Graduate Texts in Mathematics 28

Springer
New York
Berlin
Heidelberg
Barcelona
Budapest
Hong Kong
London
Milan
Paris
Santa Clara
Singapore
Tokyo

Graduate Texts in Mathematics

continued after index

Oscar Zariski   Pierre Samuel

Commutative Algebra

Volume 1

With the cooperation of

I. S. Cohen

Springer

Oscar Zariski
† Deceased

Pierre Samuel
Université de Paris-Sud
Mathématique/Bâtiment 425
91405 Orsay, France

Mathematics Subject Classification (1991):
13-01, 13AXX, 13BXX, 13CXX, 13E05, 13F05, 13F10

Library of Congress Cataloging in Publication Data

Zariski, Oscar, 1899–
Commutative algebra.
(Graduate texts in mathematics; v. 28)
Reprint of the 1958–1960 ed. published by Van Nostrand, Princeton, N.J., in series: The University series in higher mathematics, edited by M. H. Stone, D. C. Spencer, H. Whitney, and O. Zariski.
Includes index.
1. Commutative algebra I. Samuel, Pierre, 1921– joint author. II. Series: Graduate texts in mathematics; v. 28–
QA251.3.Z37 1975 512′.24 75-17751

Printed in the United States of America.

9 8 7 6 5 4

ISBN 0-387-90089-6 Springer-Verlag New York Berlin Heidelberg
ISBN 3-540-90089-6 Springer-Verlag Berlin Heidelberg New York SPIN 10531922

PREFACE

Le juge: Accusé, vous tâcherez d'être bref.
L'accusé: Je tâcherai d'être clair.

—G. COURTELINE

This book is the child of an unborn parent. Some years ago the senior author began the preparation of a Colloquium volume on algebraic geometry, and he was then faced with the difficult task of incorporating in that volume the vast amount of purely algebraic material which is needed in abstract algebraic geometry. The original plan was to insert, from time to time, algebraic digressions in which concepts and results from commutative algebra were to be developed in full as and when they were needed. However, it soon became apparent that such a parenthetical treatment of the purely algebraic topics, covering a wide range of commutative algebra, would impose artificial bounds on the manner, depth, and degree of generality with which these topics could be treated. As is well known, abstract algebraic geometry has been recently not only the main field of applications of commutative algebra but also the principal incentive of new research in commutative algebra. To approach the underlying algebra only in a strictly utilitarian, auxiliary, and parenthetical manner, to stop short of going further afield where the applications of algebra to algebraic geometry stop and the general algebraic theories inspired by geometry begin, impressed us increasingly as being a program scientifically too narrow and psychologically frustrating, not to mention the distracting effect that repeated algebraic digressions would inevitably have had on the reader, vis-à-vis the central algebro-geometric theme. Thus the idea of a separate book on commutative algebra was born, and the present book—of which this is the first of two volumes—is a realization of this idea, come to fruition at a time when its parent—a treatise on abstract algebraic geometry—has still to see the light of the day.

In the last twenty years commutative algebra has undergone an intensive development. However, to the best of our knowledge, no systematic account of this subject has been published in book form since the appearance in 1935 of the valuable *Ergebnisse* monograph "Idealtheorie" of

W. Krull. As to that monograph, it has exercised a great influence on research in the intervening years, but the condensed and sketchy character of the exposition (which was due to limitation of space in the *Ergebnisse* monographs) made it more valuable to the expert than to the student wishing to study the subject. In the present book we endeavor to give a systematic and—we may even say—leisurely account of commutative algebra, including some of the more recent developments in this field, without pretending, however, to give an encyclopedic account of the subject matter. We have preferred to write a self-contained book which could be used in a basic graduate course of modern algebra. It is also with an eye to the student that we have tried to give full and detailed explanations in the proofs, and we feel that we owe no apology to the mature mathematician, who can skip the details that are not necessary for him. We have even found that the policy of trading empty space for clarity and explicitness of the proofs has saved us, the authors, from a number of erroneous conclusions at the more advanced stages of the book. We have also tried, this time with an eye to both the student and the mature mathematician, to give a many-sided treatment of our topics, not hesitating to offer several proofs of one and the same result when we thought that something might be learned, as to methods, from each of the proofs.

The algebro-geometric origin and motivation of the book will become more evident in the second volume (which will deal with valuation theory, polynomial and power series rings, and local algebra; more will be said of that volume in its preface) than they are in this first volume. Here we develop the elements of commutative algebra which we deem to be of general and basic character. In chapter I we develop the introductory notions concerning groups, rings, fields, polynomial rings, and vector spaces. All this, except perhaps a somewhat detailed discussion of quotient rings with respect to multiplicative systems, is material which is usually given in an intermediate algebra course and is often briefly reviewed in the beginning of an advanced graduate course. The exposition of field theory given in chapter II is fairly complete and follows essentially the lines of standard modern accounts of the subject. However, as could be expected from algebraic geometers, we also stress treatment of transcendental extensions, especially of the notions of separability and linear disjointness (the latter being due to A. Weil). The study of maximally algebraic subfields and regular extensions has been postponed, however, to Volume II (chapter VII), since that study is so closely related to the question of ground field extension in polynomial rings.

Chapter III contains classical material about ideals and modules in arbitrary commutative rings. Direct sum decompositions are studied in detail. The last two sections deal respectively with tensor products of rings and free joins of integral domains. Here we introduce the notion of quasi-linear disjointness, and prove some results about free joins of integral domains which we could not readily locate in the literature.

With chapter IV, devoted to noetherian rings, we enter commutative algebra proper. After a preliminary section on the Hilbert basis theorem and a side trip to the rings satisfying the descending chain condition, the first part of the chapter is devoted mostly to the notion of a primary representation of an ideal and to applications of that notion. We then give a detailed study of quotient rings (as generalized by Chevalley and Uzkov). The end of the chapter contains miscellaneous complements, the most important of which is Krull's theory of prime ideal chains in noetherian rings. An appendix generalizes some properties of the primary representation to the case of noetherian modules.

Chapter V begins with a study of integral dependence (a subject which is nowadays an essential prerequisite for almost everything in commutative algebra) and includes the so-called "going-up" and "going-down" theorems of Cohen-Seidenberg and the normalization theorem. (Other variations of that theorem will be found in Volume II, in the chapter on polynomial and power series rings.) With Matusita we then define a Dedekind domain as an integral domain in which every ideal is a product of prime ideals and derive from that definition the usual characterization of Dedekind domains and their properties. An important place is given to the study of finite algebraic field extensions of the quotient field of a Dedekind domain, and the degree formula $\Sigma e_i f_i = n$ is derived under the usual (and necessary) finiteness assumptions concerning the integral closure of the given Dedekind domain in the extension field. This study finds its natural refinement in the Hilbert ramification theory (sections 9 and 10) and in the properties of the different and discriminant (section 11). The chapter closes with some classical number-theoretic applications and a generalization of the theorem of Kummer. The properties of Dedekind domains give us a natural opportunity of introducing the notion of a valuation (at least in the discrete case) but the reader will observe that this notion is introduced by us quite casually and parenthetically, and that the language of valuations is not used in this chapter. We have done that deliberately, for we wished to emphasize the by now well-known fact that while ideals and valuations cover substantially the same ground in the classical case (which, from a geometric point of view, is the case of dimension 1), the domain in which valuations become really significant belongs to the theory of function fields of dimension greater than 1.

The preparation of the first volume of this book began as a collaboration between the senior author and our former pupil and friend, the late Irving S. Cohen. We extend a grateful thought to the memory of this gifted young mathematician.

We wish to acknowledge many improvements in this book which are due to John Tate and Jean-Pierre Serre. We also wish to thank heartily Mr. T. Knapp who has carefully read the manuscript and the galley proofs and whose constructive criticisms have been most helpful.

Thanks are also due to the Harvard Foundation for Advanced Research whose grant to the senior author was used for typing part of the manuscript. Last but not least, we wish to extend our thanks to the D. Van Nostrand Company for having generously cooperated with our wishes in the course of the printing of the book.*

PREFACE TO THE SPRINGER EDITION

In this edition the most important change is the formulation and strengthening of Theorem 29 (p. 303) and corresponding changes in the proof of that theorem (pp. 304–305). The chief purpose of this change is to have in this volume an explicit statement of the very useful formula $f'_y = \mathfrak{C}\mathfrak{D}_x$ for extensions of Dedekind domains and the full proof of this theorem. Besides this, several minor misprints have been corrected.

* The work on this volume was supported in part by a research project at Harvard University, sponsored by the Office of Ordnance Research, United States Army, under Contract DA-19-020-ORD-3100.

TABLE OF CONTENTS

I. INTRODUCTORY CONCEPTS

§ **1. Binary operations.** Let G be an arbitrary set of elements $a, b, c, \cdots$. By a *binary operation in* G is meant a rule which associates with each *ordered* pair (a, b) of elements of G a unique element c of the same set G. A binary operation can therefore be thought of as a single-valued function whose domain is the set of all ordered pairs (a, b) of elements of G and whose range is either G itself or some subset of G. We point out explicitly that if a and b are *distinct* elements of G, then the elements of G which are associated with the ordered pairs (a, b) and (b, a) may very well be distinct.

In group theory, and in algebra generally, it is customary to denote by $a \cdot b$ or ab the element which is associated with (a, b) under a given binary operation. The element $c = ab$ is then called the *product* of a and b, and the binary operation itself is called *multiplication*. When the term "multiplication" is used for a binary operation, it carries with it the implication that "*if* $a \in G$ (read: a is an element of G) *and* $b \in G$, *then also* $ab \in G$." We shall often express this property by saying that *G is closed under the given multiplication.*

Let G be a set on which there is given a binary operation, which we write as multiplication. The operation is said to be *associative* if $(ab)c = a(bc)$ for any three elements a, b, c of G. Two elements a and b of G are said to *commute* if $ab = ba$, and the operation is said to be *commutative* if any two elements of G commute.

We assume henceforth that the operation in question is associative. It is then a simple matter to define inductively the powers of an element of G and to prove the usual rules of exponents. Namely, if $a \in G$ and if n is a positive integer, we define $a^1 = a$; if $n > 1$, $a^n = a^{n-1}a$. We then have for any positive integers m and n:

$$a^m a^n = a^{m+n}; \tag{1}$$

$$(a^m)^n = a^{mn}. \tag{2}$$

For fixed m, one can proceed by induction on n, observing that these rules hold by definition for $n = 1$. Moreover, if a and b are two elements of G which commute, then so do any powers of a and b, and

$$(3) \qquad (ab)^n = a^n b^n.$$

An *identity element* in G is an element e in G such that $ea = ae = a$ for all a in G. If G has an identity e, then it has no other. For if e' is also an identity, then $e = ee' = e'$. Moreover, we can now define a^0 to be e, and the foregoing three rules trivially hold for arbitrary non-negative exponents.

We now assume that G has an identity e. If $a \in G$, an *inverse* of a is an element a' in G such that $a'a = aa' = e$. If a'' is also an inverse of a, then $a'' = a''e = a''(aa') = (a''a)a' = ea' = a'$. Thus the inverse of a (if it exists at all) is unique. If a possesses an inverse a', then negative powers of a can also be defined. Namely, we observe that

$$a^m = a^{m+1}a'$$

for all non-negative m, and we take this as an inductive definition for negative m. Thus $a^m a = a^{m+1}$ for *all* m. The rule (1) above is then true for any fixed m (positive or negative), provided $n = 1$; it can be proved for arbitrary positive n by induction from $n - 1$ to n and for negative n by induction from $n + 1$ to n. Since, therefore, $a^m a^{-m} = e = a^{-m}a^m$, we observe that a^m has a^{-m} as inverse, so that $(a^m)^n$ is defined for every n. Rule (2) can now be proved by the two inductions used for (1). From the definition we have that $a^{-1} = a'$, and we shall always use a^{-1} for the inverse of a (if it exists). If a and b both have inverses, then so does ab, and $(ab)^{-1} = b^{-1}a^{-1}$. If, moreover, a and b commute, then so do any powers of a and b, and (3) holds for arbitrary n.

The product of n elements $a_1, \cdots, a_n$ of G is inductively defined as follows:

$$\prod_{i=1}^{n} a_i = a_1 \quad \text{if } n = 1; \qquad \prod_{i=1}^{n} a_i = \left(\prod_{i=1}^{n-1} a_i\right)a_n \quad \text{if } n > 1.$$

This product will be denoted also by $a_1a_2 \cdots a_n$. From the associativity of multiplication in G, we can prove the following general associative law, which states that the value of a product is independent of the grouping of the factors:

Let $n_0, n_1, \cdots, n_r$ be integers such that $0 = n_0 < n_1 < \cdots < n_r = n$. Then

$$\prod_{j=1}^{r}\left(\prod_{k=n_{j-1}+1}^{n_j} a_k\right) = \prod_{i=1}^{n} a_i.$$

This is clear for $n = 1$; hence we assume it proved for $n - 1$ and prove it for n factors. The formula being trivial for $r = 1$, we may assume $r > 1$. Then

$$\prod_{j=1}^{r}\left(\prod_{k=n_{j-1}+1}^{n_j} a_k\right)$$

$$= \left[\prod_{j=1}^{r-1}\left(\prod_{k=n_{j-1}+1}^{n_j} a_k\right)\right]\left[\left(\prod_{k=n_{r-1}+1}^{n_r-1} a_k\right)a_n\right] \quad \text{(by definition)}$$

$$= \left\{\left[\prod_{j=1}^{r-1}\left(\prod_{k=n_{j-1}+1}^{n_j} a_k\right)\right]\left[\prod_{k=n_{r-1}+1}^{n_r-1} a_k\right]\right\}a_n \quad \text{(by associativity)}$$

$$= \left\{\prod_{i=1}^{n-1} a_i\right\}a_n \quad \text{(by definition and induction hypothesis)}$$

$$= \prod_{i=1}^{n} a_i \quad \text{(by definition).}$$

This computation is valid unless $n_{r-1} = n - 1$; the modification necessary in this case is left to the reader.

If all $a_i = a$, then $\prod_{i=1}^{n} a_i = a^n$, and (1) and (2) are consequences (for positive exponents) of the general associative law.

§ 2. Groups

DEFINITION. *A set G which is closed under a given multiplication is called a* GROUP *if the following conditions* (GROUP AXIOMS) *are satisfied*:

G_1. *The set G is not empty.*

G_2. *If $a, b, c \in G$, then $(ab)c = a(bc)$* (ASSOCIATIVE LAW).

G_3. *There exists in G an element e such that*

(1) *For any element a in G, $ea = a$.*

(2) *For any element a in G there exists an element a' in G such that $a'a = e$.*

In view of axiom G_2 and the general associativity law proved above, we can write the product of any (finite) member of elements of G without inserting parentheses.

We proceed to show that e is an identity in G, and that for every element a has an inverse. If a is given, then by G_3 (2), there exists an a' such that $a'a = e$, and there exists an a'' such that $a''a' = e$. Then $aa' = e(aa') = (a''a')(aa') = a''(a'a)a' = a''ea' = e$; this, together with $a'a = e$, shows that a' is an inverse of a, *provided* that e is an identity. But this is immediate, for $ea = a$ by G_3 (1), and $ae = a(a'a) = (aa')a = ea = a$.

Since e is an identity in G and a' an inverse of a, it follows that both are uniquely determined. As mentioned in the preceding section, the inverse of a will be denoted by a^{-1}.

If a and b are elements of a group G, then each of the equations $ax = b$, $xa = b$, has one and only one solution. Consider, for instance, the equation $ax = b$. Multiplication on the left by a^{-1} yields $x = a^{-1}b$ as the only possible solution, and direct substitution shows that $a^{-1}b$ is indeed a solution. Similarly it can be seen that $x = ba^{-1}$ is the only solution of the equation $xa = b$.

An immediate consequence of the uniqueness of the solution of each of the above equations is the (right or left) *cancellation law*: *if $ax = ax'$ or if $xa = x'a$, then $x = x'$.*

The solvability of both equations $ax = b$, $xa = b$ is equivalent, *in the presence of* G_1 *and* G_2, to axiom G_3. For if we assume the solvability of the foregoing equations and if we assume furthermore G_1 and G_2, then we can prove G_3 as follows:

We fix an element c in G and we denote by e a solution of the equation $xc = c$. If now a is any element of G, let b be a solution of the equation $cx = a$. We will have then $ea = e(cb) = (ec)b = cb = a$, which establishes G_3 (1). As to G_3 (2), it is an immediate consequence of the solvability of the equation $xa = e$.

In practice, when testing a given set G against the group axioms, it is sometimes the case that the solvability of the equations $ax = b$, $xa = b$ follows more or less directly from the nature of the given binary operation in G. The task of proving that G is a group can therefore sometimes be simplified by using the solvability condition just stated, rather than axiom G_3.

A group which contains only a finite number of elements is called a *finite group*. By the *order* of a finite group is meant the number of elements in the group.

It may happen that a group G consists entirely of elements of the form a^n, where a is a fixed element of G, and n is an arbitrary integer, $\gtreqless 0$. If this is the case, G is called a *cyclic* group, and the element a is said *to generate* G.

§ 3. Subgroups. Given two groups G and H, denote by $\cdot$ and $\circ$ the group operations in G and in H respectively. We say that H is a *subgroup* of G if (1) H is a subset of G and (2) $a \cdot b = a \circ b$ for any pair of elements a, b in H.

Let H be a subgroup of G and let e and e' be the identity elements of G and H respectively. We have $e' \cdot e' = e' \circ e' = e'$ and $e' \cdot e = e'$.

Hence $e' \cdot e' = e' \cdot e$, and therefore, by the cancellation law which holds in G, $e' = e$. We thus see that *the identity element of a group G belongs to any subgroup H of G* (and is necessarily also the identity of H).

If H is a subgroup of G we shall *not* use different symbols (such as and o) to denote the group operations in G and H respectively. Both operations will be denoted by the same symbol, say, $\cdot$ *or* o.

Given a group G and a *non-empty* subset H_0 of G, there is a very simple criterion for H_0 to be the set of elements of a subgroup of G. Namely, we have the following necessary and sufficient condition: *if* $a, b \in H_0$, *then* $ab^{-1} \in H_0$. This condition is obviously necessary. On the other hand, if this condition is satisfied, then we have in the first place that H_0 contains the identity e of G (if a is any element of the *non-empty* set H_0, then $e = a \cdot a^{-1} \in H_0$). It follows that if $a \in H_0$, then also $a^{-1} \in H_0 (a^{-1} = e \cdot a^{-1} \in H_0)$, and if $a, b \in H_0$, then $a \cdot b = a \cdot (b^{-1})^{-1} \in H_0$. Thus H_0 is indeed a group H with respect to the group operation in G, and this group H is a subgroup of G.

Let G be an arbitrary group and let H be a subgroup of G. If a is any element of G, we denote by Ha the set of elements of G which are of the form ha, $h \in H$, and we call this set *a right coset of H*. In a similar fashion, we can define *left cosets aH* of H. If multiplication in G is commutative (§ 1), then any right coset is also a left coset: Ha and aH are identical sets.

Let Ha and Hb be two right cosets of H in G, and suppose that these two cosets have an element c in common: $c = h_1 a = h_2 b$; $h_1, h_2 \in H$. Then $b = h_2^{-1} h_1 a$, and for any element h of H we have $hb = (hh_2^{-1} h_1) a \in Ha$ (since H is a subgroup of G and hence $hh_2^{-1} h_1 \in H$). Thus $Hb \subset Ha$; and similarly we can show that $Ha \subset Hb$. *Therefore* $Ha = Hb$.

It follows that two right cosets Ha and Hb are either *disjoint* (that is, have no elements in common) or *coincide*. A similar result holds for left cosets. Note that $a \in Ha$, for H contains the identity of G. Hence every element of G belongs to some right (or left) coset.

H is said to be a *normal* (or *invariant*) subgroup of G if $Ha = aH$ for every a in G. An equivalent property is the following: for every a in G and every h in H, the element $a^{-1}ha$ belongs to H.

Suppose now that G is a finite group of order n, and let m be the order of H. Every right coset Ha of H contains then precisely m elements (if $h_1, h_2 \in H$ and $h_1 \neq h_2$, then $h_1 a \neq h_2 a$). Since every element of G belongs to one and only one right coset, it follows that m must be a divisor of n and that n/m is the number of *right* cosets of H. We have therefore proved that *if G is a finite group, then the order m of any* *subgroup H of G divides the order n of G.* The quotient n/m is called *the index* of H *in* G.

If a is an arbitrary element of a group G, the elements a^n, n any integer $\gtreqless 0$, clearly form a subgroup H of G. We call H *the cyclic subgroup generated by the element* a. If this subgroup H is finite, say of order m, then m is called *the order* of the element a; otherwise, a is said to be of *infinite order.*

Let a be an element of G, of finite order m. There exist then pairs of distinct integers n, n' such that $a^n = a^{n'}$ (otherwise the cyclic group generated by a would be infinite). From $a^n = a^{n'}$ follows $a^{n-n'} = 1$, whence there exist positive integers ν such that $a^\nu = 1$. Let μ be the smallest of these integers. Then $1, a, a^2, \cdots, a^{\mu-1}$ are distinct elements, while if n is any integer and if, say, $n = q\mu + n'$, $0 \leqq n' < \mu$, then

$$(1) \qquad a^n = a^{q\mu+n'} = (a^\mu)^q \cdot a^{n'} = a^{n'}.$$

It follows that the cyclic group generated by a consists precisely of the μ elements $1, a, a^2, \cdots, a^{\mu-1}$, *and hence* $\mu = m$. Thus the order of a is also the smallest positive integer m such that $a^m = 1$.

From (1) it follows that $a^n = 1$ if and only if $n' = 0$, that is, if and only if n is a multiple of $m(= \mu)$.

It is clear that if G is a finite group, then every element a of G has finite order, and that the order of a divides the order of G.

§ **4. Abelian groups.** Let G be a set with an associative multiplication. As defined in § 1, the multiplication is said to be commutative if $ab = ba$ for any elements a, b in G. In such a case it is permissible to change freely the order of the factors in a product $a_1 a_2 \cdots a_n$. That is to say, we have the general commutative law, which can be formally stated as follows:

Let φ be a permutation of the integers $\{1, 2, \cdots, n\}$. Then

$$\prod_{i=1}^{n} a_i = \prod_{i=1}^{n} a_{\varphi(i)}.$$

The proof is by induction and may be left to the reader.

A *group* G in which the group operation is commutative is said to be *commutative* or *abelian.* The group operation is then often written additively; that is, we write $a + b$ instead of ab and $\sum a_i$ instead of $\prod a_i$. The element $a + b$ is called the *sum* of a and b. The identity element is denoted by 0 (*zero*) and the inverse of a by $- a$. Correspondingly one writes na instead of a^n, and the rules for exponents take the form

$$(1) \qquad ma + na = (m + n)a,$$

(2) $$m(na) = (mn)a,$$

(3) $$n(a + b) = na + nb,$$

(4) $$-(na) = (-n)a.$$

The last equation is a paraphrase of the statement (in the multiplicative notation) that the inverse of a^n is a^{-n}. The equation $xa = b$, which in the abelian case is equivalent to the equation $ax = b$, assumes then the form $x + a = b$. Its unique solution $b + (-a)$ is denoted by $b - a$ and is called *the difference of b and a.* The binary operation which associates with the ordered pair (a, b) the difference $b - a$ is called *subtraction.*

§ 5. Rings

DEFINITION. *A set R in which two binary operations, $+$ (addition) and $\cdot$ (multiplication), are given is called a* RING *if the following conditions* (RING AXIOMS) *are satisfied:*

R_1. *R is an abelian group with respect to addition.*

R_2. If $a, b, c \in R$, *then* $a(bc) = (ab)c$.

R_3. *If $a, b, c \in R$, then $a(b + c) = ab + ac$ and $(b + c)a = ba + ca$ (distributive laws).*

In conformity with the additive notation for abelian groups (§ 4) the identity element of R (regarded as an additive group) is denoted by 0, and the (additive) inverse of an element a is denoted by $-a$. Therefore the following relations hold in any ring R:

$$0 + a = a + 0 = a,$$
$$a + (-a) = (-a) + a = 0,$$
$$-(-a) = a,$$
$$a + (b + c) = (a + b) + c,$$
$$a + b = b + a.$$

The abelian group which, according to the ring axiom R_1, any ring R forms with respect to addition is called the *additive group of the ring.*

A ring R is called *commutative* if multiplication is commutative in R: $ab = ba$ for any elements a, b in R.

The distributive laws hold also for subtraction:

(1) $$a(b - c) = ab - ac; \qquad (b - c)a = ba - ca.$$

To prove, for instance, the first of these two relations, we have to show that $a(b - c) + ac = ab$. This, however, follows directly from the first distributive law R_3, since $(b - c) + c = b$.

For $b = c$, relations (1) yield the following important property of the element 0:

$$a0 = 0a = 0, \tag{2}$$

for all a in R. If we put in (1) $b = 0$ we find

$$a(-c) = -ac; \quad (-c)a = -ca,$$

and if in the first of these relations we replace a by $-a$ we obtain $(-a)(-c) = -(-a)c = -(-ac)$, whence

$$(-a)(-c) = ac. \tag{3}$$

An element a of R is called a *left* (or *right*) *zero divisor* if there exists in R an element b *different from zero* such that $ab = 0$ (or $ba = 0$). By (2) the element 0 is always both a left and right zero divisor whenever R contains elements different from zero. However, it is convenient to regard 0 as a zero divisor also in the trivial case of a ring R which consists only of the element zero (*nullring*). By a *proper* zero divisor is meant a zero divisor which is different from 0. Hence a ring R has proper zero divisors if and only if it is possible to have in R a relation $ab = 0$ *with both a and b different from zero.* In the sequel we shall call *R a ring without zero divisors* if R has no *proper* zero divisors. An element of R which is not a zero divisor will be called a *regular* element. In particular, the element 0 is not a regular element.

§ 6. **Rings with identity.** If there exists in the ring R an element which is an identity *with respect to multiplication*, then, by a remark made in § 1, this element is uniquely determined. *If R is not a nullring*, we shall refer to this element as the *identity* of the ring and we shall denote it by the symbol 1. In such a ring, multiplicative inverses are referred to simply as inverses. Hence an inverse of a is an element a' such that $a'a = 1$ and $aa' = 1$; it is unique according to § 1 and will be denoted by a^{-1}.

The element 1 is its own inverse. Similarly it follows from (3) that -1 is its own inverse.

The elements 0 *and* 1 *are distinct elements of R.* For we have agreed that R is not a nullring, and if $a \neq 0$, then $a0 = 0$ and $a1 = a \neq 0$, whence $0 \neq 1$. From this it follows that *the element* 0 *has no inverse,* since for any element a in R we have $a0 = 0a = 0 \neq 1$. *Consequently a ring* (which is not a nullring) *is definitely not a group with respect to multiplication.*

An element of R is called a *unit* if it has an inverse. The elements 1 and -1 are units. The ring of integers is the simplest example of a commutative ring in which 1 and -1 are the only units. If a and b are units, we have $a^{-1}a = aa^{-1} = 1$ and $(b^{-1}a^{-1})ab = ab(b^{-1}a^{-1}) = 1$, and this shows that also a^{-1} and ab are units. It follows that *in a ring R with identity the units form a group with respect to multiplication.*

If an element a has an inverse a^{-1}, then from $ab = 0$ follows $a^{-1}ab = 0$, $1b = 0$, that is, $b = 0$. Therefore a is not a left zero divisor. Similarly it can be shown that a is not a right zero divisor. Thus *no unit in R is a zero divisor.*

A *commutative* ring with identity and having no proper zero divisors is called an *integral domain.*

§ 7. Powers and multiples. If R is an arbitrary ring and $a \in R$, then a^n is defined for all positive integers n, in accordance with § 1, and, moreover, relations (1) and (2) of that section are valid. If R is commutative, (3) also holds. If R has an element 1, then the definition in § 1 gives $a^0 = 1$, and if in addition a^{-1} exists, then a^n is defined for *all* integers n, and (1) and (2) are valid for arbitrary powers. In the commutative case, if a and b have inverses, then (3) holds for any integer n.

Since R is a group with respect to addition, the *multiples na* are defined for any integer n and any a in R. In addition to the rules for multiples given in § 4 we have the rules

$$n(ab) = (na)b = a(nb). \tag{1}$$

These follow from the general distributive laws

$$b\sum_{i=1}^{n} a_i = \sum_{i=1}^{n} ba_i, \quad \left(\sum_{i=1}^{n} a_i\right)b = \sum_{i=1}^{n} a_ib,$$

which in turn are easily proved by induction.

We point out that the associative law of multiplication has nothing to do with (1) above or with (2) of § 4, nor have the distributive laws anything to do with (1) and (3) of § 4. More generally, we note that *the symbol na should not be regarded as the product of n and a.* Not only would such an interpretation of the symbol na be ill-founded (na was defined as the *sum* of n elements, all equal to a), but it would also be meaningless, since the *integer n* is in general not even an element of R. However, if R has an identity, then using the distributive law R_3—or simply (1) above—we can write:

$$na = 1a + 1a + \cdots + 1a \ (n \text{ times}) = (1 + 1 + \cdots + 1)a = (n1)a,$$

and this time na is therefore indeed a product, namely, the product of $n1$ and a. But also in this case the factor $n1$ (which is an element of R)

should not be confused offhand with the integer n, just as the element 1 of R is not to be identified with the integer 1. We shall see in a later chapter (II, § 4) under what conditions and in what sense is the identification "$n \cdot 1 = n$" permissible.

In this book we shall study exclusively the theory of COMMUTATIVE *rings.* Since no other rings will be considered, a "ring" will mean from now on a "commutative ring."

§ 8. Fields

DEFINITION. *A ring F is called a* FIELD *if the following conditions* (FIELD AXIOMS) *are satisfied:*

F_1. *F has at least two elements.*

F_2. *F has an identity.*

F_3. *Every element of F different from zero has an inverse.*

The three field axioms can be replaced by a single axiom: *the elements of F which are different from zero form a group with respect to multiplication.* This group shall be referred to as the *multiplicative group of F.*

In a field, every element different from 0 is a unit. Therefore a field has no proper zero divisors (§ 6) and is an integral domain (in view of F_2).

If we apply the general group-theoretic considerations of § 2 to the multiplicative group of F, especially the considerations concerning the equation $ax = b$, we see that given any two elements a and b of F, both different from zero, it is possible to *divide* b by a, that is, form *the quotient* b/a. This quotient is the unique solution of the equation $ax = b$. We observe, however, that also if $b = 0$, but $a \neq 0$, then the resulting equation $ax = 0$ still has a unique solution $x = 0$, since a is not a zero divisor. For this reason we define: $0/a = 0$ $(a \neq 0)$. Hence *division by any element a different from zero is always permissible in a field.* On the other hand, if $a = 0$, then there results an equation $0 \cdot x = b$ which either has no solution (if $b \neq 0$; whence $b/0$ does not exist) or is satisfied by every element of F (if $b = 0$; whence $0/0$ is indeterminate.)

The ring of natural integers is an example of an integral domain that is not a field. Examples of fields: (a) the set of all rational numbers; (b) the set of all real numbers; (c) the set of all complex numbers.

§ 9. Subrings and subfields. A ring R' is called *a subring of R if* (a) R' is a subset of R and (b) the ring operations $+$ and $\cdot$ in R' are the same as those induced in the set R' by the corresponding ring operations $+$ and $\cdot$ in R. It follows that a subring R' of R, regarded as an additive group, must be in the first place a subgroup of the additive group of R. Hence R' must be a non-empty set and it must satisfy the following condition (§ 3):

(a) If $a, b \in R'$, then $a - b \in R'$.

Furthermore, R' must be closed under the given multiplication in R:

(b) If $a, b \in R'$, then $ab \in R'$.

Conditions (a) and (b) (together with the trivial condition that R' be a non-empty set) are also sufficient to make R' a subring of R (the associative, commutative, and distributive laws automatically hold in R' because they hold in R).

If R has an identity 1 and if this element 1 also belongs to R', then 1 is, of course, the identity of R'. In this case, we shall call R' a *unitary subring* of R (or R a *unitary overring* of R'.) However, it may well happen that while R has an identity, R' does not (for example: $R =$ ring of integers, $R' =$ ring of even integers). Less trivial possibilities are the following: (a) both R and R' have an identity, but the identity of R does not belong to R'; (b) R' has an identity but R does not (see Example 2 below). *In both cases* (a) *and* (b) *the identity of R' is necessarily a zero divisor of R.* For let $1'$ denote the identity of R' and let us assume that $1'$ is not an identity of R. There exists then in R an element a such that $1'a = b \neq a$. We have $1'b = (1' \cdot 1')a = 1'a = b$, hat is, $1'a = 1'b$, or $1'(a - b) = 0$. Since $a \neq b$, it follows that $1'$ is a zero divisor in R.

By a *subfield* of a field F we mean any subset F' of F which is a field with respect to the given field operations ($+$ and $\cdot$) in F. From the remarks just made concerning rings with identity it follows that the element 1 of F is necessarily the identity of F'. This also follows from the fact that the multiplicative group of F' must be a subgroup of the multiplicative group of F. This last condition, together with the condition that F' be a subgroup of the additive group of F, characterizes the concept of a subfield. Hence (§ 3) F' is a subfield of F if and only if the following two conditions are satisfied: (a) if $a, b \in F'$, then $a - b \in F'$; (b) if $a, b \in F'$ and $b \neq 0$, then $ab^{-1} \in F'$.

Examples. (1) If a and b are distinct elements of a field F, we may define a new addition $\oplus$ and a new multiplication $\odot$ in F as follows: $x \oplus y = x + y - a$, $x \odot y = a + (x - a)(y - a)/(b - a)$. (In geometric terms: we change the origin and the scale.) It is easily seen that the elements of F form a field also with respect to these new operations. We denote this new field by F'. It is clear that a subset of F which is a subring of F' will not in general be a subring of F. Note that a and b are respectively the zero and the identity of F'.

(2) Let A and B be two rings and let R be the set of all ordered pairs (a, b), where $a \in A$ and $b \in B$. If we define addition and multiplication in R by setting $(a, b) + (a', b') = (a + a', b + b')$, $(a, b) \cdot (b, b') = (aa'\ bb')$, then R is a ring, and the subset R' of R consisting of the elements $(a, 0)$ is a subring of R. If A has an identity, say, e_A, then $(e_A, 0)$ is the identity of R'. The ring R has an identity if and only if both A and B have identities e_A and e_B, and in that case (e_A, e_B) is the identity of R. In the present example the identities of R and R' are therefore necessarily distinct.

§ **10. Transformations and mappings.** We shall use the symbol $\subset$ for set inclusion. Thus, if S and S' are sets, then $S' \subset S$ shall mean that S' is a subset of S. If $S' \subset S$, and $S' \neq S$, we shall say that S' is a *proper* subset of S and we shall write $S' < S$.

Let S and $\bar{S}$ be arbitrary sets of elements. By *a transformation of* S into $\bar{S}$ we mean a rule which associates with every element a of S some subset of $\bar{S}$. This subset, which may be empty, will be denoted by aT. If $\bar{a}$ is an element of aT, we say that $\bar{a}$ *corresponds to* a (under the given transformation T), or that $\bar{a}$ is *a transform of* a, or that $\bar{a}$ is *a T-image* of a. It may be that to certain (or even all) elements of S there correspond no elements of $\bar{S}$.

If A is an arbitrary non-empty subset of S, the union of all T-images of all elements of A shall be referred to as *the transform of A* (under T) and shall be denoted by AT. We have $AT = \cup\, aT$, $a \in A$, where the symbol $\cup$ indicates set-theoretic addition (union of sets) and where a varies in A. We make the convention that if A is empty, then the symbol AT stands for the empty set. We say that T is a transformation of S *onto* $\bar{S}$ if $ST = \bar{S}$.

Let T be a transformation of S into $\bar{S}$, and let S' be a subset of S. Then T induces in a natural way a transformation T' of S' into $\bar{S}$: if $a \in S'$, we define $aT' = aT$. T' is called *the restriction* of T to S'.

If T is a transformation of S into $\bar{S}$ and T' is a transformation of $\bar{S}$ into some other set S', then *the product of T and T'* is the transformation of S into S' which associates with every element a of S the subset $(aT)T'$ of S'. This transformation shall be denoted by TT'. Thus, by definition, $a(TT') = (aT)T'$, and it follows that we have for any subset A of S: $A(TT') = (AT)T'$. If S_1, S_2, S_3, S_4 are sets and $T_i (i = 1, 2, 3)$ is a transformation of S_i into S_{i+1}, then clearly $(T_1T_2)T_3 = T_1(T_2T_3)$.

For a transformation T of S into $\bar{S}$, the *inverse transformation* T^{-1} of $\bar{S}$ into S is defined as follows: If $\bar{a} \in \bar{S}$, then $\bar{a}T^{-1}$ is the set of all elements of S having $\bar{a}$ as T-image; that is, $a \in \bar{a}T^{-1}$ if and only if $\bar{a} \in aT$. Clearly T is the inverse of T^{-1}.

A transformation T of S into $\bar{S}$ will be called a *mapping* of S into $\bar{S}$ if it is everywhere defined on S and is single-valued, that is, if for every element a of S, the set aT contains *one and only one* element. This element will also be denoted by aT. As with transformations in general, a mapping T of S into $\bar{S}$ is said to be a mapping *onto* $\bar{S}$ if $ST = \bar{S}$. A mapping of S into $\bar{S}$ is *univalent* if $aT = bT$ implies $a = b$ for any a and b in S. A mapping of S into $\bar{S}$ will be called *one to one*—in symbols, (1, 1)—if it is *both onto and univalent*. It is clear that, T being a mapping of S into $\bar{S}$, T^{-1} is a mapping of $\bar{S}$ into S if and only if T is one to one; and in that case, also T^{-1} is one to one.

The *identity mapping* I of a set S is defined by $aI = a$ for all a in S. If S and $\bar{S}$ are two sets, I and $\bar{I}$ their respective identity mappings, then a transformation T of S into $\bar{S}$ is a one to one mapping of S if and only if there exists a transformation $\bar{T}$ of $\bar{S}$ into S such that $T\bar{T} = I$, $\bar{T}T = \bar{I}$; and in that case $\bar{T} = T^{-1}$.

If T is a mapping of S into $\bar{S}$, and T' a mapping of $\bar{S}$ into a set S', then the product transformation TT' of S into S' is itself a mapping.

A mapping of S into $\bar{S}$ is, in fact, a single-valued function f on S to $\bar{S}$, since it associates with each element of S a unique element of $\bar{S}$. We shall frequently use the functional notation $f(a)$ to denote the element of $\bar{S}$ which corresponds to an element a of S. If f is a mapping from S into $\bar{S}$, and g a mapping from $\bar{S}$ into S', we shall write, in the usual way, $g(f(a))$ for the element of S' corresponding to a under the product of the mappings f and g.

A mapping T of a set S into a set S' is sometimes denoted by a notation of the type $a \to E(a)$, where $E(a)$ is a formula giving the value of the image aT of any element a of S.

§ 11. Group homomorphisms. From the foregoing general set-theoretic definitions we now pass to the case in which the given sets are groups. In this case one is interested in mappings of a particular type. Let G and $\bar{G}$ be two arbitrary groups. We use the multiplicative notation for the group operation in each group. By a *homomorphism*, or *homomorphic mapping*, of G into (or onto) $\bar{G}$ we mean a mapping T of G into (or onto) $\bar{G}$ which satisfies the following condition: *if a and b are any two elements of G, then*

$$(ab)T = (aT)(bT).$$

Thus a homomorphism of a group G into another group $\bar{G}$ is a mapping characterized by the condition that *the image of a product is the product of the images*: if to a there corresponds $\bar{a}$ and to b there corresponds $\bar{b}$ ($a, b \in G$; $\bar{a}, \bar{b} \in \bar{G}$), then to the product ab there corresponds the product $\bar{a}\bar{b}$, that is, we have $\overline{ab} = \bar{a}\bar{b}$.

If both groups G, $\bar{G}$ are abelian and if the group operation in both groups is written additively, then the foregoing homomorphism condition $(ab)T = (aT)(bT)$ becomes

$$(a + b)T = aT + bT.$$

A *univalent* homomorphic mapping of G into (or onto) $\bar{G}$ is called an *isomorphism*, or *an isomorphic mapping*, of G into (or onto) $\bar{G}$. It is clear that an isomorphism of G onto $\bar{G}$ is a homomorphism of G into $\bar{G}$ which is at the same time a one-to-one mapping.

Given two groups G, $\bar{G}$, we say that $\bar{G}$ is a *homomorphic* or *isomorphic image* (or *map*) of G according as there exists a homomorphism or an isomorphism of G *onto* $\bar{G}$. If T is an isomorphism of G onto $\bar{G}$, then it is clear that T^{-1} is an isomorphism of $\bar{G}$ onto G. Hence if $\bar{G}$ is an isomorphic image of G, then also G is an isomorphic image of $\bar{G}$. We say then that G and $\bar{G}$ are *isomorphic groups*. In particular, a homomorphism of a group G into *itself* is called an *endomorphism* of G; and an isomorphism of G *onto* itself is called an *automorphism* of G.

If T is a homomorphism of G into $\bar{G}$ and if T' is a homomorphism of $\bar{G}$ into a group G', then TT' is a homomorphism of G into G'. If both T and T' are homomorphisms onto, then also TT' is a homomorphism onto (of G onto G'). It follows that a homomorphic image of a homomorphic image of a group G is itself a homomorphic image of G.

If T is a homomorphism of a group G into a group $\bar{G}$, we mean by the *kernel* of T the set of all elements of G which are mapped into the identity element of $\bar{G}$.

THEOREM 1. *If T is a homomorphism of a group G into a group $\bar{G}$ and if e and $\bar{e}$ denote respectively the identity elements of G and of $\bar{G}$, then $eT = \bar{e}$. If $a \in G$ and if $aT = \bar{a}$, then $a^{-1}T = \bar{a}^{-1}$. The set GT is a subgroup of $\bar{G}$, and the kernel H of T is a normal subgroup of G.*

PROOF. From $ee = e$ follows $(eT)(eT) = eT$, and on the other hand we have $\bar{e}(eT) = eT$. Hence $(eT)(eT) = \bar{e}(eT)$, and since the cancellation law holds in any group, it follows that $eT = \bar{e}$.

From $aa^{-1} = e$ follows $(aT)(a^{-1}T) = eT = \bar{e}$, whence $a^{-1}T = \bar{a}^{-1}$, where $\bar{a} = aT$.

If $\bar{a} = aT$ and $\bar{b} = bT$ are any two elements of GT ($a, b \in G$), then $\bar{a}(\bar{b})^{-1} = (aT)(bT)^{-1} = (aT)(b^{-1}T) = (ab^{-1})T$, and therefore $\bar{a}(\bar{b})^{-1} \in GT$. This shows that GT is a subgroup of $\bar{G}$ (§ 4).

The kernel of T is a non-empty subset of G, since $eT = \bar{e}$, hence $e \in H$. If $a, b \in H$, that is, $aT = bT = \bar{e}$, then $(ab^{-1})T = (aT)(bT)^{-1} = \bar{e}$, hence $ab^{-1} \in H$, and this shows that H is a subgroup of G. If a is any element of the kernel H and if x is any element of G, we have $(x^{-1}ax)T = (xT)^{-1}(aT)(xT) = \bar{e}$, and therefore $x^{-1}ax \in H$. This shows that H is a normal subgroup of G.

The following theorem is used very frequently in testing whether a given group homomorphism is an isomorphism:

THEOREM 2. *A homomorphism T of a group G into a group $\bar{G}$ is an isomorphism if and only if the kernel H of T contains only the identity e of G.*

PROOF. In the first place it is obvious that if T is an isomorphism—hence a univalent mapping—then e is the only element of G which is mapped into the identity element $\bar{e}$ of $\bar{G}$. Conversely, let us assume that the kernel H of T contains only the identity e of G and let a and b be elements of G having the same T-image: $aT = bT$. Then $(ab^{-1})T = aT\cdot(bT)^{-1} = \bar{e}$, $ab^{-1} \in H$, $ab^{-1} = e$, $a = b$, and hence T is a univalent mapping, that is, T is an isomorphism.

As was stated in Theorem 1, the kernel of any homomorphism of a group G is a normal subgroup of G. Now, conversely, let H be a given invariant subgroup of G. The right cosets of H and G coincide then with the left cosets of H, and we can define multiplication of cosets as follows: $Ha\cdot Hb = Hab (a, b \in G)$. The product $Ha\cdot Hb$ depends only on the cosets Ha, Hb and not on the choice of representatives a and b of these cosets. For if $Ha' = Ha$ and $Hb' = Hb$, we have $a' = h_1a$ and $b' = h_2b$, where h_1 and h_2 are elements of H, and hence $Ha'\cdot Hb' = Hh_1\cdot ah_2\cdot b = Hh_1h_3\cdot ab = Hab$, where $h_3 = ah_2a^{-1} \in H$. One sees immediately that with respect to this definition of multiplication of cosets, the cosets of H form a group, the coset H being the identity of that group, and that the mapping $a \to Ha$ is a homomorphism of G onto the group of H-cosets, with kernel H. The group of cosets of the normal subgroup H is called the *factor group*, or the *quotient group*, *of G with respect to H*, and is denoted by G/H. The mapping $a \to Ha$ is called the *canonical* or *natural homomorphism* of G onto G/H.

The following situation occurs frequently in applications: we are given a group G, a set $\bar{G}$ in which a binary operation (multiplication) is defined, and a mapping T of G *onto* $\bar{G}$ which has the usual homomorphism property $(ab)T = (aT)(bT)$. We may express these conditions by saying that *the set $\bar{G}$ is a homomorphic image of the group G.*

LEMMA 1. *The homomorphic image $\bar{G}$ of a group G is a group. If G is commutative, so is $\bar{G}$.*

PROOF. We first prove the associative law in $\bar{G}$. Let $\bar{a}$, $\bar{b}$, $\bar{c}$ be arbitrary elements of $\bar{G}$; they are images of certain elements a, b, c of G, since T maps G *onto* $\bar{G}$. We have $(ab)c = a(bc)$. We have $[(ab)c]T = [(ab)T]cT = [(aT)(bT)]cT = (\bar{a}\bar{b})\bar{c}$. In a similar fashion we find that $[a(bc)]T = \bar{a}(\bar{b}\bar{c})$, and hence $(\bar{a}\bar{b})\bar{c} = \bar{a}(\bar{b}\bar{c})$. One shows then, as in the proof of Theorem 1, that $\bar{G}$ has an identity, namely, eT, where e is the identity of G, and that every element $\bar{a}$ of $\bar{G}$ has an inverse, namely, if $\bar{a} = aT$, then $\bar{a}^{-1} = (a^{-1})T$. Thus $\bar{G}$ is a group. The second assertion of the lemma is obvious.

Another situation which occurs frequently in connection with group homomorphisms is the following:

We are given two groups G and $\bar{G}$ and a transformation T of G into $\bar{G}$. It is also given that

(A) for any element a in G the set aT is non-empty;
(B) if $\bar{a} \in aT$ and $\bar{b} \in bT$, then $\bar{a}\bar{b} \in (ab)T$.

It is not given *a priori* that T is a mapping (that is, single-valued). Were this given too, then it would follow at once that T is a homomorphism of G into $\bar{G}$. The following lemma reduces the test of single-valuedness of T to the test of single-valuedness of T at the identity element e of G.

LEMMA 2. *Let T be a transformation of a group G into a group $\bar{G}$ such that conditions* (A) *and* (B) *are satisfied. If the set eT contains only one element (e denoting the identity of G), then T is a mapping, hence a homomorphism, of G into $\bar{G}$.*

PROOF. We have, by condition (B), $eT \cdot eT \in (e \cdot e)T = eT$; hence eT is the identity $\bar{e}$ of $\bar{G}$. Let a be any element of G and let us fix an element $\bar{b}$ in $(a^{-1})T$. If $\bar{a}$ is any element in aT, we have, by (B), $\bar{a}\bar{b} \in (aa^{-1})T = eT = \bar{e}$, that is, $\bar{a}\bar{b} = \bar{e}$. This shows that aT consists of the single element $\bar{b}^{-1}$. Q.E.D.

§ 12. **Ring homomorphisms.** A mapping T of a ring R into a ring $\bar{R}$ is called a *ring homomorphism*, or simply a *homomorphism*, or a *homomorphic mapping*, if T satisfies the following conditions:

$$(a + b)T = aT + bT, \tag{1}$$

$$(ab)T = (aT)(bT), \tag{2}$$

for any pair of elements a and b in R. Condition (1) signifies that T is a homomorphism of the additive group of R into the additive group of $\bar{R}$. Condition (2) is the analogue of (1) for multiplication.

A ring homomorphism which is a univalent mapping is called an *isomorphism.*

If T is a homomorphism or isomorphism of R onto $\bar{R}$, then we say that $\bar{R}$ is respectively a *homomorphic* or *isomorphic image* of R. If $\bar{R}$ is an isomorphic image of R, then also R is an isomorphic image of $\bar{R}$ (in virtue of the mapping T^{-1}), and the two rings R, $\bar{R}$ are said to be *isomorphic rings*, or R is said to be *isomorphic with* $\bar{R}$.

We use the standard notation

$$R \sim \bar{R}$$

to indicate that $\bar{R}$ is a homomorphic image of R (that is, that *there exists* a homomorphism of R *onto* $\bar{R}$) and we write

$$T{:}R \sim \bar{R}$$

to indicate that a *given* mapping T of R *onto* $\bar{R}$ is a homomorphism.

The corresponding notation for isomorphic rings is

$$R \cong \bar{R},$$
$$T: R \cong \bar{R}.$$

The same notation is used also in group theory for group homomorphism and group isomorphisms respectively.

An isomorphic mapping of a ring R (or of a group) *onto itself* is called an *automorphism.* In an automorphism $T: R \cong \bar{R}$ the two rings (or groups) R, $\bar{R}$ coincide (not merely as sets but also as rings, or groups).

By the *kernel* of a homomorphism T of a ring R into a ring $\bar{R}$ we mean the set of elements a in R such that $aT = \bar{0}$, where $\bar{0}$ denotes the zero element of $\bar{R}$.

THEOREM 3. *If T is a homomorphism of a ring R into a ring $\bar{R}$, then*

(a) $0T = \bar{0}$ *and* $(-a)T = -(aT)$, *for any element a in R;*
(b) *RT is a subring of $\bar{R}$;*
(c) *the kernel N of T is a subring of R;*
(d) *if R has an identity element* 1 *and if RT is not a nullring, then* $1T$ *is the identity element of RT, and if a^{-1} exists, then $a^{-1}T$ is the inverse of aT in the ring RT.*

PROOF

(a) This follows from Theorem 1 of § 11 as applied to the additive group of R.

(b) If $\bar{a}, \bar{b} \in RT$, then $\bar{a} = aT$, $\bar{b} = bT$, where $a, b \in R$, and $\bar{a}\bar{b} = (ab)T \in RT$. Hence RT is closed under multiplication. Since, by Theorem 1, RT is a subgroup of the additive group of $\bar{R}$, it follows (§ 9) that RT is a subring of $\bar{R}$.

The proof of (c) and (d) is equally straightforward and is left to the reader.

COROLLARY. *If T is a homomorphism of R onto $\bar{R}$ and if R has an identity element* 1, *then also $\bar{R}$ has an identity element (provided $\bar{R}$ is not a nullring) and this element is* $1T$.

It has already been pointed out that the kernel N of the homomorphism T contains at least the element 0 of R. From Theorem 2 of § 11, as applied to the additive group of R, it follows that a *homomorphism T of a ring R into a ring $\bar{R}$ is an isomorphism if and only if the kernel N of T contains only the element* 0 *of R.*

We have shown in the proof of Theorem 3 that the kernel N is closed under multiplication. Actually N has the following much stronger property: *If one of the factors a, b of a product ab belongs to N, then the product itself belongs to N.* For if, say, $a \in N$, then $(ab)T = (aT)(bT) = \bar{0}(bT) = \bar{0}$, hence $ab \in N$, as asserted. This property of the kernel N is fundamental in the formulation of the concept of an ideal, and we shall return to it in chapter III.

From a formal algebraic standpoint, isomorphic rings are not essentially distinct rings, because it is clear that an isomorphic mapping of a ring R preserves the *algebraic* properties of R (that is, those properties of R which can be formally expressed in terms of the ring operations $+$ and $\cdot$). Thus, for instance, an isomorphic image of an integral domain or of a field is again respectively an integral domain or a field.

On the other hand, a homomorphism which is not an isomorphism may affect some algebraic properties of a ring. For instance, a homomorphic image of an integral domain need not be an integral domain, and a ring which is not an integral domain may have an integral domain as a homomorphic image, (see III, § 9).

The situation for groups, which is covered by Lemma 1 of the preceding section, arises also for rings and leads to a similar lemma. Assume that we have a ring R, a set $\bar{R}$ in which two binary operations $+$ and $\cdot$ are defined, and a mapping T of R *onto* $\bar{R}$ having the usual homomorphism properties: $(a + b)T = aT + bT$, $(ab)T = aT \cdot bT$. We express these conditions by saying that the set $\bar{R}$ is *a homomorphic image of the ring R.*

LEMMA. *A homomorphic image of a ring is again a ring.*

The proof is similar to that of Lemma 1 of the preceding section and may be left to the reader.

As to Lemma 2 of the preceding section, it is automatically applicable to rings when we regard rings as additive groups.

COROLLARY. *An isomorphic image of an integral domain or of a field is again respectively an integral domain or a field.*

If T is a homomorphism of a ring R into a ring $\bar{R}$ and if R_0 is a subring of R, then the restriction T_0 of T to R_0 is a homomorphism of R_0 into $\bar{R}$. If T is an isomorphism, then also the induced homomorphism T_0 of R_0 is an isomorphism (but not conversely).

An important special case is the following: R_0 is a *common subring* of R and $\bar{R}$, and the induced homomorphism of R_0 is the *identity* (that is, the automorphism T_0 of R_0 defined by $aT_0 = a$, for all a in R_0). In this case we say that T is *a relative homomorphism* of R *over* R_0, or briefly: T is an R_0*-homomorphism* (or an R_0*-isomorphism*, if T is an isomorphism). For instance, the automorphism of $a + ib \rightarrow a - ib$ of the field of complex numbers (a, b real) is a relative automorphism over the field of real numbers.

If R_0 is a common subring of two rings R and $\bar{R}$, we say that $\bar{R}$ is an R_0*-homomorphic image* of R if there exists an R_0-homomorphism of R *onto* $\bar{R}$; and that $\bar{R}$ is an R_0*-isomorphic image* of R (or that R and $\bar{R}$ are R_0*-isomorphic*) if there exists an R_0-isomorphism of R *onto* $\bar{R}$.

If T is a homomorphism of a ring R into a ring $\bar{R}$ and T_1 is a homomorphism of a subring R_1 or R into the same ring $\bar{R}$, we shall say that T is an *extension* of T_1 if T_1 is the restriction of T to R_1. If only R, $\bar{R}$, R_1 and T_1 are given, then we say that T_1 can be extended to a homomorphism of R (into $\bar{R}$) if there exists a homomorphism T of R into $\bar{R}$ such that T is an extension of T_1.

§ **13. Identification of rings.** As an application of the concept of isomorphism extension, we shall now discuss a certain standard procedure of ring identification which is frequently used in algebra.

Given two rings R and S' we say that R *can be imbedded in* S' if there exists a ring S which contains the ring R as a subring (§ 9) and which is isomorphic with S'. It is clear that if R can be imbedded in S', then S' must contain a subring which is an isomorphic image of R. We shall prove now that this condition is also sufficient. We give the sufficiency condition in the following sharp formulation:

LEMMA. *If R and S' are rings and if T_0 is a given isomorphism of R onto a subring R' of S', then there exists a ring S which contains R as a subring and which is such that T_0 can be extended to an isomorphism T of S onto S'.*

PROOF. We shall first assume that R and S' have no elements in common. We replace in S' every element r' of R' by the corresponding element $r'T_0^{-1}$ of R. The result is a set S which is the union of the two *disjoint* sets $S' - R'$ and R, where $S' - R'$ denotes the set of elements of S' which are not in R' (the *complement* of R' in S'). We extend the one to one mapping T_0 of R onto R' to a one to one mapping T of S onto S' in the following obvious fashion: $aT = aT_0$, if $a \in R$; $aT = a$ if $a \in S - R$. The mapping T is indeed one to one since $S' - R'$ and R are disjoint. We now define addition $\oplus$ and multiplication $\odot$ in S as follows: if $a, b \in S$, then $a \oplus b = (aT + bT)T^{-1}, a \odot b = (aT \cdot bT)T^{-1}$. With this definition of the ring operations in S it follows directly from Lemma 1 of § 12 that *S is a ring and that T is an isomorphism of S onto S'.* Since T_0 is an *isomorphism* of R onto R' and T coincides with T_0 on R, it follows from the very definition of the ring operations in S that if $a, b \in R$, then $a \oplus b = a + b$ and $a \odot b = a \cdot b$, where $+$ and $\cdot$ refer to the ring operations in R. Hence *the ring R is a subring of S.* Moreover, T is, by definition, an extension of T_0.

This completes the proof if R and S' are disjoint. In case R and S' have elements in common, we first replace S' by an isomorphic ring S'_1, which is disjoint from R. For this purpose, we make use of the following elementary fact from set theory: If S' and R are arbitrary sets, there exists a set S'_1 and a mapping H of S' onto S'_1 such that S'_1 is disjoint from R and H is one to one. By means of H the ring operations can be carried over from S' to S'_1 (as they were in the preceding paragraph from S' to S by means of T), S'_1 becomes a ring, and H becomes an isomorphism of S' on S'_1. If $R'_1 = R'H$, then R'_1 is a subring of S'_1 and T_0H defines an isomorphism of R onto R'_1. Since S'_1 and R are disjoint we may apply the present lemma and obtain a ring S containing R and an isomorphism T_1 of S onto S'_1 which coincides with T_0H on R. Then T_1H^{-1} is an isomorphism of S onto S' which coincides with T_0 on R. The lemma is thereby proved.

A typical situation which will occur frequently in this book and in which we shall tacitly make use of the foregoing lemma is the following: R will be a ring (as a rule, a field) which is fixed throughout the discussion, while S' may be any ring of a certain class of rings, but in each ring S' there will be a subring R' isomorphic with R. Since we shall not be concerned with the particular nature of the elements of S' but only with S' regarded as an *abstract ring*, we are free to replace S' by an isomorphic ring S containing the fixed ring R as a subring, according to the scheme indicated in the above lemma. Actually we shall seldom carry out explicitly this cumbersome substitution of S for S'. We shall, as a rule, simply say that we *identify R' with our fixed ring R*, and we shall, therefore, without further ado regard R as a subring of S'.

§ **14. Unique factorization domains.** We first give some definitions concerning divisibility concepts in an arbitrary (commutative) ring R with identity. *The zero element of R is excluded from the considerations which follow below.*

If a and b are elements of R, we say that b *divides* a (or b is *a divisor of* a) and that *a is divisible by b* (or a is *a multiple* of b) if there exists in R an element c such that $a = bc$. Notation: $b|a$, or $a \equiv 0 \pmod{b}$. It is clear that the units of R are those and only those elements of R which are divisors of 1.

If $a = b\epsilon$ and ϵ is a unit, then a and b are called *associate elements*, or simply *associates*. We have then that $b = a\epsilon^{-1}$, and hence not only does b divide a but also a divides b. Conversely, if a and b are elements of R such that $b|a$ *and* $a|b$, and *if R is an integral domain*, then a and b are associates. For we have $a = bc$ and $b = ac'$, whence $a = ac'c$, $c'c = 1$, that is, c is a unit.

A unit ϵ divides any element a of R: $a = \epsilon \cdot \epsilon^{-1}a$. The associates of an element a and the units in R are referred to as *improper* divisors of a.

An element a is called *irreducible* if it is not a unit and if every divisor of a is improper.

DEFINITION. *An integral domain R is a* UNIQUE FACTORIZATION DOMAIN (*or briefly, a* UFD) *if it satisfies the following conditions*:

UF1. *Every non-unit of R is a finite product of irreducible factors.*

UF2. *The foregoing factorization is unique to within order and unit factors.*

More explicitly, UF2 means the following: If $a = p_1p_2 \cdots p_m = q_1q_2 \cdots q_n$, where p_i and q_j are irreducible, then $m = n$, and on renumbering the q_j, we have that p_i and q_i are associates, $i = 1, 2, \cdots, m$.

Examples of unique factorization domains: (a) the ring of integers; (b) euclidean domains (see § 15, Theorem 5); (c) the ring of polynomials in any number of indeterminates, with coefficients in a field (see § 17, Theorem 10).

THEOREM 4. *For integral domains R satisfying* UF1, *condition* UF2 *is equivalent to the following condition*:

UF3. *If p is an irreducible element in R and if p divides a product ab then p divides at least one of the factors a, b.*

PROOF. Let $ab = pc$ and let

$$a = \prod_i p'_i, \quad b = \prod_j p''_j, \quad c = \prod_k q_k$$

be factorizations of a, b, and c into irreducible factors (UF1). We have

$\prod_i p'_i \cdot \prod_j p''_j = p \cdot \prod_k q_k$, and hence if we assume that UF2 holds, then p differs from one of the factors p'_i, p''_j by a unit factor, and this proves UF3.

Conversely, assume that R satisfies conditions UF1 and UF3. Since UF2 is obvious for factorizations of irreducible elements, we shall assume that UF2 holds for any element of R which can be factored into s irreducible factors and we shall prove then that UF2 holds for any element a which can be factored into $s + 1$ irreducible factors. Let

$$(1) \qquad a = \prod_{i=1}^{s+1} p_i = \prod_{j=1}^{\sigma} p'_j,$$

be two factorizations of a into irreducible factors, one of which involves exactly $s + 1$ factors. We have that p_1 divides the product of the p'_j, and hence, by UF3, p_1 must divide one of the elements $p'_1, p'_2, \cdots, p'_\sigma$. Let, say, p_1 divide p'_1. Since p'_1 is irreducible, it follows that p_1 and p'_1 are associates. Then $p'_1 = \epsilon p_1$, where ϵ is a unit, and after cancellation of the common factor p_1, (1) yields

$$(2) \qquad \prod_{i=2}^{s+1} p_i = \epsilon \prod_{j=2}^{\sigma} p'_j.$$

On the left there is a product of s irreducible factors. Hence by our assumption, the two factorizations in (2) differ only in the order of the factors and by unit factors. Since we have already shown that p'_1 differs from p_1 by a unit factor, everything is proved.

In a unique factorization domain any pair of elements a, b has a *greatest common divisor* (GCD), that is, an element d, denoted by (a, b), which is defined as follows: (1) *d is a common divisor of a and b*; (2) *if c is a common divisor of a and b, then c divides d.* The GCD of a and b is uniquely determined to within an arbitrary unit factor. The proofs of existence and uniqueness of (a, b) are straightforward and can be left to the reader.

If $(a, b) = 1$, the elements a and b are said to be *relatively prime*. The following are important but straightforward properties of relatively prime elements:

(1) *If $(a, b) = 1$ and b divides a product ac, then b divides c.*
(2) *If $(a, b) = 1$ and if $a|c$ and $b|c$, then $ab|c$.*

§ **15. Euclidean domains.** An important class of unique factorization domains is given by the so-called *euclidean domains* or rings admitting a *division algorithm*. These rings are defined as follows:

DEFINITION. *A euclidean domain E is an integral domain in which with every element a there is associated a definite integer $\varphi(a)$, provided the function φ satisfies the following conditions:*

E1. *If b divides a, then $\varphi(b) \leqq \varphi(a)$.**

E2. *For each pair of elements a, b in E, $b \neq 0$, there exist elements q and r in E such that $a = bq + r$ and $\varphi(r) < \varphi(b)$.*

The ring of integers is a euclidean ring if we set for every integer n: $\varphi(n) = |n| =$ absolute value of n. Then for any two integers a and b the ordinary division algorithm yields integers q (quotient) and r (remainder) satisfying E2. Similarly the ring $F[X]$ of polynomials in one indeterminate X, with coefficients in a field F (see § 17, Theorem 9, Corollary 3) is a euclidean ring if for any polynomial $f(X)$ in $F[X]$ we set: $\varphi(f) =$ degree of f if $f \neq 0$; $\varphi(0) = -1$.

We proceed to derive a number of consequences from the conditions E1 and E2.

a. *If $b \neq 0$, then $\varphi(0) < \varphi(b)$.* For if in E2 the element a is the element zero, then $r = -bq$. If r were different from zero, then we would have $b|r$ and hence, by E1, $\varphi(b) \leqq \varphi(r)$, in contradiction with E2. Hence $r = 0$ and $\varphi(0) < \varphi(b)$, as asserted. We note that the function $\varphi_1 = \varphi - \varphi(0)$ also satisfies conditions E1 and E2. This new "normalized" function is such that $\varphi_1(0) = 0$ and $\varphi_1(a) > 0$ if $a \neq 0$. This normalization of the function φ can therefore always be assumed *ab initio*, if desired, but it plays no particular role in the proofs given below. As a matter of fact, we could have phrased the definition of euclidean rings in such a way as to leave out the element 0 altogether. Namely, it would have been sufficient to assume that φ is defined only for elements a different from zero, provided the requirement $\varphi(r) < \varphi(b)$ in E2 had been replaced by the alternative: either $r = 0$ or $\varphi(r) < \varphi(b)$.

b. *If a and b are associates, then $\varphi(a) = \varphi(b)$.* This follows directly from E1.

c. *If a divides b and $\varphi(b) = \varphi(a)$, then a and b are associates.* Under the assumption $\varphi(b) = \varphi(a)$, condition E2 yields: $\varphi(r) < \varphi(a)$. On the other hand, if r were different from zero then from $r = a - bq$ and $a|b$ it would follow that a divides r, whence $\varphi(a) \leqq \varphi(r)$, a contradiction. Hence $r = 0$, that is, also b divides a, and therefore a and b are associates.

* In this condition the elements a and b are automatically different from zero, since the divisibility concepts introduced in the preceding section have been restricted to elements different from zero.

d. *If* ϵ *is a unit, then* $\varphi(\epsilon) = \varphi(1)$, *and conversely.* The direct statement follows from b. and the converse from c.

THEOREM 5. *A euclidean domain is a unique factorization domain.*

PROOF. We shall show that a euclidean domain E satisfies UF1 and UF3 (see § 14, Theorem 4).

VERIFICATION OF UF1. Let a be an arbitrary non-unit. Then UF1 is vacuously true for a if $\varphi(a) = \varphi(1)$ (since this equality is in fact impossible if a is a non-unit). Hence we can use induction with respect to the value of $\varphi(a)$. We shall therefore assume that UF1 is satisfied for all elements a' such that $\varphi(a') < \varphi(a)$ and we proceed to show that UF1 is then satisfied also for the given element a. If a is irreducible, there is nothing to prove. In the contrary case we have $a = bc$, where neither b nor c is an associate of a. It follows then from E1 and c that $\varphi(b) < \varphi(a)$ *and* $\varphi(c) < \varphi(a)$. Therefore, by our induction hypothesis, both b and c are finite products of irreducible factors, and consequently also a is such a product.

VERIFICATION OF UF3. We shall first prove the following lemma:

LEMMA. *Any two elements* a, b *of* $E(a, b \neq 0)$ *have a* GCD d, *and* d *is a linear combination of* a *and* b, that is, $d = \alpha a + \beta b$, $\alpha \in E$, $\beta \in E$.

Let I denote the set of all elements of E which are linear combinations, $Aa + Bb$ of a and b $(A, B \in E)$. Among the elements of I *other than zero* we select an element d for which $\varphi(d)$ is minimum. We have $d = \alpha a + \beta b(\alpha, \beta \in E)$, and on the other hand, by E2, we can find elements s and t in E such that $a = ds + t$, $\varphi(t) < \varphi(d)$. We have then $t = a - ds = a(1 - \alpha s) + b(-\beta s) \in I$ and $\varphi(t) < \varphi(d)$. Consequently, $t = 0$, that is, d *divides* a. Similarly it can be shown that d divides b, and hence d is a *common divisor* of a and b. Moreover, since d is of the form $\alpha a + \beta b$, every common divisor of a and b is also a divisor of d. Hence d is a GCD of a and b. Q.E.D.

The verification of UF3 is now immediate. For let an irreducible element p of E divide a product ab, and let us assume that p *does not divide* a. Then the GCD of p and a is 1, and hence, by the lemma, we can write $1 = \alpha a + \beta p$. Hence $b = b \cdot 1 = \alpha ab + \beta bp$, and since $p \mid ab$ it follows that $p \mid b$. This completes the proof of the theorem.

§ 16. Polynomials in one indeterminate. Given a ring R, we shall consider sequences

$$f = \{a_0, a_1, a_2, \cdots\}, \quad a_i \in R,$$

such that all but a finite number of the a_i are zero. Let S denote the set of all such sequences. If $f, g \in S$,

$$g = \{b_0, b_1, b_2, \cdots\},$$

then we define:

(1) $$f + g = \{a_0 + b_0, a_1 + b_1, a_2 + b_2, \cdots\},$$

(2) $$fg = \{a_0b_0, a_0b_1 + a_1b_0, a_0b_2 + a_1b_1 + a_2b_0, \cdots\} = \{c_k\},$$

where

(3) $$c_k = \sum_{i+j=k} a_ib_j, \quad k = 0, 1, 2, \cdots.$$

It is immediately seen that with these definitions of addition and multiplication the set S becomes a ring. The elements of this ring S will be called *polynomials over R or polynomials with coefficients in R.*

The zero element of S is the sequence $\{0, 0, 0, \cdots\}$, and we have

$$-f = \{-a_0, -a_1, -a_2, \cdots\}.$$

If R has an identity 1, *then also S has an identity* $1'$, namely, $1' = \{1, 0, 0, \cdots\}$. The converse is also true, as can be seen by writing $\{a, 0, 0, \cdots\}.1' = \{a, 0, 0, \cdots\}$, $a \in R$ (complete the proof).

If $f = \{a_i\}$ is a non-zero polynomial (that is, if not all a_i are zero) and if n is the greatest integer such that $a_n \neq 0 (n \geqq 0)$, then n is called *the degree* of f. The degree of f will be denoted by ∂f. We do not assign any degree to the zero polynomial. If $\partial f = n$, then $a_0, a_1, \cdots, a_n$ will be called the *coefficients* of f, and a_n will be called the *leading coefficient* of f. If R has an identity and $a_n = 1$, then the polynomial f will be called *monic.*

It is clear that if $\partial f \leqq \partial g$, then $\partial(f + g) \leqq \partial g$, with equality if $\partial f < \partial g$. If $\partial f = n$ and $\partial g = m$, then it follows directly from (3) that $c_{m+n} = a_nb_m$ and $c_k = 0$ if $k > m + n$. *Hence either $a_nb_m \neq 0$, in which case $fg \neq 0$, $\partial(fg) = m + n$, and the leading coefficient of fg is a_nb_m; or $a_nb_m = 0$,* and *then either $fg = 0$ or $\partial(fg) < m + n$.* The first alternative (that is, $a_nb_m \neq 0$) certainly holds if one of a_n and b_m is not a zero divisor, in particular if either (1) R has an identity and one of f and g is monic or (2) if R is an integral domain.

The natural mapping $a \rightarrow \{a, 0, 0, \cdots\}$ is an isomorphism of R onto a subring R' of S. Hence R can be imbedded in S. However, rather than replace S by some unspecified isomorphic ring S' which contains R as a subring (see § 13), we prefer in the present case to deal with the ring S itself, since our concrete definition of a polynomial as a sequence is most convenient. It must then be emphasized that we cannot regard in all cases our original ring R as a subring of S, since, in the absence of any information about the nature of the elements of R, it cannot be excluded that R and $S - R'$ have common elements, that is, that some elements of R are in fact finite sequences of other elements of R. To avoid all unnecessary notational complications, we agree from now on to replace R by some isomorphic ring for which the above set-theoretic difficulty does not arise and to regard therefore R as a subring of S.

Summarizing, we have the following

THEOREM 6. *The polynomials with coefficients in R form a ring S in which R can be imbedded as a subring. S has an identity if and only if R has an identity; and if that is so, then* $(1, 0, 0, \cdots)$ *is the identity of S, where 1 is the identity of R. If f and g are two non-zero polynomials in S, then either $fg = 0$ or $\partial(fg) \leqq \partial f + \partial g$, and we have $\partial(fg) = \partial f + \partial g$ if and only if the product $a_n b_m$ of the leading coefficients of f and g is not zero; and if that is so, then $a_n b_m$ is the leading coefficient of fg. If R is an integral domain, so is S, and the units of S arise from the units of R under the mapping $a \to (a, 0, 0, \cdots)$.*

If—as will be the case from now on—R is regarded as a subring of S, then the element 1 of R is also the identity of S, and if R is an integral domain, then the units of R are the only units of S.

We shall now assume that R has an identity 1 and denote by X the polynomial $(0, 1, 0, \cdots)$. We find at once that if $a \in R$ and m is a non-negative integer, then $aX^m = \{c_i\}$, where $c_i = 0$ if $i \neq m$, $c_m = a$. It follows that if $f = \{a_i\}$ is a polynomial of degree n, then

$$(4) \qquad f = a_0 + a_1X + a_2X^2 + \cdots + a_nX^n, \quad a_i \in R, \quad a_n \neq 0,$$

which yields the familiar expression of a "polynomial in X". We shall call X *an indeterminate* and we shall refer to the polynomials in S as *polynomials in one indeterminate* (over R). The ring S itself will be denoted by $R[X]$ and will be referred to as a *polynomial ring in one indeterminate over R*.

The polynomials in one indeterminate, which we have defined so far in a purely formal fashion, have an important functional connotation which we proceed to elucidate. Let Δ be any unitary overring of R and let $f = a_0 + a_1X + \cdots + a_nX^n$ be any polynomial in $R[X]$. If $y \in \Delta$, we set $f(y) = a_0 + a_1y + \cdots + a_ny^n$. Then $f(y) \in \Delta$. We say that $f(y)$ is the result of *substituting y for X in the expression $f(X)$ of f*. In particular, we have, then, $f(X) = f$ (taking for Δ the ring $R[X]$ itself).

If Δ is a unitary overring of R and if y is a fixed element of Δ, the mapping $f \to f(y)$ is a R-homomorphism of $R[X]$ into Δ. This statement follows from a comparison of (1), (2), (3) with the easily proved formulas

$$\sum a_i y^i + \sum b_i y^i = \sum (a_i + b_i) y^i, \tag{5}$$

$$(\sum a_i y^i)(\sum b_j y^j) = \sum c_k y^k, \quad \text{where} \quad c_k = \sum_{i+j=k} a_i b_j. \tag{6}$$

Thus if $f(X)$ and $g(X)$ are two polynomials in X and if we set

$$h(X) = f(X) \pm g(X),\ k(X) = f(X)g(X),$$

then

$$h(y) = f(y) \pm g(y),\ k(y) = f(y)g(y).$$

For f fixed, the transformation $y \to f(y)$, $y \in \Delta$, is a mapping of Δ into itself, that is, a function of Δ to Δ. We denote this function by f_Δ. Thus with every polynomial f in $R[X]$ and with every ring Δ, unitary over R, we have associated a function f_Δ on Δ to Δ. If Δ is a subring of another ring Δ_1, which is unitary over Δ, then $f_{\Delta_1} = f_\Delta$ on Δ. It is therefore apparent that any polynomial in $R[X]$ can be thought of as the symbol of a well-defined operation which can be applied to any element y of any given ring Δ unitary over R and which, if so applied, yields a well-defined function on Δ to Δ. This operation is performed by substituting y for X in the given polynomial f, or $f(X)$. From this point of view the symbol X appears indeed as an indeterminate, or "variable," which can take values in any ring containing R.

We point out that for a *given* ring Δ containing R it may very well happen that distinct polynomials in $R[X]$ give rise to the same function on Δ. This is equivalent to saying that there may exist a non-zero polynomial f such that $f(y) = 0$ for all y in Δ. This will certainly happen if $\Delta = R$ and R contains only a finite number of elements, say, $c_1, c_2, \cdots, c_n$. For then we may set $f = (X - c_1)(X - c_2) \cdots (X - c_n)$, and obviously $f(y) = 0$ for all y in R. On the other hand, there exist rings Δ containing R such that $f_\Delta \neq g_\Delta$ *whenever* $f \neq g$. The simplest example of such a ring is the ring $R[X]$ itself, for we have $f(X) = f \neq g = g(X)$. Any ring S' containing R which is R-isomorphic with $R[X]$ (see § 12), and *a fortiori*, any ring Δ which contains such a ring S' as a subring, will share with $R[X]$ the above-mentioned property.

If $f = a \in R$, then the function f_Δ is constant: $f_\Delta(y) = a$, for all $y \in \Delta$. For this reason the elements of R regarded as polynomials will be called *constants*. In view of what was said in the preceding paragraph, it may well happen that f_Δ is constant even though $f \notin R$. Nevertheless *only* those polynomials which are in R will be called constants.

§ 17. **Polynomial rings.** We consider again a ring Δ unitary over R, and we fix an element x in Δ. We then have a mapping $f \to f(x)$ of $R[X]$ into Δ and we have seen that this mapping is a homomorphism. If f is a constant, $f = a \in R$, then $f(x) = a$, whence we are dealing with an R-homomorphism of $R[X]$ (§ 12). The image of $R[X]$ under this homomorphism is a subring of Δ (Theorem 3, b, § 12). We denote this subring by $R[x]$. This subring of Δ is uniquely determined by R and x: it consists of all elements of Δ which are of the form $a_0 + a_1x + \cdots + a_nx^n$, $a_i \in R$. It can also be characterized as the least subring of Δ containing x and all the elements of R.

DEFINITION. *We shall say that x is algebraic over R if the mapping $f \to f(x)$ is a proper homomorphism (that is, not an isomorphism). In other words* (§ 11, Theorem 2), *x is algebraic over R if and only if there exists a non-zero polynomial $g(X)$ such that $g(x) = 0$. An element x of Δ is said to be transcendental over R if it is not algebraic over R.*

It follows that if x is transcendental over R, then $R[x]$ and $R[X]$ are R-isomorphic rings, the mapping $f(X) \to f(x)$ being an R-isomorphism of $R[X]$ onto $R[x]$.

Since all rings $R[x]$, where x is transcendental over R, are R-isomorphic with $R[X]$, it is natural to call all such rings *polynomial rings.* We give therefore the following

DEFINITION. *Let R be a ring with identity and let S' be a ring unitary over R. Then S' is called a polynomial ring over R if there exists at least one R-isomorphism of $R[X]$ onto S'. In other words, S' is a polynomial ring over R if S' contains at least one element x which is transcendental over R and which is such that $S' = R[x]$. Any such element x is called a generator of S' over R.*

If S' is a polynomial ring over R, and x is a generator of S' over R, we shall also say that S' is a polynomial ring over R *in the element x.* As an example, let R be the field of rational numbers, Δ the field of real numbers, π the ratio of circumference to diameter (or any other transcendental real number). Then the subring $R[\pi]$ of Δ is a polynomial ring over R in the element π.

From the very definition of polynomial rings it follows that all polynomial rings over a given ring R are R-isomorphic. We further elaborate this fact in the following

THEOREM 7. *Let S' be a polynomial ring over a ring R in an element x; let $\bar{R}$ be a ring with identity, Δ a unitary overring of $\bar{R}$, and y an element of Δ. If T_0 is a homomorphism of R onto $\bar{R}$, then T_0 can be extended in one and only one way to a homomorphism T of S' onto $\bar{R}[y]$ such that $xT = y$.*

Moreover, T will be an isomorphism if and only if T_0 is an isomorphism and y is transcendental over $\bar{R}$.

PROOF. We observe that if T exists at all, then we have

$$(\sum a_i x^i)T = \sum (a_i T)(xT)^i = \sum (a_i T_0) y^i, \; a_i \in R,$$

so that T is uniquely determined. We make use of this formula to define T. Since x is transcendental over R, every element of S' can be *uniquely* expressed in the form $\sum a_i x^i (a_i \in R)$; thus T is single-valued. It is surely a mapping of S' *onto* $\bar{R}[y]$, since T_0 is a mapping onto $\bar{R}$. Obviously $aT = aT_0$ for $a \in R$, and $xT = y$. That T is a homomorphism follows from (5) and (6) of § 16, applied to elements of $R[x]$ and $\bar{R}[y]$.

Suppose T_0 is an isomorphism and y is transcendental over $\bar{R}$. If $(\sum a_i x^i)T = 0$, then $\sum (a_i T_0) y^i = 0$. Since y is transcendental over $\bar{R}$, each $a_i T_0$ is 0; since T_0 is an isomorphism, $a_i = 0$. Thus T is an isomorphism. The converse is similarly proved.

COROLLARY. *Let S' and $\bar{S}$ be polynomial rings over a ring R in the elements x and y respectively. Then there is a unique R-isomorphism of S' onto $\bar{S}$ which maps x into y.*

We now turn to the study of a fixed polynomial ring S in an element x over a ring R with identity. The notion of degree and leading coefficient of a polynomial is carried over in an obvious fashion from the ring $R[X]$ to the given ring S. Thus, if y is any element of S, $y \neq 0$, then $y = f(x)$, where $f = f(X)$ is a uniquely determined non-zero polynomial in $R[X]$. Then the degree and leading coefficient of f will be, by definition, the degree and leading coefficient of the element y *regarded as a polynomial in x.* It must be emphasized that the degree and leading coefficient of any given element y of S are not intrinsically related to y but depend also on the choice of the generator x. We can, however, state the following

THEOREM 8. *Let R be an integral domain and let S be a polynomial ring over R in an element x. Let x' be a non-zero element of S, of degree $n > 0$ in x (that is, n = degree of x' regarded as a polynomial in x) and let $f(X)$ be any polynomial in an indeterminate X, of degree m. Then $f(x')$ is of degree mn in x. A necessary and sufficient condition that x' be a generator of S over R is that x' be linear in x (that is, $n = 1$) and with leading coefficient a unit in R. In this case the degree of an element of S relative to x' will be equal its degree relative to x.*

PROOF. Let $x' = g(x)$ and let a and b denote the leading coefficients of g and f respectively. Then the leading term of $f(x')$ is $ba^m x^{mn}$, whence the first statement of the conclusion. If x' is a generator of S over R, then $x = f(x')$ for an appropriate f, hence $mn = 1$, $ba = 1$, so that x' has the indicated form. Conversely, if x' has this form, then $x \in R[x']$, hence $S = R[x']$. Furthermore, if $f(X)$ is of degree m, $f \neq 0$, then $f(x')$ is also of degree m in x (since $n = 1$), and hence $f(x') \neq 0$. Hence x' is a transcendental over R. This completes the proof.

COROLLARY. *If T is an R-automorphism of a polynomial ring $R[x]$ (R an integral domain), then $xT = a_0 + a_1x$, where a_1 is a unit in R. Conversely, if $x' = a_0 + a_1x$ and a_1 is a unit in R, then there exists a unique R-automorphism T of $R[x]$ such that $xT = x'$.*

The first part of the corollary follows from the fact that under the assumptions made we must have $R[x] = R[xT]$. The second part follows directly from the present theorem and from the corollary to Theorem 7.

If R has zero divisors, then it is still true that elements x' of the indicated form are generators, but the other statements of this theorem need not be true. Indeed it is possible that S is a polynomial ring in an element x' whose degree in x is greater than 1. For example, let R be a ring with identity, and suppose that R contains an element $a \neq 0$ such that $a^2 = 0$. Then, if $x' = x + ax^2$, we have $x' - ax'^2 = x$, whence $R[x'] = R[x]$.

Of particular importance are the polynomial rings over a field. These will be seen to be euclidean domains as a result of

THEOREM 9. *Let R be a ring with identity and $R[x]$ a polynomial ring over R in x. Let $f(x)$ and $g(x)$ be two polynomials in $R[x]$ of respective degrees m and n, let $k = max(m - n + 1, 0)$ and let a be the leading coefficient of $g(x)$. Then there exist polynomials $q(x)$ and $r(x)$ such that*

$$a^k f(x) = q(x)g(x) + r(x),$$

and $r(x)$ is either of degree less than n or is the zero polynomial. Moreover, if a is regular in R, then $q(x)$ and $r(x)$ are uniquely determined.

PROOF. If $m < n$, then $k = 0$, and we may take $q(x) = 0$, $r(x) = f(x)$. For $m \geqq n - 1$, $k = m - n + 1$, and we prove the first part of the theorem by induction on m, observing it to be true if $m = n - 1$. Hence let $m \geqq n$. Then $af(x) - bx^{m-n}g(x)$ has degree at most $m - 1$, where b is the leading coefficient of f. By induction hypothesis there exist polynomials $q_1(x)$ and $r_1(x)$ such that

$$a^{(m-1)-n+1}(af(x) - bx^{m-n}g(x)) = q_1(x)g(x) + r_1(x), \ \partial r_1 < n$$

or

$$r_1 = 0.$$

We need now only take $q(x) = ba^{m-n}x^{m-n} + q_1(x)$, $r(x) = r_1(x)$.

Now suppose a is regular and that we have also $a^k f = q'g + r'$, $\partial r' < n$. Then $(q - q')g = r' - r$. If $q - q' \neq 0$, then the left side has degree at least n, since the leading coefficient of $g(x)$ is regular. But this is impossible since $\partial(r' - r) < n$. Hence $q - q' = 0$, $r' - r = 0$.

COROLLARY 1. *Using the notation of the theorem let $f(x)$ be in $R[x]$ and a in R. Then $f(a) = 0$ if and only if $x - a$ is a divisor of $f(x)$ in $R[x]$.*

Since $x - a$ is of degree 1, there exist $q(x) \in R[x]$ and $b \in R$ such that $f(x) = q(x)(x - a) + b$; then $f(a) = b$, whence the corollary.

If X is an indeterminate, an element a of R such that $f(a) = 0$ will be called, as usual, a *root* of $f(X)$.

COROLLARY 2. *Let $f(X)$ be in the polynomial ring $R[X]$ in one indeterminate, over an integral domain R. If $a_1, \cdots, a_m$ are distinct roots of $f(X)$ in R, then $(X - a_1) \cdots (X - a_m)$ divides $f(X)$ in $R[X]$. If $f(X) \neq 0$, the number of roots of $f(X)$ in R is at most equal to the degree of $f(X)$.*

The first statement is true for $m = 1$; hence assume it for $m - 1$ roots, so that $f(X) = (X - a_1) \cdots (X - a_{m-1})q(X)$. Then $f(a_m) = (a_m - a_1) \cdots (a_m - a_{m-1})q(a_m)$. Since there are no zero divisors, $q(a_m) = 0$, so that $X - a_m$ divides $q(X)$, whence the first statement of the theorem. The second statement follows from considerations of degree.

If R has zero divisors, Corollary 2 need not be true. Indeed a non-zero polynomial may have infinitely many roots. For example, suppose that an element a of R, different from zero, is an *absolute zero-divisor*, that is, that $ab = 0$ for all b in R. Then every element of R is a root of the polynomial aX, which therefore has infinitely many roots (if R has infinitely many elements).

Another example (in which R will have an element 1) is the following: Let A and B be two rings with identities e_A and e_B and let R be the ring of ordered pairs (a, b) defined in Example 2 of § 9. If we set $a = (e_A, 0)$, every element of the form $(0, b)$, $b \in B$ is a root of the polynomial aX, which therefore has infinitely many roots if we take for B an infinite ring.

COROLLARY 3. *A polynomial ring $F[x]$ over a field F is a euclidean domain. Every polynomial of positive degree can be factored in the form $a \prod_{i=1}^{m} f_i(x)$, where $a \in F$ and $f_i(x)$ is a monic irreducible polynomial; this factorization is unique except for order.*

If $f(x) \in F[x]$, let $\varphi(f) = \partial f$, if $f \neq 0$; let $\varphi(0) = -1$. Condition E1 of the definition of euclidean domain (§ 15) is clearly satisfied;

condition E2 follows from the theorem. Hence $F[x]$ is a unique factorization domain. Since every polynomial in $F[x]$ has a monic associate and since associates can differ only by a non-zero factor in F, the remainder of the corollary follows.

Since a field is trivially a unique factorization domain, the following theorem, which is of the greatest importance, may be regarded as a partial generalization of the preceding corollary.

THEOREM 10. *If R is a unique factorization domain, then so is any polynomial ring over R in one transcendental.*

PROOF. Throughout this proof one should bear in mind the various assertions of Theorem 6 of § 16.

We call a polynomial *primitive* if its coefficients have no common divisors (other than units). We then observe that it is possible to write any (non-zero) polynomial $f(x)$ of $R[X]$ in the form $f(x) = cf_1(x)$, where $c \in R$, and $f_1(x)$ is primitive: namely, let c equal a GCD of the coefficients of $f(x)$. Any element c satisfying the stated condition is necessarily a GCD of the coefficients of $f(x)$ and hence is determined to within a unit factor. The factor c is called the *content* of $f(x)$ and is denoted by $c(f)$. We observe that $f(x)$ is primitive if and only if $c(f)$ is a unit in R.

We can now prove that every element of $R[x]$ factors into irreducible ones. It is clear that an element of R is irreducible (or a unit) in $R[x]$ if and only if it is irreducible (or a unit) in R. From this it follows (since R is a UFD) that every polynomial of $R[x]$ of degree zero factors into irreducibles. Suppose $f(x)$ has positive degree n and that factorization has been proved for polynomials of lower degree. We write $f(x) = cf_1(x)$, where $c = c(f) \in R$ and $f_1(x)$ is primitive, and we need only prove that $f_1(x)$ is a product of irreducibles. If $f_1(x)$ is irreducible, there is nothing to prove. Otherwise, $f_1(x) = g(x)h(x)$, where $g(x)$, $h(x) \in R[x]$, and neither is a constant since $f_1(x)$ is primitive. Hence both have degree less than n, therefore they factor into irreducible polynomials, by induction assumption, and hence so does $f_1(x)$.

We complete the proof by verifying UF3: *If $p(x), f(x), g(x) \in R[x]$, $p(x)$ irreducible, and $p(x)$ divides $f(x)g(x)$, then $p(x)$ divides either $f(x)$ or $g(x)$.* The proof must be separated into two cases, depending on whether the degree of $p(x)$ is zero or positive, and each case is covered by one of the following two lemmas.

LEMMA 1. (LEMMA OF GAUSS) *If $f(x)$, $g(x) \in R[X]$, then $c(fg) = c(f)c(g)$. In particular, the product of two primitive polynomials is primitive.*

PROOF. If $c = c(f)$, $d = c(g)$, then $f(x) = cf_1(x)$, $g(x) = dg_1(x)$, and f_1 and g_1 are primitive. Since $fg = (cd)f_1g_1$, we need only prove that

f_1g_1 is primitive—that is, it is enough to prove the second assertion of the lemma. If f_1g_1 is not primitive, let p be an irreducible element of R which divides all the coefficients of f_1g_1. If $f_1(x) = \sum a_i x^i$, $g_1(x) = \sum b_j x^j$, $a_i, b_j \in R$, let a_s, b_t be the first coefficients of f_1 and g_1 respectively which are not divisible by p (these exist since f_1 and g_1 are primitive). The coefficient of x^{s+t} in $f_1(x)g_1(x)$ is

$$\cdots + a_{s-1}b_{t+1} + a_sb_t + a_{s+1}b_{t-1} + \cdots.$$

Since R is a unique factorization domain, p does not divide a_sb_t. Since it divides all terms of the above sum which precede and follow a_sb_t, it does not divide the sum itself, a contradiction. Hence $f_1(x)g_1(x)$ is primitive, as asserted.

LEMMA 2. *If $g(x)$ divides $bf(x)$, where $b \in R$ and $g(x)$ is primitive, then $g(x)$ divides $f(x)$.*

PROOF. We have $bf(x) = g(x)h(x)$, where $h(x) \in R[x]$. By Lemma 1, $b \cdot c(f) = c(g) \cdot c(h) = c(h)$. Thus b divides $c(h)$ and hence also $h(x)$, so that $g(x)$ divides $f(x)$.

We can now prove UF3 for $R[x]$. Suppose, then, that $p(x)$ divides $f(x)g(x)$, where $p(x)$ is irreducible. If the degree of $p(x)$ is zero, so that $p(x) = p \in R$, then p divides $c(fg) = c(f)c(g)$, hence (say) $p/c(f)$ (by UF3 in R), so that $p/f(x)$.

If, on the other hand, the degree of $p(x)$ is positive, we proceed as follows. Suppose $p(x)$ does not divide $f(x)$; then we show that it divides $g(x)$. Consider* the set M of all polynomials $A(x)p(x) + B(x)f(x)$, where $A(x), B(x) \in R[x]$. Among all the non-zero polynomials of M, let $\varphi(x)$ be one of least degree, and let a be its leading coefficient. According to Theorem 9, there exists a non-negative integer k and polynomials $h(x)$ and $r(x)$, such that $a^kf = \varphi h + r$, where either $r = 0$ or $\partial r < \partial\varphi$. Since $\varphi \in M$, $\varphi = Ap + Bf$, hence $r = a^kf - \varphi h = (-Ah)p + (a^k - Bh)f$, so that $r \in M$. Hence $\partial r < \partial\varphi$ is impossible, and so $r = 0$, $a^kf = \varphi h$. We write $\varphi(x) = c\varphi_1(x)$, where $c = c(\varphi)$ and φ_1 is primitive. By Lemma 2, φ_1 divides f. Similarly φ_1 divides p. Since p is irreducible and does not divide $f(x)$, it follows that φ_1 is a unit in $R[x]$, hence is in R. Hence $\varphi \in R$; that is, the set M contains a constant $\varphi \neq 0$. From $\varphi = Ap + Bf$ we obtain $\varphi g = Apg + Bfg$, so that p divides φg. Since p is irreducible and of positive degree, it is primitive, and so Lemma 2 implies that $p(x)$ divides $g(x)$.

This completes the proof of Theorem 10. We shall use the two above lemmas on various other occasions.

* It will be noticed that this proof is very much like that of the Lemma of § 15 (p. 27), the modifications being due to the fact that our ring $R[x]$ is not euclidean but is "nearly so" (in virtue of Theorem 9).

§ 18. **Polynomials in several indeterminates.** In § 16 we have defined polynomials in one indeterminate over a given ring R and have seen that each such polynomial can be expressed in the usual form $\sum a_i X^i$. By *a polynomial in n indeterminates* we have in mind a finite sum

$$\sum a_{i_1 \cdots i_n} X_1^{i_1} \cdots X_n^{i_n},$$

where the i_j are non-negative integers and $a_{i_1 \cdots i_n} \in R$, and we seek to formalize this concept. We observe that a polynomial is determined when its coefficients $a_{i_1 \cdots i_n}$ are known, that is, when to each ordered n-tuple $(i_1, \cdots, i_n)$ of non-negative integers is assigned an element $a_{i_1 \cdots i_n}$ of R. This, in effect, will be our definition.

Let I be the set of non-negative integers, I_n the set of ordered n-tuples $(i) = (i_1, \cdots, i_n)$ of elements of I, that is, each element of I_n is a sequence of n non-negative integers. If $(j) = (j_1, \cdots, j_n)$ is in I_n, we define $(i) + (j) = (i_1 + j_1, \cdots, i_n + j_n)$.

DEFINITION. *Let R be a ring with identity, n a positive integer. A polynomial over R in n indeterminates is a mapping f of I_n into R such that $(i)f = 0$ for all but a finite number of n-tuples (i). If f and g are two such polynomials, define $h = f + g$ and $k = f \cdot g$* by

$$(i)h = (i)f + (i)g,$$

$$(i)k = \sum_{(j)+(j')=(i)} (j)f(j')g.$$

If $n = 1$, we have mappings of I into R—that is, in effect, sequences of elements of R. Thus the present definition is consistent with that of § 16.

If S denotes the set of all polynomials over R in n indeterminates it is easily seen that S is a ring. For each element a in R we define a polynomial f_a by

$$(i)f_a = a \quad \text{if } (i) = (0, \cdots, 0),$$

$$(i)f_a = 0 \quad \text{otherwise.}$$

It is immediate that f_0 is the zero of S and that, moreover, S has an identity, which is given by f_1 (1 being the identity of R). It is readily verified that

$$f_a + f_b = f_{a+b}, f_a \cdot f_b = f_{ab},$$

so that the mapping $a \to f_a$ is an isomorphism of R onto the subring of S consisting of all f_a. We shall replace each f_a by the corresponding a, so that henceforth we consider S to contain R as a subring.

If ν is a fixed integer between 1 and n, let $(j^{(\nu)})$ denote the n-tuple which has the integer 1 in the ν-th place and the integer 0 elsewhere.

We define X_ν to be the element of S which assigns the identity element of R to the n-tuple $(j^{(\nu)})$ and the zero of R to every other n-tuple. If $a \in R$, and $i_1, i_2, \cdots, i_n$ are non-negative integers, then it is easily seen that $aX_1^{i_1}X_2^{i_2} \cdots X_n^{i_n}$ is the element of S which associates a with the n-tuple $(i) = (i_1, i_2, \cdots, i_n)$ and 0 with every other one. Thus every element f of S is a sum of a finite number of special polynomials of the form

$$(1) \qquad a_{(i)}X_1^{i_1}X_2^{i_2} \cdots X_n^{i_n},$$

called *monomials*, and f is the zero element of S if and only if all the *coefficients* $a_{(i)}$ are zero. Here $i_1, i_2, \cdots, i_n$ are any non-negative integers and $a_{(i)}$ is any element of R. The ring S will be denoted by $R[X_1, \cdots, X_n]$.

By the *degree* of the monomial (1) we mean the sum of the exponents $i_1 + i_2 + \cdots + i_n$. By the *degree* ∂f of any non-zero polynomial f we mean the maximum of the degrees of the monomials of which f is the sum. If all the monomials in this sum have the same degree, then f is said to be *homogeneous* or to be *a form*. If f and g are forms, then fg is clearly either zero or a form of degree $\partial f + \partial g$.

A polynomial f of degree m can be expressed uniquely in the form

$$f = f_0 + f_1 + \cdots + f_m,$$

where each f_i is either zero or a form of degree i, and $f_m \neq 0$. From this it is clear that if $f, g \in S$ and $fg \neq 0$, then $\partial(fg) \leqq \partial f + \partial g$.

We may now state

THEOREM 11. *Let R be a ring with identity. The polynomials in n indeterminates with coefficients in R form a ring S which is unitary over R. If f and g* are *non-zero polynomials in S, then either $fg = 0$ or $\partial(fg) \leqq \partial f + \partial g$. If R is an integral domain, then so is S and then $\partial(fg) = \partial f + \partial g$.*

PROOF. All has been proved but the last statement. Suppose, then, that f and g are non-zero polynomials in S of respective degrees p and q. We write

$$f = f_0 + f_1 + \cdots + f_p, \quad g = g_0 + g_1 + \cdots + g_q,$$
$$f_p \neq 0, \quad g_q \neq 0,$$

where f_i and g_j are either zero or forms of degrees i and j respectively. Now

$$fg = \sum_{k=0}^{p+q} h_k, \qquad h_k = \sum_{i+j=k} f_i g_j.$$

Since h_k is either zero or a form of degree k, the last statement of the theorem is proved if we show that $h_{p+q} = f_p g_q$ is not zero. In other words it is sufficient to show that S is an integral domain.

For this purpose we order the monomials of a given degree ν lexicographically: $X_1^{i_1}X_2^{i_2}\cdots X_n^{i_n} < X_1^{j_1}X_2^{j_2}\cdots X_n^{j_n}$ if $i_s < j_s$, where s is the least integer $(1 \leqq s \leqq n)$ such that $i_s \neq j_s$. With respect to this ordering, and for $\nu = p$, let $aX_1^{\alpha_1}X_2^{\alpha_2}\cdots X_n^{\alpha_n}$ be the first of the monomials which actually occur in $f_p(a \neq 0)$. Similarly, let $bX_1^{\beta_1}X_2^{\beta_2}\cdots X_n^{\beta_n}$ be the first monomial of degree q which actually occurs in g_q $(b \neq 0)$. Then it is immediately seen that

$$abX_1^{\alpha_1+\beta_1}X_2^{\alpha_2+\beta_2}\cdots X_n^{\alpha_n+\beta_n}$$

is the first monomial in the product f_pg_q, and since $ab \neq 0$ it follows that $f_pg_q \neq 0$.

Often theorems on polynomials in n indeterminates are proved by induction with respect to n. We shall now put in evidence this inductive aspect of polynomial rings.

Consider the set S' of those polynomials f in $R[X_1, X_2, \cdots, X_n]$ in which the indeterminate X_n does not occur at all, or—as we shall say—which are *independent of* X_n. By these polynomials we mean those mappings f of I_n into R which satisfy the following condition: $(i_1, i_2, \cdots, i_n)f = 0$ if $i_n \neq 0$. It is clear that these mappings f in S' are in (1, 1) correspondence with the mappings of I_{n-1} into R, for any such mapping f is uniquely determined by its effect on the n-tuples of the form $(i_1, i_2, \cdots, i_{n-1}, 0)$. We can therefore identify the polynomials f in $R[X_1, X_2, \cdots, X_n]$, which are independent of X_n, with corresponding polynomials in $R[X_1, X_2, \cdots, X_{n-1}]$. It is immediately seen that the ring operations in $S = R[X_1, X_2, \cdots, X_n]$ and $S_1 = R[X_1, X_2, \cdots, X_{n-1}]$ are consistent with this identification. *Hence we can (and shall) regard* $R[X_1, X_2, \cdots, X_{n-1}]$ *as a subring of* $R[X_1, X_2, \cdots, X_n]$. We now assert that *this latter ring* S *is a polynomial ring in* X_n *over the ring* S_1, in the sense of the definition of § 17. For in the first place, every subring of S which contains S_1 and X_n contains all the monomials $aX_1^{i_1}X_2^{i_2}\cdots X_n^{i_n}$ and hence contains S. In the second place, it is obvious that X_n is a transcendental over S_1. Hence $S = S_1[X_n]$.

From this last fact, and from Theorem 6 of § 16, we can conclude by induction that S *is an integral domain if* R *is.*

Much of the discussion of § 16 and § 17 can be extended to the case of polynomials in n indeterminates.

Polynomials in $R[X_1, \cdots, X_n]$ can be construed as "functions of n variables." Let Δ be any ring unitary over R and let $x_1, \cdots, x_n$ be elements of Δ. If

$$f = \sum a_{(i)}X_1^{i_1}\cdots X_n^{i_n}$$

is any polynomial in S, we define

(2) $$f(x_1, \cdots, x_n) = \sum a_{(i)} x_1^{i_1} \cdots x_n^{i_n}.$$

Then $f(x_1, \cdots, x_n)$ is in Δ and is called the *result of substituting* x_1 *for* $X_1, \cdots, x_n$ *for* X_n in f. In particular, according to this definition, $f(X_1, \cdots, X_n)$ is f itself.

For $x_1, \cdots, x_n$ fixed, $f \to f(x_1, \cdots, x_n)$ is an R-homomorphism of $R[X_1, \cdots, X_n]$ into Δ. The image in Δ of $R[X_1, \cdots, X_n]$ will be denoted by $R[x_1, \cdots, x_n]$. It is a subring of Δ and consists of all elements of the form (2); it may also be described as the smallest subring of Δ containing $x_1, \cdots, x_n$ and R.

DEFINITION 1. *The elements* $x_1, \cdots, x_n$ *will be called* ALGEBRAICALLY DEPENDENT OVER R *if the mapping* $f \to f(x_1, \cdots, x_n)$ *is a proper homomorphism. Otherwise they will be called* ALGEBRAICALLY INDEPENDENT OVER R.

Thus, $x_1, \cdots, x_n$ are algebraically dependent over R if and only if there exists a non-zero polynomial $g(X)$ such that $g(x) = 0$.

DEFINITION 2. *Let* R *be a ring with identity and* S' *a ring unitary over* R. *Then* S' *is called a polynomial ring over* R *if there exist elements* $x_1, \cdots, x_n$ in S' *which are algebraically independent over* R *and such that* $S' = R[x_1, \cdots, x_n]$. *Any such set* $\{x_1, \cdots, x_n\}$ *will be called a generating set. More specifically we say that* S' *is a polynomial ring over* R *in* $x_1, \cdots, x_n$.

Thus S' is a polynomial ring over R if and only if there is an R-isomorphism of $R[X_1, \cdots, X_n]$ onto S' for some n. In particular $R[X_1, \cdots, X_n]$ is itself a polynomial ring under this definition. Before proving the analogue of Theorem 7 of § 17, we first state the following lemma.

LEMMA. *Let* R *be a ring with identity,* S' *a unitary overring,* $x_1, \cdots, x_n$ *elements of* S', *with* $n > 1$. *Let* $R_1 = R[x_1, \cdots, x_{n-1}]$. *Then* S' *is a polynomial ring over* R *in* $x_1, \cdots, x_n$ *if and only if* R_1 *is a polynomial ring over* R *in* $x_1, \cdots, x_{n-1}$ *and* S *is a polynomial ring over* R_1 *in* x_n.

This lemma is essentially a restatement of the inductive property of polynomial rings in n indeterminates, given earlier in this section. The proof may be left to the reader.

THEOREM 12. *Let* S' *be a polynomial ring over a ring* R *in the elements* $x_1, \cdots, x_n$, *let* $\bar{R}$ *be a ring with an identity and* Δ *a unitary overring of* $\bar{R}$; *let* $y_1, \cdots, y_n$ *be elements of* Δ. *If* T_0 *is a homomorphism of* R *onto* $\bar{R}$, *then* T_0 *can be extended in one and only one way to a homomorphism* T *of* S' *onto* $\bar{R}[y_1, \cdots, y_n]$ *such that* $x_i T = y_i$, $i = 1, \cdots, n$. *Moreover,* T

will be an isomorphism if and only if T_0 is an isomorphism and $y_1, \cdots, y_n$ are algebraically independent over $\bar{R}$.

In view of the lemma, this theorem follows from Theorem 7 of § 17.

COROLLARY 1. *Let S' and $\bar{S}$ be polynomial rings over R in $x_1, \cdots, x_n$ and in $y_1, \cdots, y_n$ respectively. Then there is a unique R-isomorphism T of S' onto $\bar{S}$ such that $x_iT = y_i$, $i = 1, \cdots, n$.*

COROLLARY 2. *Let S be a polynomial ring over R in $x_1, \cdots, x_n$, and let $\{h_1, h_2, \cdots, h_n\}$ be a permutation of the integers $\{1, 2, \cdots, n\}$. Then there is a unique R-automorphism T of S such that $x_iT = x_{h_i}$, $i = 1, \cdots, n$.*

THEOREM 13. ***If R is a UFD and S is a polynomial ring over R in n elements, then S is also a unique factorization domain.***

This follows by induction from the lemma and Theorem 10 of § 17.

THEOREM 14. *Let R be an integral domain, and $f(X_1, \cdots, X_n)$ a non-zero polynomial over R in n indeterminates. Let Q be a subset of R containing infinitely many elements. Then there exist elements $a_1, \cdots, a_n$ in Q such that $f(a_1, \cdots, a_n) \neq 0$.*

PROOF. This is true for $n = 1$, by Corollary 2 to Theorem 9 of § 17. Assuming it true for $n - 1$ indeterminates, let us write $f(X_1, \cdots, X_n) = \sum_{i=0}^{k} f_i(X_1, \cdots, X_{n-1})X_n^i$, where $f_i(X_1, \cdots, X_{n-1}) \in R[X_1, \cdots, X_{n-1}]$, and $f_k(X_1, \cdots, X_{n-1}) \neq 0$. By induction hypothesis, there exist $a_1, \cdots, a_{n-1} \in Q$ such that $f_k(a_1, \cdots a_{n-1}) \neq 0$. Since $f(a_1, \cdots, a_{n-1}, X_n) \neq 0$, the quoted corollary guarantees the existence of an $a_n \in Q$ such that $f(a_1, \cdots, a_{n-1}, a_n) \neq 0$.

From this theorem it follows that if R has infinitely many elements and if $f(a, \cdots, a_n) = 0$ for *all* $a_1, \cdots, a_n \in R$, then $f(X_1, \cdots, X_n) = 0$. On the other hand, this is obviously not true if R has but a finite number of elements, as was pointed out toward the end of § 16 in the case $n = 1$.

We now turn to the study of a fixed polynomial ring S over R in n elements $x_1, \cdots, x_n$. The notion of the degree of a polynomial in S is carried over in an obvious fashion from the ring $R[X_1, \cdots, X_n]$. As in the case $n = 1$, we point out that the degree of a polynomial f in S depends on the particular generating elements $x_1, \cdots, x_n$ and not merely on the ring S. Indeed, if $n > 1$, the degree of f may actually be different if a different set of indeterminates is used, even if R is an integral domain (or even a field; see § 17, Theorem 8). For example, let $n = 2$, and let $y_1 = x_1, y_2 = x_2 + x_1^2$. Then S is clearly also a polynomial ring in y_1, y_2, but the degree of y_2 is two as a polynomial in x_1, x_2. We shall not attempt to determine all sets of elements $y_1 \cdots, y_m$ with respect to which S is a polynomial ring over R. However, we do show that the *number* of indeterminates is invariant:

THEOREM 15. *Let S be a polynomial ring in elements $x_1, \cdots, x_n$, over a ring R, and let $y_1, \cdots, y_m$ be elements of S such that $S = R[y_1, \cdots, y_m]$. Then $m \geqq n$, and equality holds in case S is a polynomial ring in $y_1, \cdots, y_m$.**

PROOF: Since $y_j \in R[x_1, \cdots, x_n]$, we may write $y_j = b_j + y'_j$, where y'_j is a polynomial in $x_1, \cdots, x_n$ without constant term, and $b_j \in R$. Now $S = R[y'_1, \cdots, y'_m]$, and $y'_1, \cdots, y'_m$ are algebraically independent over R if and only if $y_1, \cdots, y_m$ are. Hence it is sufficient to prove the theorem with the y'_j replacing the y_j; in other words, we may assume $b_j = 0$. Then we have

(3) $$y_j = b_{j1}x_1 + \cdots + b_{jn}x_n + B_j, \quad j = 1, 2, \cdots, m,$$

where $b_{j1}, \cdots, b_{jn} \in R$, and B_j is a sum of monomials in $x_1, \cdots, x_n$ of degree two or greater. Since $x_i \in R[y_1, \cdots, y_m]$,

(4) $$x_i = a_{i0} + a_{i1}y_1 + \cdots + a_{im}y_m + A_i, \quad i = 1, 2, \cdots, n,$$

where $a_{i0}, a_{i1}, \cdots, a_{im} \in R$, and A_i is a sum of monomials in $y_1, \cdots, y_m$ of degree two or greater. Substituting in (4) the expressions for the y_j from (3) we have

$$x_i = a_{i0} + \sum_{k=1}^{n} \left(\sum_{j=1}^{m} a_{ij}b_{jk} \right) x_k + \text{terms in } x_1, \cdots, x_n \text{ of degree} \geqq 2, \quad i = 1, 2, \cdots, n.$$

Since $x_1, \cdots, x_n$ are algebraically independent over R, $a_{i0} = 0$, and

(5) $$\sum_{j=1}^{m} a_{ij}b_{jk} = 1 \text{ or } 0 \text{ according as } i = k \text{ or } i \neq k; i, k = 1, 2, \cdots, n.$$

If, now, we assume $m < n$, then each of the determinants

$$\begin{vmatrix} a_{11} \cdots a_{1m} & 0 \cdots 0 \\ \vdots \quad\quad \vdots & \vdots \quad \vdots \\ a_{n1} \cdots a_{nm} & 0 \cdots 0 \end{vmatrix}, \quad \begin{vmatrix} b_{11} \cdots b_{1n} \\ \vdots \quad\quad \vdots \\ b_{m1} \cdots b_{mn} \\ 0 \ \cdots \ 0 \\ \vdots \quad\quad \vdots \\ 0 \ \cdots \ 0 \end{vmatrix}$$

has value zero. On the other hand, in view of (5), the multiplication rule for determinants implies that the product of these two determinants is 1. This contradiction shows that $m \geqq n$. The second statement is now obvious.

* See II, §12, Theorem 25, for another proof of this theorem using the concept of the degree of transcendence.

It is possible to define also polynomials in infinitely many indeterminates. If the number of indeterminates is to be countable, we may simply construct a sequence

$$R[X_1] \subset R[X_1, X_2] \subset \cdots \subset R[X_1, X_2, \cdots, X_n] \subset \cdots,$$

where each ring of this sequence is considered a subring of its successor in the manner described earlier. The set-theoretic union of these rings, which can be made into a ring in an obvious way, may be called a polynomial ring in the sequence of indeterminates $X_1, X_2, \cdots, X_n, \cdots$. We could use transfinite induction to obtain an uncountable number of indeterminates.

It is better, however, to proceed by analogy with the procedure for n variables. To construct a polynomial ring whose indeterminates shall be in (1, 1) correspondence with the elements of a given set E, we let I_E be the collection of all systems $(i) = (i_\alpha)$, where $\alpha \in E$, i_α is a non-negative integer which is zero for almost all α in E, that is, I_E is the collection of all mappings

$$(i) : \alpha \to i_\alpha$$

of E into I such that $i_\alpha = 0$ for all but a finite number of α in E. (Thus in case E consists of the integers $1, 2, \cdots, n$, (i) becomes essentially an ordered n-tuple and $I_E = I_n$). If $(j) = (j_\alpha)$ we define $(i) + (j) = (i_\alpha + j_\alpha)$.

If R is a given ring with identity, let S be the set of all mappings f of I_E into R such that $(i)f = 0$ for all but a finite number of (i) in I_E. If $f \in S$ and $g \in S$, let $h = f + g$ and $k = fg$ be defined by

$$(i)h = (i)f + (i)g$$
$$(i)k = \sum_{(j)+(j')=(i)} [(j)f(j')g].$$

It is easily seen that S is a ring and that R can be identified with a subring of S in an obvious way.

If β is a fixed element of E, let $(j^{(\beta)})$ denote that mapping of E into I such that under $j^{(\beta)}$, $\beta \to 1$, and $\alpha \to 0$ for $\alpha \neq \beta$. We may say that $(j^{(\beta)}) = (j_\alpha^{(\beta)})$ has the integer 0 in every place but the β-th, where it has the integer 1. We then define X_β to be that element of S which assigns the identity of R to $(j^{(\beta)})$ and the zero of R to every other member of I_E. If $\beta_1, \cdots, \beta_n$ are distinct members of E, consider the subset I' of I_E consisting of those (i) such that $i_\alpha = 0$ unless α is one of $\beta_1, \cdots, \beta_n$; I' is in (1, 1) correspondence with I_n in an obvious fashion. Now consider the set S' of those f in S such that $(i)f = 0$ for (i) not in I'. Such f are completely determined by what they assign to the members (i) of I' and are thus seen to be in (1, 1) correspondence with the members of the ring of polynomials over R in n indeterminates. This correspondence is easily seen to be an isomorphism. This can be shown by direct verification. Another method starts with the observation that the elements of S' are finite sums of terms of the form

$$aX_{\beta_1}{}^{h_1} \cdots X_{\beta_n}{}^{h_n},$$

where $a \in R$ and $h_1, \cdots, h_n$ are non-negative integers, so that

$$S' = R[X_{\beta_1}, \cdots, X_{\beta_n}]. \tag{6}$$

Now it can readily be checked that $X_{\beta_1}, \cdots, X_{\beta_n}$ are algebraically independent over R, so that S' is indeed isomorphic to the ring of polynomials over R in n variables.

If f is any fixed polynomial in S, then $(i)f = 0$ for all but a finite number of (i) in I_E. For each such (i), all but a finite number of i_α are 0. Taking all (i) such that $(i)f \neq 0$, and for each such (i) all α in E such that $i_\alpha \neq 0$, we get a finite number of elements $\beta_1, \cdots, \beta_n$ of E. Then it is seen that f is in the ring (6). Thus it may be said that every single f in S is really a polynomial in only a finite number of variables, and that S is the union of all its subrings of the type of (6).

In view of the observation just made, many properties of ordinary polynomial rings can be extended to the case of polynomial rings in infinitely many variables. For example, concepts like *degree* and *homogeneity* can be defined, and theorems analogous to Theorem 11, 12 and 13 can be proved.

§ **19. Quotient fields and total quotient rings.** Let K be a field and let R be a ring contained in K. We assume that R is not the null-ring. The intersection of all the subfields of K which contain R is again a subfield of K containing R. This field, which we shall denote by F, is therefore the *smallest subfield of K which contains R* (it is not to be excluded that F coincides with R). If $a, b \in R$ and $b \neq 0$, then $a, b \in F$ since $R \subset F$, and also $a/b \in F$, since F is a field. Hence F contains all the quotients of elements of R. On the other hand, the following relations

$$\frac{a}{b} \pm \frac{c}{d} = \frac{ad \pm bc}{bd}, \tag{1}$$

$$\frac{a}{b} \cdot \frac{c}{d} = \frac{ac}{bd}, \tag{2}$$

$$\left(\frac{b}{d}\right)^{-1} = \frac{d}{b}, \tag{3}$$

$$a = ab/b, \tag{4}$$

hold for any elements a, b, c, d of K, provided $b \neq 0$ and $d \neq 0$. If we take these elements to be in R and we use the assumption that R is a ring, not the nullring, we conclude at once that *the set of all quotients* a/b, *such that* $a, b \in R$, $b \neq 0$, *is already a subfield of* K *containing* R, *and therefore coincides with* F. We shall refer to F as *the quotient field of* R *in* K.

Now suppose that a ring R has been given in advance. One may, then, inquire whether R can be at all imbedded in some field K. If R is not the nullring, an obvious necessary condition is that R have no proper zero divisors. *We shall see in a moment that this condition is also sufficient.* If, then, we assume that R has no proper zero divisors, there will exist fields K containing R as a subring. In each such field K, the given ring will have a quotient field F. We shall see that *the various fields* F *thus obtained are all* R*-isomorphic.* Any one of these R-isomorphic fields may then be referred to as a *quotient field* of R. (See the definition given below.)

Actually, we shall not confine the discussion to rings which are free from proper zero divisors, but shall prove analogous results for a much wider class of rings. Let, first, R be an arbitrary ring, not the nullring. We have agreed in § 5 (p. 8) to refer to an element of R which is not a zero divisor as a *regular* element of R. Let K be a ring *with identity* containing R as a subring. Naturally, no zero divisor of R can have an inverse in K. If b is a regular element of R, b *may* have an inverse in K. If b does have an inverse in K, then K contains also the quotients a/b, where a is any element of R. We shall assume that R *contains at least one regular element which has an inverse in* K. Under this assumption, the ring K will contain all the quotients a/b such that $a, b \in R$ and b is invertible in K. Let F denote the set of all these quotients. From the fact that R contains at least one invertible element of K, it follows that F contains R [see (4)]. Furthermore, since the product of invertible elements of K is invertible, and since relations (1) to (3) hold for any elements a, b, c, d of K, provided b and d are units in K, we conclude at once that F *is a ring* (since R is a ring). We call this ring F *the quotient ring of* R *in* K.

We note the following properties of F:

(a) F *has an identity.*

For if b is an element of R which is invertible in K and 1 is the identity of K, then $1 = b/b \in F$.

(b) R *is a subring of* F.

(c) *If an element of* R *has an inverse in* K, *that inverse is in* F.

For if $b \in R$ and $b^{-1} \in K$, then $b^{-1} = b/b^2 \in F$.

(d) *Every element of F is of the form a/b, where $a, b \in R$, and b is regular in R.*

We are thus led to make the following

DEFINITION. *If R is a ring which contains at least one regular element, then a total quotient ring of R is any ring F satisfying the above conditions* (a), (b), (d), *and the following condition* (c′), *which is stronger than* (c):

(c′) *Every regular element of R has an inverse in F.*

Before proceeding to the theorems on the uniqueness (to within R-isomorphism) and the existence of a total quotient ring of R, we list below, as corollaries, a number of consequences of the above definition. *It is always assumed that R has at least one regular element.* In the following corollaries F denotes a total quotient ring of R. The letters $a, b, c, \cdots$ stand for elements of R, and any element of R which occurs in a denominator is assumed to be a regular element of R.

COROLLARY 1. *An element a/b of F is regular in F if and only if a is regular in R. Every regular element of F has an inverse in F.* IN PARTICULAR, IF R HAS NO PROPER ZERO DIVISORS, THEN F IS A FIELD.

For if a/b is regular in F, then it is obvious that a is regular in R, and therefore $b/a \in F$. The rest of the proof is obvious.

For rings R without proper zero divisors we shall therefore use the term "*quotient field*" instead of "total quotient ring."

COROLLARY 2. *If R has an identity and if every regular element of R has an inverse in R, then $F = R$. In particular, a total quotient ring of any ring R is always its own total quotient ring.*

The first part of this corollary is an immediate consequence of the definition of total quotient rings. The second part follows from Corollary 1.

COROLLARY 3. *If K is any ring which satisfies conditions* (a), (b) *and* (c′) (*with F replaced by K*), *then the quotient ring F_1 of R in K is a total quotient ring of R, and F_1 is the smallest subring of K which satisfies conditions* (a), (b) *and* (c′) (*with F replaced by F_1*). *Furthermore, F_1 is the only subring of K which is a total quotient ring of R* (in view of condition (d)).

We now proceed to the two basic theorems on the uniqueness and the existence of the total quotient ring of R.

THEOREM 16. *Let R and R' be two isomorphic rings, each containing at least one regular element, let T_0 be an isomorphism of R onto R', and let F and F' be respective total quotient rings. Then T_0 can be extended in a unique manner to an isomorphism T of F onto F'.*

PROOF. Suppose $a/b \in F$, where a and b are in R, and b is regular in R; thus bT_0 is regular in R', since T_0 is an isomorphism. If T exists at all, then from $a = b(a/b)$ we conclude $aT_0 = aT = bT \cdot (a/b)T = bT_0 \cdot (a/b)T$, so that

$$\left(\frac{a}{b}\right)T = \frac{aT_0}{bT_0}. \tag{5}$$

Thus T is uniquely determined by T_0, if it exists at all. We prove its existence by defining it according to this formula.

By this formula, T is not defined, *a priori*, as a mapping (that is, as a single-valued transformation) because an element of F may have several representations of the form a/b. However, (5) does define T as a transformation of F into F', and it is easily verified that the conditions (A) and (B) referred to in Lemma 2 of § 11, are satisfied. Moreover, if $a/b = 0$, then $a = 0$, $aT_0 = 0$, and hence $(a/b)T = 0$. It follows, therefore, by Lemma 2, that T is a homomorphism of F into F'. Since T_0 is a mapping onto R' and since F' is a total quotient ring of R', we conclude that T maps F *onto* F'. If b is regular in R and a is any element of R, then $a = ab/b$, so that $aT = (ab)T_0/bT_0 = aT_0 \cdot bT_0/bT_0 = aT_0$, *so that T is an extension of T_0.* Finally, if $(a/b)T = 0$, then $aT_0/bT_0 = 0$, $aT_0 = 0$, hence $a = 0$ (for T_0 is an isomorphism), and $a/b = 0$; since only the zero of F maps into the zero of F', T is an isomorphism (§ 11, Theorem 2). This completes the proof of the theorem.

THEOREM 17. *If R is a ring containing at least one regular element, then R possesses a total quotient ring, which is unique to within isomorphisms over R.*

PROOF. The uniqueness follows from the preceding theorem; for if F and F' are two total quotient rings of R, apply the theorem with T_0 equal to the identity automorphism of R.

We now proceed to the existence proof by *constructing* a total quotient ring of R. For this purpose we consider ordered pairs (a, b) of elements a, b of R, *in which the element b is regular*; such pairs will be called *permissible*. In the sequel, only permissible pairs will be considered.

We shall say that two (permissible) pairs (a, b) and (c, d) are equivalent—and we shall write $(a, b) \equiv (c, d)$—if $ad = cb$. In particular, $(a, b) \equiv (ac, bc)$ for any permissible pair (a, b) and any regular element c in R. It is obvious that the relation $\equiv$ is reflexive and symmetric; that is, $(a, b) \equiv (a, b)$, and if $(a, b) \equiv (c, d)$, then $(c, d) \equiv (a, b)$. This relation is also transitive; that is, if $(a, b) \equiv (c, d)$, and if $(c, d) \equiv (e, f)$, then $(a, b) \equiv (e, f)$. Namely, we have by assumption that $ad = cb$ and $cf = ed$. Multiplying the first relation by f and the second by b, we find

$adf = cbf$, $cfb = edb$, whence $afd = ebd$. Since d is not a zero divisor, $af = eb$, that is, $(a, b) \equiv (e, f)$.

It follows that the permissible pairs fall into mutually exclusive equivalence classes, each class consisting of equivalent pairs, with non-equivalent pairs belonging to different classes. We denote by $\{a, b\}$ the equivalence class which contains a given permissible pair (a, b) and we then have:

$$\{a, b\} = \{c, d\} \text{ if and only if } ad = cb.$$

Let F' denote the set of all equivalence clases $\{a, b\}$. Addition and multiplication in F' are defined as follows:

$$\{a, b\} + \{c, d\} = \{ad + cb, bd\}$$
$$\{a, b\} \cdot \{c, d\} = \{ac, bd\}.$$

Since b and d are regular, so is bd, so that the right sides of these two formulas are meaningful. We must show that the equivalence classes $\{ad + cb, bd\}$ and $\{ac, bd\}$ depend only on the classes $\{a, b\}$, $\{c, d\}$, and not on the particular pairs used to represent them. Let, then, $(a, b) \equiv (a_1, b_1)$ and $(c, d) \equiv (c_1, d_1)$. From $ab_1 - a_1b = cd_1 - c_1d = 0$ it follows that

$$(ad + cb)b_1d_1 - (a_1d_1 + c_1b_1)bd = (ab_1 - a_1b)dd_1 + (cd_1 - c_1d)bb_1 = 0,$$

and hence $(ad + cb, bd) \equiv (a_1d_1 + c_1b_1, b_1d_1)$, as asserted. Similarly, $(ac, bd) \equiv (a_1c_1, b_1d_1)$.

With these definitions of addition and multiplication in F' it is a straightforward matter to verify the commutative laws, the associative laws, and the distributive law.

Let b_0 be a fixed regular element of R. We then see that $0' = \{0, b_0\}$ is the zero element of F', moreover $\{c, d\} = 0'$ if and only if $c = 0$. If $\{a, b\} \in F'$, then $\{a, b\} + \{-a, b\} = 0'$. It is thus proved that F' is a ring. Clearly $1' = \{b_0, b_0\}$ is the identity of F'; moreover $\{c, d\} = 1'$ if and only if $c = d$.

It is easily verified that the set R' of elements of the form $\{ab_0, b_0\}$, where a is arbitrary in R, is a subring of F' and that the mapping

$$T_0 : a \to \{ab_0, b_0\}$$

is an isomorphism of R on R'. *We assert that F' is a total quotient ring of R'.* We must, then, verify conditions (c′) and (d) of the definition. For (c), let $\{ab_0, b_0\}$ be regular in R'; then clearly a is regular in R, so that the ordered pair (b_0, b_0a) is permissible, and $\{b_0, b_0a\}$ is the inverse of $\{ab_0, b_0\}$. For (d), let $\{a, b\}$ be arbitrary in F'; then

$$\{a, b\} = \{ab_0, b_0\} \cdot \{b_0, bb_0\} = \{ab_0, b_0\}/\{bb_0, b_0\}.$$

We thus have: F' is a total quotient ring of R', and T_0 is an isomorphism of R onto R'. By the Lemma of § 13 (*p.* 19), there exists a ring F containing R such that T_0 can be extended to an isomorphism of F onto F'. This obviously implies that F is a total quotient ring of R, and the proof is thus complete.

§ 20. Quotient rings with respect to multiplicative systems. Let R be a ring. A *multiplicative system* (abbreviation: m.s.) in R is a non-empty subset M of R which does not contain the zero of R and which is closed under multiplication—that is, if $m_1 \in M$, $m_2 \in M$, then $m_1m_2 \in M$. Let us make the additional requirement that *all the elements of M are regular in R.* Thus R contains regular elements and hence has a total quotient ring F. Since M is closed under multiplication, the set of all quotients a/m, where $a \in R$, $m \in M$, is a subring of F containing R. It will be denoted by R_M and will be called the *quotient ring of R with respect to the system M.* Note the following extreme cases.

(1) R has an identity, and M is the set of all units of R. In this case $R_M = R$.

(2) M is the set of all regular elements of R. Then $R_M = F$.

Let S be an arbitrary set of regular elements of R. The set of all finite products of elements of S is a m.s. M. We shall say that this system M is *generated* by S; it is the least m.s. containing S. The proof of the following statement is straightforward and may be left to the reader: *if M_1 and M_2 are two m.s. in R (both consisting only of regular elements) and if M is the m.s. in R generated by the union $M_1 \cup M_2$, then R_M is the least subring of F which contains the rings R_{M_1} and R_{M_2}.*

We note that M consists of the elements of M_1, the elements of M_2 and the products $m_1m_2 (m_i \in M_i, i = 1, 2)$. We also note that, quite generally, the least subring of a ring F which contains two given subrings R_1 and R_2 of F consists of the elements of R_1, R_2 and all finite sums $\sum a_ib_i$ of products of elements of R_1 with elements of $R_2 (a_i \in R_1, b_i \in R_2)$.

For a given m.s. M in R, let M' be the set of all elements of R which are units in R_M. It is clear that M' is a m.s., that every element of M' is regular in R, and that M is a subset of M'. Hence $R_M \subset R_{M'}$. On the other hand, if $b' \in M'$ and $a \in R$, then $a/b' = a \cdot 1/b' \in R_M$, since b' is a unit in R_M. Hence $R_{M'} \subset R_M$, whence $R_M = R_{M'}$. If M_1 is any m.s. in R such that $R_M = R_{M_1}$, then the elements of M_1 are units in R_M and therefore $M_1 \subset M'$. We have therefore shown that *M' is the greatest m.s. in R such that $R_M = R_{M'}$.*

The m.s. M' can also be characterized as follows: *M' is the set of elements of R which divide some element of M.* For if b' is any element of M', then $1/b' \in R_M$, that is, $1/b' = a/b$, where $a \in R$, $b \in M$, and this shows that b' is a divisor of b. Conversely, if an element b' of R divides some element b of M, say $b = ab'$, $a \in R$, then b' is regular (otherwise b would be a zero divisor) and $1/b' = a/b \in R_M$; thus b' is a unit in R_M, and hence $b' \in M'$.

The following special case is noteworthy: R is an integral domain and every element of R which is not a unit is a finite product of irreducible elements of R (that is, R satisfies UF_1, § 14, p. 21). Let S denote the set of all irreducible elements of R which divide elements of M. For the purposes of the considerations that follow, associate elements will not be regarded as distinct elements of S. Let M_0 be the m.s. *generated* by S. It is clear that M_0 is a subset of M'. It may be a proper subset of M', but since every element of M' is the associate of some element of M_0, it follows that $R_{M'} = R_{M_0}$. We note that S is uniquely determined by M', since S is also the set of all irreducible elements of R which divide elements of M'. Hence S is also uniquely determined by the given quotient ring R_M. On the other hand, given an arbitrary set S of irreducible elements of R, S generates a m.s. M_0 and thus determines a quotient ring R_{M_0}. *We conclude that there is a* (1, 1) *correspondence between the quotient rings of* R (in F), *with respect to multiplicative systems in* R, *and the sets of irreducible elements of* R.

We point out the following consequence: *If R is a unique factorization domain with quotient field F, then a necessary and sufficient condition that R and F be the only quotient rings of R with respect to m.s. in R is that any two irreducible elements of R be associates.* For if we exclude the trivial case $R = F$, then the assumption that the set of all quotient rings R_M of R contains only two elements (which are then necessarily R and F) is equivalent to the assumption that the set of all irreducible elements of R contain only two distinct subsets (one of which is the empty set; this corresponds to the case $R_M = R$). Hence there is only one irreducible element p in R (apart from associates of p).

THEOREM 18. *If M is a m.s. in a ring R and $\bar{M}$ is a m.s. in the ring $\bar{R} = R_M$, then $\bar{R}_{\bar{M}}$ is the quotient ring of R with respect to a suitable m.s. in R (all the m.s. under consideration are assumed to contain only regular elements).*

PROOF. We may assume that $\bar{M}$ is the maximal m.s. in $\bar{R}$ with respect to which $\bar{R}$ has the given quotient ring $\bar{R}_{\bar{M}}$. Then $\bar{M}$ contains all the units of $\bar{R}$, and therefore $\bar{M} \supset M$. Let $M_1 = \bar{M} \cap R$. Then M_1 is a m.s. in R, $M_1 \supset M$, and we have $R_{M_1} \subset \bar{R}_{\bar{M}}$. On the other hand, let $\alpha = \dfrac{a/b}{a_1/b_1}$ be any element of $\bar{R}_{\bar{M}}$, where a, $a_1 \in R$, b, $b_1 \in M$ and

$a_1/b_1 \in \bar{M}$. We have $a_1 = a_1/b_1 \cdot b_1 \in \bar{M}$, since $b_1 \in M \subset \bar{M}$ and since $\bar{M}$ is a m.s. Hence $a_1 \in M_1$. Since also $b \in M \subset M_1$, it follows that $\alpha = ab_1/a_1b \in R_{M_1}$. This shows that $R_{M_1} = \bar{R}_{\bar{M}}$.

EXAMPLE 1. Let J be the ring of integers, and let M be the set of all integers which are not divisible by a given prime number p. Then the corresponding quotient ring, which we may denote by J_p, consists of all rational numbers of the form a/b, when a and b are integers and $b \not\equiv 0(p)$. The ring J_p has only one irreducible element (to within associates), namely, p itself, and hence its only quotient ring, other than J_p, is the entire field of rationals.

According to general considerations given above, every quotient ring of J can be obtained by choosing arbitrarily a (finite or infinite) set S of prime numbers and by considering all rational numbers a/b such that all prime factors of the denominator b are in S. The ring R' thus obtained is the quotient ring of J with respect to the m.s. generated in J by S. It is easily seen that the prime numbers which do *not* belong to S are the only irreducible elements of R' (apart from their associates in R'). It is a straightforward matter to verify that also R' is a UF-domain.

An interesting remark is the following: *every ring between the ring of integers J and the field of rationals F is a quotient ring of J.* For let R' be a ring between J and F and let M denote the set of all integers b such that R' contains an element of the form a/b, $(a, b) = 1$. Since $(a, b) = 1$, there exist integers λ and μ such that $\lambda a + \mu b = 1$. Hence if $a/b \in R'$, then also $1/b \in R'$, since $1/b = \lambda a/b + \mu$. From this it follows at once that M is a m.s. in J and that $R' = J_M$, as asserted.

It is clear that the foregoing proof is valid for any euclidean domain R. We have then the following result: *any ring between a euclidean domain R and the quotient field of R is a quotient ring of R with respect to some suitable m.s. in R.*

EXAMPLE 2. Let $R = k[X]$ be a polynomial ring in one indeterminate over a field k. If a is any element in k, then the polynomials $f(X)$ such that $f(a) \neq 0$ form a m.s. M, and the corresponding quotient ring R_M consists of all rational functions $g(X)/f(X)$ which have a finite value at $x = a$.

As in the preceding case of the ring of integers, so also in the present case, every ring between the ring $k[X]$ and its quotient field is a quotient ring of $k[X]$, since $k[X]$ is a euclidean domain.

EXAMPLE 3. R is a polynomial ring $k[X_1, X_2, \cdots, X_n]$ in n indeterminates X_i, over a field. If G is an arbitrary set of points $(a_1, a_2, \cdots, a_n)$ in the n-dimensional space *over* $k(a_i \in k)$, then the set of polynomials $f(X_1, X_2, \cdots, X_n)$ such that $f(a_1, a_2, \cdots, a_n) \neq 0$ for all points (a) in G is a m.s. M. The corresponding quotient ring R_M consists of all rational functions $f(X)$ which are finite at *each* point of G.

§ 21. Vector spaces

DEFINITION. *Let F be a field. A set V is called a vector space over F if*

(a) *V is a commutative group* (the group operation will be written additively) *and if*

(b) *with every ordered pair (a, x) $(a \in F, x \in V)$ there is associated a unique element of V, to be denoted by ax, such that the following relations hold for any elements a, b of F and any elements x, y of V:*

(1) $$a(x + y) = ax + ay;$$

(2) $$(a + b)x = ax + bx;$$

(3) $$(ab)x = a(bx);$$

(4) $$1 \cdot x = x.$$

The elements of a vector space V are sometimes called *vectors*, the best-known example of a vector space being the three-dimensional vector space of ordinary geometry. The element ax is sometimes called the *product* of a and x. As in § 5, it is easily proved that $a0 = 0x = 0$ (we denote by the same symbol 0 the element zero of F and the element zero of V) and that $(-1)x = -x$. Notice also that *the relation $ax = 0$ implies $a = 0$ or $x = 0$*: in fact, if $a \neq 0$, a admits an inverse a^{-1}, whence $x = 1x = (a^{-1}a)x = a^{-1}(ax) = 0$.

Given a vector space V over a field F, a *non-empty* subset W of V is called a *subspace*, or a *vector subspace*, of V if the relations $x,y \in W$ imply $x - y \in W$ (whence W is a subgroup of the group V), and if the relations $a \in F$, $x \in W$ imply $ax \in W$. A subspace W of V is also a vector space over F, if we define the product of $a \in F$ and $x \in W$ to be ax.

It is clear that any intersection of subspaces of a vector space V is itself a subspace. Thus, given any subset X of V, there exists a *least subspace* containing V, namely, the intersection of all subspaces containing X. This subspace is called the *subspace generated*, or *spanned*, by X, or *the span of X*. We shall denote it by $s(X)$. Note that our definition of $s(X)$ implies that if X is the empty set then $s(X)$ consists of the zero vector only. It is clear that $s(X)$ consists of all the linear combinations $\sum_{i=1}^{n} a_i x_i$, where $\{x_i\}$ is any finite family of elements of X and $\{a_i\}$ any finite family of elements of F. (We adopt the convention that if $\{x_i\}$ is an empty set then zero is a linear combination of the x_i.)

We shall now put into evidence five properties of the operation s of "span," from which all the other elementary properties of vector space may be deduced. This axiomatic treatment has the advantage that it also applies to the study of algebraic dependence in field theory (cf. II, § 12).

THEOREM 19. *The operation s is a mapping of the set of all subsets of V into itself which has the following properties:*

(S_1) *If $X \subset Y$, then $s(X) \subset s(Y)$.*

(S_2) *If x is an element of V and X a subset of V such that $x \in s(X)$, then there exists a finite subset X' of X such that $x \in s(X')$.*

(S_3) *For every subset X of V we have $X \subset s(X)$.*

(S_4) *For every subset X of V we have $s(s(X)) = s(X)$.*

(S_5) *The relations $y \in s(X, x)$ and $y \notin s(X)$ imply $x \in s(X, y)$* ("exchange property"). (Here $s(X, x)$ stands for $s(X \cup \{x\})$.)

PROOF. Properties (S_1) and (S_3) are evident. Property (S_2) follows from the fact that every element of $s(X)$ is a linear combination of a finite number of elements of X. Since the span of a subspace W is W itself, (S_4) holds. Finally the relation $y \in s(X, x)$ means that there exist elements a, b_i of F and x_i of X such that $y = ax + \sum_{i=1}^{n} b_i x_i$. We have $a \neq 0$ since $y \notin s(X)$. Whence $x = a^{-1}y - \sum_{i=1}^{n} a^{-1}b_i x_i$, and therefore $x \in S(X, y)$.

From now on we consider a set V with a mapping s of the set of all subsets of V into itself which satisfies conditions (S_1), (S_2), (S_3), (S_4), (S_5). A subset X of V is called a *system of generators of* V if $s(X) = V$. A subset X of V is said to be *free* if for every x in X, we have $x \notin s(X - x)$, where $X - x$ denotes the complement of $\{x\}$ in X. A *basis* of V is a subset X which is at the same time free and a system of generators. Note that if X is a free set, every subset of X is free.

CASE OF VECTOR SPACES. A system X of generators of a vector space V is a subset of V such that every element of V is a linear combination of elements of X. For X to be a free subset of V it is necessary and sufficient that the following condition holds:

(I) *Every relation $\sum_{i=1}^{n} a_i x_i = 0$ ($a_i \in F$, $x_i \in X$) implies that $a_i = 0$ for every i.*

In fact, if X is free, a relation $\sum_{i=1}^{n} a_i x_i = 0$, with, say, $a_1 \neq 0$, implies $x_1 = -\sum_{i=2}^{n} a_1^{-1} a_i x_i$, whence $x_i \in s(X - x_1)$, in contradiction with the hypothesis. Conversely, if (I) holds, a relation $x \in s(X - x)(x \in X)$ gives a non-trivial linear relation between the elements of X, in contradiction with (I).

The elements of a free subset of a vector space V are said to be *linearly independent*; notice that they must then be all distinct and all $\neq 0$. As a partial converse we notice that, if a vector x is $\neq 0$, then the subset $\{x\}$ is free according to (I), since $ax = 0(a \in F)$ implies $a = 0$.

A basis X of V is then a subset of V such that every element X of V can be expressed in one and only one way, as a linear combination of elements of X (the assertion of uniqueness is an immediate consequence of the assumption that X is free).

We now return to the axiomatic situation.

THEOREM 20. *Let X be a subset of V. The three following assertions are equivalent*:

(a) *X is a minimal system of generators of V.*
(b) *X is a maximal free subset of V.*
(c) *X is a basis of V.*

PROOF. We give a cyclic proof. Let us first prove that (a) implies (c). We have to prove that X is free. Assume the contrary to be true. There exists then an element x in X such that $x \in s(X - x)$. Since we have $X - x \subset s(X - x)$ (by (S_3)), it follows that $X \subset s(X - x)$, and therefore $V = s(X) \subset s(s(X - x))$ (by (S_1)) $= s(X - x)$ (by (S_4)). Thus $X - x$ is a system of generators, in contradiction with the hypothesis that no proper subset of X is a system of generators.

We now prove that (c) implies (b). We know that X is free. For every x in V, $x \notin X$, we have $x \in s(X)$ since X is a system of generators, whence $X \cup \{x\}$ cannot be free. Thus no subset of V properly containing X can be free, and this proves (b).

Finally we show that (b) implies (a). Let us first show that X is a system of generators. In fact, for every x in V such that $x \notin X$, $X \cup \{x\}$ is not free, whence we have, either $x \in s(X)$, or $y \in s(X - y, x)$ for some y in S. In the second case the hypothesis that X is free implies that $y \notin s(X - y)$, whence $x \in s(X - y, y) = s(X)$ by (S_5). Hence in either case we have $x \in s(X)$ for every $x \notin X$, and also for every $x \in X$ by (S_3). Therefore $s(X) = V$, and X is a system of generators. If X were not a minimal system of generators, there would exist x in X such that $V = s(X - x)$, whence $x \in s(X - x)$, in contradiction with the fact that X is free. Q.E.D.

REMARK. In the last part of the proof we have shown that, if X is free and if $x \notin s(X)$, then $X \cup \{x\}$ is free.

THEOREM 21. *Let L be a free subset of V, and S a finite system of generators of V. There exists a subset S' of S with the following properties: $L \cup S'$ is a basis of V and $L \cap S'$ is empty.*

PROOF. There exist subsets S'' of S such that $L \cup S''$ is free and $L \cap S''$ is empty (for example, the empty set). Thus, among the subsets S'' of S such that $L \cup S''$ is free and $L \cap S''$ is empty, we may choose a maximal one, S' (for example, one with the greatest possible number of elements). We need only to show now that $V = s(L \cup S')$. By (S_4) this is equivalent to showing that $S \subset s(L \cup S')$, or that, for every element x of S such that $x \notin S'$, we have $x \in s(L \cup S')$. This, however, follows from the fact that the relation $x \notin s(L \cup S')$ would imply that $L \cup S' \cup \{x\}$ is free, according to the remark made above, and this contradicts the maximality of S'. Q.E.D.

COROLLARY. *If V admits a finite system S of generators, it admits a basis $B \subset S$.*

In fact, we take for L the empty set.

Theorem 21 and its corollary remain valid if S is not a finite set. Namely, if S is any system of generators of V, one uses Zorn's lemma for proving the existence of a maximal subset S' of S such that $L \cup S'$ is free and $L \cap S'$ is empty. We shall discuss the general case in II, § 12, in connection with infinite transcendental extensions of fields.

THEOREM 22. *If V admits a finite basis B of n elements then every basis B' of V is finite and has exactly n elements.*

PROOF. Let m be the number of common elements of B and B'. If $m = n$, that is, if $B \subset B'$, then $B = B'$ by Theorem 20 (b) and the theorem is proved. We shall now assume that $m < n$ and we shall proceed by induction from $m + 1$ to m. Let $B = \{x_1, x_2, \cdots, x_n\}$. We may assume that $x_1, x_2, \cdots, x_m$ are the common elements of B and B'. The set $B - x_{m+1}$ cannot be a set of generators of V, by Theorem 20 (a). Then $s(B - x_{m+1}) \neq V$, while $s(B') = V$, and this implies that $B' \not\subset s(B - x_{m+1})$, since $s(s(B - x_{m+1})) = s(B - x_{m+1})$. Let then y be an element of B' which does not belong to $s(B - x_{m+1})$. By the remark made above, the set $B_1 = (B - x_{m+1}) \cup \{y\}$ is free. From $y \notin s(B - x_{m+1})$ and $y \in s((B - x_{m+1}), x_{m+1})$ $(= s(B) = V)$ follows by the "Exchange property" (S_5) that $x_{m+1} \in s(B_1)$. Hence $B \subset s(B_1)$, $V = s(B) \subset s(B_1)$, showing that B_1 is a system of generators of V. Thus B_1 is a base of V. Also B_1 has n elements, but B_1 and B' have the $m + 1$ elements $x_1, x_2, \cdots, x_m, y$ in common. Hence, by our induction hypothesis, B' has exactly n elements.

CASE OF VECTOR SPACES. Let V be a vector space over a field F. If V admits a finite system of generators, then V admits a *finite basis*, and any two bases of V have the same number of elements. This number is called *the dimension of V over F*, and is denoted by $[V:F]$ or by dim (V). A vector space which admits a finite basis is said to be *finite-dimensional*. If a vector space V does not admit any finite basis, we say that V is *infinite-dimensional*, and we set $[V:F] = \infty$ in this case.

We conclude this section by giving some useful results about finite-dimensional vector spaces. Given two vector spaces V, W over the same field F, we say that a mapping T of V into W is a *homomorphism* (or a *linear transformation*) if $(x + y)T = xT + yT$ for every x and y in V, and if $(ax)T = a(xT)$ for every x in V and every a in F. Then T is, in particular, a homomorphism of the additive group of V into that of W (§ 11). It is easily seen, as in Theorem 1 of § 11, that the *kernel* of T is a vector subspace of V, and that the *image* VT of V is a vector subspace of W. A homomorphism of V into W which is univalent (that is, whose kernel is (0)) is called an *isomorphism* of V into W. A homomorphism of V into itself is called an *endomorphism*; an endomorphism of V which is univalent and onto is called an *automorphism* of V.

THEOREM 23. *Let V be a finite-dimensional vector space over a field F, and T a homomorphism of V into another vector space W. Then the kernel K of T and the image VT of V are finite-dimensional vector spaces, and we have*

$$[V:F] = [K:F] + [VT:F].$$

PROOF. The fact that K is finite-dimensional is included in the following lemma:

LEMMA. *Let V be a finite-dimensional vector space and V' a subspace of V. Then V' is finite-dimensional. For every basis $(x_1, \cdots, x_p) = B$ of V' there exists a basis $(x_1, \cdots, x_p, x_{p+1}, \cdots, x_q)$ of V which extends B. (It follows that if V' is a proper subspace of V, then* dim $V' <$ dim V.)

If V' were not finite-dimensional, then no finite free subset of V' could be maximal (Theorem 20 (b)); we could then construct by induction a strictly increasing infinite sequence $X_1 < X_2 < X_3 < \cdots$ of finite free subsets of V'. Their union X is obviously free, both in V' and in V. Then Theorem 21 guarantees the existence of a basis of V containing the infinite set X, in contradiction with Theorem 22. Thus V' is finite-dimensional. Then a basis B of V' is a free subset of V, and Theorem 21 proves that it can be included in a basis of V. This proves the lemma.

This being so, let $\{x_1, \cdots, x_p\}$ be a basis of K, and let us extend it to a basis $\{x_1, \cdots, x_p, x_{p+1}, \cdots, x_q\}$ of V. We assert that $\{x_{p+1}T, \cdots, x_qT\}$ is a basis of VT. In fact, every element of VT may be written in the form $\left(\sum_{i=1}^{q} a_i x_i\right)T = \sum_{i=1}^{q} a_i(x_iT) = \sum_{j=p+1}^{q} a_j(x_jT)$, since $x_iT = 0$ for $i = 1, \cdots, p$. Thus $\{x_{p+1}T, \cdots, x_qT\}$ is a system of generators of VT. On the other hand, this system is free in VT, since a relation $\sum_{j=p+1}^{q} a_j(x_jT) = 0$ implies $\sum_{j=p+1}^{q} a_jx_j \in K$, that is, $\sum_{j=p+1}^{q} a_jx_j = \sum_{i=1}^{p} a_ix_i$ for suitable elements a_i of F. The linear independence of the vectors x_i, x_j implies that $a_j = 0$ for $j = p + 1, \cdots, q$. This proves that $[VT : F] = q - p$. Since $[V : F] = q$ and $[K : F] = p$, Theorem 23 is proved.

COROLLARY. *Let V be a finite-dimensional vector space. For an endomorphism T of V to be univalent, it is necessary and sufficient that it be onto.*

In fact the assertion that T is univalent means that its kernel K is (0), that is, that $[K : F] = 0$. The assertion that T is onto means that $VT = V$, that is, that $[VT : F] = [V : F]$ according to the lemma.

II. ELEMENTS OF FIELD THEORY

§ 1. **Field extensions.** Let k and K be two fields such that k is a subfield of K. We say then that K is an *extension* of k. If $x_1, x_2, \cdots$ x_n are fixed elements of K, then K contains the ring $k[x_1, x_2, \cdots, x_n]$ (the least subring of K which contains k and the elements $x_1, \cdots, x_n$; see I, § 18, p. 37). This ring is an integral domain (since K is a field).

If $f(X_1, X_2, \cdots, X_n)$ and $g(X_1, X_2, \cdots, X_n)$ are two polynomials in $k[X_1, X_2, \cdots, X_n]$ and if $g(x_1, x_2, \cdots, x_n) \neq 0$ [whence, *a fortiori*, $g(X_1, X_2, \cdots, X_n) \neq 0$], then the quotient $f(x_1, x_2, \cdots, x_n)/g(x_1, x_2, \cdots, x_n)$ belongs to K (since K is a field), and the set of all such quotients is a field; in fact, it is the least subfield of K which contains k and the elements $x_1, x_2, \cdots, x_n$. This field, which is merely the quotient field in K of the integral domain $k[x_1, x_2, \cdots, x_n]$ (I, § 19), shall be denoted by $k(x_1, x_2, \cdots, x_n)$. It shall be referred to as *the field generated over k by* $x_1, x_2, \cdots, x_n$, or the field obtained *by adjoining to k* the elements $x_1, x_2, \cdots, x_n$.

An extension K of k is said to be *finitely generated over k*, if $K = k(x_1, x_2, \cdots, x_n)$, where the x_i are suitable elements of K. We say that K is a *simple extension* of k if K can be obtained from k by the adjunction of a single element x.

If K and K' are two extensions of k, we say in accordance with the terminology introduced in I, § 12, that the two fields K and K' are *k-isomorphic*, or *isomorphic over k*, or *isomorphic extensions of k*, if there exists a k-isomorphism σ of K onto K'.

§ 2. **Algebraic quantities.** Let the field K be an extension of k and let x be an element of K which is algebraic over k (I, § 17, p. 28). Let $f(X)$ be a polynomial in $k[X]$ of least degree such that $f(x) = 0$.

THEOREM 1. *The polynomial $f(X)$ is irreducible over k (that is, $f(X)$ is an irreducible element of $k[X]$; see I, § 14). If $g(X)$ is any other polynomial such that $g(x) = 0$, then $f(X)$ divides $g(X)$ (in $k[X]$).*

PROOF. Suppose that $f(X) = f_1(X)f_2(X), f_i(X) \in k[X]$. Then $f_1(x)f_2(x) = 0$, and since K is a field (and hence has no proper zero divisors), either $f_1(x) = 0$ or $f_2(x) = 0$. Let, say, $f_1(x) = 0$. Since $\partial f_1 \leqq \partial f$, and since $f(X)$ is a polynomial of least degree such that $f(x) = 0$, we must have $\partial f_1 = \partial f$, and hence f_2 is of degree zero, that is, f_2 is a unit in $k[X]$. This shows that $f(X)$ is irreducible.

Let $g(X)$ be a polynomial in $k[X]$ such that $g(x) = 0$. Since $k[X]$ is a euclidean domain (I, § 17, Theorem 9), division by $f(X)$ yields: $g(X) = q(X)f(X) + r(X)$, where either $r(X) = 0$ or $\partial r < \partial f$. Substituting x for X we have $g(x) = r(x)$, whence $r(x) = 0$. Therefore we cannot have $\partial r < \partial f$, and hence $r(X) = 0$, and $f(X)$ divides $g(X)$. This completes the proof.

An immediate consequence is the following

COROLLARY. *There is one and—apart from an arbitrary unit factor $c \neq 0, c \in k$—only one irreducible polynomial $f(X)$ in $k[X]$ such that $f(x) = 0$. There is exactly one such polynomial which is monic.*

The monic irreducible polynomial in $k[X]$ of which x is a root will be called the *minimal polynomial of x in $k[X]$, or over k.*

THEOREM 2. *If x is algebraic over k, then the field $k(x)$ coincides with the ring $k[x]$. Moreover, if the minimal polynomial of x over k is of degree n, then any element of $k(x)$ has a unique expression of the form $c_0x^{n-1} + c_1x^{n-2} + \cdots + c_{n-1}, c_i \in k$.*

PROOF. Let $f(X)$ be the minimal polynomial of x over k, and let $\dfrac{h(x)}{g(x)}$ be any element of $k(x)$. Since $g(x) \neq 0, f(X)$ does not divide $g(X)$ and hence $f(X)$ and $g(X)$ are relatively prime (since $f(X)$ is irreducible, by Theorem 1). Hence 1 is a highest common divisor of $f(X)$ and $g(X)$, and we have an identity of the form $1 = A(X)f(X) + B(X)g(X)$, where $A(X)$ and $B(X)$ belong to $k[X]$. Substituting x for X, we have $1 = B(x)g(x)$, that is, $g(x)$ is a unit in $k[x]$. This implies that $h(x)/g(x) \in k[x]$, which proves the first part of the theorem.

Now let $y = g(x)$ be any element of $k(x)$, where $g(X) \in k[X]$. By the division algorithm in $k[X]$ we find as in the proof of Theorem 1 that $y = r(x) = c_0x^{n-1} + c_1x^{n-2} + \cdots + c_{n-1}$, where n is the degree of f, and the c_i are in k. If $r_1(X)$ is any other polynomial in $k[X]$, of degree $\leqq n - 1$, such that $y = r_1(x)$, then x is a root of the polynomial $r(X) - r_1(X)$, and since this polynomial is either zero or of degree $< n$ it must be the zero polynomial. This completes the proof.

COROLLARY. *If x is algebraic over k, then the field $k(x)$, regarded as a vector space over k, is of dimension n* (see I, § 21), *where n is the degree of the minimal polynomial of x over k. The elements* $1, x, x^2, \cdots, x^{n-1}$ *form a basis of $k(x)$ over k.*

THEOREM 3. *Let K and K' be two extensions of k and let x and x' be* *elements of K and K' respectively which are algebraic over k. If x and x' have the same minimal polynomial $f(X)$ in $k[X]$, then there exists a k-isomorphism of $k(x)$ onto $k(x')$ which carries x into x', and conversely.*

PROOF. Assume that x and x' are roots of one and the same irreducible polynomial $f(X)$ in $k[X]$. By Theorem 2, we obtain a (1, 1) mapping of $k(x)$ onto $k(x')$, if we let correspond to each element $c_0x^{n-1} + c_1x^{n-2} + \cdots + c_{n-1}$ of $k(x)$ the element $c_0x'^{n-1} + c_1x'^{n-1} + \cdots + c_{n-1}$ of $k(x')$. Let this mapping be denoted by σ. It is clear that σ transforms each element of k into itself and that $x\sigma = x'$. So it remains to show that σ is an isomorphism. It is obvious that $(\xi + \eta)\sigma = \xi\sigma + \eta\sigma$ for any ξ and η in $k(x)$. We now prove that $(\xi\eta)\sigma = \xi\sigma \cdot \eta\sigma$. This will complete the proof of the direct part of the theorem. Let $\xi = r(x)$, $\eta = s(x)$ and $\xi\eta = t(x)$, where $r(X)$, $s(X)$ and $t(X)$ are polynomials in $k[X]$, of degrees $\leqq n - 1$. We have then: $\xi\sigma = r(x')$, $\eta\sigma = s(x')$ and $(\xi\eta)\sigma = t(x')$. Since x is a root of $r(X)s(X) - t(X)$, we must have $r(X)s(X) - t(X) = A(X)f(X)$, where $A(X) \in k[X]$ (Theorem 1). Since also $f(x') = 0$, it follows that $r(x')s(x') = t(x')$, that is, $\xi\sigma \cdot \eta\sigma = (\xi\eta)\sigma$, as asserted.

Conversely, if there exists a k-isomorphism σ of $k(x)$ onto $k(x')$ such that $x\sigma = x'$ and if $f(X)$ is the minimal polynomial of x over k, then we have $f(x)\sigma = 0$, and since $f(x)\sigma = f(x')$ it follows that $f(x') = 0$. The consideration of σ^{-1} shows at once that not only $f(x') = 0$ but that $f(X)$ is also the minimal polynomial of x' over k.

Another proof of the direct part of the theorem is the following:

For any $F(X)$ in $k[X]$ we set $F(x)\sigma = F(x')$. Then σ is a transformation (*a priori* not necessarily single-valued) of $k[x]$ onto $k[x']$ which satisfies the homomorphism conditions for sums and products. If $F(x) = 0$, then $f(X)$ divides $F(X)$ in $k[X]$, and since also $f(x') = 0$ it follows that $F(x') = 0$, that is, $F(x)\sigma = 0$. By Lemma 2 of I, § 11, it follows that σ is a homomorphism. By the same token also σ^{-1} is single-valued. Hence σ is an isomorphism.

DEFINITION. *Two elements x and y of one and the same extension field K of k are conjugate over k if they are algebraic over k and have the same minimal polynomial over k.*

COROLLARY. *If the minimal polynomial of x over k is of degree n, then the number of conjugates of x over k in K is at most n. Moreover, if x and y are conjugates, then the fields $k(x)$ and $k(y)$ are isomorphic extensions of k.*

The first part of the corollary follows from the fact that a polynomial $f(X)$ in $K[X]$, of degree n, can have at most n roots in K (see, for instance, I, § 17, Theorem 9, Corollary 2). The second part of the corollary follows from Theorem 3.

Theorem 3 shows that if k is a field and $f(X)$ is an irreducible polynomial in $k[X]$, then there exists—up to k-isomorphisms—*at most one* simple extension $k(x)$ of k such that x is a root of $f(X)$. We prove now the following

THEOREM 3'. *If $f(X)$ is a non-constant irreducible polynomial in $k[X]$, there exists a simple extension $k(x)$ of k such that x is a root of $f(X)$.*

PROOF. It will be sufficient to prove the theorem for monic polynomials $f(X)$. Let n be the degree of $f(X)$, $n \geqq 1$. By Theorem 2, if there exists an extension $k(x)$ such that x is a root of $f(X)$, then the elements of $k(x)$ are all expressible in the form $c_0x^{n-1} + c_1x^{n-2} + \cdots + c_{n-1}$, $c_i \in k$. This suggests the following procedure for a proof of our theorem.

Consider the subset Δ of $k[X]$ consisting of the zero of $k[X]$ and of all polynomials in $k[X]$ which are of degree $\leqq n - 1$. This subset Δ is a subgroup of the additive group of $k[X]$. It is, however, not closed under multiplication in $k[X]$. We shall make the additive group Δ into a *field* by introducing in Δ a new multiplication, which we shall denote by $\circ$, and we shall show that the field thus obtained is the field whose existence is asserted in the theorem.

Let $g(X), h(X) \in \Delta$. To define the new product $g(X) \circ h(X)$ we multiply $g(X)$ and $h(X)$ in $k[X]$ and we divide the resulting polynomial by $f(X)$, getting as remainder a polynomial $r(X)$ which is either zero or is of degree $\leqq n - 1$:

$$g(X)h(X) = q(X)f(X) + r(X). \tag{1}$$

The polynomial $r(X)$ belongs to Δ and is uniquely determined by $g(X)$ and $h(X)$ ($f(X)$ being fixed).

We set

$$g(X) \circ h(X) = r(X). \tag{2}$$

It is immediately seen that this multiplication in Δ is associative, commutative and satisfies the distributive law. For instance, to prove the associative law,

$$[g(X) \circ h(X)] \circ l(X) = g(X) \circ [h(X) \circ l(X)],$$

we show that either product is equal to the remainder $r'(X)$ obtained by dividing $g(X)h(X)l(X)$ by $f(X)$. Let us show, for instance, that

$$[g(X) \circ h(X)] \circ l(X) = r'(X).$$

By (1), $g(X)h(X) - r(X)$ is divisible by $f(X)$. Hence $g(X)h(X)l(X) - r(X)l(X)$ is also divisible by $f(X)$. Since also $g(X)h(X)l(X) - r'(X)$ is divisible by $f(X)$, it follows that $r(X)l(X) - r'(X)$ is divisible by $f(X)$. Since $r'(X) \in \Delta[X]$, $r'(X)$ is the remainder of the division of $r(X)l(X)$ by

$f(X)$, and hence, according to our definition of the multiplication $\circ$, we have $r(X) \circ l(X) = r'(X)$, that is,

$$[g(X) \circ h(X)] \circ l(X) = r'(X).$$

Thus we have now Δ defined as a commutative ring. The identity 1 of $k[X]$ is also the identity of Δ. *We now prove that Δ is a field.* Let $g(X)$ be any element of Δ, different from zero. Since $g(X)$ is of degree less than n and $f(X)$ is irreducible, the two polynomials $g(X), f(X)$ are relatively prime. Hence there exist polynomials $h(X)$ and $A(X)$ such that $h(X)g(X) + A(X)f(X) = 1$. In this identity we may assume that $h(X)$ is of degree $\leqq n - 1$, since we may write $h(X) = B(X)f(X) + h_1(X)$, with $\partial h_1 \leqq n - 1$, and then we find $h_1(X)g(X) + A_1(X)f(X) = 1$, where $A_1(X) = A(X) + B(X)g(X)$. Hence $h(X)$ belongs to Δ. In the case of the two polynomials $g(X)$ and $h(X)$ under consideration, we find that (1) holds with $q(X) = -A(X)$ and $r(X) = 1$, and hence Δ is a field.

If $g(X)$ and $h(X)$ are elements of Δ such that the (old) product $g(X)h(X)$ is a polynomial $F(X)$ of degree $< n$, then from our definition of multiplication in Δ it follows that $g(X) \circ h(X) = F(X)$. Hence if $c \in k$ and m is any integer $< n$, then the element cX^m of Δ is actually the circle product $c \circ X \circ X \circ \cdots \circ X$ of c and m factors X. Since addition in Δ is the same as addition in $k[X]$, *we conclude that X is a generator of Δ over k.*

At this stage, it will be convenient to denote the element X of $k[X]$, when this element is regarded as an element of the *field* Δ, by some letter other than X, say, by x. When that is done, then, we can dispense with the symbol $\circ$, used for multiplication in Δ, without introducing any ambiguity in our notation. We therefore write $g(x)h(x)$ for $g(X) \circ h(X)$. Our last conclusion, to the effect that X is a generator of Δ over k, can now be expressed, without ambiguity, by writing: $\Delta = k(x)$.

Let now $f(X) = X^n + f_1(X)$, where $f_1(X)$ has degree $\leqq n - 1$. We have $X^{n-1} \cdot X = f(X) - f_1(X)$, hence $x^n = x^{n-1} \cdot x = -f_1(x)$, by definition (2). Therefore $x^n + f_1(x) = 0$, that is, $f(x) = 0$. This completes the proof of the theorem.

Corollary. *If k is a field and $f(X) = a_0X^n + a_1X^{n-1} + \cdots + a_n$, $a_0 \neq 0$, is an arbitrary non-constant polynomial in $k[X]$, there exists an extension field K of k such that $f(X)$ factors completely in linear factors in $K[X]$:*

$$(3) \qquad f(X) = a_0(X - x_1)(X - x_2) \cdots (X - x_n), \quad x_i \in K.$$

For $n = 1$, there is nothing to prove. We use induction with respect to n. We fix an irreducible factor $\varphi(X)$ of $f(X)$ and we consider some simple extension $k_1 = k(x_1)$ of k such that $\varphi(x_1) = 0$. Then $f(x_1) = 0$, and therefore $f(X)$ is divisible by $X - x_1$ in $k_1[X]$: $f(X) = (X - x_1)f_1(X)$, $f_1(X) \in k_1[X]$. Since $f_1(X)$ is of degree $n - 1$, there exists, by our induction hypothesis, an extension K of k_1 such that $f_1(X) = a_0(X - x_2)(X - x_3) \cdots (X - x_n)$, $x_i \in K$, and from this (3) follows.

§ 3. Algebraic extensions

DEFINITION 1. *If $K \supset k$, then K is an algebraic extension of k if every element of K is algebraic over k. Extensions which are not algebraic are called transcendental extensions.*

The simplest example of an algebraic extension is the field $k(x)$, x algebraic over k. That not only x but every element of this field is algebraic over k will follow from the theorem below and from the fact that $k(x)$ is a finite dimensional vector space over k (§ 2, Theorem 2, Corollary).

THEOREM 4. *If $K \supset k$ and if the dimension of K (regarded as a vector space over k) is finite, say n, then K is an algebraic extension of k, and every element x of K satisfies an equation of degree $\leqq n$ over k (whence the minimal polynomial of x in $k[X]$ is of degree $\leqq n$; see § 2, Theorem* 1).

PROOF. $1, x, x^2, \cdots, x^n$ are linearly dependent over k.

DEFINITION 2. *The dimension n of K over k is called the degree of K over k and is denoted by $[K:k]$. We set $[K:k] = \infty$ if K, regarded as a vector space over k, has infinite dimension. If $[K:k]$ is finite, then K is said to be a finite extension of k, or also that K/k is a finite extension.*

COROLLARY. *If K is an extension of k and $x \in K$, then x is algebraic over k if and only if $k(x)$ is a finite extension of k. In that case, if $n = [k(x):k]$, the minimal polynomial of x in $k[X]$ is of degree n.*

This follows at once from the preceding theorem and from Theorem 2 of § 2.

Let k, K and L be fields such that $k \subset K \subset L$ and let $[K:k] = n$, $[L:K] = m$.

THEOREM A. *If $\omega_1, \omega_2, \cdots, \omega_n$ is a basis of K/k and $\xi_1, \xi_2, \cdots, \xi_m$ is a basis of L/K, then the mn products*

(1) $$\omega_i\xi_j, \quad i = 1, 2, \cdots, n; j = 1, 2, \cdots, m,$$

form a basis of L/k.

PROOF. If ζ is any element of L, then $\zeta = \sum_{j=1}^{m} A_j\xi_j$, $A_j \in K$. Furthermore, we have $A_j = \sum_{i=1}^{n} a_{ij}\omega_i$, $a_{ij} \in k$. Hence $\zeta = \sum_{i=1}^{n}\sum_{j=1}^{m} a_{ij}\omega_i\xi_j$. This shows that L, regarded as a vector space over k, is spanned by the mn

vectors $\omega_i\xi_i$. It remains to show that these mn vectors are linearly independent over k. Let $\sum_{i=1}^{n}\sum_{j=1}^{m} c_{ij}\omega_i\xi_j = 0$, $c_{ij} \in k$. We set $C_j = \sum_{i=1}^{n} c_{ij}\omega_i$. Then $\sum_{j=1}^{m} C_j\xi_j = 0$, $C_j \in K$, and since the ξ's form a basis of L over K, we must have $C_j = 0, j = 1, 2, \cdots, m$. From $\sum_{i=1}^{n} c_{ij}\omega_i = 0$ and from the fact that the ω's form a basis of K over k, we conclude that all the c_{ij} are zero. This completes the proof.

An immediate consequence of the foregoing theorem is the following relation:

$$[L:k] = [L:K]\cdot[K:k]. \tag{2}$$

THEOREM B. *If $x_1, x_2, \cdots, x_n$ are in an extension field K of k and are algebraic over k, then $k(x_1, x_2, \cdots, x_n)$ is an algebraic extension of k, of finite degree.*

PROOF. Each x_i, being algebraic over k, is *a fortiori* algebraic over $k(x_1, x_2, \cdots, x_{i-1})$. Hence $k(x_1, x_2, \cdots, x_i)$ is a *simple algebraic* extension of $k(x_1, x_2, \cdots, x_{i-1})$, and therefore $[k(x_1, x_2, \cdots, x_i): k(x_1, x_2, \cdots, x_{i-1})] = m_i =$ a finite integer $\geqq 1$ (by the corollary of Theorem 4). It follows then from (2) that $[k(x_1, x_2, \cdots, x_n): k] = m_1m_2 \cdots m_n$, and Theorem 4 is applicable.

COROLLARY. *If K is an extension field of k, the elements of K which are algebraic over k form a field.*

THEOREM C. *If K is an algebraic extension of k and L is an algebraic extension of K, then L is an algebraic extension of k.*

PROOF. Assume first that the degree $[K:k]$ is finite, and let x be any element of L. Since x is algebraic over K, the field $K(x)$ has finite degree over K. Hence by (2), $K(x)$ has also finite degree over k, and *a fortiori* $k(x)$ has finite degree over k. This implies that x is algebraic over k. In the general case, let $X^n + A_1X^{n-1} + \cdots + A_n$ be some polynomial in $K[X]$ which has x as a root (for instance, the minimal polynomial of x over K), and let $K' = k(A_1, A_2, \cdots, A_n)$. Then x is already algebraic over K', and since K' is finitely generated over k, the relative degree $[K':k]$ is finite, by Theorem B. The assertion that x is algebraic over k now follows from the preceding case.

COROLLARY. *Let K be an extension field of k and let k_0 be the subfield of K consisting of the elements of K which are algebraic over k (see Corollary of Theorem B). Then every element of K which is algebraic over k_0 belongs to k_0.*

We express this property of the field k_0 by saying k_0 is *algebraically closed in K*. We refer to k_0 as *the algebraic closure of k in K*.

§ 4. **The characteristic of a field.** Let k be a field and let e be the identity of k. The integral multiples ne of $e(n \gtreqless 0)$ form a subring E of k [in view of the relations $(n \pm m)e = ne \pm me$, $(nm)e = ne \cdot me$ (I, § 4)], in fact the least subring of k containing e.

Let Δ be the quotient field of E in k (I, § 19). Any subfield of k contains the ring E and hence must also contain the field Δ. Hence Δ is the *smallest subfield* of k, and is in fact the intersection of all the subfields of k.

DEFINITION 1. *A field which does not contain any proper subfields is called a prime field.*

It follows from this definition that the above subfield Δ of k is a prime field. Since every subfield of k contains Δ, Δ is the only prime subfield of k. *Thus every field k contains a unique prime field.*

We consider the mapping

$$n \rightarrow ne \tag{1}$$

of the ring J of integers onto E. This mapping is a homomorphism (in view of the relations given above). Two cases are possible: (a) either (1) is an isomorphism, or (b) it is a proper homomorphism.

If (1) is an isomorphism, we say that k has *characteristic zero.* In this case, we have $ne \neq 0$ if $n \neq 0$, and the ring E is an infinite ring, isomorphic to the ring J of integers. The quotient field Δ of E in k is then isomorphic to the field of rational numbers, the isomorphism between the former and the latter being given by $\frac{n}{m} \cdot e \leftrightarrow \frac{n}{m}$, $m \neq 0$ (see I, § 19, Theorem 16). It is clear that if a field k is of characteristic zero, then every subfield of k is of characteristic zero, and that if one subfield of k is of characteristic zero, then k itself is of characteristic zero.

We also note that—as has just been shown—*any prime field of characteristic zero is isomorphic to the field of rational numbers.*

We now consider the case in which the homomorphic mapping (1) is not an isomorphism. In this case, the kernel N of (1), that is, the set of all n such that $ne = 0$, contains at least one integer n which is different from 0 (I, § 11, Theorem 2). Since $ne = 0$ implies $-ne = 0$, the kernel contains also positive integers. Let p be the least positive integer in N. We have then

$$pe = 0 \tag{2}$$

and

$$re \neq 0, \text{ if } 0 < r < p. \tag{3}$$

Since N is a subring of J (I, § 12, Theorem 3, c), N contains also all the multiples mp of p. On the other hand, if n is an arbitrary integer, we can write $n = qp + r$, where $0 \leqq r < p$, and then we find that $ne = qpe + re = re$, since $qpe = 0$. We have therefore,

$$(4) \qquad ne = re, \qquad 0 \leqq r < p,$$

and hence, by (3), $ne \neq 0$ if *n is not divisible by p*, for in that case $0 < r$. The kernel N of the homomorphsim (1) consists therefore of all the multiples of p.

Since $1 \cdot e = e \neq 0$, p is greater than 1. We assert that *p is a prime number.* For if $p = n_1 n_2$, then $0 = (n_1 n_2)e = (n_1 e)(n_2 e)$, and hence either $n_1 e = 0$ or $n_2 e = 0$ (since k is a field and has no proper zero divisors), that is, either n_1 or n_2 is equal to p. The prime number p is called the *characteristic of the field k.* Every field k has therefore a well-defined characteristic p which is either zero or a prime number ($p > 1$).

We continue with the case $p \neq 0$. Relation (4) shows that *the ring E is finite and consists of the elements*:

$$0, e, 2e, \cdots, (p - 1)e.$$

These p elements are *distinct*, in view of (3).

The ring E is a field. For let ne be any non-zero element of E. Since n is not divisible by p, n and p are relatively prime and hence there exist integers m and q such that $mn - qp = 1$. We have then $(me)(ne) = (mn)e = (qp)e + e = e$, and so ne has the inverse me, which proves that E is a field. (Note the similarity of this reasoning to that employed in the proof of Theorem 2 in § 2.)

Let k' be any other field *of the same characteristic* $p \neq 0$ *as* k, and let E' be the set of integral multiples ne' of the identity e' of k'. It is then immediately seen that the transformation $ne \to ne'$ is an isomorphic mapping of E onto E'. We thus see that if there exist at all fields of a given characteristic $p \neq 0$, then there also exist prime fields of characteristic p, and any two prime fields of the same characteristic p are isomorphic. Using the ring J of integers, we can now construct fields of any characteristic $p \neq 0$. The construction is quite similar to that of simple algebraic extensions of a field k, used in the proof of Theorem 3′ in § 2. The role of the irreducible polynomial $f(X)$ is now played by the prime number p. We denote, namely, by J_p the set of integers $0, 1, 2, \cdots, p - 1$. If m and n are any elements of J_p, we define addition $+$ and multiplication $\circ$ in J_p as follows: $m + n$ is the remainder of the division of $m + n$ by p and $m \circ n$ is the remainder of the division of mn by p. Using arguments similar to those used in the proof of Theorem 3′ in § 2, one proves that J_p is a field. Since every element of

$\mathcal{J}_p$ is clearly an integral multiple of 1 (that is, we have $m = 1 + 1 + \cdots + 1$, m times, for all m such that $1 \leqq m \leqq p - 1$), $\mathcal{J}_p$ is a prime field. Since $\mathcal{J}_p$ contains p elements, p is the characteristic of $\mathcal{J}_p$.

The following identities hold in any field k of characteristic p:

$$pa = 0, \tag{5}$$

$$(b \pm c)^p = b^p \pm c^p, \tag{6}$$

where a, b and c are elements of k. The first of these relations follows from $pa = p(ea) = (pe)a$. The second relation is obtained by observing that since p is a prime number, all the binomial coefficients of $(a \pm b)^p$, except the first and the last, are divisible by p. Hence applying (5), we have $(b \pm c)^p = b^p + (\pm 1)^p c^p$. If $p \neq 2$, p is odd and (6) follows. If $p = 2$, we have $(b - c)^2 = b^2 + c^2$, but this time we have $c^2 = -c^2$ since $2c^2 = 0$.

The identity (6) leads to an important consequence. Let k be a field of characteristic p *different from zero* and let us denote by k^p the set of all elements of k which are of the form a^p, $a \in k$. By (6), the set k^p is closed under addition and subtraction. Since we also have for any b and c in k: $b^p c^p = (bc)^p$ and—if $c \neq 0$—$b^p/c^p = (b/c)^p$, k^p is also closed under multiplication and division. *Hence k^p is a subfield of k.* We consider the mapping

$$x \rightarrow x^p, \quad x \in k. \tag{7}$$

Clearly, we have $xy \rightarrow x^p \cdot y^p$. This, in conjunction with (6), implies that the mapping (7) is a homomorphism. Since $x^p = 0$ implies $x = 0$, *it follows that* (7) *is an isomorphism of k onto k^p.*

DEFINITION 2. *A field k is called perfect if it is either of characteristic zero or is of characteristic $p \neq 0$ and coincides with its subfield k^p.*

It follows that if k is of characteristic $p \neq 0$, it is perfect if and only if for every element x in k there exists another element y in k such that $x = y^p$. This element y is uniquely determined by x, since (7) is one to one. This element y is denoted by $\sqrt[p]{x}$.

If k is of characteristic $p \neq 0$ and is not perfect, there exist elements in k which are not p-th powers of elements of k. If x is such an element, there we agree to indicate this property of x by the notation: $\sqrt[p]{x} \notin k$.

An example of a perfect field of characteristic p is the prime field $\mathcal{J}_p$. To see this, we shall prove a more general result.

DEFINITION 3. *A Galois field is a field containing only a finite number of elements.*

It is clear that the characteristic of a Galois field must be different from zero, for any field of characteristic zero contains the (infinite) field of rational numbers.

Now suppose that k is a Galois field and let p be the characteristic of k. Since (7) is an *isomorphism* of k *onto* k^p, the two fields have the same (finite) number of elements. Since $k^p \subset k$, it follows that $k = k^p$. We have thus proved

THEOREM 5. *Every Galois field is perfect.*

§ 5. **Separable and inseparable algebraic extension.** Let k be a field and let $k[X]$ be the polynomial ring in the indeterminate X over k. If

$$f(X) = a_0X^n + a_1X^{n-1} + \cdots + a_n,\ a_i \in k,\ a_0 \neq 0,$$

is any polynomial in $k[X]$, of degree n, we define the *derivative* $f'(X)$ *of* $f(X)$ in the usual fashion:

$$f'(X) = na_0X^{n-1} + (n-1)a_1X^{n-2} + \cdots + a_{n-1}.$$

The derivative $f'(X)$ is again a polynomial in $k[X]$. If the characteristic of k is zero, then a coefficient $(n-i)a_i$ of $f'(X)(i = 0, 1, \cdots, n-1)$ can be zero if and only if a_i is zero. *Hence* $f'(X) \neq 0$ if $n > 0$.

Suppose, however, that k has characteristic $p \neq 0$. In that case, $(n-i)a_i$ is zero if *either* $a_i = 0$ *or* $n - i$ is divisible by p. In particular, since $a_0 \neq 0$, we have $na_0 = 0$ only if n is divisible by p. It follows that $f'(X) = 0$ if and only if n is divisible by p *and* all those coefficients a_i of $f(X)$ are zero for which $n - i$ is not divisible by p. When that is so, the terms a_iX^{n-i} which actually occur in $f(X)$ are such that the exponent $n - i$ is divisible by p. *That signifies that* $f(X)$ *is a polynomial in* X^p, that is, $f(X) \in k[X^p]$. This, then, is a necessary and sufficient condition for the vanishing of $f'(X)$.

DEFINITION 1. *An irreducible polynomial* $f(X)$ *in* $k[X]$ *is separable or inseparable according as* $f'(X) \neq 0$ *or* $f'(X) = 0$. *An arbitrary polynomial* $f(X)$ in $k[X]$ *is separable if all its irreducible factors are separable; otherwise* $f(X)$ *is inseparable.*

If k is of characteristic zero, every polynomial in $k[X]$, of positive degree, is separable. For fields of characteristic $p \neq 0$ we have the following

THEOREM 6. *A field k of characteristic $p \neq 0$ is perfect if and only if every polynomial in $k[X]$ of positive degree is separable.*

PROOF. Assume k perfect. It will be sufficient to show that every irreducible polynomial in $k[X]$ of positive degree is separable. Now if $f(X)$ is an arbitrary polynomial in $k[X]$ such that $f'(X) = 0$, then $f(X) \in k[X^p]$, that is, we have $f(X) = \Sigma b_jX_j^p = (\Sigma \beta_jX^j)^p$, where $\beta_j = \sqrt[p]{b_i} \in k$ (since k is perfect), and hence $f(X)$ is not irreducible.

Conversely, assume that k is not perfect. There exists then at least one element a in k such that a is not the p-th power of an element of k. Set $f(X) = X^p - a$. We have $f'(X) = 0$, and hence the proof will be complete *if we show that* $X^p - a$ *is irreducible in* $k[X]$. We shall prove the following more general result:

THEOREM 7. *If* $a \in k$, $\sqrt[p]{a} \notin k$ *and* e *is an integer* $\geqq 0$, *then* $X^{p^e} - a$ *is irreducible in* $k[X]$.

PROOF. The theorem is trivial if $e = 0$ for in that case we have $X^{p^e} - a = X - a$ (the condition $\sqrt[p]{a} \notin k$ is in this case irrelevant). We now proceed by induction with respect to e. Let $\varphi(X)$ be a monic irreducible factor of $X^{p^e} - a$ in $k[X]$ and let $[\varphi(X)]^h$ be the highest power of $\varphi(X)$ which divides $X^{p^e} - a$:

(1) $$X^{p^e} - a = [\varphi(X)]^h\psi(X), \quad (\varphi(X), \psi(X)) = 1.$$

Taking derivatives of both sides★ and dividing by $[\varphi(X)]^{h-1}$, we have the identity

$$h\varphi'(X)\psi(X) + \varphi(X)\psi'(X) = 0,$$

and hence $\psi(X)$ divides the product $\varphi(X)\psi'(X)$. Since $\psi(X)$ and $\varphi(X)$ are relatively prime, we must have $\psi'(X) = 0$, for in the contrary case $\psi'(X)$ would be a non-zero polynomial of smaller degree than $\psi(X)$, and $\psi(X)$ could not divide $\psi'(X)$. We must therefore have simultaneously: $h\varphi'(X) = 0$, $\psi'(X) = 0$. The second of these relations implies that $\psi(X) \in k[X^p]$, say, $\psi(X) = \psi_1(X^p)$, where $\psi_1(X) \in k[X]$. The first implies that the derivative of $[\varphi(X)]^h$ is zero, whence also $[\varphi(X)]^h \in k[X^p]$, say $[\varphi(X)]^h = \varphi_1(X^p)$, $\varphi_1(X) \in k[X]$. Hence, by (1), we have $X^{p^e} - a = \varphi_1(X^p)\psi_1(X^p)$, or—replacing X^p by $X : X^{p^{e-1}} - a = \varphi_1(X)\psi_1(X)$. Since $X^{p^{e-1}} - a$ is irreducible in $k[X]$ (by our induction hypothesis) and since $\varphi_1(X)$ is of positive degree, it follows that $\psi_1(X)$ is of degree zero, and hence $\psi_1(X) = 1$ since both polynomials $X^{p^{e-1}} - a$ and $\varphi_1(X)$ are monic. We have therefore $X^{p^{e-1}} - a = \varphi_1(X)$, $X^{p^e} - a = [\varphi(X)]^h$. Were h a multiple of p, $X^{p^e} - a$ would be a power of $[\varphi(X)]^p$, and since the coefficients of $[\varphi(X)]^p$ belong to k^p it would then follow that also the coefficients $X^{p^e} - a$ all belong to k^p. This, however, is in contradiction with our assumption that $\sqrt[p]{a} \notin k$. Hence h is not divisible by p. Since $h\varphi'(X) = 0$, it follows now that $\varphi'(X) = 0$, $\varphi(X) \in k[X^p]$, and this implies at once that $h = 1$ for otherwise the relation $X^{p^e} - a = [\varphi(X)]^h$ would imply that $X^{p^{e-1}} - a$ is reducible in $k[X]$. Hence $X^{p^e} - a = \varphi(X)$, Q.E.D.

★ We use here the familiar rule for the derivative of a product. This rule is a straightforward consequence of our purely formal definition of the derivative of a polynomial.

A shorter proof of the above theorem can be given by making use of the existence of an algebraic extension k' of k such that $\varphi(X)$ has a root α in k' (see Theorem 3', § 2). We have $a = \alpha^{p^e}$ and hence $X^{p^e} - a = (X - \alpha)^{p^e}$. It is now easy to see that $X^{p^e} - a$ is necessarily a power of $\varphi(X)$. For assume this is not the case, and let $\psi(X)$ be an irreducible factor of $X^{p^e} - a$ such that $(\varphi(X), \psi(X)) = 1$. Then we have an identity of the form $A(X)\varphi(X) + B(X)\psi(X) = 1$, where $A(X)$, $B(X) \in k[X]$. Since $\psi(X)$ divides $(X - \alpha)^{p^e}$ in $k'[X]$, α is a root of $\psi(X)$, and hence the substitution $X \rightarrow \alpha$ in the above identity leads to a contradiction $(0 = 1)$. Since therefore $\varphi(X)$ is necessarily a power of $X - \alpha$, it follows that $X^{p^e} - a = [\varphi(X)]^{p^\rho}$, $\rho \geqq 0$, and the fact that $\sqrt[p]{a} \notin k$ yields at once $\rho = 0$.

Let $f(X)$ be a polynomial in $k[X]$. If $f(X) \in k[X^p]$, then $f(X) = f_1(X^p)$, and the degree of $f(X)$ is divisible by p. If also $f_1(X) \in k[X^p]$, then $f_1(X) = f_2(X^p)$ and $f(X) = f_2(X^{p^2})$, and the degree of $f(X)$ is divisible by p^2. Since the degree of $f(X)$ is finite, there exists an integer $e \geqq 0$ such that $f(X) \in k[X^{p^e}]$, $f(X) \notin k[X^{p^{e+1}}]$. We set then $f(X) = f_0(X^{p^e})$, $n =$ degree of $f(X)$, $n_0 =$ degree of $f_0(X)$. Here $f_0(X) \notin k[X^p]$ and

$$n = n_0 p^e.$$

DEFINITION 2. *The integer n_0 is called the reduced degree of $f(X)$, or the degree of separability of $f(X)$, while e and p^e are called respectively the exponent of inseparability and the degree of inseparability of $f(X)$.*

It is clear that an *irreducible* polynomial $f(X)$ is separable if and only if $n = n_0$.

Let K be an extension field of a field k and let x be an element of K which is algebraic over k.

DEFINITION 3. *The element x is separable or inseparable over k according as the minimal polynomial $f(X)$ of x in $k[X]$ is separable or inseparable.*

It follows that if k is of characteristic zero or is a perfect field of characteristic $p \neq 0$, then every algebraic quantity over k is necessarily separable.

COROLLARY 1. *If x is algebraic over k and $f(X)$ is the minimal polynomial of x over k, then x is inseparable over k if and only if $f'(x) = 0$. If x is inseparable over k and $g(X)$ is any polynomial in $k[X]$ such that $g(x) = 0$, then $g'(x) = 0$.*

For if $f(x) = f'(x) = 0$, $f(X)$ irreducible, then necessarily $f'(X) = 0$ since $f'(X)$ is of smaller degree than $f(X)$; and hence x is inseparable over k. Conversely, if x is inseparable over k, then $f'(X) = 0$

and hence $f'(x) = 0$. If $g(x) = 0$, then $f(X)$ divides $g(X)$, $g(X) = A(X)f(X)$, $g'(x) = A(x)f'(x)$, and hence $g'(x) = 0$ if x is inseparable over k.

Let $g(X)$ be any polynomial in $k[X]$ such that $g(x) = 0$. Since $x \in K$ and $g(X)$ is also a polynomial in $K[X]$, $X - x$ must divide $g(X)$ in $K[X]$. Let $(X - x)^s$ be the highest power of $X - x$ which divides $g(X)$ in $K[X]$:

$$g(X) = (X - x)^s g_1(X), \tag{2}$$

where $g_1(X) \in K[X]$ and $g_1(x) \neq 0$. Since x belongs also to the subfield $k(x)$ of K, a similar argument is applicable to the field $k(x)$ (instead of to K), and hence if $(X - x)^\rho$ is the highest power of $X - x$ which divides $g(X)$ in $k(x)[X]$, then $g(X) = (X - x)^\rho g_2(X)$, where $g_2(X) \in k(x)[X] \subset K[X]$ and $g_2(x) \neq 0$. The identity $(X - x)^s g_1(X) = (X - x)^\rho g_2(X)$, together with the inequalities $g_1(x) \neq 0$, $g_2(x) \neq 0$, implies that $\rho = s$, $g_1(X) = g_2(X)$. Hence the integer s depends only on x and $g(X)$, and not on the choice of the extension field K of k containing x. This integer s is called *the multiplicity of the root x of $g(X)$*. We say that x is a *simple* root or a *multiple* root of $g(X)$ according as $s = 1$ or $s > 1$.

Taking derivatives of both sides of (2) we find that if $s = 1$ then $g'(x) = g_1(x) \neq 0$, and if $s > 1$ then $g'(x) = 0$. We can therefore re-state Corollary 1 in the following form:

COROLLARY 2. *If x is algebraic over k and $f(X)$ is the minimal polynomial of x in $k[X]$, then x is inseparable over k if and only if x is a multiple root of $f(X)$. If $g(X)$ is any polynomial in $k[X]$ such that $g(x) = 0$ and if x is inseparable over k, then x is a multiple root of $g(X)$.*

DEFINITION 4. *An algebraic extension K of k is a separable extension of k if every element of K is separable over k. In the contrary case, K is called an inseparable extension of k.*

From now on we shall assume in this section that the characteristic p of k is $\neq 0$.

DEFINITION 5. *An element $x \in K$ is purely inseparable over k if some p^e-th power of x belongs to k, $e \geqq 0$.* (In particular, if $e = 0$, that is, if $x \in k$, then x is purely inseparable over k.) *K is a purely inseparable extension of k if every element of K is purely inseparable over k.*

COROLLARY 3. *If K is a finite purely inseparable extension of k, then the degree $[K:k]$ is a power of p.*

By formula (2) of § 3 it is sufficient to prove the corollary under the assumption that K is a simple extension of k, say, $K = k(x)$. Let $e \geqq 0$ be the smallest integer such that $x^{p^e} \in k$, and let $x^{p^e} = a$. Then a

is not a p-th power of an element of k, and hence the polynomial $X^{p^e} - a$ is irreducible in $k[X]$ (Theorem 7). Since x is a root of this polynomial, $X^{p^e} - a$ is the minimal polynomial of x over k and hence (Theorem 2, Corollary, § 2) $[k(x) : k] = p^e$.

LEMMA 1. *If x is both separable and purely inseparable over k, then $x \in k$.*

PROOF. If e is the least non-negative exponent such that $x^{p^e} \in k$ and if $x^{p^e} = a$, then the proof of the Corollary 3 shows that $X^{p^e} - a$ is the minimal polynomial of x in $k[X]$. Since x is separable over k, it follows that $f'(X) \neq 0$, and this is possible only if $e = 0$. Hence $x \in k$, as asserted.

LEMMA 2. *If K is a separable algebraic extension of k and L is any field between k and K $(k \subset L \subset K)$, then K is a separable algebraic extension of L.*

PROOF. Let x be any element of K and let $f(X)$ be the minimal polynomial of x in $k[X]$. Since x is separable over k, x is a simple root of $f(X)$, by the first part of Corollary 2. Since $f(X)$ is also a polynomial in $L[X]$, it follows then from the second part of Corollary 2 that x is also separable over L.

If L is any subset of K, we denote by $k(L)$ the subfield of K obtained by adjoining to k all the elements of L, that is, $k(L)$ is the set of all elements of K which are of the form $f(x_1, x_2, \cdots, x_n)/g(x_1, x_2, \cdots, x_n)$, where $f(X_1, X_2, \cdots, X_n), g(X_1, X_2, \cdots, X_n) \in k[X_1, X_2, \cdots, X_n], x_i \in L$ $(i = 1, 2, \cdots, n)$, $g(x_1, x_2, \cdots, x_n) \neq 0$ and n is an arbitrary integer. We denote by $k[L]$ the ring consisting of all polynomials $f(x_1, x_2, \cdots, x_n)$ such as above. Then $k(L)$ is the smallest subfield of K which contains k and L, $k[L]$ is the smallest subring of K containing k and L, and $k(L)$ is the quotient field of $k[L]$ in K.

We shall denote by kL the set of all finite sums of products of elements of k by elements of L. This set is, in general, not a ring, unless L is a ring, and in the latter case we have $kL = k[L]$.

If L is a field and if every element of L is algebraic over k, then $k(L) = kL$. For in that case we have for any elements $x_1, x_2, \cdots, x_n$ of $L : k(x_1, x_2, \cdots, x_n) = k[x_1, x_2, \cdots, x_n]$ (Theorem 2, § 2), hence $k(x_1, x_2, \cdots, x_n) \subset k[L] = kL$.

We now proceed to prove the following criterion for separable algebraic extensions:

THEOREM 8. *If K is a separable algebraic extension of k, then $kK^p = K$. Conversely, if K is an extension of k such that $kK^p = K$ and if the extension K/k is* **finite**, *then K is a separable extension of k.*

PROOF. From a preceding observation it follows that kK^p is a field.

Moreover, every element of K is purely inseparable over kK^p, for $K^p \subset kK^p$. If K is a separable extension of k, it follows then from $k \subset kK^p \subset K$ and from Lemmas 1 and 2 that $kK^p = K$, which proves the first part of the theorem.

Assume now that $kK^p = K$ and that K is a finite (hence algebraic) extension of k. Let $[K:k] = n$. Let $\omega_1, \omega_2, \cdots, \omega_h$ be any elements of K which are linearly independent over k. *We assert that* $\omega_1{}^p, \omega_2{}^p, \cdots, \omega_h{}^p$ *are also linearly independent over* k. For the proof of this assertion, extend the set $\omega_1, \omega_2, \cdots, \omega_h$ to a basis $\omega_1, \omega_2, \cdots, \omega_n$ of K/k. We have $K = \sum_{i=1}^{n} k\omega_i$, $K^p = \sum_{i=1}^{n} k^p\omega_i{}^p$, $K = kK^p = \sum_{i=1}^{n} k\omega_i{}^p$. This shows that $\omega_1{}^p, \omega_2{}^p, \cdots, \omega_n{}^p$ also form a basis of K over k, which proves our assertion. Now let x be any element of K, let $f(X)$ be the minimal polynomial of x over k, and let m be the degree of $f(X)$. Assume for a moment that x is inseparable over k, and let m_0 be the reduced degree of $f(X)$ (see Definition 2), so that $m_0 < m$. Then $1, x, x^2, \cdots, x^{m_0}$ are linearly independent over k, but $1, x^{p^e}, x^{2p^e}, \cdots, x^{m_0p^e}$ are linearly dependent over k, a contradiction. Hence x is separable over k. Q.E.D.

COROLLARY. *If x is separable over k, then $k(x) = k(x^p)$, and therefore $k(x)$ is a separable extension of k. Conversely, if $k(x) = k(x^p)$, then x is separable over k.*

For if we set $K = k(x)$, then $K^p = k^p(x^p)$ and $kK^p = k(x^p)$, and since $k(x)$ is a finite extension of k, it follows, by the theorem just proved, that x is separable over k if $k(x) = k(x^p)$. On the other hand, if x is separable over k, x is both separable (Lemma 2) and purely inseparable over $k(x^p)$, and hence $x \in k(x^p)$, $k(x) \subset k(x^p) \subset k(x)$, i.e., $k(x) = k(x^p)$. Thus $K = kK^p$, and K is a separable extension of k.

THEOREM 9. *If L is a separable extension of k and K is a separable extension of L, then K is a separable extension of k.*

PROOF. Since every element x of K is separable algebraic over $k(L_0)$, where L_0 is a suitable finite subset of L (depending on x), it is sufficient to prove the theorem for *finite* extensions L/k, K/L. We have, by Theorem 8, $L = kL^p$, $K = LK^p = kL^pK^p \subset kK^p$, hence $K = kK^p$. Since K/k is a finite extension, K is separable over k (second half of Theorem 8).

THEOREM 10. *If $x_1, x_2, \cdots, x_n$ are elements of K which are separable over k, then $k(x_1, x_2, \cdots, x_n)$ is a separable extension of k.*

PROOF. Set $K_i = k(x_1, x_2, \cdots, x_i)$. We know that K_1 is a separable extension of k (Theorem 8, Corollary). Assume that K_i is a separable extension of k. Since x_{i+1} is separable over K_i (Lemma 2), it follows that K_{i+1} is a separable extension of K_i, whence K_{i+1} is also a separable extension of k (Theorem 9). This completes the proof.

Let K be an arbitrary extension of k and let $\bar{k}$ be the algebraic closure of k in K (see end of § 3, p. 61). Let k_0 be the set of all elements of K which are separable algebraic over k. Then $k \subset k_0 \subset \bar{k}$, and k_0 is a field (Theorem 10). *We shall refer to k_0 as the maximal separable extension of k in K.* We say that k is *quasi-algebraically closed* in K if $k = k_0$.

Let x be any element of $\bar{k}$ and let $f(X)$ be the minimal polynomial of x over k. Let p^e be the degree of inseparability of $f(X)$ ($e = 0$, if and only if $x \in k_0$). Then x^{p^e} is separable over k, and therefore $x^{p^e} \in k_0$. *Consequently x is purely inseparable over k_0.* This holds for any element x of $\bar{k}$; *consequently $\bar{k}$ is a purely inseparable extension of k_0.* It follows that any algebraic extension $\bar{k}$ of k can be obtained in two steps: a separable extension $k \to k_0$ followed by a purely inseparable extension $k_0 \to \bar{k}$.

Let K be a *finite algebraic* extension of k. In this case $\bar{k} = K$. Let $n_0 = [k_0 : k]$. The degree $[K : k_0]$ is a power of p, since K is a purely inseparable extension of k_0 (see Definition 5, Corollary). Let $[K : k_0] = p^e$. Then $[K : k] = n = n_0 p^e$. The integers n_0 and p^e are called respectively *the separable* and the *inseparable factor* of the degree $[K : k]$, or also the *degree of separability* and the *degree of inseparability* of K/k. In symbols

$$n_0 = [K : k]_s, \quad p^e = [K : k]_i, \tag{3}$$

whence

$$[K : k] = [K : k]_s \cdot [K : k]_i. \tag{4}$$

We consider now the special case in which $K = k(x) =$ *a simple algebraic* extension of k. Let $f(X)$ be the minimal polynomial of x over k. Let n_0 be the reduced degree of $f(X)$ and let p^e be the degree of inseparability of $f(X)$, so that $n = n_0 p^e$, where n is the degree of $f(X)$. *It is not difficult to see that n_0 and p^e are equal respectively to $[k(x) : k]_s$ and $[k(x) : k]_i$.* For let $y = x^{p^e}$. Then y is separable over k, and $k(y)$ is a separable extension of k. Moreover $[k(y) : k] = n_0$, since the minimal polynomial of y over k has degree n_0. The element x is purely inseparable over $k(y)$, and hence *any* element of $k(x)$ is *purely inseparable* over $k(y)$ (from $x^{p^e} \in k(y)$ follows $z^{p^e} \in k(y)$ for all z in $k(x)$). It follows that every element of $k(x)$ which is separable over k [and hence also over $k(y)$] belongs to $k(y)$ (Lemma 1). Hence $k(y) = k_0$, and $n_0 = [k(x) : k]_s$. We have $n_0 \cdot p^e = n = [k(x) : k] = [k(x) : k]_s \cdot [k(x) : k]_i = n_0 \cdot [k(x) : k]_i$, and therefore $p^e = [k(x) : k]_i$, as asserted.

§ 6. Splitting fields and normal extensions. We have shown in § 2 that if $f(X)$ is any polynomial in $k[X]$, then there exists an extension K of k such that $f(X)$ factors completely in $K[X]$ into linear factors:

$$(1) \qquad f(X) = a_0(X - x_1)(X - x_2) \cdots (X - x_n), \quad x_i \in K.$$

Here a_0 is the leading coefficient of $f(X)$. If $f(X)$ is irreducible in $k[X]$, then the n quantities x_i are distinct [and hence each x_i is a simple root of $f(X)$], if and only if $f(X)$ is a separable polynomial (see § 5, Definition 3, Corollary 2). If $f(X)$ is reducible in $k[X]$, then the x_i are distinct if and only if $f(X)$ is a separable polynomial *and* has no multiple factors in $k[X]$. This follows from the fact that two distinct irreducible monic polynomials in $k[X]$ cannot have a common root in any extension field of k (a quantity x which is algebraic over k has a unique minimal polynomial in $k[X]$).

THEOREM 11. *If $f(X)$ is an irreducible inseparable polynomial in $k[X]$, of reduced degree n_0 and exponent of inseparability e, then each linear factor in* (1) *appears exactly p^e times.*

PROOF. We have $f(X) = \varphi(X^{p^e})$, where $\varphi(X)$ is an irreducible separable polynomial in $k[X]$. Each element $x_i^{p^e}$ is a root of $\varphi(X)$, necessarily a simple root, and hence $\varphi(X) = (X - x_i^{p^e})\varphi_i(X)$, where $\varphi_i(X) \in K[X]$ and $\varphi_i(x_i^{p^e}) \neq 0$. We have, then, $f(X) = (X - x_i)^{p^e} f_i(X)$ where $f_i(x_i) = \varphi_i(x_i^{p^e}) \neq 0$. This shows that $X - x_i$ is exactly a p^e-fold factor of $f(X)$, as asserted.

Let K be an extension field of k in which $f(X)$ factors completely into linear factors and let (1) be the factorization of $f(X)$ in $K[X]$. The field $k(x_1, x_2, \cdots, x_n)$ is clearly the smallest subfield of K which contains k and in which $f(X)$ factors completely into linear factors.

DEFINITION 1. *The field $k(x_1, x_2, \cdots, x_n)$ is called a splitting field over k of the polynomial $f(X)$.*

A splitting field of $f(X)$, over k, is therefore any extension field L of k in which $f(X)$ factors completely into linear factors *and* which is generated over k by the roots of $f(X)$ in L.

We have proved in § 2 (see Theorem 3) that if $f(X)$ is an irreducible polynomial in $k[X]$ and x, x' are roots of $f(X)$ in some extension fields K and K' of k respectively, then the fields $k(x)$ and $k(x')$ are k-isomorphic extensions of k. Our next object is to prove the following analogous result for splitting fields: *if $f(X)$ is an arbitrary polynomial in $k[X]$* (not necessarily irreducible), *any two splitting fields of $f(X)$ over k are k-isomorphic extensions of k.* Before we do that, we restate Theorem 3 of § 2 in a slightly more general form:

LEMMA 1. *Let τ be an isomorphism between two fields k and $\bar{k}$ and let*

$\varphi(X) = a_0X^n + a_1X^{n-1} + \cdots + a_n$ *be an irreducible polynomial in* $k[X]$. *Let* $\bar{\varphi}(X) = [\varphi(X)]\tau$ *be the corresponding polynomial in* $\bar{k}[X]$, *that is, let* $\varphi(X) = \bar{a}_0X^n + \bar{a}_1X^{n-1} + \cdots + \bar{a}_n$, *where* $\bar{a}_i = a_i\tau$. *Let, moreover,* x *be a root of* $\varphi(X)$ *in some extension field of* k *and let* $\bar{x}$ *be a root of* $\bar{\varphi}(X)$ *in some extension field of* $\bar{k}$. *Then the isomorphism* τ *can be extended to an isomorphism* ρ *of* $k(x)$ *onto* $\bar{k}(\bar{x})$ *such that* $x\rho = \bar{x}$, *and the extension is unique.*

If $k = \bar{k}$ and τ is the identity, then the lemma coincides with Theorem 3 of § 2. In the general case, the proof of the lemma is similar to the proof of Theorem 3 and may be left to the reader.

The uniqueness theorem on splitting fields, which we propose to prove, is the following:

THEOREM 12. *Let* k, $\bar{k}$ *and* τ *have the same meaning as in the preceding lemma, and let* $f(X)$ *be an arbitrary monic polynomial in* $k[X]$, *of degree* n. *Let* $\bar{f}(X) = [f(X)]\tau$ *be the corresponding polynomial in* $\bar{k}[X]$ *and let* k', $\overline{k'}$ *be splitting fields over* k *and* $\bar{k}$ *of* $f(X)$ *and* $\bar{f}(X)$ *respectively. Then the isomorphism* τ *can be extended to an isomorphism* ρ *of* k' *onto* $\overline{k'}$, *and any such extension* ρ *sends each root of* $f(X)$ *in* k' *into a root of* $\bar{f}(X)$ *in* $\overline{k'}$ *(and similarly* ρ^{-1} *sends each root of* $\bar{f}(X)$ *into a root of* $f(X)$.

PROOF. The theorem is trivial for $n = 1$. We shall therefore proceed by induction from $n - 1$ to n. Let $\varphi(X)$ be an irreducible factor of $f(X)$ in $k[X]$ and *let* $\bar{\varphi}(X) = [\varphi(X)]\tau$ be the corresponding irreducible factor of $\bar{f}(X)$ in $\bar{k}[X]$. Then both $\varphi(X)$ and $\bar{\varphi}(X)$ have roots in k' and $\overline{k'}$ respectively. We fix a root x_1 of $\varphi(X)$ in k' and a root $\overline{x_1}$ of $\bar{\varphi}(X)$ in $\overline{k'}$. By Lemma 1, there exists an isomorphism τ_1 between $k(x_1)$ and $\bar{k}(\bar{x}_1)$ which is an extension of τ and which sends x_1 into $\bar{x}_1$. Let $f(X) = (X - x_1)f_1(X)$ and $\bar{f}(X) = (X - \bar{x}_1)\bar{f}_1(X)$. The polynomials $f_1(X)$ and $\bar{f}_1(X)$ have coefficients in $k(x_1)$ and $\bar{k}(\bar{x}_1)$ respectively, and are of degree $n - 1$. It is clear that they are corresponding polynomials under the isomorphism τ_1 between $k(x_1)$ and $\bar{k}(\bar{x}_1)$. Furthermore, the fields k' and $\overline{k'}$ are respectively splitting fields of $f_1(X)$ over $k(x_1)$ and of $\bar{f}_1(X)$ over $\bar{k}(\bar{x}_1)$. It follows from our induction hypothesis that the isomorphism τ_1 can be extended to an isomorphism ρ of k' onto $\overline{k'}$. Then ρ is an extension of τ, and since the last statement in the theorem is self-evident, the proof is complete.

COROLLARY. *Let* k' *be a splitting field over* k *of a polynomial* $f(X)$ *whose coefficients belong to a certain subfield* k_0 *of* k. *Then any* k_0-*isomorphism of* k *into* k' *can be extended to an automorphism of* k'.

For, in the present case, the isomorphic field $\bar{k}$ is contained in k', and, on the other hand, the polynomial $\bar{f}(X)$ coincides with $f(X)$. Hence we can take for $\bar{k}'$ the field k' itself.

The preceding theorem has several important consequences. We recall (§ 2, *p.* 57) that two algebraic elements x and y of an extension field K of k are said to be conjugate over k if they are the roots of one and the same irreducible polynomial in $k[X]$. It was shown earlier in this section that if an element x of K is a root of an irreducible polynomial $f(X)$ in $k[X]$, of reduced degree n_0 and exponent of inseparability e, then x is a p^e-fold root of $f(X)$. Hence x has at most n_0 conjugate elements in K (including x itself). If the number of conjugate elements of x contained in K is exactly n_0, or—what is the same thing—if $f(X)$ factors completely in $K[X]$ into linear factors, then we shall say that "*K contains all the conjugates of x over k.*"

DEFINITION 2. *An extension K of k is said to be normal over k, or a normal extension of k, if K is an algebraic extension of k and if every irreducible polynomial $f(X)$ in $k[X]$ which has a root in K factors completely in $K[X]$ into linear factors, or—what is the same thing—if K contains then a splitting field of $f(X)$ over k.*

It is clear that this definition is equivalent to the following: *K is a normal extension of k if K is an algebraic extension of k and contains with every element x also all the conjugates of x over k.*

COROLLARY 1. *If K is a finite normal extension of k, then K is a splitting field of some polynomial $f(X)$ in $k[X]$.*

For let $K = k(\alpha_1, \alpha_2, \cdots, \alpha_m)$ be a finite normal extension of k, and let $f_i(X)$ be the minimal polynomial of α_i in $k[X]$. Since K is normal over k, K contains a splitting field of $f_i(X)$ over k. Then K also contains a splitting field, over k, of the product $f(X)$ of the m polynomials $f_i(X)$. Since K is generated over k by roots of $f(X)$ (namely by $\alpha_1, \alpha_2, \cdots, \alpha_m$), it follows that K itself is a splitting field of $f(X)$ over k.

COROLLARY 2. *If K is a finite normal extension of k and α, β are any two elements of K which are conjugate over k, then there exists a k-automorphism of K which sends α into β.*

For, by Corollary 1, K is a splitting field of some polynomial $f(X)$ in $k[X]$. Then K is also a splitting field of $f(X)$ over $k(\alpha)$ and also over $k(\beta)$. Since there exists a k-isomorphism of $k(\alpha)$ onto $k(\beta)$ which sends α into β, our corollary follows at once from Theorem 12.

COROLLARY 3. *Let K be a finite normal extension of k. If an element α of K is left invariant under all k-automorphisms of K, then α is purely inseparable over k.*

For α must then coincide with all its conjugates over k, by Corollary 2, and hence the minimal polynomial of α in $k[X]$ has reduced degree 1.

COROLLARY 4. *If K is a finite normal extension of k and if L is a field between k and K, then any k-isomorphism of L into K can be extended to an automorphism of K.*

Apply the corollary of Theorem 12, taking for k', k, and k_0 the fields K, L, and k respectively.

We shall have occasion to use the following lemma:

LEMMA 2. *Let $k \subset L \subset \Delta \subset K$ be successive finite algebraic extensions of k, where K is a normal extension of k. If Δ possesses n L-isomorphisms into K, then every k-isomorphism of L into K has exactly n extensions which are isomorphisms of Δ into K.*

PROOF. Let G be the group of all k-automorphisms of K and let $G(L)$ (respectively, $G(\Delta)$) be the subgroup of G consisting of those automorphisms of K which leave fixed every element of L (respectively, of Δ). It is clear that $G(\Delta)$ is a subgroup of $G(L)$. Let

$$G(L) = \bigcup_{i=1}^{n} G(\Delta)\varphi_i, \tag{1}$$

$$G = \bigcup_{j=1}^{m} G(L)\psi_j, \tag{2}$$

be the decomposition of $G(L)$ into right $G(\Delta)$-cosets and that of G into right $G(L)$-cosets. Then the mn $G(\Delta)$-cosets are distinct and

$$G = \bigcup_{i,j} G(\Delta)\varphi_i\psi_j \tag{3}$$

is the decomposition of G into right $G(\Delta)$-cosets.

It is clear that the m automorphisms ψ_j have distinct restrictions to L and that the restriction of any element ψ of G to L coincides with the restriction of one of the ψ_j. Since by Corollary 4 to Definition 2 every k-isomorphism of L into K is the restriction of some automorphism of K, it follows that L has exactly m k-isomorphisms into K and that these are given by the restrictions of $\psi_1, \psi_2, \cdots, \psi_m$ to L.

In a similar fashion it follows from (3) that Δ has exactly mn k-isomorphisms into K and that these are given by the restrictions of the mn products $\varphi_i\psi_j$ to Δ. Now, since each φ_i reduces to the identity on L and since ψ_j and $\psi_{j'}$ have distinct restrictions to L if $j \neq j'$, it follows that each k-isomorphism of L into K, say the isomorphism represented by the restriction of ψ_j, has exactly n extensions to Δ which are k-isomorphisms of Δ into K, namely the restrictions of $\varphi_1\psi_j, \varphi_2\psi_j, \cdots, \varphi_n\psi_j$ to Δ. In particular, the identical isomorphism of L into K has also n such extensions to Δ, that is, Δ possesses exactly n L-isomorphisms into K. This completes the proof of the lemma.

We now prove the converse of Corollary 1 to Definition 2.

THEOREM 13. *Any splitting field over k of a polynomial $f(X)$ in $k[X]$ is a finite normal extension of k.*

PROOF. Let K be a splitting field, over k, of a polynomial $f(X)$ in $k[X]$ and let $\varphi(X)$ be any irreducible polynomial in $k[X]$ which has a root α in K. We fix a splitting field K' of $\varphi(X)$ *over* K. Let β be any root of $\varphi(X)$ in K'. Since $\varphi(X)$ is irreducible over k, we have a k-isomorphism τ between $k(\alpha)$ and $k(\beta)$ which sends α into β. This isomorphism leaves $f(X)$ invariant (since the coefficients of f are in k), and on the other hand the fields K and $K(\beta)$ are splitting fields of $f(X)$ respectively over $k(\alpha)$ and $k(\beta)$ [since $k(\alpha) \subset K$, $k(\beta) \subset K(\beta)$ and $K = k(x_1, x_2, \cdots, x_n)$]. Hence, by Theorem 12, the isomorphism τ of $k(\alpha)$ onto $k(\beta)$ can be extended to an isomorphism ρ of K onto $K(\beta)$. We are dealing here with an isomorphism ρ of K into a field *containing* K, namely into K'. Since ρ is also a k-isomorphism and since the polynomial $f(X)$, whose coefficients are *in* k, factors completely in $K[X]$ into linear factors, it follows that ρ must transform *onto* itself the set of roots of $f(X)$ in K. Since the roots of $f(X)$ in K generate K over k, *it follows that ρ is an automorphism of K.* Since $\alpha \in K$ and $\alpha\rho = \beta$, we have $\beta \in K$. We have thus proved that K contains all the roots of $\varphi(X)$ in K' (whence K actually coincides with K'). This shows that K is a normal extension of k and completes the proof of the theorem.

Let $K = k(\alpha_1, \alpha_2, \cdots, \alpha_m)$ be a finite extension of k and let $f_i(X)$ be the minimal polynomial of α_i in $k[X]$. We set $f(X) = f_1(X)f_2(X) \cdots f_m(X)$ and we consider a splitting field K' of $f(X)$ *over* K. Since K is generated over k by roots of $f(X)$ (namely by $\alpha_1, \alpha_2, \cdots, \alpha_m$) it follows that K' is generated over k (and not only over K) by the roots of $f(X)$, whence K' is also a splitting field of $f(X)$ *over* k. By Theorem 13, K' is then a normal extension of k. We have therefore constructed an over-field K' of K which is normal (and finite) over k. If K_1 is any field between K and K' which is normal over k, then each of the m polynomials $f_i(X)$ must factor completely in $K_1[X]$ into linear factors (since $f_i(X)$ is irreducible in $k[X]$ and has a root in $K_1[X]$, namely α_i). *This shows that K' coincides with K_1.* Hence K' is a least normal extension of k which contains K as a subfield. We note furthermore that if K'' is any normal extension of k which contains K as a subfield, then K'' must contain a splitting field of $f(X)$ over k (since this must be so for each irreducible factor $f_i(X)$ of $f(X)$) and the latter field will of course contain K. In particular, then, if K'' is a least normal extension of k containing K as a subfield, then K'' must be itself a splitting field of

$f(X)$ over K, and hence K' and K'' are K-isomorphic (Theorem 12). We have therefore proved the following theorem:

THEOREM 14. *If K is a finite extension of k then there exists a least normal extension of k containing K, and any two such extensions are K-isomorphic.*

An almost immediate consequence of this theorem and of Corollary 2 of Definition 2 is the following theorem which gives a *characteristic property of finite normal extensions*:

THEOREM 15. *A finite extension K of k is normal over k if and only if it satisfies the following condition: if K' is any extension of K then any k-isomorphism of K into K' is necessarily a k-automorphism of K.*

PROOF. That any (finite *or* infinite) normal extension K of k satisfies the condition of the theorem is obvious, since a k-isomorphism of K into K' sends any element of K into a conjugate element over k. Conversely, assume that a finite extension K of k satisfies that condition. We fix a finite extension K' of K which is normal over k, for instance a least normal extension of k containing K (Theorem 14). If γ is any element of K, then K' contains all the conjugates of γ over k. If γ' is one of these conjugate elements, then there exists a k-automorphism ρ of K' which sends γ into γ' (Definition 2, Corollary 2). Then ρ induces a k-isomorphism of K into K', and by our assumption this induced k-isomorphism is necessarily a k-automorphism of K. Hence $\gamma' = \gamma\rho \in K$ (since $\gamma \in K$). We have thus shown that K contains all the conjugates of γ over k. Since γ is an arbitrary element of K it follows that K is normal over k. Q.E.D.*

As a final application of the preceding results, we shall now investigate the following question: *if K is a finite normal extension of k, how many k-automorphisms does K admit?* We incorporate the answer to this question in the following more general result:

THEOREM 16. *Let L be a finite algebraic extension of k and let K be an extension field of L which is normal over k. If n_0 is the separable factor of the degree $[L:k]$, then there exist precisely n_0 distinct k-isomorphisms of L into K.*

PROOF. In the proof we may assume that K is a finite extension of k, in fact we may even assume that K is a least normal extension of k containing L, for every k-isomorphism of L into K necessarily maps L into the least normal extension of k which contains L and is contained in K.

* It is clear (and follows also directly from the proof) that Theorem 15 remains true also if only finite extensions K' of K are allowed in the statement of the theorem; in fact the theorem remains true if we take for K' a *fixed* normal extension of k.

The theorem is obvious in the case $n_0 = 1$, for in that case every element of L is purely inseparable over k and therefore is left invariant by every k-isomorphism of L into K; that is, the identity is the only k-isomorphism of L into K. The theorem is also obvious if L is a simple extension of k, say $L = k(\alpha)$. For in that case, n_0 is also the reduced degree of the minimal polynomial of α in $k[X]$ (see end of § 5, p. 71). Hence α has exactly n_0 conjugate elements in K (that is, conjugate over k). If $\alpha_1, \alpha_2, \cdots, \alpha_{n_0}$ are these conjugate elements ($\alpha_1 = \alpha$), then there exists a unique k-isomorphism τ_i of $k(\alpha)$ onto $k(\alpha_i)$ which sends α into α_i. It is clear that the n_0 isomorphisms τ_i ($\tau_1 =$ the identity) are the only k-isomorphisms of $k(\alpha)$ into K, since any k-isomorphism of $k(\alpha)$ into K must send α into a conjugate element of α over k, that is, into one of the elements α_i.

After these preliminary remarks, we proceed to prove our theorem by induction on n_0. We assume namely that the theorem is true for all finite algebraic extensions of k for which n_0 is less than a given integer m, $m > 1$. Let $n_0 = m$ for the given field L. Since $m > 1$, there exist elements in L, not in k, which are separable over k. We fix one such element, say, α. Let s be the degree $[k(\alpha) : k]$. Since α is separable over k, the maximal separable extension of k in L coincides with the maximal separable extension of $k(\alpha)$ in L. It follows that if we denote by r the separable factor of the degree $[L : k(\alpha)]$, then $m = sr$. Since $s > 1$, we have $r < m$. By our induction hypothesis, the theorem is therefore valid for L if we replace k by $k(\alpha)$. Hence there exist exactly r distinct $k(\alpha)$-isomorphisms of L into K (note that K, being normal over k, is *a fortiori* normal over $k(\alpha)$). Let $\tau_1, \tau_2, \cdots, \tau_r$ be the $k(\alpha)$-isomorphisms of L into K. Since K is normal over k, K contains all the conjugates of α over k, say $\alpha_1, \alpha_2, \cdots, \alpha_s$. For each $j = 1, 2, \cdots, s$, we fix a k-automorphism σ_j of K which sends α into α_j (Definition 2, Corollary 2) and we set $\rho_{ij} = \tau_i\sigma_j$, $i = 1, 2, \cdots, r$; $j = 1, 2, \cdots, s$. Then each ρ_{ij} is a k-isomorphism of L into K. *The $m(= rs)$ isomorphisms ρ_{ij} are distinct.* For we have $\alpha\rho_{ij} = \alpha\sigma_j$ (since α is left invariant by τ_i), and hence if $\rho_{ij} = \rho_{i'j'}$ then $\alpha\sigma_j = \alpha\sigma_{j'}$, that is, $\alpha_j = \alpha_{j'}$. This implies $j = j'$, and from this it follows at once that $\tau_i = \tau_{i'}$ (since the σ_j, as automorphisms of K, are univalent mappings of K). Hence $i = i'$, and this proves the assertion that the m isomorphisms ρ_{ij} are distinct. Now let ρ be an arbitrary k-isomorphism of L into K. The element α is transformed by ρ into one of its conjugate elements $\alpha_1, \alpha_2, \cdots, \alpha_s$. Let, say, $\alpha\rho = \alpha_j$. Then $\rho\sigma_j^{-1}$ is a k-isomorphism of L into K which leaves α fixed, that is, $\rho\sigma_j^{-1}$ is a $k(\alpha)$-isomorphism of L into K. Hence $\rho\sigma_j^{-1}$ coincides with one of the isomorphisms

$\tau_1, \tau_2, \cdots, \tau_r$, say, with τ_i, and hence $\rho = \tau_i\sigma_j = \rho_{ij}$. This completes the proof.

COROLLARY 1. *Let L be a finite algebraic extension of k and let n_0 be the separable factor of the degree $[L:k]$. Then L possesses at most n_0 k-automorphisms, and the maximum n_0 is reached if and only if L is a normal extension of k.*

The first part of the corollary is an immediate consequence of the Theorem 16 and of the existence of finite extensions K of L which are normal over k. If L is a normal extension of k, we can identify, in Theorem 16, the field K with L and we deduce then that L possesses n_0 k-automorphisms. Conversely, if L possesses n_0 k-automorphisms, then it follows from the above theorem that if K' is any extension field of L, every k-isomorphism of L into K' is necessarily an automorphism of L. Hence, by Theorem 15, L is a normal extension of k.

COROLLARY 2. *If $k \subset L \subset \Delta$ are successive finite algebraic extensions of k, then*

$$[\Delta : k]_s = [\Delta : L]_s \cdot [L : k]_s, \tag{4}$$

$$[\Delta : k]_i = [\Delta : L]_i \cdot [L : k]_i. \tag{5}$$

It is sufficient to prove (4) since the product of the right-hand sides of (4) and (5) is equal to the product of the left-hand sides, in view of relation (2) of § 3 and relation (4) of § 5. Let $m_0 = [L:k]_s$, $n_0 = [\Delta : L]_s$. Then n_0 is the number of L-isomorphisms of Δ into K, where K is some extension of Δ which is normal over k (for instance, the least normal extension of k containing Δ), and m_0 is the number of k-isomorphisms of L into K. By Lemma 2, the product $m_0 n_0$ is the number of k-isomorphisms of Δ into K, and since this number is equal to $[\Delta : k]_s$ relation (4) is proved.

Another proof of (4) can be based on the following property of finite separable extensions K/k established in the course of the proof of Theorem 8 of § 5: *if $x_1, x_2, \cdots, x_n$ are elements of K which are linearly independent over k, then for any integer $e \geqq 0$ also the elements $x_1^{p^e}, x_2^{p^e}, \cdots, x_n^{p^e}$ are linearly independent over k.* Let L_0, Δ_0, and Δ'_0 be respectively the maximal separable extension of k in L, of L in Δ, and of k in Δ. We have $k \subset L_0 \subset L \subset \Delta_0 \subset \Delta$, $k \subset L_0 \subset \Delta'_0 \subset \Delta_0 \subset \Delta$, and $[\Delta : k]_s = [\Delta'_0 : k] = [\Delta'_0 : L_0] \cdot [L_0 : k] = [\Delta'_0 : L_0] \cdot [L : k]_s$. Hence to prove (4) we have to show that

$$[\Delta'_0 : L_0] = [\Delta_0 : L]. \tag{6}$$

Let $x_1, x_2, \cdots, x_n$ be elements of Δ'_0 which are linearly independent over L_0. The x_i are also in Δ_0; we assert that they are linearly independent over L. For let $\sum u_i x_i = 0$, $u_i \in L$. Since L is a purely inseparable extension of L_0, we have $u_i^{p^e} \in L_0$ for some integer $e \geqq 0$, and also $\sum u_i^{p^e} x_i^{p^e} = 0$. From this relation and from the separability of the extension Δ'_0/L_0 it follows that $u_i^{p^e} = 0$, $u_i = 0$, and this proves our assertion. We have therefore shown that $[\Delta'_0 : L_0] \leqq [\Delta_0 : L]$. On the other hand, let now $x_1, x_2, \cdots, x_n$ be elements of Δ_0 which are linearly independent over L. Since Δ_0 is a purely inseparable extension of Δ'_0, there is an integer $e \geqq 0$ such that the p^e-th powers of the x_i belong to Δ'_0. In view of the separability of the extension Δ_0/L, the p^e-th powers of the x_i are still linearly independent over L, and hence also over the subfield L_0 of L. We have thus found n linearly independent elements of Δ'_0 over L_0. This shows that $[\Delta_0 : L] \leqq [\Delta'_0 : L_0]$ and establishes (6).

§ **7. The fundamental theorem of Galois theory.** If K is any field, then the automorphisms of K clearly form a group (of transformations). If K contains a subfield k, then also the k-automorphisms of K form a group. If K is a finite normal extension of k, the group of k-automorphisms of K is called *the Galois group of K with respect to k*. We shall denote this group by $G(K/k)$. By Theorem 16 of § 6, $G(K/k)$ is a finite group.

Let K be a finite normal extension of k. If H is any subgroup of $G(K/k)$, then it is easily seen that the elements of K which are left invariant under all the automorphisms belonging to H form a subfield of K. We denote this subfield by $F(H)$ (*the fixed field of H*). On the other hand, if L is any subfield of K such that $k \subset L$, then K is also a normal extension of L, and the Galois group of K with respect to L is clearly a subgroup of $G(K/k)$; it consists precisely of those automorphisms in $G(K/k)$ which leave invariant every element of L.

The fundamental theorem of Galois theory asserts the following:

THEOREM 17. *If K is a finite normal separable extension of k, then there is a one-to-one correspondence between the subgroups H of $G(K/k)$ and the subfields L of K which contain k, corresponding elements H and L being such that $L = F(H)$ and $H = G(K/L)$.*

PROOF. The correspondence $L \rightarrow G(K/L)$ defines a mapping of the set of all subfields L of K which contain k into the set of all subgroups of $G(K/k)$. If L is a given subfield of K containing k and if $H = G(K/L)$, then it follows from the separability and normality of K/L and from § 6, Definition 2, Corollary 3, that $L = F(H)$. *Hence the above mapping $L \rightarrow G(K/L)$ is univalent.* To complete the proof of the theorem, it remains to show that the mapping is *onto* the set of all subgroups of $G(K/k)$. Let H be any subgroup of $G(K/k)$ and let $L = F(H)$. We shall show that H is the Galois group of K with respect to L. The proof of this assertion will complete the proof of the theorem.

It is clear that $H \subset G(K/L)$. Let n denote the order of the group H. Suppose that it has already been proved that

$$(1) \qquad [K:L] \leqq n.$$

Since K is a normal *and* separable extension of L, we have, by Theorem 16, Corollary (§ 6), that the order of $G(K/L)$ is equal to $[K:L]$, *hence is* $\leqq n$, by (1). On the other hand, H is a subgroup of $G(K/L)$ and has order n. It follows at once that $H = G(K/L)$, as asserted.

It remains to prove the inequality (1). Let $\alpha_1, \alpha_2, \cdots, \alpha_{n+1}$ be arbitrary $n + 1$ elements of K. We have to show that these elements are linearly dependent over L. In the proof we may assume that no α_i is zero. Let $\tau_1, \tau_2, \cdots, \tau_n$ be the elements of the group H. We find a set of $n + 1$ elements c_j in K, *not all zero*, such that the following system of n homogeneous equations is satisfied:*

$$(2) \qquad \sum_{j=1}^{n+1} c_j(\alpha_j\tau_i) = 0, \quad i = 1, 2, \cdots, n.$$

Among all such sets $\{c_1, c_2, \cdots, c_{n+1}\}$ we choose one with the smallest number of non-zeros. We assume that $\{c_1, c_2, \cdots, c_{n+1}\}$ has already been chosen in this fashion. Let, say, $c_1, c_2, \cdots, c_r \neq 0$, $c_{r+1} = c_{r+2} = \cdots = c_{n+1} = 0$. Then $r \geqq 2$, for if $r = 1$, then $\alpha_1\tau_i = 0$, $\alpha_1 = 0$ [since $\{\tau_1, \tau_2, \cdots, \tau_n\}$ is a *non-empty* set of *automorphisms* of K (the identity belongs to the set)]. We have then

$$(3) \qquad \sum_{j=1}^{r} c_j(\alpha_j\tau_i) = 0, \quad i = 1, 2, \cdots, n,$$

and, in particular, taking for τ_i the identity of H, we have

$$(4) \qquad \sum_{j=1}^{r} c_j\alpha_j = 0.$$

We may assume that $c_1 = 1$. *We claim then that $c_2, \cdots, c_r$ belong to L,* whence by (4) $\alpha_1, \cdots, \alpha_{n+1}$ are indeed linearly dependent over L, as was asserted.

We have to prove that $c_j\tau_i = c_j$, $i = 1, 2, \cdots, n$ (since L is the fixed

* We presuppose here the knowledge of the theory of simultaneous linear homogeneous equations, with coefficients in a field K (see, for instance, G. Birkhoff and S. MacLane, *A Survey of Modern Algebra*, Chapter X). The existence of a non-trivial solution $(c_1, c_2, \cdots, c_{n+1})$ of (2) follows from the theory of vector spaces which was developed in I, § 21 [the set of all n-tuples $(x_1, x_2, \cdots, x_n)$, $x_i \in K$, is an n-dimensional vector space over K, and hence the $n + 1$ vectors $v_j = (\alpha_j\tau_1, \alpha_j\tau_2, \cdots, \alpha_j\tau_n)$ are linearly dependent over K].

field of H). Let us prove for instance that $c_j\tau_1 = c_j$. If we apply to (3) the automorphism τ_1 of K we find

$$\sum_{j=1}^{r} (c_j\tau_1)(\alpha_j\tau_i\tau_1) = 0, \quad i = 1, 2, \cdots, n.$$

The n products $\tau_i\tau_1$ give again *all* the elements of the *finite* group H. Hence we have

$$\sum_{j=1}^{r} (c_j\tau_1)(\alpha_j\tau_i) = 0, \quad i = 1, 2, \cdots, n. \tag{5}$$

Subtracting (5) from (3) and taking into account that $c_1 = c_1\tau_1 = 1$, we find

$$\sum_{j=2}^{r} (c_j\tau_1 - c_j)(\alpha_j\tau_i) = 0, \quad i = 1, 2, \cdots, n.$$

Here we have a set of n relations similar to (2), but the number of terms in each of these relations is *less* than r. Hence, by our choice of the set $\{c_1, c_2, \cdots, c_r, 0, 0, \cdots, 0\}$, we must have $c_j\tau_1 = c_j, j = 2, 3, \cdots, r$. In a similar fashion we can prove that $c_j\tau_i = c_j$, $j = 2, 3, \cdots, r$, $i = 1, 2, \cdots, n$, and this completes the proof of the theorem.

COROLLARY. *If $k \subset L \subset K$, then L is a normal extension of k if and only if $G(K/L)$ is an invariant subgroup of $G(K/k)$, and when that is so, then the Galois group $G(L/k)$ is isomorphic to the factor group $G(K/k)/G(K/L)$.*

Let $H = G(K/L)$. If τ is any fixed element of $G(K/k)$, it is immediately seen that the elements of the form $x\tau$, $x \in L$, form a subfield of K, which we shall denote by $L\tau$, and that $\tau^{-1}H\tau = G(K/L\tau)$. If L is a normal extension of k, then $L\tau = L$ (Theorem 15, § 6) and hence $\tau^{-1}H\tau = H$, and H is an invariant subgroup of $G(K/k)$. Conversely, if H is an invariant subgroup of $G(K/k)$, then we have $H = \tau^{-1}H\tau = G(K/L)$, that is, $G(K/L) = G(K/L\tau)$. Hence, by the theorem just proved above, $L = L\tau$. This holds for all elements τ of the Galois group $G(K/k)$, and therefore L is a normal extension of k (see footnote at the end of proof of Theorem 15). Furthermore, the mapping $\tau \to$ restriction of τ to L ($\tau \in G$) is a homomorphism of $G(K/k)$ into $G(L/k)$, with kernel H. From Corollary 4 to Definition 2 of § 6, it follows that this homomorphism is onto $G(L/k)$, and this establishes the last part of the corollary.

§ 8. Galois fields. Let K be a Galois field of characteristic p (see Definition 3, § 4) and let J_p be the prime field contained in K (§ 4). In view of the finiteness of K, it follows at once that K is a finite algebraic extension of J_p (see, for instance, Theorem 4, § 3). Let n be the degree $[K:J_p]$ and let $\{x_1, x_2, \cdots, x_n\}$ be a basis of K over J_p. Then every element of K has a unique expression of the form $a_1x_1 + a_2x_2 + \cdots + a_nx_n$, $a_i \in \mathcal{J}_p$. Since each coefficient a_i can take independently p values ($\mathcal{J}_p$ being a field containing exactly p elements), *it follows that the number of elements in K is p^n.* Thus the number of elements of a Galois field of characteristic p is always a power of p.

We note that a similar argument can be applied to obtain the following results: *if k is a Galois field consisting of m elements and if K is a finite extension of k, of degree n, then K consists of m^n elements* (and is therefore also a Galois field).

The elements of K, other than 0, form a multiplicative group, of order $h = p^n - 1$. We have therefore $x^h = 1$ for all elements x of this group, and consequently $x^{p^n} - x = 0$ for all elements x of K (including 0). Since the degree of the polynomial $X^{p^n} - X$ is the same as the number of elements of K, we conclude that *the polynomial $X^{p^n} - X$ factors completely into linear factors in $K[X]$ and that we have*

$$X^{p^n} - X = \prod_{i=1}^{p^n} (X - \alpha_i), \tag{1}$$

where $\alpha_1, \alpha_2, \cdots, \alpha_{p^n}$ are all the elements of K. It follows also that K is a splitting field, over $\mathcal{J}_p$, of the polynomial $X^{p^n} - X$, and is therefore a normal extension $\mathcal{J}_p$ (§ 6, Theorem 13). Hence, by Theorem 12 (§ 6), *any two Galois fields with the same number of elements* (and consequently of the same characteristic p) *are isomorphic.*

The Galois field having p^n elements is denoted by $GF(p^n)$. That there exist fields $GF(p^n)$ for any prime number p and any positive integer n follows from the existence of splitting fields (§ 6). Namely, it is easily shown that any splitting field of the polynomial $X^{p^n} - X$, over $\mathcal{J}_p$, is in fact a field $GF(p^n)$. The proof is as follows:

Let K be a splitting field of $X^{p^n} - X$, over $\mathcal{J}_p$, and let (1) be the factorization of $X^{p^n} - X$ into linear factors in $K[X]$. Since the derivative of $X^{p^n} - X$ is -1, it follows that each α_i is a simple root of $X^{p^n} - X$ (§ 5, Definition 3, Corollaries 1 and 2). Hence the p^n elements α_i are distinct. If α_i and α_j are any two roots of $X^{p^n} - X$ in K, then $(\alpha_i - \alpha_j)^{p^n} = \alpha_i^{p^n} - \alpha_j^{p^n} = \alpha_i - \alpha_j$, $(\alpha_i\alpha_j)^{p^n} = \alpha_i^{p^n}\alpha_j^{p^n} = \alpha_i\alpha_j$, and if furthermore $\alpha_j \neq 0$ then also $(\alpha_j^{-1})^{p^n} = \alpha_j^{-1}$. In other words: $\alpha_i - \alpha_j$, $\alpha_i\alpha_j$ and—if $\alpha_j \neq 0$—also α_j^{-1} are roots of $X^{p^n} - X$ in K and therefore belong to the set $\{\alpha_1, \alpha_2, \cdots, \alpha_{p^n}\}$. Consequently *this set is a subfield K' of K*, and K' is a Galois field of p^n elements. Clearly $\mathcal{J}_p \subset K'$, and hence $K' = \mathcal{J}_p(\alpha_1, \alpha_2, \cdots, \alpha_{p^n}) = K$, as asserted.

THEOREM 18. *The multiplicative group of a Galois field $GF(p^n)$ is cyclic.*

PROOF. Let $h = q_1^{r_1}q_2^{r_2} \cdots q_m^{r_m}$ be the decomposition into prime factors of the order h of the multiplicative group of $GF(p^n)(h = p^n - 1)$, and let $h_i = h/q_i$. The polynomial $X^{h_i} - 1$ has at most h_i roots in $GF(p^n)$, and since $h_i < h$ it follows that there exist elements $\neq 0$ in $GF(p^n)$ which are not roots of this polynomial. We fix such an element β_i for each $i = 1, 2, \cdots, m$ and we set $y_i = \beta_i^{h/q_i^{r_i}}$, $y = y_1y_2 \cdots y_m$. We have $y_i^{q_i^{r_i}} = 1$, whence the order of y_i is a divisor of $q_i^{r_i}$ (see I, § 3) and is therefore a power $q_i^{s_i}$ of q_i, $s_i \leqq r_i$. On the other hand, $y_i^{q_i^{r_i-1}} = \beta_i^{h_i} \neq 1$. Hence y_i is exactly of order $q_i^{r_i}$. *We claim that h is precisely the order of y.* For assume the contrary. Then the order of y is a proper divisor of h and is therefore a divisor of at least one of the m integers h/q_i, say of h/q_1. We have then $1 = y^{h/q_1} = y_1^{h/q_1}y_2^{h/q_1} \cdots y_m^{h/q_1}$. Now if $2 \leqq i \leqq m$, then $q_i^{r_i}$ divides h/q_1, and hence $y_i^{h/q_i} = 1$. Therefore $y_1^{h/q_1} = 1$. This implies that the order of y_1 must divide h/q_1, which is impossible since the order of y_1 is $q_1^{r_1}$.

The cyclic subgroup of the multiplicative group of $GF(p^n)$, generated by the element y, is therefore of order h = order of the multiplicative group of $GF(p^n)$. Hence y is a generator of this latter group. This completes the proof.

§ 9. The theorem of the primitive element. Let Δ be an algebraic extension of a field k. An element α of Δ is a *primitive element of* K/k if $K = k(\alpha)$.

THEOREM 19. *Every finite separable extension Δ of k has a primitive element* (and hence every such extension Δ is a simple extension).

PROOF. We shall prove here this theorem by the "method of indeterminates," a method due to Kronecker. We shall give the proof only in the case in which k has infinitely many elements. If k is a finite field, then also Δ is a finite field (see § 8), and in that case we know from the preceding section that every non-zero element of Δ is the power of a single element θ. This element θ is then a primitive element.

Let $\Delta = k(\alpha_1, \alpha_2, \cdots, \alpha_n)$. We adjoin to Δ $n + 1$ "indeterminates" $X, X_1, X_2, \cdots, X_n$, that is, we consider the polynomial ring $\Delta[X, X_1, \cdots, X_n]$ and its quotient field $\Delta(X, X_1, \cdots, X_n)$. We set $k^\star = k(X_1, X_2, \cdots, X_n)$, $\Delta^\star = \Delta(X_1, X_2, \cdots, X_n)$. We have then $\Delta^\star = k^\star(\alpha_1, \alpha_2, \cdots, \alpha_n)$, and $\Delta^\star$ is a finite algebraic separable extension of $k^\star$ since the α_i, being separable over k, are also separable over $k^\star$ (see § 5, Lemma 2). We consider in $\Delta^\star$ the element

(1) $$\alpha^\star = X_1\alpha_1 + X_2\alpha_2 + \cdots + X_n\alpha_n.$$

Let $F(X)$ be the minimal polynomial of $\alpha^\star$ in $k^\star[X]$. The coefficients of $F(X)$ are rational functions of $X_1, X_2, \cdots, X_n$, with coefficients in k. Let $g(X_1, X_2, \cdots, X_n)$ be a common denominator of these rational functions, where $g(X_1, X_2, \cdots, X_n)$ is then an element in $k[X_1, X_2, \cdots, X_n]$. Then

$$g(X_1, X_2, \cdots, X_n)F(X) = f(X, X_1, X_2, \cdots, X_n) \in k[X, X_1, X_2, \cdots, X_n],$$

and we have

$$f(\alpha^\star, X_1, X_2, \cdots, X_n) = 0. \tag{2}$$

Let

$$G(X_1, X_2, \cdots, X_n) = f(X_1\alpha_1 + X_2\alpha_2 + \cdots + X_n\alpha_n, X_1, X_2, \cdots, X_n). \tag{3}$$

Then $G(X_1, X_2, \cdots, X_n)$ is a polynomial in $X_1, X_2, \cdots, X_n$, with coefficients in Δ, and we have, by (2): $G(X_1, X_2, \cdots, X_n) = 0$. Therefore also the partial derivatives $\partial G/\partial X_i$, $i = 1, 2, \cdots, n$, are all zero. By (3), we have, then:

$$\alpha_i f'(\alpha^\star, X_1, X_2, \cdots, X_n) + f_i(\alpha^\star, X_1, X_2, \cdots, X_n) = 0, \quad i = 1, 2, \cdots, n, \tag{4}$$

where

$$f'(X, X_1, X_2, \cdots, X_n) = \frac{\partial f(X, X_1, X_2, \cdots, X_n)}{\partial X},$$

$$f_i(X, X_1, X_2, \cdots, X_n) = \frac{\partial f(X, X_1, X_2, \cdots, X_n)}{\partial X_i}.$$

The left-hand side in each of the equations (4) is, by (1), a polynomial in $\Delta[X_1, X_2, \cdots, X_n]$, and hence is the zero polynomial. Consequently, the equations (4) remain valid if we substitute for $X_1, X_2, \cdots, X_n$, any elements of k. On the other hand, we have $f'(X, X_1, X_2, \cdots, X_n) = g(X_1, X_2, \cdots, X_n)F'(X)$, and hence $f'(\alpha^\star, X_1, X_2, \cdots, X_n) \neq 0$, since $\alpha^\star$ is separable over $k^\star$ and therefore $F'(\alpha^\star) \neq 0$. Hence $f'(\alpha^\star, X_1, X_2, \cdots, X_n)$ is a non-zero polynomial in $\Delta[X_1, X_2, \cdots, X_n]$. Since $k \subset \Delta$ and k is an *infinite* field, we can find elements $c_1, c_2, \cdots, c_n$ in k such that $(c_1, c_2, \cdots, c_n)$ is not a zero of that polynomial (I, § 18, Theorem 14). We have then, setting

$$\alpha = c_1\alpha_1 + c_2\alpha_2 + \cdots, + c_n\alpha_n,$$

that

$$f'(\alpha, c_1, c_2, \cdots, c_n) \neq 0 \tag{5}$$

and

$$\alpha_i f'(\alpha, c_1, c_2, \cdots, c_n) + f_i(\alpha, c_1, c_2, \cdots, c_n) = 0, \quad i = 1, 2, \cdots, n. \tag{6}$$

Equation (6) and the inequality (5) imply that $\alpha_i \in k(\alpha)$, and since $\alpha \in \Delta$, it follows that $\Delta = k(\alpha)$. This completes the proof of the theorem.

REMARK. Theorem 16 (§ 6) is also an immediate consequence of the above theorem of the primitive element, since—as has been pointed out in the beginning of the proof of Theorem 16—that theorem is obvious if L is a simple extension of k.

§ 10. Field polynomials. Norms and traces. Let K be a finite algebraic extension of a field k, of degree n, and let x be any element of K. If we fix a basis $\omega_1, \omega_2, \cdots, \omega_n$ of K/k, we can write:

$$(1) \qquad x\omega_i = \sum_{j=1}^{n} a_{ij}\omega_j,\ a_{ij} \in k,\ i = 1, 2, \cdots, n,$$

or, in matrix notations:

$$(1') \qquad x\Omega = A\Omega,$$

where A is the matrix $\|a_{ij}\|$ and Ω is the 1-column matrix

$$\begin{Vmatrix} \omega_1 \\ \omega_2 \\ \vdots \\ \omega_n \end{Vmatrix}.$$

The elements a_{ij}, and hence the matrix A, are uniquely determined by the element x and by the basis Ω. We shall denote by $|B|$ the determinant of a square matrix B. Then it follows from (1) that

$$(2) \qquad |xE - A| = 0,$$

where E is the unit n-rowed matrix.

The polynomial $|XE - A|$ is monic, of degree n, and its coefficients are in k. Equation (2) signifies that x is a root of this polynomial. *It is not difficult to see that for a given element x of K this polynomial does not depend on the choice of the basis* $\{\omega_1, \omega_2, \cdots, \omega_n\}$. For let $\omega'_1, \omega'_2, \cdots, \omega'_n$ be another basis of K/k. We have then $\omega'_i = \sum_{j=1}^{n} b_{ij}\omega_j$, $\omega_i = \sum_{j=1}^{n} b'_{ij}\omega'_{ij}$, $i = 1, 2, \cdots, n$, where the b_{ij} and b'_{ij} are elements of k. If Ω' denotes the one-column matrix

$$\begin{Vmatrix} \omega'_1 \\ \omega'_2 \\ \vdots \\ \omega'_n \end{Vmatrix}.$$

and B, B' denote the square matrices $\|b_{ij}\|$ and $\|b'_{ij}\|$, respectively, then the above relations can be written in matrix notation as follows:

(3) $$\Omega' = B\Omega, \qquad \Omega = B'\Omega'.$$

From (3) it follows that $\Omega = C\Omega$ where C is the matrix $B'B$. Since the elements of Ω are linearly independent over k, C is necessarily the unit matrix E, *whence B is a non-singular matrix and $B' = B^{-1}$.* Now, dealing with the basis $\{\omega'_1, \omega'_2, \cdots, \omega'_n\}$, we have relations similar to (1'): $x\Omega' = A'\Omega'$. Hence, by (3): $xB\Omega = A'B\Omega$, or $Bx\Omega = A'B\Omega$, and therefore, by (1'): $BA\Omega = A'B\Omega$. Again using the fact that $\omega_1, \omega_2, \cdots, \omega_n$ are linearly independent over k, we see that the relation $BA\Omega = A'B\Omega$ implies that $BA = A'B$, that is, $A' = BAB^{-1}$. We have therefore that the matrix $XE - A'$, which is the analogue of $XE - A$, relative to the basis Ω', is given by $XE - BAB^{-1}$. Since XE commutes with every n-rowed square matrix, we have therefore $XE - A' = B^{-1}XEB - BAB^{-1} = B(XE - A)B^{-1}$, and hence

$$|XE - A'| = |B| \cdot |XE - A| \cdot |B^{-1}| = |XE - A|,$$

which proves our assertion.

The polynomial $|XE - A|$ is called *the field polynomial of x, relative to k, or over k.* We emphasize that the field polynomial of x, over k, depends not only on x but also on the field K. This dependence on K is already obvious from the fact that the degree of the field polynomial is always equal to the degree n of K/k. In particular, the field polynomial of x is not necessarily the minimal polynomial of x over k.

We note that if K is regarded as a vector space over k then in terms of linear algebra the field polynomial of x is the *characteristic polynomial* of the linear transformation in K defined by $z \to zx$, $z \in K$.

Let

$$X^n + a_1X^{n-1} + \cdots + a_n$$

be the field polynomial of x over k. Expanding the determinant $|XE - A|$, we find

(4) $$a_1 = -\sum_{i=1}^{n} a_{ii},$$

(5) $$a_n = (-1)^n|A|.$$

We set

(6) $$\text{Norm}_{K/k}x = N_{K/k}(x) = (-1)^n a_n = |A|,$$

(7) $$\text{Trace}_{K/k}x = T_{K/k}(x) = -a_1 = \sum_{i=1}^{n} a_{ii}.$$

The index K/k will frequently be omitted when there is no possibility of confusion.

Norms and traces obey the following laws:

a) $N(xy) = N(x)\cdot N(y)$.
b) If $x \in k$, then $N(x) = x^n$.
c) $T(x + y) = T(x) + T(y)$.
d) $T(cx) = cT(x)$, $c \in k$.
e) If $x \in k$, then $T(x) = nx$.

PROOF. If, for a given basis Ω of K/k, we have $x\Omega = A\Omega$ and $y\Omega = B\Omega$, then $(x + y)\Omega = (A + B)\Omega$ and $xy\Omega = BA\Omega$. In view of the definition of traces and norms, relations a) and c) follow immediately. If $x \in k$, then A is the diagonal matrix xE_n, and this implies relations b) and e). Property d) follows directly from (4) and (7).

Also the norm and trace of an element x of K depend not only on x and k but also on the extension field K.

Let Δ be a finite extension of K, of degree m, and let x be any element of K. If we regard x as an element of Δ, we can consider the trace $T_{\Delta/k}(x)$ and norm $N_{\Delta/k}(x)$, and as was pointed out above, these are to be distinguished from $T_{K/k}(x)$ and $N_{K/k}(x)$. We shall now prove the following relations:

$$N_{\Delta/k}(x) = [N_{K/k}(x)]^m, \tag{8}$$

$$T_{\Delta/k}(x) = m[T_{K/k}(x)]. \tag{9}$$

For the proof, we fix a basis $\{\omega_1, \omega_2, \cdots, \omega_n\}$ of K/k and a basis $\{\xi_1, \xi_2, \cdots, \xi_m\}$ of Δ/K. Then the mn products $\omega_i\xi_j$ form a basis of Δ/k (see § 3, Theorem A, p. 60). We order these products, as follows: $\omega_i\xi_j$ precedes $\omega_{i'}\xi_{j'}$ if $j < j'$ or if $j = j'$ and $i < i'$, and we denote these products, in this order, by $\zeta_1, \zeta_2, \cdots, \zeta_N$, $N = mn$. We denote by Ω and Z the one-column matrices which have, respectively, $\omega_1, \omega_2, \cdots, \omega_n$ and $\zeta_1, \zeta_2, \cdots, \zeta_N$ as elements. Let $x\Omega = A\Omega$ and $xZ = CZ$, so that A and C are square matrices, with elements in k, having respectively n and mn rows. Now, we observe that if $A = \|a_{ij}\|$, whence $x\omega_i = \sum_{j=1}^{n} a_{ij}\omega_j$, then $x\omega_i\xi_\alpha = \sum_{j=1}^{n} a_{ij}\omega_j\xi_\alpha$. Hence if $\omega_i\xi_\alpha = \zeta_\mu (1 \leqq \mu \leqq N)$ then $x\zeta_\mu = \sum_{\nu=1}^{N} c_{\mu\nu}\zeta_\nu$, where $c_{\mu\nu} = a_{ij}$ if n divides *both* $\mu - i$ and $\nu - j$, *and* the absolute value of the difference $\mu - \nu$ is $< n$ (or—equivalently—if $\mu - i = \nu - j \equiv 0 \pmod{n}$), while all the other elements $c_{\mu\nu}$ of the matrix C are zero. This signifies that C has the following form: it is obtained from the m-rowed unit matrix E_m on replacing each diagonal element 1 by the matrix A and each other element of E_m by the zero n-rowed matrix; in symbols: $C = A^{(m)}$. It follows at once that the sum of the diagonal elements of C is the m-fold of the sum of the diagonal elements of A and that $|C| = |A|^m$. This establishes (8) and (9).

Another proof of (8) and (9) will be found at the end of this section.

Let $f(X)$ be the field polynomial of x over k, when x is regarded as an element of K, and let $F(X)$ be the field polynomial of x over k, when x is regarded as an element of Δ. From the preceding proof, we have that $F(X) = |(XE - A)^{(m)}|$ $(= |XE - C|)$, and hence

$$(10) \qquad F(X) = [f(X)]^m.$$

As a consequence of (10) we can now prove the following theorem:

THEOREM 20. *If $g(X)$ is the minimal polynomial of x over k, then $f(X)$ is a power of $g(X)$, and $f(X) = g(X)$ if and only if x is a primitive element of K over k (that is, if $K = k(x)$;* see § 9).

PROOF. Let $g(X)$ be of degree s, and let $g_1(X)$ be the field polynomial of x when x is regarded as an element of $k(x)$. Since $[k(x):k] = s$, it follows that $g_1(X)$ is also of degree s. Since x is a root of $g_1(X)$ and $g(X)$ is the minimal polynomial of x in $k[X]$, it follows (see § 2, Theorem 1) that $g(X) = g_1(X)$ (both g and g_1 being monic polynomials). We have thus shown that if x is regarded as an element of $k(x)$, then the minimal polynomial of x in $k[X]$ coincides with the field polynomial of x over k. This proves the first part of the theorem [apply (10) after replacing K by $k(x)$ and Δ by K] and also the "if" part of the second half of the theorem. The "only if" follows from observing that, by (10), and from the fact that $g(X)$ is the field polynomial of x over k, when x is regarded as an element of $k(x)$, it follows that $f(X) = [g(X)]^m$, where $m = [K:k(x)]$. Hence if $f(X) = g(X)$, then $m = 1$ and hence $K = k(x)$.

The field polynomial $f(X)$ of x over k $(x \in K)$ can itself be interpreted as a norm. For that purpose, we consider the field $K(X)$ and we observe that the algebraic closure of $k(X)$ in $K(X)$ contains K (since K is an algebraic extension of k) and X [since $X \in k(X)$], hence coincides with $K(X)$. In other words: $K(X)$ is an algebraic extension of $k(X)$. Furthermore, since $K = \sum_{i=1}^{n} k \cdot \omega_i = k(\omega_1, \omega_2, \cdots, \omega_n)$, we have $K(X) = k(X)(\omega_1, \omega_2, \cdots, \omega_n)$, and therefore (see § 2, Theorem 2) $K(X) = k(X)[\omega_1, \omega_2, \cdots, \omega_n] = k(X)K = \sum_{i=1}^{n} k(X) \cdot \omega_i$. This implies that $\omega_1, \omega_2, \cdots, \omega_n$ is also a basis of $K(X)$ over $k(X)$, provided we show that the ω's are linearly independent over $k(X)$. But this follows immediately from the linear independence of the ω's over k and from the fact that X is a transcendental over $k(\omega_1, \omega_2, \cdots, \omega_n)$. We have therefore proved that $K(X)$ is a finite extension of $k(X)$, that $[K(X):k(X)] = [K:k] = n$, and that $\{\omega_1, \omega_2, \cdots, \omega_n\}$ is a basis of $K(X)$ over $k(X)$. Now, we have $(X - x)\Omega = X\Omega - A\Omega = (XE - A)\Omega$, where Ω is the one-column matrix

$$\left\| \begin{matrix} \omega_1 \\ \omega_2 \\ \cdot \\ \cdot \\ \cdot \\ \omega_n \end{matrix} \right\|.$$

It follows that $N_{K(X)/k(X)}(X - x) = |XE - A| = f(X)$, that is, *the field polynomial of x over k, when x is regarded as an element of K, is the norm of $X - x$ over $k(X)$, when $X - x$ is regarded as an element of $K(X)$.*

We shall conclude this section with the derivation of an expression for the trace and norm of x in terms of the conjugates of x (in some normal extension of k containing K). In view of (8) and (9) it will be sufficient to deal with the case in which $K = k(x)$. Let $f(X) = X^n + a_1X^{n-1} + \cdots + a_n$ be the minimal polynomial of x over k. We consider some normal extension K' of k containing $k(x)$ [for instance, the least normal extension of k containing $k(x)$]. Let

$$f(X) = \prod_{i=1}^{n} (X - x_i),$$

where $x_i \in K'$ $(x_1 = x)$, whence

$$(11) \qquad a_1 = -\sum_{i=1}^{n} x_i,$$

$$(12) \qquad a_n = (-1)^n \prod_{i=1}^{n} x_i.$$

Since we know already that $f(X)$ is also the field polynomial of x over k (Theorem 20), we find, by (6) and (7):

$$(13) \qquad N(x) = \prod_{i=1}^{n} x_i,$$

$$(14) \qquad T(x) = \sum_{i=1}^{n} x_i.$$

If x is separable over k, then $x_1, x_2, \cdots, x_n$ are distinct and so we have that *the norm and trace of x are equal respectively to the product and sum of the conjugates of x* (in K'). If x is inseparable over k, and if n_0 and p^e are respectively the separable and inseparable factors of the degree n of $f(x)$, then (§ 6, Theorem 11)

$$f(X) = \prod_{i=1}^{n_0} (X - x_i)^{p^e},$$

and x has only n_0 distinct conjugates. It follows from (13) and (14) that

$$N(x) = \left(\prod_{i=1}^{n_0} x_i\right)^{p^e}, \tag{15}$$

$$T(x) = p^e \cdot \left(\sum_{i=1}^{n_0} x_i\right) = 0. \tag{16}$$

COROLLARY. *If K is a finite extension of k and x is an element of K which is inseparable over k, then $T_{K/k}(x) = 0$.*

This follows at once from (16) and (9).

We shall now derive another expression of $N_{K/k}(x)$ and $T_{K/k}(x)$, where K is a finite algebraic extension of k and x is an element of K. Let $m = [K:k]$, $m_0 = [K:k]_s$, $p^f = [K:k]_i$ and let n, n_0 and p^e be the corresponding degrees for $k(x)$ instead of K. Let $K^\star$ be the least normal extension of k containing K and let $\{\varphi_i; i = 1, 2, \cdots, m_0\}$ be the set of k-isomorphisms of K into $K^\star$. Let $\{x_j; j = 1, 2, \cdots, n_0\}$ be the set of distinct conjugates of x in $K^\star$ (one of the x_j, say x_1, being x itself). By Lemma 2 of § 6 each of the n_0 k-isomorphisms of $k(x)$ into $K^\star$ has exactly m_0/n_0 extensions among the φ_i. Hence each of the conjugates x_j of x occurs m_0/n_0 times in the set $\{x\varphi_1, x\varphi_2, \cdots, x\varphi_{m_0}\}$. Therefore

$$\prod_{i=1}^{m_0} x\varphi_i = \left(\prod_{j=1}^{n_0} x_j\right)^{m_0/n_0}, \tag{17}$$

$$\sum_{i=1}^{m_0} x\varphi_i = m_0/n_0 \cdot \sum_{j=1}^{n_0} x_j. \tag{18}$$

By (8) and (9), with K and Δ replaced by $k(x)$ and K respectively, we have:

$$N_{K/k}(x) = (N_{k(x)/k}(x))^{m_0 p^f/n_0 p^e},$$

$$T_{K/k}(x) = m_0 p^f/n_0 p^e \cdot T_{k(x)/k}(x),$$

and hence, in view of (15) and (16) we find

$$N_{K/k}(x) = \left(\prod_{i=1}^{m_0} x\varphi_i\right)^{p^f}, \tag{19}$$

$$T_{K/k}(x) = p^f \sum_{i=1}^{m_0} x\varphi_i. \tag{20}$$

These are the desired expressions of the norm and trace of x; they are generalizations of (15) and (16) from the case $K = k(x)$ to the case of an arbitrary finite algebraic extension of k.

Using the expressions (19) and (20) we can derive the following *transitivity law* for norms and traces: *if* $k \subset L \subset \Delta$ *are successive finite algebraic extensions of* k *and* x *is an element of* Δ, *then*

$$N_{\Delta/k}(x) = N_{L/k}(N_{\Delta/L}(x)), \tag{21}$$

$$T_{\Delta/k}(x) = T_{L/k}(T_{\Delta/L}(x)). \tag{22}$$

For the proof we shall use the notations of the proof of Lemma 2 of § 6. We may assume that ψ_1 is the identity automorphism of K. We have by (19) and (20):

$$N_{\Delta/L}(x) = \left(\prod_{i=1}^{n} x\varphi_i\right)^{p^\alpha}, \quad p^\alpha = [\Delta : L]_i,$$

$$N_{L/k}(N_{\Delta/L}(x)) = \left(\prod_{j=1}^{m} (N_{\Delta/L}(x)\psi_j)\right)^{p^\beta}, \quad p^\beta = [L : k]_i,$$

or

$$N_{L/k}(N_{\Delta/L}(x)) = \left(\prod_{j=1}^{m} \prod_{i=1}^{n} x\varphi_i\psi_j\right)^{p^{\alpha+\beta}} \tag{23}$$

Now, we know from the proof of Lemma 2 of § 6, that the restrictions of the products $\varphi_i\psi_j$ to Δ are distinct and give all the k-isomorphisms of Δ into K. Furthermore, by Corollary 2 to Theorem 16 of § 6, we know that $p^{\alpha+\beta} = [\Delta : k]_i$. Hence (21) follows from (23) in view of the expression of the norm obtained in (19) (and applied to the field Δ instead of to K). The proof of (22) is quite similar.

We note that relations (8) and (9) can be derived as consequences of (21) and (22). In (8) and (9) the element x belongs to a finite algebraic extension K of k, and Δ is a finite algebraic extension of K, of degree m. The norm $N_{\Delta/K}(x)$ and trace $T_{\Delta/K}(x)$ are equal to x^m and mx respectively, since x belongs to K. Hence $N_{\Delta/k}(x) = N_{K/k}(x^m) = (N_{K/k}(x))^m$ and $T_{\Delta/k}(x) = T_{K/k}(mx) = mT_{K/k}(x)$.

§ 11. The discriminant. Let K be an algebraic extension of k, of degree n, and let $\{\omega_1, \omega_2, \cdots, \omega_n\}$ be a basis of K/k.

DEFINITION. *The determinant*

$$d = |T(\omega_i \cdot \omega_j)| \tag{1}$$

is called the discriminant of the basis $\{\omega_1, \omega_2, \cdots, \omega_n\}$.

The discriminant of a basis $\{\omega_1, \omega_2, \cdots, \omega_n\}$ of K/k will also be denoted by $d\{\omega_1, \omega_2, \cdots, \omega_n\}$ or by $d_{K/k}\{\omega_1, \omega_2, \cdots, \omega_n\}$.

If $\{\omega'_1, \omega'_2, \cdots, \omega'_n\}$ is another basis of K/k, then

$$\omega'_i = \Sigma_j a_{ij}\omega_j, \; a_{ij} \in k, \; |A| = |a_{ij}| \neq 0,$$

and

$$T(\omega'_i \cdot \omega'_j) = \Sigma_{\alpha,\beta} a_{i\alpha} a_{j\beta} T(\omega_\alpha \cdot \omega_\beta).$$

Hence if d' denotes the discriminant $d\{\omega'_1, \omega'_2, \cdots, \omega'_n\}$ of the new basis, then by the rule of multiplication of determinants we have the following relations:

$$d' = d \cdot |A|^2. \tag{2}$$

COROLLARY. *If the discriminant of one basis is zero, then the discriminant of every basis is zero.*

The statement "*the field discriminant of K/k is zero (is not zero)*" has therefore a meaning. We mean here by the *field discriminant of K/k*, the discriminant of any basis of K/k. By (2), the field discriminant of K/k is only determined to within a factor which is the square of an arbitrary non-zero element of k (arbitrary, because if a is any element of k, $a \neq 0$, and if we set $\omega'_1 = a\omega_1$, $\omega'_i = \omega_i$, $i = 2, 3, \cdots, n$, then $\omega'_1, \omega'_2, \cdots, \omega'_n$ is a basis of K/k, and in this case we have $|A| = a$).

THEOREM 21. *The field discriminant of K/k is zero if and only if $T(\xi) = 0$ for all ξ in K.*

PROOF. The "if" part is obvious. Assume now that $d = 0$. We can then find n elements $c_1, c_2, \cdots, c_n$, *not all zero*, such that $\Sigma_j c_j T(\omega_i \omega_j) = 0$, for $i = 1, 2, \cdots, n$. We set $z = \Sigma_j c_j \omega_j$. Then $z \neq 0$, and we have $T(\omega_i z) = 0, i = 1, 2, \cdots, n$. From this it follows that $T(yz) = 0$ for all y in K. If $\xi \in K$, we take $y = \xi/z$ and we find $T(\xi) = 0$. Q.E.D.

COROLLARY. *If the field discriminant of K/k is zero, then k is of characteristic $p \neq 0$, and n is a multiple of p.*

For $T(1) = n$.

In order to derive further results on the discriminant, we go back to the notion of a field polynomial, developed in § 10. Let K_0 be the maximal separable extension of k contained in K (§ 5) and let $n = n_0 p^e$ where $n_0 = [K_0 : k]$. If ξ is any element of K_0, then we have $T_{K/k}(\xi) = p^e T_{K_0/k}(\xi)$ [see (9), § 10]. If K is an inseparable extension of k, that is, if $e \geqq 1$, this implies that $T_{K/k}(\xi) = 0 (\xi \in K_0)$. If ξ is in K but not in K_0, then ξ is inseparable over k, and hence we have again $T_{K/k}(\xi) = 0$, by the corollary on p. 91. We have thus proved that *if K is an inseparable extension of k, then $T_{K/k}(\xi) = 0$ for all ξ in K*, and hence, by Theorem 21 above, *the field discriminant of K/k is zero.*

We now consider the case in which K is a separable extension of k. Let Δ be a least normal extension of K containing K (§ 6) and let $\tau_1 (= 1), \tau_2, \cdots, \tau_n$ be the distinct k-isomorphisms of K into Δ (§ 6, Theorem 16). Let x be any element of K and let $x_i = x\tau_i$. Each element x_i is a conjugate of x over k, and every conjugate of x in K coincides with one of the elements x_i (§ 6, Theorem 15). The n

elements x_i are not, however, necessarily distinct. If there are ν isomorphisms τ_i which leave x invariant, then x itself, and also every conjugate element of x, occurs exactly ν times in the set $\{x_1, x_2, \cdots, x_n\}$. Then x has only m distinct conjugate elements over k, where $m = n/\nu$, and m is, then, also the degree of the minimal polynomial $g(X)$ of x over k. Now if $f(X) = 0$ is the field polynomial of x over k when x is considered as an element of K, then we know from § 10, formula (10), that $f(X) = [g(X)]^\nu$. *It follows that*

$$f(X) = \prod_{i=1}^{n} (X - x_i).$$

From this we conclude at once that

$$T_{K/k}(x) = \sum_{i=1}^{n} x_i, \tag{3}$$

$$N_{K/k}(x) = \prod_{i=1}^{n} x_i. \tag{4}$$

These formulas are similar to (13) and (14) of § 10 which were obtained in the special case $K = k(x)$.

We shall now apply (3) as follows:

Let $\{\omega_1, \omega_2, \cdots, \omega_n\}$ be a basis of K/k and let $\omega_i^{(\alpha)} = \omega_i\tau_\alpha$, $\alpha = 1, 2, \cdots, n$. Then $T(\omega_i \cdot \omega_j) = \sum_{\alpha=1}^{n} \omega_i^{(\alpha)}\omega_j^{(\alpha)}$, and from this, by the rule of multiplication of determinants, we obtain the following expression of $d(\omega_1, \omega_2, \cdots, \omega_n)$:

$$d(\omega_1, \omega_2, \cdots, \omega_n) = \begin{vmatrix} \omega_1^{(1)}, \omega_2^{(1)}, \cdots, \omega_n^{(1)} \\ \omega_1^{(2)}, \omega_2^{(2)}, \cdots, \omega_n^{(2)} \\ \cdots\cdots\cdots\cdots \\ \omega_1^{(n)}, \omega_2^{(n)}, \cdots, \omega_n^{(n)} \end{vmatrix}^2. \tag{5}$$

Since K is separable over k, there exists a primitive element of K/k (§ 9). Let x be a primitive element. Then $\{1, x, x^2, \cdots, x^{n-1}\}$ is a basis of K/k, and (5) yields:

$$d(1, x, x^2, \cdots, x^{n-1}) = \begin{vmatrix} 1, x_1, x_1^2, \cdots, x_1^{n-1} \\ 1, x_2, x_2^2, \cdots, x_2^{n-1} \\ \cdots\cdots\cdots\cdots \\ 1, x_n, x_n^2, \cdots, x_n^{n-1} \end{vmatrix}^2, \tag{6}$$

where $x_i = x\tau_i$. The Vandermonde determinant on the right-hand side of (6) is different from zero since $x_1, x_2, \cdots, x_n$ are distinct elements (x being separable over k and n being the degree of $k(x)$ over k). Hence $d(1, x, x^2, \cdots, x^{n-1}) \neq 0$.

We have therefore proved the following

THEOREM 22. *The field discriminant of K/k is zero if and only if K is an inseparable extension of k.*

COROLLARY. *K is a separable extension of k if and only if there exists an element x in K such that $T_{K/k}(x) \neq 0$.* This is an immediate consequence of Theorems 21 and 22.

Note. If x is a primitive element of the separable extension K/k, of degree n as above, and if $g(X)$ is the minimal polynomial of x over k, then $g(X) = \prod_{i=1}^{n} (X - x_i)$, $g'(x_i) = \prod_{j \neq i} (x_i - x_j)$ and $N(g'(x)) = \prod_i \prod_{j,\, i<j} (x_i - x_j)^2$. Hence, by computation of the Vandermonde determinant (6):

$$(7) \qquad d(1, x, x^2, \cdots, x^{n-1}) = \prod_i \prod_{j,\, i<j} (x_i - x_j)^2 = N(g'(x)).$$

§ **12. Transcendental extensions.** An extension K of a field k is *transcendental* if it is not algebraic, that is, if K contains elements which are transcendental over k. An example of a transcendental extension of k is *the field of rational functions in n indeterminates over k*, that is, the quotient field $k(X_1, X_2, \cdots, X_n)$ of the polynomial ring $k[X_1, X_2, \cdots, X_n]$ in n indeterminates ($n \geqq 1$) over k; or also any k-isomorphic image $k(x_1, x_2, \cdots, x_n)$ of $k(X_1, X_2, \cdots, X_n)$, where therefore $x_1, x_2, \cdots, x_n$ are algebraically independent over k and $k[x_1, x_2, \cdots, x_n]$ is a polynomial ring over k (I, § 18). It is clear that any extension of a transcendental extension of k is itself a transcendental extension of k.

The definition of algebraic independence over k, given in I, § 18, can be extended to infinite sets of elements. If K is an extension of k and L is a subset of K, then the elements of L are said to be algebraically independent (a.i.) over k if each finite subset of L consists of elements which are algebraically independent over k. Such a set L will be called a *transcendence set* (over k).

We shall use the terminology and notation introduced in II, § 1 and § 5. If K can be obtained from k by the adjunction of the elements of some transcendence set L, then K is said to be *a pure transcendental extension of k*. An example of a pure transcendental extension of k is the field $k(X_1, X_2, \cdots, X_n)$ of rational functions in n indeterminates over k. For a given integer n, any two fields which can be obtained from k by the adjunction of n algebraically independent elements (and which are therefore pure transcendental extensions of k) are k-isomorphic (see I, § 18, Theorem 12, Corollary 1; and I, § 19, Theorem 16).

Let L be a transcendence set in K/k and let x be an element of K not in L. Let L' be the set consisting of x and the elements of L.

LEMMA. *L' is a transcendence set if and only if x is a transcendental over $k(L)$.*

PROOF. Suppose that x is transcendental over $k(L)$ and let $x_1, x_2, \cdots, x_n$ be any elements of L'. If all the x_i are already in L, they are a.i. over k. Assume that $x_n = x$ and let $f(X_1, X_2, \cdots, X_n)$ be a polynomial, with coefficients in k, such that $f(x_1, x_2, \cdots, x_n) = 0$. Then x is a root of the polynomial $f(x_1, x_2, \cdots, x_{n-1}, X)$ in $k(x_1, x_2, \cdots, x_{n-1})[X]$. This polynomial must be zero since x is transcendental over $k(L)$. Hence, if $f(X_1, X_2, \cdots, X_n) = A_0(X_1, X_2, \cdots, X_{n-1})X_n^g + \cdots A_g(X_1, X_2, \cdots, X_{n-1})$, then we must have $A_i(x_1, x_2, \cdots, x_{n-1}) = 0, i = 0, 1, \cdots, g$. Since $x_1, x_2, \cdots, x_{n-1}$ are a.i. over k, the polynomials $A_i(X_1, X_2, \cdots, X_{n-1})$ must be zero. This implies that also $f(X_1, X_2, \cdots, X_n)$ is the polynomial zero. Conversely, assume that L' is a transcendence set. Let $F(X)$ be a polynomial in $k(L)[X]$ such that $F(x) = 0$. Since the number of coefficients of $F(X)$ is finite, there exists a finite subset L_1 of L such that these coefficients belong already to $k(L_1)$. Let $x_1, x_2, \cdots, x_n$ be the elements of L_1. If $F(X) = a_0X^g + a_1X^{g-1} + \cdots + a_g$, then we can write the a_i as quotients of polynomials in $k[x_1, x_2, \cdots, x_n]$, with the same denominator:

$$a_i = \frac{A_i(x_1, x_2, \cdots, x_n)}{B(x_1, x_2, \cdots, x_n)}, \quad i = 0, 1, \cdots, g.$$

If we set

$$\begin{aligned} f(X_1, X_2, \cdots, X_n, X) = {} & A_0(X_1, X_2, \cdots, X_n)X^g \\ & + A_1(X_1, X_2, \cdots, X_n)X^{g-1} + \cdots + A_g(X_1, X_2, \cdots, X_n), \end{aligned}$$

then f is a polynomial with coefficients in k, and from $F(x) = 0$ follows that $f(x_1, x_2, \cdots, x_n, x) = 0$. Since L' is a transcendence set, the elements $x_1, x_2, \cdots, x_n, x$ are a.i. over k, and hence $f(X_1, X_2, \cdots, X_n, X) = 0$. This implies that $A_i(X_1, X_2, \cdots, X_n) = 0, i = 0, 1, \cdots, g$, and hence also $a_i = 0, i = 0, 1, \cdots, g$, that is, $F(X) = 0$. We have therefore proved that x is transcendental over $k(L)$, and this completes the proof of the lemma.

DEFINITION 1. *A transcendence set L in K is called a transcendence basis of K/k if it is maximal, that is, if L is not a proper subset of another transcendence set.*

From the preceding considerations, it follows at once that *a transcendence set L is a transcendence basis of K/k if and only if K is an algebraic extension of $k(L)$.*

At this stage we shall incorporate our further conclusions concerning algebraic dependence in the general axiomatic treatment of dependence as developed in I, § 21. This is possible since we can now define the "span" $s(X)$ of a subset X of K as the algebraic closure of $k(X)$ in K (the algebraic closure of k in K, if X is empty). Then it is immediately seen that the conditions (S_1)–(S_5) of Theorem 19 of I, § 21 are satisfied. In fact, it is obvious that

(S_1). If $X \subset Y$, then $s(X) \subset s(Y)$.

(S_2). If $x \in s(X)$, then there exists a finite subset Y of X such that $x \in s(Y)$.

(S_3). $X \subset s(X)$ for all subsets X of K.

(S_4). $s(s(X)) = s(X)$ (this simply expresses the transitivity of algebraic dependence).

We shall now verify the condition (S_5):

(S_5). The relations $y \in s(X, x)$ and $y \notin s(X)$ imply $x \in s(X, y)$.

There exists, by (S_2), a finite set of elements $x_1, x_2, \cdots, x_n$ in X such that y is algebraic over $k(x_1, x_2, \cdots, x_n, x)$. There exists then a polynomial $f(X_1, X_2, \cdots, X_n, Z, Y)$, with coefficients in k, such that $f(x_1, x_2, \cdots, x_n, x, Y) \neq 0$ and $f(x_1, x_2, \cdots, x_n, x, y) = 0$. We write $f(X_1, X_2, \cdots, X_n, Z, Y) = \sum_{i=0}^{g} A_i(X_1, X_2, \cdots, X_n, Y)Z^i$, and we observe that the $g + 1$ polynomials $A_i(X_1, X_2, \cdots, X_n, Y)$ are not all zero, since $f(X_1, X_2, \cdots, X_n, Z, Y)$ is not the zero polynomial. Since $y \notin s(X)$ (that is, since y is a transcendental over $k(X)$), it follows therefore that not all the elements $A_i(x_1, x_2, \cdots, x_n, y)$ of $k(x_1, x_2, \cdots, x_n, y)$ are zero. Therefore $f(x_1, x_2, \cdots, x_n, Z, y) \neq 0$, and since $f(x_1, x_2, \cdots, x_n, x, y) = 0$, it follows that x is algebraic over $k(x_1, x_2, \cdots, x_n, y)$, that is, $x \in s(X, y)$, as was asserted.

We now generalize Theorems 21 and 22 of I, § 21, to the case of sets V which do not necessarily admit a finite system of generators. We recall from I, § 21, that it is assumed that we are given a mapping s of the set of all subsets of V into itself, and that this mapping satisfies conditions (S_1) to (S_5). We recall also that a set X is called a generating system of V if $s(X) = V$, that X is a free set if for any x in X we have $x \notin s(X - x)$, and that X is called a basis of V if it is both a generating system of V and a free set. In our case of an extension field K of k, X is a generating system of K if K is an algebraic extension of $k(X)$; X is a free set if it is a transcendence set (in view of the lemma proved above) and X is a basis of K/k if it is a transcendence basis.

For the purpose of the generalization of Theorems 21 and 22 of I, § 21, we must give some preliminary definitions concerning partially ordered sets.

A set S is said to be partially ordered if there is given in S a binary relation $\prec$ which is defined for *certain* pairs (a, b) of elements of S (it is not necessary that the relation $\prec$ be defined for all pairs (a, b) of elements of S) and which satisfies the following conditions: (1) $a \prec a$ for any element a of S; (2) if $a \prec b$ and $b \prec a$, then $a = b$; (3) if $a \prec b$ and $b \prec c$, then $a \prec c$. A subset S_1 of S is *totally ordered* if, given any two elements a, b of S_1, at least one of the relations $a \prec b$ or $b \prec a$ holds.

Let S_1 be a subset of a partially ordered set S. An element c of S is called *an upper bound* of S_1 if $a \prec c$ for all a in S_1. An element a_0 of S is a *maximal* element of S if $a_0 \prec a$ implies $a_0 = a$.

A partially ordered set S is said to be *inductive* if every totally ordered subset of S has an upper bound in S.

ZORN'S LEMMA. *If a partially ordered set S is inductive, then there exist maximal elements in S.**

We now begin with the following generalization of Theorem 21 of I, § 21.

THEOREM 23. *Let L be a free subset of V and S a system of generators of V. There exists a subset S' of S such that $L \cup S'$ is a basis of V and $L \cap S'$ is empty.*

PROOF. We partially order, by set-theoretic inclusion, the set M of all subsets S_α of S such that $L \cap S_\alpha$ is empty and $L \cup S_\alpha$ is a free set. The set M is non-empty since the empty subset of S belongs to M. It is clear that M is an inductive set (since from (S_2) it follows that any ascending chain of free sets has a limit (union) which is also a free set). Let S' be a maximal element of M. Then $L \cap S'$ is empty and $L \cup S'$ is a free set. We shall show that $L \cup S'$ is a generating system of V, hence a basis of V, and this will complete the proof of the theorem. Since $s(S) = V$, it will be sufficient to show that $S \subset s(L \cup S')$, in view of (S_1) and (S_4) Let x be any element of S. If $x \in L \cup S'$, then $x \in s(L \cup S')$, by (S_3) Assume that $x \notin L \cup S'$ and let $(S', x) = S''$. Then S'' is a subset of S such that $L \cap S''$ is empty. Since S' is a proper subset of S'', it follows by the maximality of S', that $L \cup S''$, that is, $(L \cup S', x)$ is not a free set. Since $L \cup S'$ is a free set, it follows that $x \in s(L \cup S')$ (see Remark at the end of the proof of Theorem 20, I, § 21). This completes the proof.

* For a proof of Zorn's lemma see, for instance, John L. Kelley, *General Topology*, p. 33 (University Series in Higher Mathematics, Van Nostrand Co., Inc., Princeton, N.J., 1955).

COROLLARY 1. *If L is a transcendence set in K/k and S is a subset of K such that K is an algebraic extension of $k(S)$, then there exists a subset S' of S such that $L \cap S'$ is empty and $L \cup S'$ is a transcendence basis of K/k.*

COROLLARY 2. *Any subset S of K such that K is an algebraic extension of $k(S)$ contains a transcendence basis of K/k.*

We have only to apply Corollary 1 to the case in which L is the empty set.

COROLLARY 3. *There exist transcendence bases of K/k.*

We apply Corollary 2 for the case $S = K$.

NOTE. In the case of a vector space V over a field k, Theorem 23 guarantees the existence of a *basis* (or *vector basis*) of V over k.

The following is a generalization of Theorem 22 in I, § 21:

THEOREM 24. *Any two bases of V have the same cardinal number.*

PROOF. This theorem has been proved in I, § 21 under the assumption that there exists at least one finite basis of V. We shall therefore assume now that every basis of V is infinite.

Let B be a basis of V and let x be any element of V. By (S_2), there exist finite subsets E of B such that $x \in s(B)$. We assert that there exists a smallest finite subset E_x of B such that $x \in s(E_x)$ (and such that any other subset E of B with the property $x \in s(E)$ contains E_x). To see this, it is sufficient to prove the following: *if E'* and *E'' are two subsets of B such that $x \in s(E') \cap s(E'')$ and if we have $x \notin s(E'_1)$ for every proper subset E'_1* of *E', then $E' \subset E''$.* Assuming the contrary, let y be an element of E' not in E'' and let E'_1 denote the set $E' - y$. We have $x \notin s(E'_1)$ and $x \in s(E'_1, y)$. Hence, by (S_5), we have $y \in s(E'_1, x)$. Since $x \in s(E'')$ it follows that $y \in s(E'_1 \cup E'')$. This is in contradiction with the fact that $y \notin E'_1 \cup E''$ and that $E'_1 \cup E'' \cup \{y\} \subset B$ is a free set.

Now let B' be another basis of V. We consider the mapping $x \to E_x (x \in B', E_x \subset B)$, where E_x is the finite subset of B defined above. From set theory it is known that the cardinal number of B' is not less than the cardinal number of the set $\bigcup_{x \in B'} E_x$ (since each set E_x is *finite*). On the other hand, we have $B = \bigcup_{x \in B'} E_x$ since $B' \subset s(\cup E_x)$, $V = s(B') = s(\cup E_x)$, and therefore the subset $\bigcup_{x \in B'} E_x$ of B must coincide with the *basis* B. Hence the cardinal number of B' is not less than the cardinal number of B. Interchanging the roles of B and B' we conclude that B and B' have the same cardinal number. Q.E.D.

As a consequence we have the following result:

THEOREM 25. *Any two transcendence bases of K/k have the same cardinal number.*

NOTE. In the case of a vector space V over a field k, Theorem 24 leads to the notion of the *dimension* of V over k, this being the common cardinal number of all vector bases of V/k.

DEFINITION 2. *The common cardinal number of the various transcendence bases of K/k is called the transcendence degree of K/k* (abbreviation: tr. d. K/k).

It is clear that K is an algebraic extension of k if and only if tr. d. $K/k = 0$.

THEOREM 26. *Let $k \subset K \subset \Delta$ be successive extensions of k. Then* tr. d. $\Delta/k =$ tr. d. $\Delta/K +$ tr. d. K/k.

PROOF. Let L and M be transcendence bases of K/k and Δ/K respectively. It will be sufficient to prove that $L \cup M$ is a transcendence basis of Δ/k. Let $\{x_1, x_2, \cdots, x_n, y_1, y_2, \cdots, y_n\}$ be any finite subset of $L \cup M$, where we assume that the x_i are in L and the y_j are in M. Let $f(\{X\}, \{Y\}) = f(X_1, X_2, \cdots, X_m, Y_1, Y_2, \cdots, Y_n)$ be a polynomial in $m + n$ indeterminates, with coefficients in k, such that $f(\{x\}, \{y\}) = 0$. The polynomial $f(\{x\}, Y)$ in the n indeterminates Y_j has coefficients in K and must be zero since the y_j are a.i. over K. Since the x_i are a.i. over k, it follows that $f(\{X\}, \{Y\})$, regarded as a polynomial in $\{Y\}$ with coefficients in $k[\{X\}]$, must be zero. Hence $f(\{X\}, \{Y\}) = 0$ and that shows that $L \cup M$ is a transcendence set.

By assumption, K is an algebraic extension of $k(L)$. It follows that $K(M)$ is an algebraic extension of $k(L)(M) = k(L \cup M)$. But Δ is an algebraic extension of $K(M)$. Hence Δ is an algebraic extension of $k(L \cup M)$. This shows that $L \cup M$ is a transcendence basis of Δ/k.

THEOREM 27. *Let K and K' be two extensions of k, contained in some larger field Ω, and let (K, K') be the smallest subfield of Ω containing both fields K and K'. Then* tr. d. $(K, K')/K \leqq$ tr. d. K'/k, *and* tr. d. $(K, K')/k \leqq$ tr. d. $K/k +$ tr. d. K'/k.

PROOF. Let L' be a transcendence basis of K'/k. We have $(K, K') = K(K')$. Since every element of K' is algebraic over $k(L')$, it follows that (K, K') is algebraic over $K(L')$. Therefore, by Theorem 24, L' contains a transcendence basis of $(K, K')/K$. We have then: tr. d. $(K, K')/K \leqq$ tr. d. K'/k. By the preceding theorem, we have: tr. d. $(K, K')/k =$ tr. d. $(K, K')/K +$ tr. d. K/k; and this, combined with the above inequality, establishes the theorem.

We shall use the term "transcendence degree" also when dealing with *integral domains* (not necessarily fields) containing k. If R is an integral domain, $R \supset k$, and if K is the quotient field of R, then we set tr. d. $R/k =$ tr. d. K/k. Note that since $K = k(R)$, there exist transcendence bases of K/k which are *subsets of R* (see Theorem 24, Corollary 2). Such transcendence bases of K/k will be referred to in the sequel as transcendence bases of R/k.

Two k-isomorphic domains R and R' ($k \subset R$, $k \subset R'$) have naturally the same tr. d. over k. Of particular importance in applications are two theorems which are proved immediately below.

THEOREM 28. *Let R and R' be integral domains containing k. If R' is a k-homomorphic image of R, then* tr. d. $R'/k \leqq$ tr. d. R/k.

PROOF. We assume, then, that there exists a k-homomorphism τ of R onto R' and we consider a transcendence basis L' of R'/k. For every element x' of L' we fix an element x in R such that $x' = x\tau$ and we denote by L the set of all elements x obtained in this fashion.* From the fact that L' is a transcendence set, it follows at once that also L is a transcendence set. By Theorem 23, L is contained in some transcendence basis M of R/k. The cardinal numbers of L' and L are the same since for every x' in L' there is only one element x in L such that $x' = x\tau$ and since therefore the correspondence $x' \to x$ is one to one. Since L is a subset of M, the proof is complete.,

THEOREM 29. *If* tr. d. $R/k =$ tr. d. $R'/k = n$ (n finite), *then any k-homomorphism τ of R/k onto R'/k is an isomorphism.*

PROOF. We use the notations of the proof of the preceding theorem. Let $L' = \{x'_1, x'_2, \cdots, x'_n\}$ be a transcendence basis of R'/k and let $L = \{x_1, x_2, \cdots, x_n\}$, where $x'_i = x_i\tau$. This time L is not only a transcendence set but also *a transcendence basis of R/k*, since tr. d. $R/k = n$. Now let $u \in R$, $u \neq 0$. Since u is algebraically dependent on $k(x_1, x_2, \cdots, x_n)$, we have a relation of the form

$$(1) \qquad A_0(x)u^g + A_1(x)u^{g-1} + \cdots + A_g(x) = 0, \quad g \geqq 1,$$

where

$$A_i(X) = A_i(X_1, X_2, \cdots, X_n) \in k[X_1, X_2, \cdots, X_n]$$

and where $A_0(x) \neq 0$. We take g as small as possible. Then $A_g(x) \neq 0$ for otherwise we could divide (1) by u (since $u \neq 0$) and have an equation for u, of degree $< g$. Applying to (1) the homomorphism τ we find

$$(2) \qquad A_0(x')u'^g + A_1(x')u'^{g-1} + \cdots + A_g(x') = 0,$$

where $u' = u\tau$. Since $A_g(x) \neq 0$, the polynomial $A_g(X)$ is not zero, and hence $A_g(x') \neq 0$, since $x'_1, x'_2, \cdots, x'_n$ are a.i. over k. Consequently, by (2), we have $u' \neq 0$, and this shows that τ is an isomorphism.

* If L' is an infinite set, this procedure involves the axiom of choice. It can be easily replaced by another argument which is based exclusively on Zorn's lemma and which would show the existence of a subset L of R such that: (1) $\tau(L) = L'$; (2) the transformation of L onto L' induced by τ is one to one. This may be left to the reader.

REMARK. If R is a field, then Theorem 29 is trivial and is true without any condition on the transcendence degrees, since a field does not admit proper homomorphisms (that is, homomorphisms which are not isomorphisms).

§ 13. Separably generated fields of algebraic functions. Let k be a field and let K be an extension of k.

DEFINITION 1. *K is a field of algebraic functions of r independent variables over k if K is finitely generated over k and if* tr. d. $K/k = r$.

DEFINITION 2. *A transcendence basis $\{z_1, z_2, \cdots, z_r\}$ of a field K/k of algebraic functions of r independent variables is called a separating transcendence basis of K/k if K is a separable (algebraic) extension of $k(z_1, z_2, \cdots, z_r)$.*

The field K is said to be separably generated over k if there exists a separating transcendence basis of K/k.

EXAMPLE. Let $r = 1$, $K = k(x, y)$, where x is a transcendental, and let $Y^p - x^2 = 0$ be the minimal polynomial of y over $k(x)$, where p is the characteristic of K ($p \neq 2$). Then K is an inseparable extension of $k(x)$. Nevertheless K is separably generated over k since K is a separable extension of $k(y)$ (whence y by itself is a separating element of K/k, that is, the set $\{y\}$ consisting of the single element y is a separating transcendence basis of K/k).

We begin with two simple lemmas which we shall have occasion to use in this section.

LEMMA 1. *Let R be a unique factorization domain and let K be the quotient field of R. Let X be an indeterminate over K.*

(1) *If a polynomial $f(X)$ in $R[X]$, of positive degree, is irreducible in $R[X]$, it is also irreducible in $K[X]$.*

(2) *If a primitive polynomial $f(X)$ in $R[X]$* (see I, § 17) *is irreducible in $K[X]$ it is also irreducible in $R[X]$.*

(3) *If $f(X)$ and $g(X)$ are polynomials in $R[X]$ such that $g(X)$ divides $f(X)$ in $K[X]$ and if $g(X)$ is primitive, then $g(X)$ also divides $f(X)$ in $R[X]$.*

PROOF

(1) Let $F_1(X)$ be a non-unit in $K[X]$ which divides $f(X)$ in $K[X]: f(X) = F_1(X)F_2(X)$, $F_i(X) \in K[X]$, $\partial F_1(X) > 0$. Then $F_i(X) = f_i(X)/a_i$, where $f_i(X) \in R[X]$ and $a_i \in R$, and we have $a_1a_2f(X) = f_1(X)f_2(X)$. Since $f(X)$ is irreducible in $R[X]$ it follows that $f(X)$ must divide in $R[X]$ one of the two polynomials $f_i(X)$, and therefore we have either $\partial f \leqq \partial f_1$ or $\partial f \leqq \partial f_2$. On the other hand, $\partial f = \partial f_1 + \partial f_2$, and by assumption, $\partial f_1 > 0$ (since $F_1(X)$ is not a unit and has therefore

positive degree). Hence $\partial f_2 = 0$, and $F_2(X)$ is a unit in $K[X]$. Since $f(X)$ has positive degree, it is not a unit in $K[X]$. Consequently $f(X)$ is an irreducible element in $K[X]$.

(2) Assume that we have $f(X) = f_1(X)f_2(X)$ where $f_1(X)$ and $f_2(X)$ are polynomials in $R[X]$. Since $f(X)$ is irreducible in $K[X]$, one of the polynomials $f_i(X)$ must be of degree zero. Let, say, $\partial f_1 = 0$, whence $f_1(X) = a \in R$. From $f(X) = af_2(X)$ follows that a divides, in R, the content of $f(X)$, and hence a is a unit in R since $f(X)$ is primitive. This shows that $f(X)$ is an irreducible element of $R[X]$.

(3) We have, by assumption: $f(X) = g(X)\cdot h(X)/a$, where $h(X) \in R[X]$ and $a \in R$. Thus $g(X)$ divides $af(X)$ in $R[X]$. Since $g(X)$ is primitive, it follows from I, § 17, Lemma 2, that $g(X)$ divides $f(X)$ in $R[X]$.

This completes the proof of the lemma.

LEMMA 2. *Let $x_1, x_2, \cdots, x_n, x_{n+1}$ be elements of an extension field K of a field k and assume that these $n+1$ elements x_i are algebraically dependent over k but that the n elements $x_1, x_2, \cdots, x_n$ are algebraically independent over k. Then the set A of polynomials $g(X_1, X_2, \cdots, X_n, X_{n+1})$ in $k[X_1, X_2, \cdots, X_n, X_{n+1}]$ with the property that $g(x_1, x_2, \cdots, x_n, x_{n+1}) = 0$ contains a polynomial $f(X_1, X_2, \cdots, X_n, X_{n+1})$ such that every polynomial in the set is a multiple of f in $k[X_1, X_2, \cdots, X_n, X_{n+1}]$.*

PROOF. Since the $n+1$ elements x_i are algebraically dependent over k, the set A contains polynomials different from zero. Let $f(X_1, X_2, \cdots, X_n, X_{n+1})$ be a non-zero polynomial in A, of smallest possible degree q in X_{n+1}, and let us write

$$f = A_0(X_1, X_2, \cdots, X_n)X_{n+1}{}^q + A_1(X_1, X_2, \cdots, X_n)X_{n+1}{}^{q-1} + \cdots + A_q(X_1, X_2, \cdots, X_n).$$

Since $x_1, x_2, \cdots, x_n$ are algebraically independent over k, X_{n+1} must actually occur in f, and we may also assume that f is a primitive polynomial in X_{n+1} over $k[X_1, X_2, \cdots, X_n]$. If $g(X)$ is any polynomial in the set A, then by Theorem 9 of I, § 17, we can write $A_0{}^s g = Qf + R$, where s is an integer $\geqq 0$, Q and R are polynomials in $k[X_1, X_2, \cdots, X_n, X_{n+1}]$ and R is either of degree *less* than q in X_{n+1} or is the zero polynomial. It is clear that we have $R(x_1, x_2, \cdots, x_n, x_{n+1}) = 0$, that is, the polynomial R belongs to the set A. Hence by our choice of f and by Lemma 1 we have $R = 0$; this completes the proof of the lemma.

It is obvious that the polynomial $f(X)$ is irreducible over k and that it is uniquely determined to within an arbitrary non-zero factor in k; moreover, among the polynomials in the set A the polynomial $f(X)$ is *characterized* by the condition that it be irreducible. When the elements $x_1, x_2, \cdots, x_n, x_{n+1}$ satisfy the conditions stated in the lemma, we shall refer to the relation $f(x_1, x_2, \cdots, x_n, x_{n+1}) = 0$ as *the irreducible* algebraic relation between the x_i, over k.

THEOREM 30. *Let $\{x_1, x_2, \cdots, x_n\}$ be a set of generators of K/k, $K = k(x_1, x_2, \cdots, x_n)$. If K is separably generated over k, then already the set $\{x_1, x_2, \cdots, x_n\}$ contains a separating transcendence basis of K/k.* (MacLane.)

PROOF. We first prove the theorem in the case $r = 1$. By assumption, there exists in K a separating transcendental element z. We have $z \notin k(z^p)$, since z is a transcendental. Hence, by Theorem 7, § 5, the polynomial $X^p - z^p$ is irreducible over $k(z^p)$. Since z is a root of this polynomial, it follows that *z is inseparable over $k(z^p)$*. Since $z \in k(x_1, x_2, \cdots, x_n)$, we conclude by Theorem 10, § 5, that at least one of the n elements x_i must be inseparable over $k(z^p)$. Let, say, x_1 be inseparable over $k(z^p)$. *We shall now prove that x_1 is a separating transcendental of K/k.*

Let $f(X, Z)$ be an irreducible polynomial in $k[X, Z]$ such that $f(x_1, z) = 0$. By Lemma 1 it follows that $f(X, Z)$ is irreducible in $k(Z)[X]$ (since $f(X, Z)$ must be of positive degree in X). Since z is a transcendental over k, it follows that also the polynomial $f(X, z)$ is irreducible over $k(z)$ and differs therefore from the minimal polynomial of x_1 in $k(z)[X]$ by a factor which is an element of $k[z]$. Since z is a separating element, we have $f'_{x_1}(x_1, z) \neq 0$. The polynomial $f(X, Z)$ is independent of Z if and only if x_1 is algebraic over k, and if that were the case then we would have $f(X, Z) = \varphi(X)$ and $\varphi'(x_1) = f'_{x_1}(x_1, z) \neq 0$. This would imply that x_1 is separable over k, hence *a fortiori* also over $k(z^p)$ (Lemma 2, § 5), contrary to our assumption on x_1. *Hence x_1 is a transcendental over k, and $f(X, Z)$ is not independent of Z.*

It is therefore possible now to assert that z is algebraic over $k(x_1)$ and that $f(x_1, Z)$ differs from the minimal polynomial of z in $k(x_1)[Z]$ only by a factor which is an element of $k[x_1]$. *We also assert that z is separable over $k(x_1)$.* For assume the contrary. Then we must have $f(X, Z) \in k[X, Z^p]$, say $f(X, Z) = \varphi(X, Z^p)$. From $\varphi'_{x_1}(x_1, z^p) = f'_{x_1}(x_1, z) \neq 0$, it would then follow that x_1 is separable over $k(z^p)$, which is contrary to our assumption on x_1.

Since z is separable over $k(x_1)$ and since all x_i are separable over $k(z)$, it follows (Theorem 9, § 5) that x_1 is a separating element.

To complete the proof of the theorem, we now proceed by induction with respect to r. We assume then that the theorem is true for fields of algebraic functions of $r - 1$ independent variables.

Let $\{z_1, z_2, \cdots, z_r\}$ be a separating transcendence basis of K/k. If we set $k_1 = k(z_1)$, then K can be regarded, *over* k_1, as a field of algebraic functions of $r - 1$ independent variables. Moreover, we have $K = k_1(x_1, x_2, \cdots, x_n)$, and $\{z_2, z_3, \cdots, z_r\}$ is a separating transcendence basis of K/k_1. By our induction hypothesis, $r - 1$ of the elements x_i will form a separating transcendence basis of K/k_1. Let, say, $\{x_1, x_2, \cdots, x_{r-1}\}$ be a separating transcendence basis of K/k_1. If we set $k' = k(x_1, x_2, \cdots, x_{r-1})$, then $K = k'(x_r, x_{r+1}, \cdots, x_n)$ and K is a field of algebraic functions of *one* variable, *over* k'. Moreover, z_1 is a separating element of K/k'. Hence, by the case $r = 1$, one of the elements $x_r, x_{r+1}, \cdots, x_n$ will also be a separating element of K/k'. If, say, x_r is such an element, then the r elements $x_1, x_2, \cdots, x_{r-1}, x_r$ form a separating transcendence basis for K/k. This completes the proof of the theorem.

The following lemma will be followed by an application to the case of perfect ground fields k.

LEMMA 3. *If the field $K = k(x_1, x_2, \cdots, x_n)$, of transcendence degree r over k, is not separably generated over k, then for a suitable labeling of the x_i the field $k(x_1, x_2, \cdots, x_{r+1})$ is of transcendence degree r over k and is not separably generated over k.*

PROOF. If $n = r + 1$ there is nothing to prove. Assume that $n > r + 1$ and that the theorem is true for $n - 1$. We may assume that x_1 is a.d. on $x_2, x_3, \cdots, x_n$ (over k), and consequently that the field $k(x_2, x_3, \cdots, x_n)$ has transcendence degree r over k. If this field is not separably generated over k, the assertion of the lemma follows by the case $n - 1$. Assume that $k(x_2, x_3, \cdots, x_n)$ is separably generated over k. Then by the preceding theorem we may assume that $x_2, x_3, \cdots, x_{r+1}$ form a separating transcendence basis of $k(x_2, x_3, \cdots, x_n)/k$. In that case the field K is a separable extension of $k(x_1, x_2, \cdots, x_{r+1})$, and therefore this latter field enjoys the properties stated in the lemma.

The following is a straightforward application to perfect fields:

THEOREM 31. *If k is a perfect field, then $K = k(x_1, x_2, \cdots, x_n)$ is always separably generated over k* (F. K. Schmidt).

PROOF. By the above lemma, it is sufficient to prove the theorem in the case $n = r + 1$, where $r = $ tr. d. K/k. In this case, we have *the irreducible* relation $f(x_1, x_2, \cdots, x_r, x_{r+1}) = 0$ between the $r + 1$ elements x_i (see Lemma 2). If $x_1, x_2, \cdots, x_r$ do not form a separating transcendence basis, then $f(X) \in k[X_1, X_2, \cdots, X_n, X_{r+1}{}^p]$. If *no* r of the elements x_i form a separating transcendence basis, then $f(X) \in k[X_1{}^p, X_2{}^p, \cdots, X_{r+1}{}^p]$. But then $f(X)$ is the p-th power of a polynomial in $k[X]$, since k is perfect—a contradiction.

§ 14. Algebraically closed fields

DEFINITION 1. *A field k is said to be algebraically closed if it possesses no proper algebraic extensions (that is, if every algebraic extension of k coincides with k).*

It is not difficult to see that the following properties of a field k are equivalent:

(a) k is algebraically closed.
(b) Every irreducible polynomial in $k[X]$ is of degree 1.
(c) Every polynomial in $k[X]$, of positive degree, factors completely in $k[X]$ in polynomials of degree 1.
(d) Every polynomial in $k[X]$, of positive degree, has at least one root in k.

In fact, if $f(X)$ is an irreducible polynomial in $k[X]$, of degree $n \geqq 1$, then we know that there exists an algebraic extension K of k, of relative degree n over k (Theorem 3′, § 2). If k is algebraically closed we must have $K = k$, whence $n = 1$. Thus (a) implies (b).

It is clear that (b), (c), and (d) are equivalent. Finally, if K is any algebraic extension of k and x is any element of K, then the minimal polynomial of x over k is irreducible and therefore has degree 1 if (b) holds. Therefore $x \in k$, $K = k$, that is, (b) implies (a).

COROLLARY. *If k is a subfield of an algebraically closed field K, then the algebraic closure k' of k in K is an algebraically closed field.*

For if $f(X)$ is any polynomial in $k'[X]$, then $f(X)$ must have a root α in K; this root α is then also algebraic over k (by the transitivity of algebraic dependence) and therefore belongs to k'.

DEFINITION 2. *If k is a subfield of a field K, then K is said to be an algebraic closure of k if* (1) *K is an algebraic extension of k and* (2) *K is an algebraically closed field.*

COROLLARY. *If an algebraic extension K of k has the property that every polynomial $f(X)$ in $k[X]$ factors completely in linear factors in $K[X]$, then K is an algebraic closure of k.*

For if K' is any algebraic extension of K and x is an element of K', then we have for the minimal polynomial $f(X)$ of x over k a complete factorization $f(X) = \prod(X - x_i)$, $x_i \in K$. Since $f(x) = 0$, we must have $x = x_i$ for some i, whence $x \in K$, and thus $K' = K$.

The following fundamental theorem guarantees the existence and the essential unicity of an algebraic closure of a given field k:

THEOREM 32. *If k is a field, then there exists an algebraic closure of k, and any two algebraic closures of k are k-isomorphic fields.*

PROOF. Let N denote the set of all ordered pairs $(f(X), n)$, where

$f(X) \in k[X]$ and n is any non-negative integer. We agree to identify any element c of k with the pair $(X - c, 0)$. We consider the set S of all fields Σ such that (a) the elements of Σ form a subset E of N; (b) k is a subfield of Σ; (c) if $z = (f(X), n) \in \Sigma$, then $f(z) = 0$. The set S is non-empty since $k \in S$. We observe that if A denotes the set of all ordered triads (z_1, z_2, z_3) of elements of Σ such that $z_3 = z_1 + z_2$ and if M denotes the set of all ordered triads (y_1, y_2, y_3) of elements of Σ such that $y_3 = y_1y_2$, then the field Σ is uniquely determined by, and can be identified with, the ordered triad (E, A, M). Since E is a subset of N, while A and M are suitable subsets of the set product $N \times N \times N$, our set S of fields Σ is well defined from the standpoint of the theory of sets.

We partially order the set S by setting $\Sigma \prec \Sigma'$ if Σ is a subfield of $\Sigma'(\Sigma, \Sigma' \in S)$. It is clear that S is an inductive set. By Zorn's lemma, let K be a maximal element of S. We shall show that K is an algebraic closure of k.

The property (c) enjoyed by any field Σ in S implies that Σ is an algebraic extension of k. Hence K is an algebraic extension of k. We shall show that the assumption that K has proper algebraic extensions leads to a contradiction with the maximality of K in M. Assume then that there exists a proper algebraic extension K' of K. We shall now define a (1, 1) mapping φ of K' into N. If $x \in K$ we set $\varphi(x) = x$. If $x \in K'$, $x \notin K$, we consider the minimal polynomial $f(X)$ of x over k, we denote by $z_1, z_2, \cdots, z_h$ the roots of $f(X)$ in K $(h \geqq 0)$ and by $x_1, x_2, \cdots, x_g$ the roots of $f(X)$ which are in K' but not in K $(g \geqq 1; x_1 = x)$. We fix g distinct non-negative integers $n_1, n_2, \cdots, n_g$ such that $z_i \neq (f(X), n_j)$ $(i = 1, 2, \cdots, h; j = 1, 2, \cdots, g)$ and we set $\varphi(x_j) = (f(X), n_j)$, $j = 1, 2, \cdots, g$. It is then clear that φ is a (1, 1) mapping of K' into N and that φ is the identity on K. Let $E_0 = \varphi(K')$. We carry over the field structure of K' to the set E_0, by means of the mapping φ, thus getting a field K_0. From the definition of φ it follows at once that $K_0 \in S$. Since K is a proper subfield of K', it is also a proper subfield of K_0, and this contradicts the maximality of K.

The second half of the theorem will be included in the following stronger result:

THEOREM 33. *Let K' be an algebraically closed field and let K be an algebraic extension of a field k. If φ is an isomorphism of k into K' then φ can be extended to an isomorphism of K into K'.*

PROOF. We first show that Theorem 33 implies the second half of Theorem 32. If K and K' are two algebraic closures of k, we apply

Theorem 33 to the identity mapping φ of k into K', and we thus find that there exists a k-isomorphism ψ of K into K'. Since K is an algebraic closure of k, also $\psi(K)$ is an algebraic closure of k, and therefore $\psi(K) = K'$ since K' is an algebraic extension of $\psi(K)$. Thus ψ is a k-isomorphism of K *onto* K'.

We now begin the proof of Theorem 33. Let M be the set of all ordered triads (L, L', ψ), where L is any field between k and K, L' is a field between $\varphi(k)$ and K', and ψ is an isomorphism of L onto L' such that $\psi = \varphi$ on k. The set M is non-empty since $(k, \varphi(k), \varphi) \in M$. We partially order M by setting $(L, L', \psi) \prec (L_1, L'_1, \psi_1)$ if L is a subfield of L_1 and if $\psi_1 = \psi$ on L. If $N = \{(L_\alpha, L'_\alpha, \psi_\alpha)\}$ is a totally ordered subset of M, and if we set $L = \bigcup L_\alpha$, $L' = \bigcup L'_\alpha$, then L and L' are fields between k and K, and between $\varphi(k)$ and K' respectively. The mappings ψ_α determine uniquely an isomorphism ψ of L onto L' such that $\psi = \psi_\alpha$ on L_α. Thus (L, L', ψ) is an element of M which is an upper bound of the set N, and M is therefore shown to be inductive. Let (L_0, L'_0, ψ_0) be a maximal element of M. We shall show that $L_0 = K$, and this will complete the proof of the theorem.

Let x be any element of K and let $f(X)$ be the minimal polynomial of x over L_0. Let $f'(X)$ be the ψ_0-transform of the polynomial $f(X)$. Then $f'(X) \in L'_0[X] \subset K'[X]$, and since K' is algebraically closed the polynomial $f'(X)$ has a root x' in K'. By Lemma 1 of § 6, ψ_0 can be extended to an isomorphism ψ_1 of $L_0(x)$ onto $L'_0(x')$ such that $\psi_0(x) = x'$. Then $(L_0, L'_0, \psi_0) \prec (L_0(x), L'_0(x'), \psi_1) \in M$, and hence, by the maximality of (L_0, L'_0, ψ_0), we have $(L_0, L'_0, \psi_0) = (L_0(x), L'_0(x'), \psi_1)$, $L_0 = L_0(x)$, $x \in L_0$. We have therefore shown that $L_0 = K$. Q.E.D.

Let k be a field and let K be an algebraically closed field containing k. If the characteristic p of k is $\neq 0$, then the elements x of K such that $x^p \in k$ form a field containing k. This field shall be denoted by $k^{p^{-1}}$. Since K is algebraically closed, K contains a root of every polynomial of the form $X^p - a$, $a \in k$. Therefore $k^{p^{-1}}$ consists of the p-th roots of the elements of k, and we have $k = (k^{p^{-1}})^p$. It is also obvious that the fields $k^{p^{-1}}$ obtained in relation to various algebraic closures of k are always k-isomorphic to each other.

In a similar fashion we can define the fields $k^{p^{-n}}$ for $n = 1, 2, \cdots$. These fields form an ascending chain $k \subset k^{p^{-1}} \subset k^{p^{-2}} \subset \cdots$, and their union is a field between k and K which we shall sometimes denote by $k^{p^{-\infty}}$; it is the least perfect field containing k and is therefore referred to as *the perfect closure of* k. If k itself is perfect, then $k^{p^{-\infty}} = k$.

If the characteristic p of k is zero, we set $k^{p^{-n}} = k^{p^{-\infty}} = k$.

§ 15. Linear disjointness and separability. Let S be a ring containing a field k and such that the identity 1 of k is also the identity of S. Then S is a vector space over k, and in this sense we speak below of subspaces of S.

DEFINITION 1. *Two subspaces L and L' of S are said to be linearly disjoint over k if the following condition is satisfied: whenever $x_1, x_2, \cdots, x_n$ are elements of L which are linearly independent over k and $x'_1, x'_2, \cdots, x'_m$ are elements of L' which are linearly independent over k, then the mn products $x_i x'_j$ are also linearly independent over k.*

Linear disjointness of L, L' is clearly a *symmetric* relationship between the two spaces and is *relative* to the preassigned ground field k. *The following property is equivalent to linear disjointness* and will be the one most frequently used in the sequel:

(LD). *Whenever $x_1, x_2, \cdots, x_n$ are elements of L which are linearly independent over k then these elements x_i are also linearly independent over L'.*

For, assume that L and L' are linearly disjoint over k and let $u'_1x_1 + u'_2x_2 + \cdots + u'_nx_n = 0$, where the u'_i are in L' and $x_1, x_2, \cdots, x_n$ are elements of L which are linearly independent over k. Let $\{x'_1, x'_2, \cdots, x'_m\}$ be a basis of the vector space $ku'_1 + ku'_2 + \cdots + ku'_n$ over k and let $u'_i = \sum_j a_{ij}x'_j$, where $a_{ij} \in k$. Then $\sum_{i,j} a_{ij}x_ix'_j = 0$, and hence, by the linear disjointness of L and L' over k it follows that all the a_{ij} are zero, and so also all the u'_i are zero, showing that condition (LD) is satisfied. Conversely, assume property (LD) and let the two sets $\{x_1, x_2, \cdots, x_n\}$ and $\{x'_1, x'_2, \cdots, x'_m\}$ be as in the above definition. Assume that we have a relation $\sum_{i,j} a_{ij}x_ix'_j = 0$, $a_{ij} \in k$. Since, for each i, the sums $\sum_j a_{ij}x'_j$ belong to L', it follows from (LD) that $\sum_j a_{ij}x'_j = 0$, for all i, and hence all the a_{ij} are zero since the x'_j are linearly independent over k. Hence L and L' are linearly disjoint over k.

There is a close connection between the concept of linear disjointness and the concepts of separable or inseparable extensions. In this section we shall study this connection. To introduce the topic we begin by proving a theorem which is essentially a restatement of Theorem 8 (§ 5) in terms of linear disjointness.

THEOREM 34. *If a field K is an algebraic extension of a field k, then a necessary and sufficient condition that K be a separable algebraic extension of k is that K and $k^{p^{-1}}$ be linearly disjoint over k.* (We take $k^{p^{-1}}$ to be a subfield of an algebraic closure of K.)

PROOF. Assume that K is a separable algebraic extension of k. We have to show that if $u_1, u_2, \cdots, u_g$ are elements of K which are linearly independent over k, these elements are also linearly independent of $k^{p^{-1}}$, or—equivalently—that $u_1^p, u_2^p, \cdots, u_g^p$ are linearly independent over k. We consider the field $K_1 = k(u_1, u_2, \cdots, u_g)$ and we extend the set $\{u_1, u_2, \cdots, u_g\}$ to a basis $\{u_1, u_2, \cdots, u_m\}$ of K_1/k. We have, by Theorem 8 (§ 5), $K_1 = kK_1^p$, whence

$$K_1 = \sum_{i=1}^{m} ku_i^p.$$

Consequently also the elements $u_1^p, u_2^p, \cdots, u_m^p$ form a basis of K_1/k and are therefore linearly independent over k.

Assume now that K and $k^{p^{-1}}$ are linearly disjoint over k, and let x be any element of K. Let $f(X)$ be the minimal polynomial of x in $k[X]$ and let m be the degree of $f(X)$. If h is any integer $\leqq m$, then $1, x, x^2, \cdots, x^{h-1}$ are linearly independent over k. Consequently, by the linear disjointness of K and $k^{p^{-1}}$ over k, the p-th powers of these elements are also linearly independent over k. This implies that $f(X) \notin k[X^p]$, that is, x is separable over k. Q.E.D.

We shall need later on the following lemma:

LEMMA. *Let L and L' be subspaces of a ring S/k, and let $\{u_\alpha\}$ and $\{u'_\beta\}$ be bases (finite or infinite) of L/k and L'/k respectively. Then a necessary and sufficient condition that L and L' be linearly disjoint over k is that the products $u_\alpha u_\beta'$ be linearly independent over k. An equivalent condition is that the u_α be linearly independent over L'. In particular, if the dimensions of L and L' are both finite then L and L' are linearly disjoint over k if and only if* $\dim LL'/k = \dim L/k \cdot \dim L'/k$; *here LL' denotes the space spanned by the products uu', $u \in L$, $u' \in L'$.*

PROOF. It follows directly from the definition of linear disjointness that if L and L' are linearly disjoint over k, then the products $u_\alpha u'_\beta$ are linearly independent over k. It is also obvious that the linear independence of the products $u_\alpha u'_\beta$ over k is equivalent to the linear independence of the u_α over L'. Now, assume that the products $u_\alpha u'_\beta$ are linearly independent over k. We first consider the case in which L and L' have finite dimension, say $s = \dim L/k$, $t = \dim L'/k$. Since the st products $u_\alpha u'_\beta$ span the space LL', it follows that $\dim LL'/k = st$. Now, if $x_1, x_2, \cdots, x_n$ are elements of L which are linearly independent over k and $x'_1, x'_2, \cdots, x'_m$ are elements of L' which are linearly independent over k, then we extend the sets $\{x_1, x_2, \cdots, x_n\}$, $\{x'_1, x'_2, \cdots, x'_m\}$ to bases $\{x_1, x_2, \cdots, x_s\}$, $\{x'_1, x'_2, \cdots, x'_t\}$ of L/k and L'/k respectively and we observe that the st products $x_\alpha x'_\beta$ must be linearly independent over k, since they span LL'/k and since $\dim LL'/k = st$. In particular, also the

mn products $x_i x'_j$ ($i = 1, 2, \cdots, n$; $j = 1, 2, \cdots, m$) are linearly independent over k, and this proves the linear disjointness of L and L' over k. This also establishes the second half of the lemma.

In the general case, we can always find a finite subset $\{u_1, u_2, \cdots, u_s\}$ of the set $\{u_\alpha\}$ such that the elements $x_1, x_2, \cdots, x_n$ belong to the finite-dimensional space L_0 spanned by $u_1, u_2, \cdots, u_s$. Similarly, there exists an integer t such that $x'_j \in L'_0 = ku'_1 + ku'_2 + \cdots + ku'_t$, $j = 1, 2, \cdots, m$. Since L_0 and L'_0 are linearly disjoint over k, by the preceding case, the products $x_i x'_j$ are linearly independent over k. This completes the proof of the lemma.

COROLLARY 1. *Let k, K and S be fields such that $k \subset K \subset S$. If X is a (finite or infinite) set of algebraically independent elements of S over K, then the subfields K and $k(X)$ of S are linearly disjoint over k.*

It is obvious that K and $k(X)$ are linearly disjoint over k if and only if K and $k[X]$ are linearly disjoint over k. Now, the set of all monomials $x_1^{i_1} x_2^{i_2} \cdots x_n^{i_n}$, $x_\alpha \in X$, is a basis of the $k[X]$ over k, and since, by assumption, these monomials are also independent over K, the corollary follows from the lemma.

COROLLARY 2. *If a field K is a purely transcendental extension of a field k, then K and $k^{p^{-1}}$ are linearly disjoint over k.*

For, if $K = k(X)$, where X is a suitable transcendence basis of K/k, then the elements of X, being algebraically independent over k, are also algebraically independent over $k^{p^{-1}}$. Therefore, by the preceding corollary, K and $k^{p^{-1}}$ are linearly disjoint over k.

For extension fields K of k which are not algebraic over k there is no complete equivalence between the concept of separable generation and that of linear disjointness (over $k^{p^{-1}}$). However, we have the following theorem:

THEOREM 35. *Let K be an extension of k. A necessary condition for K to be separably generated over k is that K and $k^{p^{-1}}$ be linearly disjoint over k. If K is finitely generated over k, then the foregoing condition is also sufficient.*

PROOF. Assume that K is separably generated over k and let B be a separating transcendence basis of K/k. By Corollary 2 to Lemma 1, the fields $k(B)$ and $k^{p^{-1}}$ are linearly disjoint over k. Now, let $u_1, u_2, \cdots, u_g$ be elements of $k^{p^{-1}}$ which are linearly independent over k. They are then also linearly independent over $k(B)$. Since K is a separable algebraic extension of $k(B)$, it follows from Theorem 34 that $u_1, u_2, \cdots, u_g$ are also linearly independent over K, showing that K and $k^{p^{-1}}$ are linearly disjoint over k.

Assume now that K is *finitely generated over* k, say, $K = k(x_1, x_2, \cdots, x_n)$, and that K and $k^{p^{-1}}$ are linearly disjoint over k. Let r be the transcendence degree of K/k, whence $n \geqq r$. If $n = r$, there is nothing to prove. We next consider the case $n = r + 1$. In this case, let $f(X)$ be the irreducible polynomial in $k[X_1, X_2, \cdots, X_{r+1}]$ such that $f(x) = 0$ (§ 13, Lemma 2). We assert that $f(X) \notin k[X_1^p, X_2^p, \cdots, X_{r+1}^p]$. For, assume the contrary, and let $f(X) = g(X_1^p, X_2^p, \cdots, X_{r+1}^p)$, where $g(X_1, X_2, \cdots, X_{r+1}) \in k[X]$. If $\omega_1, \omega_2, \cdots, \omega_m$ are the monomials in $x_1, x_2, \cdots, x_{r+1}$ which actually occur in the expression $g(x_1, x_2, \cdots, x_{r+1})$, then the ω are linearly independent over k (since the degree of these monomials is less than the degree of $f(X)$), while $\omega_1^p, \omega_2^p, \cdots, \omega_m^p$ are linearly dependent over k since $g(x_1^p, x_2^p, \cdots, x_{r+1}^p) = 0$. This contradicts the linear disjointness of K and $k^{p^{-1}}$ over k. This contradiction shows that we have indeed $f(X) \notin k[X_1^p, X_2^p, \cdots, X_{r+1}^p]$. We therefore may assume that one of the $r + 1$ variables X_i, say, X_{r+1}, which actually occurs in $f(X)$, occurs in some term of $f(X)$ with an exponent which is not a multiple of p. The elements $x_1, x_2, \cdots, x_r$ are then necessarily algebraically independent over k, and furthermore, x_{r+1} is separable algebraic over $k(x_1, x_2, \cdots, x_r)$. Hence $\{x_1, x_2, \cdots, x_r\}$ is a separating transcendence basis of K/k.

For $n > r + 1$ we shall use induction with respect to n. The linear disjointness of $k(x_1, x_2, \cdots, x_n)$ and $k^{p^{-1}}$ over k implies the linear disjointness of $k(x_1, x_2, \cdots, x_{n-1})$ and $k^{p^{-1}}$ over k. Hence, by our induction hypothesis, $k(x_1, x_2, \cdots, x_{n-1})$ is separably generated over k. Let $\{z_1, z_2, \cdots, z_m\}$ be a separating transcendence basis of $k(x_1, x_2, \cdots, x_{n-1})$ over k. Then m is either $r - 1$ or r. The field $k(x_1, x_2, \cdots, x_n)$ is a separable algebraic extension of $K_1 = k(z_1, z_2, \cdots, z_m, x_n)$, and we have only to show now that K_1 is separably generated over k. The field K_1/k has the same transcendence degree r as K/k and it is generated over k by at most $r + 1$ elements (since $m \leqq r$). Furthermore, since $K_1 \subset K$, also K_1 and $k^{p^{-1}}$ are linearly disjoint over k. Hence, by the case $n \leqq r + 1$, K_1 is indeed separably generated over k. This completes the proof of the theorem.

The preceding theorem and the reasoning used in the proof of that theorem enable us to give a second proof of Theorem 30 (§ 13), that is, of the assertion that if $K = k(x_1, x_2, \cdots, x_n)$ is a separably and finitely generated extension of k, then already the set of generators x_i contains a separating transcendence basis of K/k. For $n = r$ there is nothing to prove. The case $n = r + 1$ has been settled in the course of the preceding proof (since K and $k^{p^{-1}}$ are linearly disjoint over k, by the first part of Theorem 35). For $n > r + 1$ we again use induction with respect to n and repeat the last part of the preceding proof, observing that this time the induction hypothesis permits us to find a separating transcendence basis $\{z_1, z_2, \cdots, z_m\}$ of $k(x_1, x_2, \cdots, x_{n-1})/k$ *within the set of generators* $\{x_1, x_2, \cdots, x_{n-1}\}$. Then the $m + 1$ generators of K_1/k are in the set $\{x_1, x_2, \cdots, x_n\}$, and Theorem 30 now follows by the case $n = r + 1$.

Also Theorem 31 (§ 13) is an immediate consequence of Theorem 35, for if k is a perfect field, that is, if $k^{p^{-1}} = k$, then K and $k^{p^{-1}}$ are linearly disjoint over k.

DEFINITION 2. *If K is an extension field of a field k of characteristic p, then K is said to be a separable extension of k if K and $k^{p^{-1}}$ are linearly disjoint over k.*

In view of Theorem 34, for *algebraic* extensions K of k, separability in the sense of this definition is equivalent with separability as defined in § 5. Similarly, for *finitely generated* extensions K of k it is true that K/k is a separable extension if and only if K is separably generated over k (Theorem 35). However, if K is not finitely generated over k, it may very well be separable over k without being separably generated over k. For instance, if k is a perfect field every extension K of k is separable over k. However, if x is a transcendental over k, then it is easily seen that the field $K = k(x, x^{1/p}, x^{1/p^2}, \cdots, x^{1/p^n}, \cdots)$ is not separably generated over k.

We note that Corollary 2 of the lemma proved earlier in this section can now be re-stated as follows: *if K is a purely transcendental extension of k, then K is separable over k.*

The transitivity of separability proved in § 5 (Theorem 9) for algebraic extensions extends to arbitrary field extensions. We have, namely, the following theorem:

THEOREM 36. *If K' is separable over k, and if K'' is separable over K', then K'' is separable over k.*

PROOF. Given a set F of elements of $k^{p^{-1}}$ which are linearly independent over k, the elements of F are also linearly independent over K' since K' is separable over k. Hence the elements of F are also linearly independent over K'' since K'' is separable over K' and since $F \subset K'^{p^{-1}}$. Thus K'' and $k^{p^{-1}}$ are linearly disjoint over k.

§ 16. Order of inseparability of a field of algebraic functions. We shall deal in this section with finitely generated extensions of a field k, that is, with fields of algebraic functions over k (see Definition 1 in § 13). For these fields we shall define a numerical character, called the *order of inseparability* of the field (relative to k); it generalizes the concept of degree of inseparability of finite algebraic extensions and plays an important role in abstract algebraic geometry. We shall also give some results concerning the behavior of the order of inseparability under extensions of the ground field k.

In the sequel, p denotes the characteristic of our fields. If $p = 0$, then all powers p^e of p are to be replaced by 1. If A and B are subfields of some field S then we denote by (A, B) the compositum of these fields (the least subfield of S containing A and B). *All the fields which are considered below are assumed to be subfields of some algebraically closed field S.*

LEMMA 1. *If L, K and k' are fields and K is a finite extension of L, then*

$$[(K, k'):(L, k')] \leqq [K:L], \tag{1}$$

$$[(K, k'):(L, k')]_s \leqq [K:L]_s, \tag{2}$$

$$[(K, k'):(L, k')]_i \leqq [K, L]_i. \tag{3}$$

If the fields K and k' are linearly disjoint over some common subfield k of L and k', then the equality sign holds in (1), (2), *and* (3).

PROOF. Let $\{u_1, u_2, \cdots, u_n\}$ be a basis of K over L. Since the elements of K are algebraic over the field (L, k'), we have $(K, k') = (L, k')[K] = (L, k')\sum_{q=1}^{n} Lu_q = \sum_{q=1}^{n} (L, k')u_q$, and this proves inequality (1).

Let K_0 be the maximal separable extension of L in K. From the separability of the extension K_0/L follows the separability of the extension $(K_0, k')/(L, k')$, while from the fact that K is purely inseparable over K_0 follows that (K, k') is a purely inseparable extension of (K_0, k'). *Hence (K_0, k') is the maximal separable extension of (L, k') in (K, k').* It follows that $[(K, k'):(L, k')]_s = [(K_0, k'):(L, k')]$ and $[(K, k'):(L, k')]_i = [(K, k'):(K_0, k')]$. Therefore (2) is obtained by replacing K by K_0 in (1), and we obtain (3) by replacing L by K_0 in (1).

We now assume that K and k' are linearly disjoint over some common subfield k of L and k'. It will be sufficient to show that the equality sign holds in (1), for then it will follow that the equality sign also holds in (2) and (3), in view of the inequalities (2) and (3) and of the product formula (5) of § 6. To show that the equality sign holds in (1) we must show that the u_q are linearly independent over (L, k'). Now, the u_q are linearly independent over L. Therefore we have to show that *the fields K and (L, k') are linearly disjoint over L.* This we now proceed to show.

Let $\{a_i\}$ be a finite set of elements of (L, k') which are linearly independent over L. We can write the α_i in the form $\alpha_i = f_i/f_0$, where f_0

and the f_i belong to $k'[L]$. The numerators are linearly independent over L and we have to show that they are also linearly independent over K. In other words, it is sufficient to show the linear disjointness of K and $k'[L]$ over L. Now the ring $k'[L]$, regarded as a vector space over L, is spanned by the field k'. Hence we can find a basis B of $k'[L]$ over L such that $B \subset k'$. The elements of B are linearly independent over L, hence *a fortiori* also over k. It follows that the elements of B are also linearly independent over K, since k' and K are linearly disjoint over k. We have therefore shown that the vector space $k'[L]/L$ has an L-basis B whose elements are linearly independent over K, and the linear disjointness of $k'[L]$ and K over L now follows from the lemma proved in § 15.

COROLLARY 1. *Let k' be a finite algebraic extension of a field k and let B be a (finite or infinite) set of elements (of the big field S) which are algebraically independent over k'. Then*

$$[k'(B) : k(B)]_s = [k' : k]_s, \tag{4}$$

$$[k'(B) : k(B)]_i = [k' : k]_i. \tag{5}$$

By Corollary 1 to the lemma of § 15, the fields $k(B)$ and k' are linearly disjoint over k. Hence the corollary follows from the foregoing lemma by setting $L = k$, $K = k'$ and $k' = k(B)$.

COROLLARY 2. *Let k' be a finite algebraic extension of k and let $k'(x) = k'(x_1, x_2, \cdots, x_n)$ be a finitely generated extension of k'. Then $[k' : k]_i \geqq [k'(x) : k(x)]_i$, or equivalently (since both sides of this inequality are powers of p):*

$$[k' : k]_i = p^s[k'(x) : k(x)]_i, \tag{6}$$

where s is a non-negative integer. Furthermore, if $\{z\} = \{z_1, z_2, \cdots, z_r\}$ is any transcendence basis of $k(x)/k$, then

$$[k(x) : k(z)]_i = p^s[k'(x) : k'(z)]_i. \tag{7}$$

The first half of this corollary follows from the inequality (1) of Lemma 1 upon replacing L, K and k' by k, k' and $k(x)$ respectively. To obtain the second half of the corollary we observe that we have, in view of relations (5), § 6:

$$[k'(x) : k(z)]_i = [k'(x) : k(x)]_i[k(x) : k(z)]_i,$$

and also

$$[k'(x) : k(z)]_i = [k'(x) : k'(z)]_i[k'(z) : k(z)]_i.$$

Comparing these two expressions of $[k'(x) : k(z)]_i$ we see that (7) follows from (5) (with $B = \{z\}$) and (6).

DEFINITION 1. *If $k(x)$ is a finitely generated extension of k, we mean by the order of inseparability of $k(x)/k$ (or of the extension $k(x)$ of k) the smallest value of the degree of inseparability $[k(x):k(z)]_i$ for all possible choices of a transcendence basis $\{z\} = \{z_1, z_2, \cdots, z_r\}$ of $k(x)/k$.*

We shall denote the order of inseparability of $k(x)/k$ by $[k(x):k]_i$. In the special case in which $k(x)$ is an algebraic extension of k, $\{z\}$ is the empty set, $k(z) = k$, and hence the order of inseparability of $k(x)/k$ is the degree of inseparability of $k(x)$ over k. Thus our notation $[k(x):k]_i$ is consistent in this case with the one already used for algebraic extensions.

THEOREM 37. *The notations being the same as in the foregoing definition, already the set of generators $x_1, x_2, \cdots, x_n$ contains a transcendence basis $\{z\}$ such that $[k(x):k(z)]_i = [k(x):k]_i$.*

PROOF. Let $\bar{k}$ be the algebraic closure of k in some algebraic closure of $k(x)$. Since $\bar{k}$ is a perfect field, $\bar{k}(x)$ is separably generated over $\bar{k}$ (Theorem 31, § 13). By Theorem 30, § 13, the set of generators $\{x_1, x_2, \cdots, x_n\}$ contains a separating transcendence basis of $\bar{k}(x)/\bar{k}$. Let $\{x_1, x_2, \cdots, x_r\}$ be such a basis. Then the x_i are separably algebraic over $\bar{k}(x_1, x_2, \cdots, x_r)$, $i = 1, 2, \cdots, n$. If we denote by k' the field obtained by adjoining to k the coefficients of the minimal polynomials $f_{r+1}(X), f_{r+2}(X), \cdots, f_n(X)$ of $x_{r+1}, x_{r+2}, \cdots, x_n$ respectively, over $\bar{k}$, then k' is a finite algebraic extension of k, *and it is clear that $\{x_1, x_2, \cdots, x_r\}$ is also a separating transcendence basis of $k'(x)/k'$* (since the $f_i(X)$ are separable polynomials). Let s be the integer defined by (6). If we identify the set $\{z_1, z_2, \cdots, z_r\}$ with $\{x_1, x_2, \cdots, x_r\}$, we have, in view of (7), $[k(x):k(x_1, x_2, \cdots, x_r)]_i = p^s$, while if $\{z\}$ is any other transcendence basis of $k(x)/k$, then we have, again by (6), that $[k(x):k(z)]_i \geqq p^s$. This establishes the theorem.

The foregoing theorem tells us that if $\{z\}$ ranges over the set of all transcendence bases which can be extracted from the set of generators $x_1, x_2, \cdots, x_n$, then $[k(x):k]_i = \min \{[k(x):k(z)]_i\}$. The theorem is therefore a generalization of Theorem 30, § 13, for if $k(x)$ is separably generated over k, then $[k(x):k]_i = 1$.

We shall now consider a certain class of ground field extensions $k \to K$ (within the big field S) and we shall derive some properties of the order of inseparability $[k(x):k]_i$ in relation to such extensions.

We first prove the following lemma:

LEMMA 2. *Let K be an algebraic extension of k and let $\{z\}$ be a transcendence basis of $k(x)/k$. If we have $[k'(x):k'(z)]_i = [k(x):k(z)]_i$ for all fields k' between k and K which are finite over k, then we have also $[K(x):K(z)]_i = [k(x):k(z)]_i$.*

PROOF. For any field k' between k and K we denote by $L(k')$ the maximal separable extension of $k'(z)$ in $k'(x)$. Let $\{\xi_1, \xi_2, \cdots, \xi_m\}$ be a basis of $k(x)$ over $L(k)$ (whence $m = [k(x) : k(z)]_i$). It is immediately seen that $k'(x) = L(k')\xi_1 + L(k')\xi_2 + \cdots + L(k')\xi_m$. By our assumption, the ξ_i must be linearly independent over $L(k')$ if k' is a finite extension of k. Since every finite set of elements of $L(K)$ is contained in some $L(k')$, where k' is a finite extension of k, it follows that the ξ_i are also linearly independent over $L(K)$, and this proves the lemma.

The extensions $k \to K$ which we shall deal with in the remainder of this section are the so-called *free extensions of k relative to $k(x)$*. We give, namely, the following definition:

DEFINITION 2. *Given two subrings R and R' of S which contain k, we say that R and R' are free over k if the following condition is satisfied: whenever $\{x_1, x_2, \cdots, x_r\}$ and $\{x'_1, x'_2, \cdots, x'_s\}$ are finite subsets of R and R' respectively such that the elements of each set are algebraically independent over k, then also the $r + s$ elements $x_1, x_2, \cdots, x_r, x'_1, x'_2, \cdots, x'_s$ are algebraically independent over k.*

It follows at once that *if either R or R' is algebraic over k, then R and R' are free over k.*

We note that if each of the two integral domains R and R' has finite transcendence degree over k (see II, § 12, p. 100), then Definition 2 is equivalent to the following: *R and R' are free over k if and only if*

$$\text{(8)} \qquad \text{tr. d. } [R, R']/k = \text{tr. d. } R/k + \text{tr. d. } R'/k,$$

where $[R, R']$ denotes, as usual, the smallest subring of S which contains both R and R'. For, clearly, if $\{x_1, x_2, \cdots, x_m\}$ and $\{x'_1, x'_2, \cdots, x'_n\}$ are transcendence bases of R/k and R'/k respectively, then every element of $[R, R']$ is algebraic over the field $k(x_1, x_2, \cdots, x_m, x'_1, x'_2, \cdots, x'_n)$ (this field is to be thought of as a subfield of the quotient field of S) and hence, if R and R' are free over k, then the $m + n$ elements x_i, x'_j are distinct and constitute a transcendence basis of $[R, R']/k$. Conversely, if (8) holds, and if $\{x_1, x_2, \cdots, x_r\}$ and $\{x'_1, x'_2, \cdots, x'_s\}$ are finite transcendence sets in R/k and R'/k respectively, then we extend these sets to transcendences bases $\{x_1, x_2, \cdots, x_r, x_{r+1}, \cdots, x_m\}$ and $\{x'_1, x'_2, \cdots, x'_s, x'_{s+1}, \cdots, x'_n\}$ of R/k and R'/k respectively (see II, § 12, p. 101). The preceding argument shows that the set of $m + n$ elements x_i, x'_j contains a transcendence basis of $[R, R']/k$, and, consequently, if (8) holds, then the $m + n$ elements x_i, x'_j (and therefore also the given $r + s$ elements $x_1, x_2, \cdots, x_r, x'_1, x'_2, \cdots, x'_s$) must be algebraically independent over k.

We note that an equivalent formulation of (8) is the following:

(9) $$\text{tr. d. } [R, R']/R = \text{tr. d. } R'/k,$$

for tr. d. $[R, R']/R +$ tr. d. $R/k =$ tr. d. $[R, R']/k$.

The preceding proof can be extended to the general case of integral domains R, R' having arbitrary transcendence degree over k and leads to the conclusion that if B is a transcendence basis of R/k and B' is a transcendence basis of R'/k, then R and R' are free over k if and only if B and B' have no elements in common and the elements of $B \cup B'$ are algebraically independent over k (whence $B \cup B'$ is a transcendence basis of $[R, R']/k$).

In applications one uses frequently the following equivalent (but asymmetric) formulation of the concept of free integral domains over k: *the integral domains R and R' are free over k if and only if any set of algebraically independent elements of R over k is also such over R'.* That this is equivalent to our original definition follows immediately from our preceding considerations.

THEOREM 38. *If $k(x)$ is a finitely generated extension of k and if K is an extension field of k such that K and $k(x)$ are free over k, then for any transcendence basis $\{z\}$ of $k(x)/k$ we have*

(10) $$[K(x) : K(z)]_i/[k(x) : k(z)]_i = [K(x) : K]_i/[k(x) : k]_i.$$

In particular, if K is a finite algebraic extension of k, then the common value of both sides of (10) *is equal to* $[K : k]_i/[K(x) : k(x)]_i$.

PROOF. If K is a finite extension of k, then (6) and (7) show that the value of the ratio on the left-hand side of (10) is independent of the choice of the transcendence basis $\{z\}$ and is equal to $[K : k]_i/[K(x) : k(x)]_i$. If we now let $\{z\}$ range over the set of all transcendence bases of $k(x)/k$ which can be extracted from the set $\{x\}$ of generators, then we deduce at once from Theorem 37 that the foregoing ratio is equal to the ratio $[K(x) : K]_i/[k(x) : k]_i$.

We next consider the case in which K is an arbitrary algebraic extension of k. We fix a transcendence basis $\{u\}$ of $k(x)$ over k. For any two fields k_1 and k_2 between k and K such that $k_1 \subset k_2$ we have $[k_1(x) : k_1(u)]_i \geqq [k_2(x) : k_2(u)]_i$, by (2). We can therefore find a field k' between k and K such that k' is a finite extension of k and such that $[k''(x) : k''(u)]_i = [k'(x) : k'(u)]_i$ for any field k'' between k' and K which is a finite extension of k'. By the finite case considered in the preceding part of the proof we have then $[k''(x) : k''(z)]_i = [k'(x) : k'(z)]_i$ for any transcendence basis $\{z\}$ of $k'(x)/k$ and for any field k'' between k and K such that k'' is a finite extension of k'. It follows then by Lemma 2 that $[K(x) : K(z)]_i = [k'(x) : k'(z)]_i$ for any transcendence basis $\{z\}$ of

$k'(x)/k'$. Hence $[K(x):K]_i = [k'(x):k']_i$. We now have, for any transcendence basis $\{z\}$ of $k(x)/k$:

$$[k(x):k(z)]_i = p^s[k'(x):k'(z)]_i = p^s[K(x):K(z)]_i,$$

where

$$p^s = [k(x):k]_i/[k'(x):k']_i = [k(x):k]_i/[K(x):K]_i,$$

and this establishes the theorem in the case in which K is an algebraic extension of k.

In the general case we consider a transcendence basis B of K/k. Since K and $k(x)$ are free over k, the elements of B are algebraically independent over $k(x)$. Hence by Corollary 1 of Lemma 1 we have

$$(11) \qquad [k(B, x):k(B, z)]_i = [k(x):k(z)]_i$$

for any transcendence basis $\{z\}$ of $k(x)/k$. Hence

$$(12) \qquad [k(B, x):k(B)]_i = [k(x):k]_i.$$

Since K is an algebraic extension of $k(B)$, we have by the preceding case:

$$[k(B, x):k(B, z)]_i = p^s[K(x):K(z)]_i,$$

where

$$p^s = [k(B, x):k(B)]_i/[K(x):K]_i,$$

and from this the theorem follows in view of (11) and (12).

COROLLARY 1. *If $\bar{k}$ is the algebraically closed field given by the algebraic closure of k in S, then*

$$(13) \qquad [k(x):k(z)]_i = [k(x):k]_i[\bar{k}(x):\bar{k}(z)]_i$$

for any transcendence basis $\{z\}$ of $k(x)/k$.

For $\bar{k}$ being a perfect field, the order of inseparability of $\bar{k}(x)$ over $\bar{k}$ is equal to 1.

COROLLARY 2. *The assumptions being as in the first part of Theorem* 38, *and if we assume furthermore that either* (a) *K and $k(x)$ are linearly disjoint over k or* (b) *K is separably generated over k, then $[K(x):K]_i = [k(x):k]_i$.*

In the case (a) the corollary is an immediate consequence of (10) and of Lemma 1 as applied to the case $L = k(z)$, $K = k(x)$ and $k' = K$. If K is a finite separable extension of k, then the corollary follows at once from (10) and from the second half of Theorem 38. If K is a purely transcendental extension of k then K and $k(x)$ are linearly disjoint over k by Corollary 1 to the lemma of § 15 (since K and $k(x)$ are free over k) and we are then in the case (a). The general case under (b) is now settled immediately by following up a purely transcendental extension of k by a separable algebraic extension and by using Lemma 2.

§ 17. Derivations

DEFINITION. *Let S be a ring and R a subring of S. A mapping D of R into S is said to be a derivation of R (with values in S) if, for every x, y in R, D satisfies the following conditions:*

(1) $D(x + y) = D(x) + D(y)$.

(2) $D(xy) = xD(y) + yD(x)$.

REMARK. The notion of a derivation of R may be generalized to the case in which S is replaced by an R-module (see III, § 1). In that case, the products $xD(y)$ and $yD(x)$ on the right-hand side of (2) are products of an element of R and an element of the R-module S, and hence are elements of S; thus formula (2) is meaningful.

Successive applications of formula (2) show that for every x in R and for every integer $n \geq 1$ we have $D(x^n) = nx^{n-1}D(x)$. In particular, if S has a unit element e which lies in R, we have $D(e) = D(e^2) = 2eD(e) = 2D(e)$, whence $D(e) = 0$. From (2) it follows that the *kernel* of the additive homomorphism D (that is, the set of all elements x in R such that $D(x) = 0$) is a *subring* of R; the relation $D(e) = 0$ now shows that this subring contains the unit element e, hence also all integral multiples $ne(n \in J)$.

A derivation D of a ring R such that $D(x) = 0$ for every element x of a subring R' of R is said to be *trivial on* R', or to be *an R'-derivation* of R.

LEMMA. *Let R be an integral domain and let D be a derivation of R with values in a field K' containing R. Then D can be extended, and in a unique way, to a derivation D' of the quotient field K of R. For $x/y \in K$ $(x, y, \in R, y \neq 0)$ we have*

$$D'(x/y) = (yD(x) - xD(y))/y^2. \tag{3}$$

PROOF. From $D(x) = D'(y \cdot (x/y)) = yD'(x/y) + (x/y)D(y)$ it follows that relation (3) holds for every derivation D' of K which extends D. This proves the uniqueness of D'. As to the existence of D', we first have to show that (3) actually defines a mapping of K, that is, that if $x/y = x'/y'$ then $(yD(x) - xD(y))/y^2 = (y'D(x') - x'D(y'))/y'^2$, and it will be sufficient to show this in the case in which $x' = zx$, $y' = zy$, $z \in R$. In this case we have $y'D(x') = zy(zD(x) + xD(z))$, $x'D(y') = zx(zD(y) + yD(z))$, whence $y'D(x') - x'D(y') = z^2(y(D(x) - xD(y))$, and this proves the above equality. A straightforward computation, which may be left to the reader, shows that the mapping D' satisfies conditions (1) and (2) and hence is a derivation.

EXAMPLES OF DERIVATIONS

EXAMPLE 1. Let R be a ring, D a derivation of R, and A the *polynomial ring* $R[X_1, \cdots, X_n]$. With every polynomial

$$f(X_1, \cdots, X_n) = \sum a_{q_1, \cdots q_n} X_1^{q_1} \cdots X_n^{q_n}$$

we associate the polynomial

$$\sum D(a_{q_1, \cdots, q_n}) X_1^{q_1} \cdots X_n^{q_n}$$

obtained from f by coefficientwise derivation, and we denote this polynomial by $f^D(X_1, \cdots, X_n)$. It is a straightforward matter to check that the mapping $f \to f^D$ is a *derivation* of A. Obviously, this derivation is an extension of D.

EXAMPLE 2. Let R' be a ring and R be the polynomial ring $R'[X_1, \cdots, X_n]$. We may formally define *the partial derivative* $f'_{X_1}(X_1, \cdots, X_n)$ (with respect to X_1) of a polynomial f in R as follows: if $f(X_1, \cdots, X_n) = \sum a_{q_1, \cdots, q_n} X_1^{q_1} \cdots X_n^{q_n}$ then $f'_{X_1}(X_1, \cdots, X_n) = \sum a_{q_1, \cdots, q_n} q_1 X_1^{q_1 - 1} X_2^{q_2} \cdots X_n^{q_n}$. The mapping $f \to f'_{X_1}$ is denoted by D_1 or by $\partial/\partial X_1$. A straightforward computation shows that D_1 is an R'-derivation of the polynomial ring R; (it is even trivial on $R'[X_2, \cdots, X_n]$). We call this derivation of R the *partial derivation with respect to* X_1. In the same way we define the partial derivations D_i for $i = 2, \cdots, n$. By the above lemma, if R' is a field these derivations can be extended, and in a unique way, to derivations of the rational function field $K = R'(X_1, \cdots, X_n)$; these derivations bear the same name and are denoted in the same way as their restrictions to the polynomial ring R. Formulae (1) and (2) show that the derivation D_i $(1 \leq i \leq n)$ is uniquely determined by the condition that it be trivial on R' and that it satisfy the following relations:

$$D_i(X_i) = 1, \quad D_i(X_j) = 0 \quad \text{for} \quad i \neq j. \tag{4}$$

EXAMPLE 3. The preceding example can be generalized to the case of polynomial rings $R = R'[\{X_\alpha\}]$ in infinitely many indeterminates X_α indexed by a set $A = \{\alpha\}$. For any $\alpha \in A$ we denote by S_α the set of indeterminates X_β, $\beta \neq \alpha$, and by R_α the polynomial ring $R'[S_\alpha]$. Then $R = R_\alpha[X_\alpha]$ is a polynomial ring in one indeterminate X_α over the ring R_α, and thus there is a unique derivation D_α of R which is trivial on R_α and such that $D_\alpha(X_\alpha) = 1$. We denote this derivation by $\partial/\partial X_\alpha$. We have then $D_\alpha(X_\beta) = 0$ if $\alpha \neq \beta$. If R' is a field, the derivations $\partial/\partial X_\alpha$ can be extended, in a unique way, to the quotient field $R'(\{X_\alpha\})$.

Let now K be a field and L an extension field of K. We are going to study *the set of all derivations of K with values in L.* If D and D' are two such derivations, the mapping $D + D'$ defined by $(D + D')(x) = D(x) + D'(x)$ is also a derivation of K with values in L: in fact formulae (1) and (2) are "linear in D." In the same way we see that, if D is a derivation of K and a is an element of L, the mapping aD defined by $(aD)(x) = aD(x)$ is also a derivation of K. Thus the derivations of K with values in L form a *vector space over* L, which we shall denote by $\mathscr{D}_K$ or $\mathscr{D}_K(L)$. If K' is a subfield of K, the derivations of K which are trivial on K' form a vector subspace of $\mathscr{D}_K$, which we shall denote by $\mathscr{D}_{K/K'}$.

REMARK. If D and D' are two derivations of K with values in L, the composite mappings DD' and $D'D$ are not, in general, derivations. However, the mapping $D'D - DD'$ [defined by $(D'D - DD')(x) = D'(D(x)) - D(D'(x))$] *is* a derivation of K. In fact, formula (1) is obvious for $D'D - DD'$. On the other hand, we have

$$\begin{aligned}(D'D - DD')(xy) &= D'(xD(y) + yD(x)) - D(xD'(y) + yD'(x))\\ &= x\cdot(D'D - DD')(y) + y\cdot(D'D - DD')(x),\end{aligned}$$

since the terms $D'(x)D(y)$ and $D'(y)D(x)$ cancel; thus $D'D - DD'$ satisfies (2). The derivation $D'D - DD'$ is sometimes denoted by $[D, D']$ and called the *bracket* of D and D'. The formulae $[D, D] = 0$, $[D, D'] = -[D', D]$ are obvious. Furthermore, it is a straightforward matter to check the "Jacobi Identity"

$$[[D, D'], D''] + [[D', D''], D] + [[D'', D], D'] = 0$$

between any three derivations D, D', D'' of K. One expresses the above properties of the bracket $[D, D']$ by saying that $\mathscr{D}_K$ is a *Lie Algebra* over L. Similarly $\mathscr{D}_{K/K'}$ is also a Lie Algebra over L.

EXAMPLE 4. Let k be a field and $K = k(X_1, \cdots, X_n)$ the rational function field in n variables over k. We are going to prove that the *partial derivations* D_i $(i = 1, \cdots, n)$ *form a basis of the vector space* $\mathscr{D}_{K/k}$ of k-derivations of K (with values in any extension field L of K). In fact, let $D \in \mathscr{D}_{K/k}$. If we set $D' = \sum D(X_i)D_i$, D' is a k-derivation of K which coincides with D at the elements $X_1, \cdots, X_n$, and hence D' coincides with D on the polynomial ring $k[X_1, \cdots, X_n]$ since the kernel of the *derivation* $D' - D$ is a ring. Consequently, $D' = D$ on the quotient field K, by formula (3) (Lemma), showing that the D_i span $\mathscr{D}_{K/k}$. We now prove that the D_i are linearly independent (over L). In fact, if $\sum a_iD_i = 0$ $(a_i \in L)$, then $0 = (\sum a_iD_i)(X_j) = a_j$ for $j = 1, \cdots, n$.

THEOREM 39. *Let K be a field, $F = K(x_1, \cdots, x_n)$ a finitely generated extension field of K, D a derivation of K with values in some extension field L of F, and $\{u_1, \cdots, u_n\}$ a set of n elements of L. In order for there to exist a derivation D' of F extending D and such that $D'(x_i) = u_i$ for $i = 1, \cdots, n$, it is necessary and sufficient that, for every polynomial $f(X_1, \cdots, X_n)$ in $K[X_1, \cdots, X_n]$ such that $f(x_1, \cdots, x_n) = 0$, we have*

$$f^D(x_1, \cdots, x_n) + \sum_{i=1}^{n} u_i\cdot(D_if)(x_1, \cdots, x_n) = 0, \tag{5}$$

and if this condition is satisfied then the derivation D' is uniquely determined. Furthermore, if $\{f_j\}$ is a set of polynomials such that $f_j(x_1, x_2, \cdots, x_n) = 0$ for all j and such that every polynomial $f \in K[X_1, \cdots, X_n]$ for which we have $f(x_1, \cdots, x_n) = 0$ may be written in the form $f = \sum_j g_j f_j$ with $g_j \in K[X_1, \cdots, X_n]$ (in other words, if $\{f_j\}$ is "a basis of the ideal of algebraic relations satisfied by $x_1, \cdots, x_n$ over K"), then for the existence of D' it is sufficient that

$$(6) \qquad f_j^D(x_1, \cdots, x_n) + \sum_{i=1}^{n} u_i \cdot (D_i f_j)(x_1, \cdots, x_n) = 0,$$

for every j.

PROOF. We first observe that if D' is any derivation of F extending D then we have, for every polynomial $g \in K[X_1, \cdots, X_n]$, the relation

$$(7) \quad D'(g(x_1, \cdots, x_n)) = g^D(x_1, \cdots, x_n) + \sum_i (D'(x_i))(D_i g)(x_1, \cdots, x_n).$$

In fact, formula (2) for the derivative of a product shows that (7) is true when g is a monomial $aX_1^{q_1} \cdots X_n^{q_n}$ $(a \in K)$; hence (7) is true for any polynomial g, by linearity. Since $D'(0) = 0$, (7) shows that relation (4) is necessary, and (7) also shows that there is at most only one derivation D' of $K[x_1, x_2, \cdots, x_n]$, and hence also at most one derivation of the quotient field F, such that D′ is an extension of D and $D'(x_i) = u_i$, $i = 1, 2, \cdots, n$. Conversely, if (5) holds for any polynomial f over K such that $f(x_1, \cdots, x_n) = 0$, we define D' by setting

$$D'(g(x_1, \cdots, x_n)) = g^D(x_1, \cdots, x_n) + \sum_i u_i \cdot (D_i g)(x_1, \cdots, x_n).$$

Then D' is a mapping, for the value $D'(g(x_1, \cdots, x_n))$ depends only on the element $g(x_1, \cdots, x_n)$ of F and not on the polynomial g [since, if $g(x_1, \cdots, x_n) = h(x_1, \cdots, x_n)$, we may apply (5) to the polynomial $g - h$, and we then find that $D'(g(x_1, x_2, \cdots, x_n)) = D'(h(x_1, x_2, \cdots, x_n))$]. It is clear that D' is an additive homomorphism (that is, D' satisfies (1)). On the other hand, we have $D'(x_i) = u_i$. Finally, condition (2) is satisfied by D' since the mappings $g \to g^D$ and $g \to u_i \cdot D_i g$ are derivations of $K[X_1, \cdots, X_n]$ (Examples 1 and 2, p. 121). Thus condition (5) is sufficient, since a derivation D' of $K[x_1, \cdots, x_n]$ can be extended to the quotient field F (Lemma). It remains to be shown that condition (6) implies condition (5) for any polynomial f such that $f(x_1, \cdots, x_n) = 0$. Since $f = \sum_j g_j f_j$, we have $f^D(x_1, \cdots, x_n) = \sum_j f_j^D(x_1, \cdots, x_n) g_j(x_1, \cdots, x_n) + \sum_j f_j(x_1, \cdots, x_n) g_j^D(x_1, \cdots, x_n) = \sum f_j^D(x_1, \cdots, x_n) g_j(x_1, \cdots, x_n)$, since $f_j(x_1, \cdots, x_n) = 0$ for every j.

In the same way we see that

$$(D_i f)(x_1, \cdots, x_n) = \sum_j g_j(x_1, \cdots, x_n) \cdot (D_i f_j)(x_1, \cdots, x_n).$$

Therefore we obtain (5) by multiplying each relation (6) by $g_j(x_1, \cdots, x_n)$ and adding the resulting relations. Q.E.D.

We shall now apply Theorem 39 in various important cases of field extensions.

COROLLARY 1. *Let K be a field and let $F = K(x)$ be a simple transcendental extension of K. If D is a derivation of K with values in some field L containing F, and if u is any element of L, then there exists one and only one derivation D' of F extending D, such that $D(x) = u$.*

In fact, 0 is the only polynomial f in $K[X]$ such that $f(x) = 0$.

COROLLARY 1′. *Let K be a field and let $F = K(S)$ be a purely transcendental extension of K; here S denotes a set of generators of F/K which are algebraically independent over K. Let $x \to u_x$ be a mapping of S into a field L containing F. If D is any derivation of K with values in L, then there exists one and only one derivation D' of F extending D, such that $D(x) = u_x$ for all x in S.*

It is clear that if D' exists it is uniquely determined on $K[S]$, whence also on F. To show the existence of D' we shall use Zorn's lemma. Let I be the set of all subsets S_α of S with the property that D admits an extension D_α to $F_\alpha = K(S_\alpha)$ such that $D_\alpha x = u_x$ for all x in S_α. The set I is non-empty since the empty set belongs to I. If $S_\alpha \subset S_\beta$ then D_β is an extension of D_α. It follows at once that the set I, partially ordered by set-theoretic inclusion, is inductive. By Zorn's lemma, there exists a maximal element S' in I. Let $F' = K(S')$ and let D' be the derivation of F' extending D and such that $D'(x) = u_x$ for all x in S'. If $S' \neq S$, then there exists an element y in S which does not belong to S', and by Corollary 1 there exists a derivation D'' of the field $F'(y) = K(S' \cup \{y\})$ which is an extension of D' and is such that $D'y = u_y$. This contradicts the maximality of S' in I. Hence $S' = S$ and the corollary is proved.

COROLLARY 2. *Let K be a field, and $F = K(x)$ a simple separable algebraic extension of K. Then every derivation D of K may be extended, in one and only one way, to a derivation of F.*

In fact, every polynomial $g \in K[X]$ such that $g(x) = 0$ is a multiple of the minimal polynomial f of x over K. Thus the set of relations (6) reduces to the single relation $f^D(x) + uf'_X(x) = 0$. Since x is separable over K we have $f'_X(x) \neq 0$, and hence the relation $f^D(x) + uf'_X(x) = 0$ is satisfied by one and only one value of u. If u_0 is that value of u then the extension D' of D to F such that $D'(x) = u_0$ is the only extension of D to F.

COROLLARY 2′. *Corollary 2 remains valid if F is an arbitrary separable algebraic extension of K.*

The proof, which is similar to and even simpler than the proof of Corollary 1′, is a straightforward application of Zorn's lemma and may be left to the reader.

COROLLARY 3. *If a field F is a separably generated extension of a field K then every derivation D of K can be extended to a derivation of F.*

This follows immediately from Corollaries 1 and 2.

COROLLARY 4. *Let K be a field of characteristic $p \neq 0$ and let $F = K(x)$ $(F \neq K)$ be a simple purely inseparable extension of K. Let e be the smallest integer $(e \geqq 1)$ such that $y = x^{p^e} \in K$. Then a derivation D of K has an extension D' to F if and only if $D(y) = 0$, and if D is such a derivation of K (with values in a field L containing F) then the value of $D'(x)$ can be assigned arbitrarily in L.*

The minimal polynomial of x over K is $X^{p^e} - y$ (Theorem 7 of § 5). Hence relation (6) reduces to $D(y) = 0$.

COROLLARY 4′. *Let K be a field of characteristic $p \neq 0$ and let F be a purely inseparable extension of K such that $F^p \subset K$. If S is a set of generators of F/K and if D is a derivation of K such that $D(x^p) = 0$ for every element x of S, then D can be extended to F.*

We consider the set I of all pairs (F', D') composed of a field F' such that $K \subset F' \subset K(S)$ and a derivation D' of F' extending D. We set $(F', D') \prec (F'', D'')$ if $F' \subset F''$ and if D'' extends D'. This relation is an order relation, for which I is obviously inductive. By Zorn's lemma there exists a maximal element (F', D') of I. If $F' \neq K(S)$, there exists an element x in S such that $x \notin F'$; we then have $x^p \in F'$ and $D'(x^p) = 0$, whence D' may be extended to $K(S)(x)$ according to Corollary 4. Since this contradicts the maximality of (F', D'), we have $F' = K(S)$. Q.E.D.

COROLLARY 5. *Let K be a field of characteristic $p \neq 0$, and F a purely inseparable extension of K such that $F^p \subset K$. If $[F : K]$ is finite, say $[F : K] = p^s$, then the vector space $\mathscr{D}_{F/K}$ of K-derivations of F has dimension s. There exists a set $\{x_1, \cdots, x_s\}$ of s elements of F and a basis $\{D_1, D_2, \cdots, D_s\}$ of $\mathscr{D}_{F/K}$, with the following properties: $F = K(x_1, \cdots, x_s)$, $x_j \notin K(x_1, \cdots, x_{j-1})$ for $j = 1, \cdots, s$, and $D_i(x_i) = 1$, $D_i(x_j) = 0$ for $i \neq j$.*

We construct by induction a sequence $x_1, \cdots, x_j, \cdots$ of elements of F such that $x_j \notin K(x_1, \cdots, x_{j-1})$. Since $x_j^p \in K(x_1, \cdots, x_{j-1})$, we have $[K(x_1, \cdots, x_j) : K(x_1, \cdots, x_{j-1})] = p$. Since $[F : K]$ is finite, this sequence must have a finite number of terms, say s, and then we have $[F : K] = p^s$. For every $j = 1, \cdots, s$ there exists, by Corollary

4, a $K(x_1, \cdots, x_{j-1})$-derivation D'_j of $K(x_1, \cdots, x_j)$ such that $D'_j(x_j) = 1$. We extend D'_j to a derivation D_j of $F = K(x_1, \cdots, x_s)$ by imposing on D_j the conditions $D_j(x_{j+1}) = \cdots = D_j(x_s) = 0$; this is possible, as follows by successive applications of Corollary 4, since $x_i^p \in K$ for $i = j+1, \cdots, s$. We thus have s derivations $D_1, \cdots, D_s$ of F over K such that $D_j(x_j) = 1$ and $D_j(x_i) = 0$ for $i \neq j$. These derivations are linearly independent: in fact, if $\sum_i a_i D_i = 0$, we have $a_j = (\sum_i a_i D_i)(x_j) = 0$ for every j. Finally, if D is any K-derivation of F, the mapping $D' = D - \sum_i D(x_i)D_i$ is a derivation of F which takes the value 0 on K and on the set $\{x_1, \cdots, x_s\}$. Hence D' takes the value 0 on $K[x_1, \cdots, x_s] = F$; this proves that the derivations $D_1, \cdots, D_s$ span $D_{F/K}$. Q.E.D.

REMARK. The formula $D(x^p) = px^{p-1}$ shows that in characteristic $p \neq 0$, every derivation of a field L is trivial on the subfield L^p. In particular, a *perfect* field of characteristic $p \neq 0$ has no non-trivial derivations. Corollaries 4 and 4′ show that L^p is exactly the set of all elements x of L such that $D(x) = 0$ for every derivation D of L. If K is a subfield of L we see, in the same way, that $K(L^p)$ is the set of all elements x in L such that $D(x) = 0$ for every K-derivation of L.

THEOREM 40. *Let K be a field and $F = K(x_1, \cdots, x_n)$ a finitely generated extension of K. For F to be separably algebraic over K, it is necessary and sufficient that 0 be the only K-derivation of F.*

PROOF. Necessity follows from Corollary 2′ to Theorem 39. Conversely let j be the largest index for which F is *not* separably algebraic over $K(x_1, \cdots, x_j)$. Then the field $K(x_1, \cdots, x_{j+1})$ is either purely transcendental or algebraic inseparable over $K(x_1, \cdots, x_j)$. At any rate there exists a non-trivial derivation D of $K(x_1, \cdots, x_{j+1})$ which is trivial on $K(x_1, \cdots, x_j)$ (Corollaries 1 and 4 to Theorem 39). Since F is separably algebraic over $K(x_1, \cdots, x_{j+1})$, D can be extended to F (Corollary 2 to Theorem 39). This contradicts our hypothesis and proves the sufficiency. Q.E.D.

COROLLARY. *Let K be a field, $\{x_1, \cdots, x_n\}$ a finite set of n elements of some extension field F of K, and $\{f_1, \cdots, f_n\}$ a finite set of n elements of $K[X_1, \cdots, X_n]$ such that $f_i(x_1, \cdots, x_n) = 0$ for $i = 1, \cdots, n$. If $\det\left(\frac{\partial f_i}{\partial X_j}(x_1, \cdots, x_n)\right) \neq 0$ then $K(x_1, \cdots, x_n)$ is separable algebraic over K. Conversely, if $K(x_1, x_2, \cdots, x_n)$ is separable algebraic over K then there exist polynomials $f_1, f_2, \cdots, f_n$ in $K[X_1, X_2, \cdots, X_n]$ such that $f_i(x_1, x_2, \cdots, x_n) = 0$ for $i = 1, 2, \cdots, n$ and such that the above determinant is different from 0.*

In fact, let D be a K-derivation of $K(x_1, \cdots, x_n)$. Since $D(f_i(x_1, \cdots, x_n)) = 0$, we have $\sum_j (\partial f_i/\partial X_j)(x_1, \cdots, x_n)\cdot D(x_j) = 0$ for $i = 1, \cdots, n$. We thus have a system of n homogeneous linear equations in $D(x_1), \cdots, D(x_n)$ with non-vanishing determinant. Hence $D(x_1) = \cdots = D(x_n) = 0$, and D is the trivial derivation.

Assume now that $K(x_1, x_2, \cdots, x_n)$ is separable algebraic over K. Then there exist no non-trivial K-derivations of $K(x_1, x_2, \cdots, x_n)$. Consider the system of homogeneous equations $\sum_{i=1}^{n} u_i(D_i f)(x_1, x_2, \cdots, x_n) = 0$ obtained by letting f vary in the set of all polynomials f in $K[X_1, X_2, \cdots, X_n]$ such that $f(x_1, x_2, \cdots, x_n) = 0$. By Theorem 39 (equation (5)) it follows that this system of equations has only the trivial solution $u_1 = u_2 = \cdots = u_n = 0$. Hence the system must contain some set of n equations with non-vanishing determinant, and this completes the proof of the corollary.

THEOREM 41. *Let K be a field and $F = K(x_1, \cdots, x_n)$ a finitely generated extension field of K. The dimension s of $\mathscr{D}_{F/K}$ is equal to the smallest number of elements $u_1, \cdots, u_t$ of F such that F is separable algebraic over $K(u_1, \cdots, u_t)$. If K has characteristic $p \neq 0$ then $p^s = [F : K(F^p)]$. The relation $s =$ tr. d. F/K characterizes separably generated extensions.*

PROOF. A K-derivation D of $F' = K(u_1, \cdots, u_t)$ is uniquely determined by $D(u_1), \cdots, D(u_t)$; hence the dimension of $\mathscr{D}_{F'/K}$ is at most equal to t. If F is separable algebraic over F', every derivation of F' has a unique extension to F (Corollary 2′ of Theorem 39), whence $s \leq t$. For proving our first assertion we now have to show the existence of s elements $u_1, \cdots, u_s$ of F such that F is separable algebraic over $K(u_1, \cdots, u_s)$. This is clear if $p = 0$ for, in that case, s is the transcendence degree of F/K (see Corollary 2′ to Theorem 39 and the example preceding Theorem 39). If $p \neq 0$ we observe that every K-derivation of F is trivial on $K(F^p)$ and that F is purely inseparable over $K(F^p)$. Therefore, if we apply Corollary 5 of Theorem 39 we find that there exist s elements $u_1, \cdots, u_s$ of F and s linearly independent K-derivations $D_1, \cdots, D_s$ of F such that $D_i(u_i) = 1$, $D_i(u_j) = 0$ for $i \neq j$. If a derivation D of F is trivial on $K(u_1, \cdots, u_s)$, we have $D = 0$, since we can write $D = \sum_i a_i D_i$, whence $0 = D(u_j) = a_j$ for every j. Hence F is separable algebraic over $K(u_1, \cdots, u_s)$ by Theorem 40, and the first part of our theorem is proved. The second assertion of the theorem is a partial repetition of Corollary 5 of Theorem 39. The third assertion follows from the first. Q.E.D.

THEOREM 42. *Let K be a field of characteristic $p \neq 0$ and F an extension field of K. For F to be separable over K* (§ 15), *it is necessary and sufficient that every derivation of K be extendable to F.*

PROOF. Suppose first that F is separable over K, that is, suppose that F and $K^{1/p}$ are linearly disjoint over K. Then, since $x \to x^p$ is an isomorphism, the fields F^p and K are linearly disjoint over K^p. Let $\{u_\alpha\}$ be a basis (finite or infinite) of F^p considered as a vector space over K^p; we have the "multiplication table" $u_\alpha u_\beta = \sum_\gamma c_{\alpha\beta\gamma} u_\gamma (c_{\alpha\beta\gamma} \in K^p)$. Then $\{u_\alpha\}$ is obviously a basis of the ring $K[F^p]$ considered as a vector space over K. Let D be a derivation of K. Every element x of $K[F^p]$ may be written, in a unique way, in the form $x = \sum_\alpha x_\alpha u_\alpha$ ($x_\alpha \in K$, $x_\alpha = 0$ except for a finite number of indices). Hence, if we set $D'(x) = \sum_\alpha D(x_\alpha)u_\alpha$, then D' is a mapping of $K[F^p]$ into some extension field of F, and this mapping obviously satisfies the condition $D'(x + y) = D'(x) + D'(y)$. To show that D' is a derivation we consider any two elements $x = \sum_\alpha x_\alpha u_\alpha$ and $y = \sum y_\beta u_\beta$ of $K[F^p]$; we have

$$
\begin{aligned}
D'(xy) &= D'\Big(\sum_{\alpha, \beta, \gamma} x_\alpha y_\beta c_{\alpha\beta\gamma} u_\gamma\Big) \\
&= \sum (D(x_\alpha)y_\beta + x_\alpha D(y_\beta))c_{\alpha\beta\gamma} u_\gamma \\
&= xD'(y) + yD'(x),
\end{aligned}
$$

since $D(c_{\alpha\beta\gamma}) = 0$ for $c_{\alpha\beta\gamma} \in K^p$. Now, the derivation D' of $K[F^p]$ (which extends D since one of the elements u_α may be taken to be 1) may be extended to a derivation D'' of the quotient field $K(F^p)$ (Lemma). Finally, since F is purely inseparable over $K(F^p)$ and since D'' takes the value 0 on F^p, by construction (if $\sum x_\alpha u_\alpha \in F^p$ then the x_α are in K^p and so $D(x_\alpha) = 0$), Corollary 4′ of Theorem 39 shows that D'' may be extended to F. Thus the necessity is proved.

Conversely, suppose that every derivation of K can be extended to F. We shall prove that if $x_1, \cdots, x_t$ are elements of F which are linearly independent over K, then the p-th powers of these elements are also linearly independent over K (a condition which is equivalent to separability; see § 15). Assuming the contrary, choose among the non-trivial linear relations with coefficients in K satisfied by the x_i^p, one with the smallest number of non-vanishing terms. By renumbering the indices we may write this relation in the form $a_1x_1^p + \cdots + a_nx_n^p = 0$ ($a_i \in K$, $a_i \neq 0$). We may also assume that $a_1 = 1$. Now, given any derivation D of K, we extend it to F and we continue to denote by D the extended derivation. Since $D(x_i^p) = 0$ and $D(a_1) = D(1) = 0$, the relation gives, by derivation, $D(a_2)x_2^p + \cdots + D(a_n)x_n^p = 0$. Since this relation has only $n - 1$ terms, all its coefficients vanish, whence $D(a_i) = 0$ for all i and all D. Since a_i is annihilated by every derivation of K, a_i belongs to K^p (remark after Corollary 5 of Theorem 39), whence $a_i = b_i^p$ with $b_i \in K$, $b_i \neq 0$. Thus the relation $\sum a_i x_i^p = 0$ gives $\sum b_i^p x_i^p = (\sum b_i x_i)^p = 0$, whence $\sum b_i x_i = 0$, in contradiction with the linear independence of $x_1, \cdots, x_t$ over K. Q.E.D.

We shall terminate this section with a brief discussion of *p-bases* of fields of characteristic $p \neq 0$. In the sequel F denotes a given field of characteristic $p \neq 0$.

Given a finite set of elements $x_1, x_2, \cdots, x_n$ of F, these elements are said to be *p-independent* if the n^p monomials $x_1^{i_1} x_2^{i_2} \cdots x_n^{i_n}$ $(0 \leqq i_q < p)$ are linearly independent over F^p. The set $\{x_1, x_2, \cdots, x_n\}$ is then said to be a *p-independent set*. An arbitrary (finite or infinite) subset S of F is said to be a p-independent set if every finite subset of S is p-independent.

A subset S of F is said to be a *p-basis* if it is a p-independent set and if it is at the same time a system of generators of F/F^p, that is, if $F = F^p(S)$.

Part of the discussion of p-independence and p-bases shall be given here as a special case of the general axiomatic notion of dependence developed in connection with vector spaces (I, § 21) and transcendental extensions (§ 12). Namely, let us introduce the following relation φ between subsets of F: if A is any subset of F then $\varphi(A) = F^p(A)$. This relation φ satisfies the five axioms given in I, § 21. The validity of the first four of these axioms is self-evident, and we have only to check the validity of the last axiom ("principle of exchange"): if A is a subset of F and x, y are elements of F such that $y \notin F^p(A)$ and $y \in F^p(A, x)$, then $x \in F^p(A, y)$. Since $F^p(A, x) = F^p(A)[x]$ and $x^p \in F^p$, we may write y in the form $y = a_0 x^{p-1} + a_1 x^{p-2} + \cdots + a_{p-2}x + a_{p-1}$, where the a_i are in $F^p(A)$. Since $y \notin F^p(A)$, the coefficients $a_0, a_1, \cdots, a_{p-2}$ are not all zero. It follows that x is both separable and purely inseparable over $F^p(A, y)$ and hence belongs to $F^p(A, y)$, and this proves the principle of exchange.

We now show that a finite subset $\{x_1, x_2, \cdots, x_n\}$ of F is free (with respect to the relation φ; see I, § 21) if and only if it is a p-independent set. Assume that the set $\{x_1, x_2, \cdots, x_n\}$ is free. To show that the elements x_i are p-independent we shall use induction with respect to n, since the case $n = 0$ is trivial. Let $f(X_1, X_2, \cdots, X_n)$ be a non-zero polynomial with coefficients in F^p and of degree *less* than p in *each* of the variables X_i. We have to show that $f(x_1, x_2, \cdots, x_n) \neq 0$. We may assume that X_n actually occurs in f. By our induction hypothesis, the polynomial $f(x_1, x_2, \cdots, x_{n-1}, X)$ in $F^p(x_1, x_2, \cdots, x_{n-1})$ $[X]$ is different from zero. Since this polynomial is of degree less than p in X, it is separable. Since x_n is purely inseparable over F^p and since it does not belong to $F^p(x_1, x_2, \cdots, x_{n-1})$ (the set $\{x_1, x_2, \cdots, x_n\}$ being free), it follows that x_n cannot be a root of the above polynomial, and this proves the p-independence of the elements $x_1, x_2, \cdots, x_n$.

Conversely, if the elements $x_1, x_2, \cdots, x_n$ do not form a free set then we may assume that $x_n \in F^p(x_1, x_2, \cdots, x_{n-1})$. Since $x_i^p \in F^p$, we may write x_n in the form $x_n = g(x_1, x_2, \cdots, x_{n-1})$, where g is a polynomial with coefficients in F^p and of degree less than p in each of the arguments. Then the relation $g(x_1, x_2, \cdots, x_{n-1}) - x_n = 0$ is a relation of linear dependence (over F^p) between the monomial x_n and the $(n-1)^p$ monomials $x_1^{i_1}x_2^{i_2} \cdots x_{n-1}^{i_{n-1}}$ $(0 \leqq i_q < p)$, showing that the elements $x_1, x_2, \cdots, x_n$ are not p-independent.

Using the general results established in I, § 21, we have therefore established the following facts:

(1) *A subset S of F is a p-basis of F if and only if it is a basis of F with respect to the relation φ, or—in other words—if and only if S is a minimal system of generators of F over F^p.*

(2) *There exist p-bases of F, and any two p-bases of F have the same cardinal number.*

The following theorem, which is in part a generalization of Corollary 5 of Theorem 39, establishes a relationship between p-independence and F^p-derivations of F:

THEOREM 43. *A subset S of F is p-independent if and only if for every element x of S there exists a derivation D_x of F/F^p such that $D_x(x) = 1$ and $D_x(y) = 0$ for every element y of S different from x.*

PROOF. Assume that S is a p-independent set. Let $\mathscr{I}$ be the set of all subsets of S having the following property: if A is any set in the family $\mathscr{I}$ then there exists for each element x of A a derivation $D_{x, A}$ of $F^p(A)$ such that $D_{x, A}(x) = 1$ and $D_{x, A}(y) = 0$ for all elements y of A which are different from x. We partially order $\mathscr{I}$ by set-theoretic inclusion and we show that $\mathscr{I}$ is inductive. Let $\mathscr{I}'$ be a totally ordered subset of $\mathscr{I}$. If $A, B \in \mathscr{I}'$ and if, say, $A \subset B$, then it is clear that for each element x of A the derivation $D_{x, B}$ is an extension of the derivation $D_{x, A}$. It follows that if C denotes the union of all the sets A belonging to $\mathscr{I}'$ then for each element x of C the various derivations $D_{x, A}$ $(x \in A \in \mathscr{I}')$ have a common extension $D_{x, C}$ to $F^p(C)$ such that $D_{x, C}(x) = 1$ and $D_{x, C}(y) = 0$ for all elements y in C which are different from x. Hence $C \in \mathscr{I}$, showing that $\mathscr{I}$ is inductive. By Zorn's lemma, let S' be a maximal element of $\mathscr{I}$. Were S' a proper subset of S, we could choose an element z in S which does not belong to S' and then, by Corollary 4 of Theorem 39, there would exist a derivation D of $F^p(S', z)$ which is trivial on $F^p(S')$ and such that $D(z) = 1$ (since $z \notin F^p(S')$, in view of the p-independence of the elements of S). This would contradict the maximality of S'. Hence $S' = S$. Now, each of the derivations $D_{x, S}$, $x \in S$, can be extended to a derivation D_x of F, by Corollary 4′ of Theorem 39. These derivations D_x satisfy the conditions stated in the theorem.

Conversely, assume that for each element x of S there exists a derivation D_x of F/F^p satisfying the conditions stated in the theorem. If $x_1, x_2, \cdots, x_n$ are arbitrary elements of S then the derivation D_{x_i} $(1 \leqq i \leqq n)$ is trivial on $F^p(x_1, x_2, \cdots, x_{i-1}, x_{i+1}, \cdots, x_n)$ while $D_{x_i}(x_i) = 1$. Hence x_i does not belong to the field $F^p(x_1, x_2, \cdots, x_{i-1}, x_{i+1}, \cdots, x_n)$. This shows that the set $\{x_1, x_2, \cdots, x_n\}$ is free (with respect to the relation φ) and hence is p-independent. Q.E.D.

We note that if F is a finitely generated extension of F^p and S is a p-basis of F, then the derivations D_x, $x \in S$, form a basis of $\mathscr{D}_{F/F^p}$; this, in fact, is the meaning of Corollary 5 of Theorem 39.

III. IDEALS AND MODULES

§ 1. Ideals and modules. In Chapter I we defined the concept of a homomorphism of a ring (I, § 12) and saw that the kernel plays an important role. We proved in I, § 12 that the kernel is not only a subring but actually *an ideal* in accordance with the following:

DEFINITION 1. *Let R be a ring. An ideal in R is a non-empty subset $\mathfrak{A}$ of R such that*

(a) *If $a_1, a_2 \in \mathfrak{A}$, then $a_1 - a_2 \in \mathfrak{A}$.*

(b) *If $a \in \mathfrak{A}$ and $b \in R$, then $ab \in \mathfrak{A}$.*

(We recall our convention, made in I, § 7, p. 10, that "ring" always means a commutative ring.) Condition (a) of this definition simply states that $\mathfrak{A}$ is a subgroup of the additive group of R. Condition (b), taken together with (a), implies that $\mathfrak{A}$ is a subring of R. But not every subring is an ideal; for example, in the ring of rational numbers the set of integers is a subring but not an ideal. Or again, if $F[x]$ is a polynomial ring over a field, then $F[x^2]$ is a subring of $F[x]$, but not an ideal.

If R is any ring and a is any element of R, then the set of all elements xa, $x \in R$, is clearly an ideal. It is called the *principal ideal determined by a* and is denoted by Ra. We note that if R has an identity, then a is a unit if and only if $Ra = R$. In a ring with identity the ideal Ra obviously is the smallest ideal in R containing a. In the general case the smallest ideal containing a is the set of all elements of R of the form $ra + na$, where $r \in R$ and n is an integer; this set is denoted by (a). If R has an identity, then $(a) = Ra$.

Another example of an ideal may be obtained by taking R to be a polynomial ring in n indeterminates $X_1, X_2, \cdots, X_n$ over a ring R_0. If $a_1, a_2, \cdots, a_n$ are fixed elements of R_0, the set of all polynomials $f(X_1, \cdots, X_n)$ in R such that $f(a_1, a_2, \cdots, a_n) = 0$ is an ideal.

Every ring R (except the nullring) has at least two ideals: the entire ring R and the set (0) consisting of 0 alone. The latter is identical with the principal ideal $R0$; if R has an identity, the former is $R1$. An ideal of R distinct from (0) and R will be called a *proper* ideal.

If R has an identity element 1 and if an ideal $\mathfrak{A}$ in R contains a unit u, then $\mathfrak{A} = R$, for $\mathfrak{A}$ contains $bu^{-1}u$ for every $b \in R$. If R is a field, then it has only the two (improper) ideals (0) and $R1$, for if $\mathfrak{A}$ is an ideal in R and $\mathfrak{A} \neq (0)$, then $\mathfrak{A}$ contains an element $a \neq 0$, hence contains $aa^{-1} = 1$, hence equals R. *Conversely, if R has an identity and has only two ideals, then R is a field*; for if $a \in R$, $a \neq 0$, then $Ra \neq R0$, hence $Ra = R$, $1 \in Ra$, so $1 = xa$ for some $x \in R$.

EXAMPLE. If G is an arbitrary abelian (additive) group then G can be made into a commutative ring by setting $ab = 0$ for all a, b in G. This ring has no multiplicative identity. Every subgroup of the *group* G is then an ideal in the *ring* G. If we take for G a finite group of prime order then we obtain an example of a ring which has no proper ideals and yet is not a field.

The above example is the most general of its kind, for we can easily prove the following result: *if a ring R has no proper ideals and is not a field, then the additive group of R is cyclic, of prime order, and we have $ab = 0$ for all a, b in R.* To prove this, we consider the set $\mathfrak{A}$ of all elements a of R such that Ra is the zero ideal. In other words, $\mathfrak{A}$ is the set of all absolute zero divisors of R. Since we have assumed that R is not a field, there exist elements a in R, different from zero, such that $Ra \neq R$. For any such element a we must have then $Ra = (0)$, for R has no proper ideals. Hence the set $\mathfrak{A}$ contains elements different from zero. On the other hand, it is obvious that $\mathfrak{A}$ is an ideal. Hence $\mathfrak{A} = R$ and we have therefore $ab = 0$ for all a, b in R. Every subgroup of the additive group of R is then an ideal of R, and therefore the additive group of R must be finite, of prime order, since (0) and R are its only subgroups.

The concept of ideal is susceptible of an immediate generalization.

DEFINITION 2. *Let S be a ring and R a subring of S. An R-MODULE in S is a non-empty subset $\mathfrak{A}$ of S such that*

(a) *If $a_1, a_2 \in \mathfrak{A}$, then $a_1 - a_2 \in \mathfrak{A}$.*
(b) *If $a \in \mathfrak{A}$ and $b \in R$, then $ba \in \mathfrak{A}$.*

Any ideal in S is clearly an R-module. In particular, the ideals of R are precisely those subsets of R which are R-modules. We note that as far as the elements of S are concerned, only the operation of addition is involved in the module concept. Multiplication enters only between an element of R and an element of S. This suggests a further (and final) generalization. We thus come to the concept of an abstract module, which will be fundamental for many of the subsequent developments in this book.

DEFINITION 3. *Let R be a ring. A set M is called a* MODULE OVER R *(or an R-*MODULE*) if*

(a) *M is a commutative group (the group operation will be written as addition).*

(b) *With every ordered pair (a, x) in which $a \in R$ and $x \in M$ there is associated a unique element of M, to be denoted by ax, such that the following relations hold:*

$$a(x + y) = ax + ay \tag{1}$$

$$(a + b)x = ax + bx \tag{2}$$

$$(ab)x = a(bx), \tag{3}$$

where a, b are any elements of R and x, y are any elements of M. The element ax will sometimes be called the *product* of a and x.

The above definition of an R-module in an overring S of R is a special case of the present definition, the element of M associated with the pair (a, x) being simply the product of a and x as elements of the ring S. In that special case, equations (1), (2), (3) are consequences of the ring axioms.

If R has an identity element, then the R-module M is said to be *unitary* if

$$1x = x, \quad \text{for all} \quad x \in M. \tag{4}$$

If R is a ring with an identity and S is a ring containing R, then clearly S is a unitary overring of R if and only if S is unitary when considered as a *module* over R.

Perhaps the best-known examples of modules are the vector spaces. In terms of modules the definition of vector spaces, as given in I, § 21, signifies that *a vector space is a unitary module over a field.* We shall see later on in this chapter (§ 12) that the elementary properties of vector spaces which we have established in I, § 21 (such as those relative to dimension, linear dependence, etc.) are also consequences of general theorems about modules.

It must be emphasized that if M is a commutative group (written additively) and if R is a ring, then M may be an R-module in more than one way. That is, given a in R and x in M there may be more than one way to define the product ax so that relations (1), (2), (3) hold. (There is always at least one way—we may define ax to be the zero of M for all a in R and $x \in M$; in this case we refer to M as a *trivial* R-module.) If we have two different definitions of the product ax then we really have two distinct R-modules, although the underlying group M is the same in both cases. If R is a subring of a ring S, then, in general, S may be thought of as an R-module in various ways. However, whenever in this book we regard S as an R-module, *we shall always mean it in the sense mentioned above* (that is, for a in R and x in S we shall mean by ax the product of a and x as elements of S), unless the contrary is explicitly stated.

As a final example, let M be any (additively written) commutative group. Then M can always be regarded as a J-module, where J denotes the ring of integers. Namely, if $n \in J$ and $x \in M$, take nx to mean what it usually does, as defined in I, § 4. Clearly M is unitary. The possibility of thus construing M as a J-module shows that any statement about modules implies a statement about general commutative groups.

We return to the general case where M is an R-module. With every element a of R we can associate a mapping T_a of M into itself, defined by

$$xT_a = ax, \; x \in M.$$

Equation (1) above states that T_a is an endomorphism of M, regarded as a group. It follows that if 0 is the zero element of M, then $a0 = 0$ for all $a \in R$. Similarly, if 0 denotes the zero element of R, then $0x$ equals the zero of M, for $0x = (0 + 0)x = 0x + 0x$, whence the assertion. From this it follows easily that $-(ax) = (-a)x = a(-x)$ for any $a \in R$ and $x \in M$. We shall use the same symbol for the zeros of M and of R; only rarely is there any possibility of ambiguity.

It has just been observed that with every element a of R may be associated an endomorphism, or, as it is sometimes called, an *operator*, T_a of M (M being regarded as a group). Sometimes the elements of R themselves are referred to as operators. The module M is often called *a group with a ring R of operators*.

The notion of group with operators may be generalized in two directions. In the first place, it will be noted that the ring property of R, in particular equations (2) and (3), will play no essential role until § 5. Hence most of the considerations up to that point could be carried through without change on the assumption that R is a set with each of whose elements is associated an endomorphism of the group M. In other words, with every $a \in R$ and $x \in M$ there is associated an element ax of M such that (1) holds; nothing more need be required. We have no occasion, however, to make use of this more general formulation.

A second generalization consists in dropping the assumption that M is commutative. The proofs which follow do not apply without some modification to the non-commutative case. A treatment of the most general case can be found in Chapter V of Jacobson's *Lectures in Abstract Algebra*, Vol. I.

§ 2. Operations on submodules

DEFINITION. *Let R be a ring and M an R-module. An R-submodule (or simply a submodule) of M is a non-empty subset N of M such that*

(a) *If $x_1, x_2 \in N$, then $x_1 - x_2 \in N$.*
(b) *If $a \in R$ and $x \in N$, then $ax \in N$.*

The first condition (together with the non-emptiness of N) states that N is a subgroup of M. If N is a submodule of M, then it is also an R-module if, for $a \in R$ and $x \in M$, we define the product of a and x to be ax, the product which is already defined in virtue of the fact that M is an R-module.

If M is a vector space over a field F, any F-submodule is called a *subspace*; it is itself a vector space over F.

Any given ring R may be regarded as a module over itself; its submodules are then simply its ideals. Any statement about submodules of a given module therefore implies a statement about the ideals of a ring.

Every module M has as submodules the set consisting of zero alone (it will be denoted by (0)) and M itself. Any other submodule is called *proper*.

If A and B are non-empty subsets of an R-module M, then the *sum* of A and B, denoted by $A + B$, is the set of all elements of the form $x + y$, where $x \in A, y \in B$. The *negative* of A, denoted by $- A$, is the set of all elements $- x$, where $x \in A$. If A consists of a single element x, then $A + B$ will be denoted by $x + B$.

It is immediately verified that the operation of addition is commutative and associative and that the set consisting of 0 alone (usually denoted by (0)) is the zero element for this addition. But the subsets of M do not form a group under this addition since $A + (- A) \neq (0)$, unless A consists of a single element. The set $A + (- B)$, which clearly consists of all elements $x - y$, where $x \in A, y \in B$, will be denoted by $A - B$.

For the most part, addition will be applied to submodules. It is easy to verify the fundamental fact that *the sum of two submodules is also a submodule*. In particular, the sum of two ideals in a ring R is also an ideal. Furthermore, if L and N are submodules of M, then $L + N$ is the *smallest* submodule of M containing L and N, in the sense that $L + N$ contains L and N, and any other submodule containing L and N must contain $L + N$. If $N_1, \cdots, N_h$ are submodules of M, then from the associativity of addition it follows that $N_1 + \cdots + N_h$ is defined. It will be denoted by $\sum_{i=1}^{h} N_i$ and clearly consists of all sums $\sum_{i=1}^{h} x_i$, where

$x_i \in N_i$. This suggests that if $\{N_\alpha\}$ is any collection (finite or infinite) of submodules of M, where the index α ranges over some set A, we define $\sum_\alpha N_\alpha$ to consist of all sums $\sum_\alpha x_\alpha$, where $x_\alpha \in N_\alpha$ and $x_\alpha = 0$ for all but a finite number of indices α. Clearly $\sum N_\alpha$ is the smallest submodule of M containing all the N_α.

A second operation on submodules is ordinary set-theoretic intersection. *If L and N are submodules of M, then so is $L \cap N$* (consisting of those elements of M common to L and N). This operation is, of course, also commutative and associative, and the module M itself acts as identity for this operation since $M \cap N = N$ for any submodule N. If $M_1, \cdots, M_r$ are submodules of M, then $M_1 \cap \cdots \cap M_r$ is also denoted by $\bigcap_{i=1}^{r} M_i$.

The two operations of addition and intersection are related by a very important identity due to Dedekind (the so-called *modular law*) between three submodules K, L, N of a module M:

(5) If $K \supset L$, then $K \cap (L + N) = L + (K \cap N)$.

It is clear that the right side is contained in the left. On the other hand, if x is contained in the left-hand side, then since $x \in L + N$, we have $x = y + z$, with $y \in L$, $z \in N$. Then $z = x - y \in K$, since $x \in K$ and $y \in L \subset K$. Thus $z \in K \cap N$, and so $x \in L + (K \cap N)$.

Let R be a ring and M an R-module; let A and L be respective non-empty subsets of R and M. The *product* of A and L, denoted by AL, is the set of all sums $\sum_{i=1}^{n} a_i x_i$, where $a_i \in A$, $x_i \in L$ and n is an arbitrary positive integer.

It is easily verified that if A is an ideal in R or if L is a submodule of M, then AL is a submodule of M. For AL is clearly closed under subtraction. If $b \in R$ and if $\sum a_i x_i \in AL$, then $b \sum a_i x_i = \sum (ba_i) x_i = \sum a_i (bx_i)$. From the first or second of these two last sums, depending on whether A is an ideal or L is a submodule, we see that $b \sum a_i x_i \in AL$.

The two conditions for a non-empty subset L of M to be a submodule may now be expressed as follows: $L - L \subset L$, $RL \subset L$.

If L consists of a single element x and A is closed under addition, then AL consists of all elements ax, where $a \in A$; it will be denoted by Ax. If A consists of a single element a and L is closed under addition, then AL consists of all elements ax, where $x \in L$; it will be denoted by aL.

If R is a given ring we have seen how it may be considered as an R-module. If we apply the above definition of product to this particular case, we have a definition for the product of two non-empty subsets A

and B of R. Namely, AB consists of all sums $\sum_{i=1}^{n} a_i b_i$, $a_i \in A$, $b_i \in B$, n arbitrary. In particular the product of two ideals in a ring R is again an ideal, and the notation Ra introduced in § 1 (p. 132) for the principal ideal determined by a is consistent with our present definitions.

It is possible to define also a quotient operation between modules, but we shall do so only for ideals (see § 7, p. 147).

§ 3. Operator homomorphisms and difference modules. In I, § 11, we have defined homomorphisms between groups. This definition applies in particular to modules but is too general, since it takes into account only the group character of M and not the fact that M admits the elements of R as operators.

DEFINITION. *Let R be a ring and let M and M' be two R-modules. An R-homomorphism of M into M' is a mapping T of M into M' such that*

$$(x + y)T = xT + yT, \quad x, y \in M, \tag{1}$$

$$(ax)T = a(xT), \quad x \in M, a \in R. \tag{2}$$

Equation (1) states that T is a homomorphism of M into M' when each is considered as a group. As to (2), let T_a and T'_a be the respective endomorphisms of M and M' determined by a; then (2) states that $T_aT = TT'_a$.

If M and M' are vector spaces over a field F, then an F-homomorphism of M into M' is called a *linear transformation* of M into M' (see I, § 21, p. 53).

When it is not desired to call attention to the specific ring R, we may refer to an *operator homomorphism* instead of an *R-homomorphism.* The terms *R-homomorphism onto, R-isomorphism, R-endomorphism, R-automorphism* are now self-explanatory (cf. I, § 11). If M, M', M'' are R-modules and T and T' are R-homomorphisms of M into M' and of M' into M'' respectively, then TT' is in an R-homomorphism of M into M''.

THEOREM 1. *Let M and M' be modules over a ring R, and let T be an R-homomorphism of M into M'. Then $0T$ is the zero of M', $(-x)T = -(xT)$ for any $x \in M$. If A and L are non-empty subsets of R and M respectively, then $(AL)T \subset A(LT)$. The kernel of T is an R-submodule of M, and T is an isomorphism if and only if the kernel is (0). If L and L' are R-submodules of M and M' respectively, then LT and $L'T^{-1}$ are R-submodules of M' and M respectively.*

PROOF. The first statement follows from Theorem 1 of I, § 11. If

$a_i \in A$ and $x_i \in L$, $i = 1, 2, \cdots, n$ then $(\sum a_i x_i)T = \sum(a_i x_i)T = \sum a_i(x_i T) \in A(LT)$, whence the second statement. For the third, let N denote the kernel of T; in accordance with the general definition given in I, § 11 (p. 14), N consists of all $x \in M$ such that $xT = 0$. By Theorem 1 of I, § 11, N is surely a subgroup of M; it is a submodule since $(RN)T \subset R(NT) \subset R0 = (0)$, hence $RN \subset N$. The second part of the third statement follows from Theorem 2 of I, § 11. Let us prove that $L'T^{-1}$ is an R-submodule of M. If $x, y \in L'T^{-1}$, then xT, $yT \in L'$, hence $(x - y)T = xT - yT \in L'$, so $x - y \in L'T^{-1}$; if $a \in R$, then $(ax)T = a(xT) \in L'$, hence $ax \in L'T^{-1}$. Similarly for LT.

If R is a ring, a clear distinction must be made between the R-homomorphisms of R (considered as an R-module) and its homomorphisms as a ring. In the former case the homomorphism T must satisfy the condition $(ax)T = a(xT)$ for any a and x in R, whereas in the latter case T satisfies the condition $(ax)T = (aT)(xT)$. For example, with every element c of R we can associate an R-endomorphism T_c, defined by $xT_c = cx$ for $x \in R$. Clearly $(x + y)T_c = xT_c + yT_c$, and $(ax)T_c = c(ax) = a(cx) = a(xT_c)$, so that T_c is indeed an R-homomorphism. But T_c is not in general a ring homomorphism, since this would require that $(ax)T_c = (aT_c)(xT_c) = (ca)(cx)$. Thus we would have $cax = c^2ax$ for all a and x in R; if, for example, R has no zero divisors, then this is possible only if $c = 0$ or 1.

It may be noted that if R has an identity, then *every* R-endomorphism T of R is of the form T_c, where $c = 1T$, for if x is any element of R, then $xT = (x1)T = x(1T) = cx = xT_c$.

Since the kernel of an R-homomorphism of an R-module is an R-submodule of M, it is natural to ask whether, conversely, every R-submodule of an R-module M is the kernel of some R-homomorphism of M. We shall now show that this question is to be answered in the affirmative. To see this, let N be any R-submodule of M. Since M is an abelian group, every subgroup of M is a normal subgroup and gives rise to a factor group. Let $\bar{M}$ be the factor group of M with respect to N. Since we use the additive notation for the group operation in M we shall denote the factor group $\bar{M}$ also by $M - N$ (instead of by M/N; see I, § 11). We denote by T the canonical homomorphism of M onto $\bar{M}$.

To make an R-module of $\bar{M}$ we specify that the product of an element $a \in R$ and an element $x + N$ of $\bar{M}$ shall be $ax + N$.* To show that this product is unambiguous, we must show that if $x + N = y + N$ (so

* This is not in general the same as the set $a(x + N)$ according to the meaning this notation acquires from § 2.

that $x - y \in N$), then $ax + N = ay + N$. But this follows from the fact $ax - ay = a(x - y)$ is in N, which in turn follows from the fact that N is a *submodule* (and not merely a subgroup) of M. It is then clear that T is an R-homomorphism. We thus have:

THEOREM 2. *If N is a submodule of an R-module M, then the group of cosets $x + N$ can be made into an R-module $\bar{M}$ which is an R-homomorphic image of M with kernel N.*

The R-module $\bar{M}$ which has just been constructed for any R-submodule N of M and which is denoted by $M - N$ * is called the *difference module* of M with respect to N. The difference module $M - N$ is also sometimes called *the factor module of M by N* and is denoted by M/N. The mapping $x \to x + N$ is called the *canonical* or *natural homomorphism* of M onto $M - N$.

If $N = (0)$, then this natural homomorphism is an isomorphism.

THEOREM 3. *If M' is any R-homomorphism image of M, with kernel N, then the elements of M' are in* (1, 1) *correspondence with those of $M - N$, and the correspondence is an R-isomorphism.*

PROOF. Let T be the natural R-homomorphism of M onto $M - N$ and let S be the given R-homomorphism of M onto M'. Consider the transformation $s = S^{-1}T$ of M' onto $M - N$. If x', y' are any two elements of M' and if $\bar{x} \in x's$, $\bar{y} \in y's$, then $\bar{x} = xT$ and $\bar{y} = yT$, where x and y are suitable elements in $x'S^{-1}$ and $y'S^{-1}$ respectively. Since $\bar{x} + \bar{y} = (x + y)T$ and $x' + y' = (x + y)S$, it follows that $\bar{x} + \bar{y} \in (x' + y')s$. Furthermore, if $0'$ is the zero of M', then $0's$ consists only of the zero of $M - N$ since S and T have the same kernel. It follows from Lemma 2 of I, § 11, that s is a homomorphism. By the same token also s^{-1} is a mapping (since in the above argument M' and $M - N$ can be interchanged). Hence s is an isomorphism. To show that s is an R-isomorphism, let x' be any element of M' and let $a \in R$. If x is any element in $x'S^{-1}$, then $x' = xS$ and $\bar{x} = x's = xT = x + N$. Since $ax' = (ax)S$ and $a\bar{x} = ax + N = (ax)T$, it follows that $a(x's) = (ax')s$, and hence s is an R-isomorphism.

We shall use the notation $M \sim M'$ to indicate that M is R-homomorphic to M' and the notation $M \cong M'$ to indicate that M and M' are R-isomorphic.

§ 4. The isomorphism theorems.

The two theorems of this section are often called the *Dedekind-Noether isomorphism theorems.*

* Although this notation conflicts with the notation $A - B$ for the set of differences introduced in § 2, it will always be clear from the context whether we are referring to one or the other meaning of the notation.

THEOREM 4. *Let T be an R-homomorphism of a module M onto a module M' with kernel N. Then there is a* (1, 1), *inclusion preserving correspondence between submodules L' of M' and submodules L of M containing N, such that if L and L' correspond then $LT = L'$, $L'T^{-1} = L$. When L and L' correspond, then T induces an R-homomorphism of L onto L', the modules $L - N$ and L' are isomorphic, and so are the modules $M - L$ and $M' - L'$.*

PROOF. If L is a submodule of M containing N, then $L' = LT$ is an R-submodule of M' by Theorem 1. That distinct L's give rise to distinct L''s follows from the fact that $(LT)T^{-1} = L$. To prove this formula we first note that $L \subset (LT)T^{-1}$ trivially; on the other hand, if $x \in (LT)T^{-1}$, then $xT \in LT$, so $xT = yT$ with $y \in L$, hence $x - y \in N \subset L$, so that $x \in L$, as required. It remains to prove that every submodule L' of M' actually arises in this way; this follows from the facts that $L'T^{-1}$ is a submodule of M (Theorem 1), that $N \subset L'T^{-1}$ (obvious), and that $(L'T^{-1})T = L'$ (since T is an *onto* mapping). Thus the first statement of the conclusion is proved. The isomorphism $L - N \cong L'$ follows from Theorem 3 since T induces an R-homomorphism of L onto L', with kernel N. To prove $M - L \cong M' - L'$, we observe that T is given as an R-homomorphism of M onto M', and that $x' \to x' + L'$ is an R-homomorphism of M' onto $M' - L'$ (the natural homomorphism). The product of these two is an R-homomorphism of M onto $M' - L'$. If x is in the kernel of this product homomorphism then $xT + L' = L'$, so that $xT \in L'$, $x \in L'T^{-1} = L$; and *vice versa*, L is contained in this kernel. Hence the isomorphism follows from Theorem 3.

COROLLARY. *Let N and L be submodules of an R-module M with $N \subset L$. Then $L - N$ is a submodule of $M - N$, and*

$$(M - N) - (L - N) \cong M - L.$$

This follows from the theorem with $M' = M - N$.

THEOREM 5. *If N and L are submodules of an R-module M, then*

$$(L + N) - N \cong L - (L \cap N). \tag{1}$$

PROOF. If T is the natural homomorphism of $L + N$ onto $(L + N) - N$, then T induces an R-homomorphism of L into $(L + N) - N$ (even though L may not contain N). We assert that T maps L *onto* $(L + N) - N$. For if $x + N$ is any element of $(L + N) - N$, with $x \in L + N$, then $x = y + z$, where $y \in L$, $z \in N$, and $x + N = y + N = yT \in LT$. Hence T induces an R-homomorphism of L onto $(L + N) - N$. Since the kernel is obviously $L \cap N$, the conclusion follows from Theorem 3.

Any element of the left side of (1) is of the form $x + N$, $x \in L + N$; any element of the right side is of the form $y + (L \cap N)$, $y \in L$. The two elements correspond in the R-isomorphism (1) if and only if $x - y \in N$.

§ 5. Ring homomorphisms and residue class rings. It was observed at the beginning of this chapter that in a homomorphism of one ring on another the kernel is an ideal. It will now be proved that any ideal in a ring is the kernel of some homomorphism of the ring. Since we are now interested in R *as a ring* and not as a module, our present considerations are not special cases of those of § 3, although some of these results can be applied.

We first make clear the implications of the preceding sections for a quite arbitrary commutative (and additively written) group M. We may then regard M as a J-module as described in § 1. In that case *every* subgroup of M is obviously a J-submodule, and *every* homomorphism of M into another group (which is similarly regarded as a J-module) is necessarily a J-homomorphism. It follows that the results of §§ 1–4 apply to arbitrary commutative groups, their subgroups, and their homomorphisms. More precisely, Theorems 1–5 are valid if in place of *module, submodule, R-homomorphism,* we write *commutative group, subgroup, homomorphism* respectively.

Now let T be a homomorphism of a ring R onto a ring R', and let N be the kernel, so that N is an ideal. The mapping T is in particular a homomorphism of the additive group of R onto the additive group of R'. From § 3 it then follows that the elements of R' are in one to one correspondence with the cosets $a + N$, $a \in R$.

The notion of congruence is convenient in the present connection. If N is an arbitrary ideal in a ring R, and a, b are elements of R, then a and b are said to be *congruent modulo* N if $a - b \in N$, and we express this by the notation

$$a \equiv b\ (N).$$

Obviously $a \equiv b\ (N)$ is another way of stating that a and b determine the same coset, that is, $a + N = b + N$. The relation of congruence is clearly reflexive, symmetric, and transitive. Moreover it is preserved under addition and multiplication. That is,

$$a \equiv b\ (N),\ c \equiv d\ (N)$$

imply

$$a + c \equiv b + d,\ ac \equiv bd\ (N).$$

For example, to prove the latter: $ac - bd = (a - b)c + b(c - d) \in N$.

It follows that if $a \equiv b\ (N)$ and if $f(X)$ is a polynomial over R, then $f(a) \equiv f(b)\ (N)$.

In number theory it is customary to write $a \equiv b \pmod{m}$ if the difference of the integers a and b is divisible by the integer m. This is equivalent to the statement $a \equiv b\ (Jm)$ in our notation, where J is the ring of integers.

If, then, T is a homomorphism of R onto R' with kernel N, we see that $aT = bT$ if and only if $a \equiv b\ (N)$.

The preceding discussion suggests, as in § 3, that if we are given an ideal N in a ring R and wish to construct a homomorphic image $\bar{R}$ with kernel N, then we should take $\bar{R}$ to be the set of cosets of N. Addition of cosets has been defined in § 3 and we know that $\bar{R}$ is a group. We define multiplication of cosets by the formula

$$(1) \qquad (a + N)(b + N) = ab + N.\star$$

We must show that this product is independent of the particular elements a and b of the cosets, that is, that if $a \equiv a_1\ (N)$ and $b \equiv b_1\ (N)$, then $ab \equiv a_1b_1\ (N)$. This has just been proved. Thus $\bar{R}$ is a set with two operations defined in it, and the mapping T, defined by

$$aT = a + N,\ a \in R,$$

is a mapping of R onto $\bar{R}$. It follows from (1) that $(ab)T = (aT)(bT)$; that $(a + b)T = aT + bT$ we know from § 3. From the lemma of I, § 12, we conclude that $\bar{R}$ is a ring, that T is a homomorphism of R onto $\bar{R}$ and N is the kernel. Hence we have:

THEOREM 6. *If N is an ideal in a ring R, then the cosets $a + N$ can be made into a ring $\bar{R}$ in such a way that the mapping $a \to a + N$ is a homomorphism of R onto $\bar{R}$ with kernel N. If R' is any homomorphic image of R with kernel N, then the elements of R' are in one to one correspondence with those of $\bar{R}$, and this correspondence is an isomorphism.*

The first statement has been proved. The proof of the second statement is similar to the proof of Theorem 3 (§ 3).

When N is an ideal in a ring the cosets $a + N$ are usually called *residue classes* and the ring $\bar{R}$ just constructed is called the *residue class ring* of R with respect to N. It is denoted by R/N. The homomorphism $a \to a + N$ of R onto R/N is called the *natural homomorphism* of R onto R/N.

THEOREM 7. *Let T be a homomorphism of a ring R onto a ring R', with*

$\star$ It must be carefully noted that $ab + N$ does *not* in general consist of sums of products of elements of $a + N$ and $b + N$, so that this product is *not* the same as the one referred to at the end of § 2. Take, for example, $R = J$, $a = b = 0$, $N = 2J$.

kernel N. Then there is a (1, 1) *inclusion preserving correspondence between the ideals* $\mathfrak{A}'$ *of* R' *and the ideals* $\mathfrak{A}$ *of R which contain N, such that if* $\mathfrak{A}$ *and* $\mathfrak{A}'$ *correspond, then*

$$\mathfrak{A}T = \mathfrak{A}', \; \mathfrak{A}'T^{-1} = \mathfrak{A}.$$

When $\mathfrak{A}$ *and* $\mathfrak{A}'$ *correspond, T induces a homomorphism of* $\mathfrak{A}$ *onto* $\mathfrak{A}'$, *and*

$$\mathfrak{A}/N \cong \mathfrak{A}', \qquad R/\mathfrak{A} \cong R'/\mathfrak{A}'.$$

PROOF. The proof parallels that of Theorem 4 (though it is not a logical consequence of it) and so will not be given in detail. Let us merely prove, as an example, that $\mathfrak{A}'T^{-1}$ is an ideal in R. If $a, b \in \mathfrak{A}'T^{-1}$, then $aT, bT \in \mathfrak{A}'$, hence $(a - b)T = aT - bT \in \mathfrak{A}'$, so that $a - b \in \mathfrak{A}'T^{-1}$. If $a \in \mathfrak{A}'T^{-1}$ and $c \in R$, then $aT \in \mathfrak{A}'$, $cT \in R'$, hence $(ca)T = (cT)(aT) \in \mathfrak{A}'$, so that $ca \in \mathfrak{A}'T^{-1}$. This shows that $\mathfrak{A}'T^{-1}$ is an ideal.

COROLLARY. *Let N and* $\mathfrak{A}$ *be ideals in a ring R with* $N \subset \mathfrak{A}$. *Then* $\mathfrak{A}/N$ *is an ideal in* R/N *and*

$$(R/N)/(\mathfrak{A}/N) \cong R/\mathfrak{A}.$$

By analogy with Theorem 5 (§ 4) we have the following

THEOREM 8. *If L is a subring of a ring R, and N is an ideal in R, then the residue class ring* $L/L \cap N$ *is isomorphic with the subring* $(L + N)/N$ *of the residue class ring* R/N.

PROOF. It is sufficient to observe that the natural homomorphism of R onto R/N induces a homomorphism of L onto the subring $L' = (L + N)/N$ of R/N and that the kernel of this induced homomorphism is $L \cap N$. Hence $L' \cong L/L \cap N$.

§ 6. The order of a subset of a module

DEFINITION. Let M be a module over a ring R and let N be any subset of M. The set of all elements a of R such that $aN = (0)$ is called the *order* (or the *annihilator*) of N.

The order is clearly an ideal in R. The cases of interest are those in which N is either a submodule or a single element. The order of the zero of M is, of course, R itself.

If M is simply a commutative group, and we regard it as a J-module, then the order of any element x of M is an ideal $\mathfrak{A}$ in J. We know that $\mathfrak{A}$ (or *any* ideal in J) consists of all the multiples of a uniquely determined non-negative integer n (cf. II, § 4, where it was shown that this property belongs to the kernel of any homomorphism of J). If $n > 0$, then it is clearly the smallest positive integer m such that $mx = 0$, and x has order n in the sense of ordinary group theory. If $n = 0$, so that the order of x is (0), elementary group theory usually speaks of x as having infinite order.*

We return to the general case of a module M over a ring R. Let $\mathfrak{A}$ be any ideal contained in the order of M itself. An important procedure in the theory of modules is the construing of M as a module over the ring $\bar{R} = R/\mathfrak{A}$. If $\bar{c} \in \bar{R}$, then we must define $\bar{c}x$ for every $x \in M$. Now $\bar{c}$ arises as an image of some element c in R, and we define $\bar{c}x = cx$. This defines $\bar{c}x$ uniquely, since if $\bar{c}$ arises also from c_1 in R, then $c_1 - c \in \mathfrak{A}$, hence $c_1x - cx = (c_1 - c)x = 0$. (Since every $\bar{c}$ in $\bar{R}$ is a residue class $c + \mathfrak{A}$, our definition amounts to placing $\bar{c}x$ equal to $(c + \mathfrak{A})x$, the latter product being intended in the sense of § 3; $(c + \mathfrak{A})x$ consists of a single element of M since $\mathfrak{A}x = (0)$). That M thus becomes an $\bar{R}$-module is trivial to verify.

Any R-submodule of M will be an $\bar{R}$-submodule, and conversely. The study of M as an R-module is thus equivalent to its study as an $\bar{R}$-module. The latter study is often the easier since $\bar{R}$ will often have a simpler structure (in some sense) than R. We may say that the ring of operators R is "too large," that it is "more natural" to use $R/\mathfrak{A}$ as the ring of operators, where $\mathfrak{A}$ is the actual order of M. This special case—where $\mathfrak{A}$ is the order of M and not merely an ideal contained in the order—is the one occurring most often in practice. In this case, the order of M regarded as an $\bar{R}$-module is clearly (0).

These considerations allow us to clarify the distinction between the residue class ring $R/\mathfrak{A}$ and the difference module $R - \mathfrak{A}$, where R is a ring and $\mathfrak{A}$ is an ideal in R. If we regard R as an R-module, then $\mathfrak{A}$ is a submodule, and we are thus enabled to define a new R-module $R - \mathfrak{A}$, as described in § 3. Now the elements of $R - \mathfrak{A}$ are, just like the elements of $R/\mathfrak{A}$, the cosets of $\mathfrak{A}$; that is, $R/\mathfrak{A}$ and $R - \mathfrak{A}$ are identical as sets. Moreover, there is an addition in each and this addition is the same for both. Hence $R/\mathfrak{A}$ and $R - \mathfrak{A}$ are identical as (additive) groups.

Since $R/\mathfrak{A}$ is a ring, multiplication is defined between its elements. Indeed, if $\bar{c}$ and $\bar{x}$ are elements of $R/\mathfrak{A}$ then $\bar{c}\bar{x}$ is the coset of $cx + \mathfrak{A}$, where c and x are elements of the cosets $\bar{c}$ and $\bar{x}$ respectively. $R/\mathfrak{A}$, like any ring, may be regarded as a module over itself.

Now the order of the R-module $R - \mathfrak{A}$ surely contains $\mathfrak{A}$ (it may well be larger; for instance, if R is the ring of even integers and $\mathfrak{A}$ is the set

* *Warning*: If M is an ordinary finite group, say, of order n, then the order of M in the sense just defined need not be Jn; for instance, the order of the additive group of any Galois field of characteristic p is a power of p (see II, § 8), but the order of this group (or its annihilator) in J is always the principal ideal Jp.

of integers which are divisible by 4, then the order of $R - \mathfrak{A}$ is the whole ring R.) Hence we may regard $R - \mathfrak{A}$ as an $R/\mathfrak{A}$-module. This means that given an element $\bar{c}$ of $R/\mathfrak{A}$ and an element $\bar{x}$ of $R - \mathfrak{A}$ there is defined a product—we denote it for the moment by $\bar{c}\cdot\bar{x}$. According to the definition given earlier in this section, $\bar{c}\cdot\bar{x} = c\bar{x}$, where c is an element of the coset $\bar{c}$. According to the definition of § 3 (p. 139), the product $c\bar{x}$ (which is meaningful since $R - \mathfrak{A}$ is an R-module) is the coset $cx + \mathfrak{A}$, where x is in the coset $\bar{x}$. Thus we have proved that $\bar{c}\cdot\bar{x} = \bar{c}\bar{x}$.

This means that not only are $R - \mathfrak{A}$ and $R/\mathfrak{A}$ identical as sets and as additive groups, but they are also identical as $R/\mathfrak{A}$-modules. We may also express this fact by saying that the identity mapping of the $R/\mathfrak{A}$-module $R - \mathfrak{A}$ onto the $R/\mathfrak{A}$-module $R/\mathfrak{A}$ is an $R/\mathfrak{A}$-homomorphism.

For this reason it is usually unnecessary to distinguish between the two, although sometimes in delicate arguments one must keep in mind that these two modules are, strictly speaking, not the same.

§ **7. Operations on ideals.** Let R be a ring. We proceed to define five fundamental operations on ideals in R. Since R is also an R-module, the definitions regarding the sum of subsets of a module (§ 2) apply also to subsets of R. Thus if A and B are subsets of R, then $A \pm B$ consists of all $a \pm b$, $a \in A$, $b \in B$.

The sum $\mathfrak{A} + \mathfrak{B}$ of two ideals $\mathfrak{A}$ and $\mathfrak{B}$ is likewise an ideal. Thus the set of ideals is closed under addition, an operation which is commutative and associative.

The second operation, as with submodules, is intersection, or common part. If $\mathfrak{A}$ and $\mathfrak{B}$ are ideals, so is $\mathfrak{A} \cap \mathfrak{B}$. This operation, too, is commutative and associative. As with submodules we have here the *modular law* connecting sum and intersection:

(1) If $\mathfrak{A} \supset \mathfrak{B}$, then $\mathfrak{A} \cap (\mathfrak{B} + \mathfrak{C}) = \mathfrak{B} + (\mathfrak{A} \cap \mathfrak{C})$,

where $\mathfrak{A}$, $\mathfrak{B}$, $\mathfrak{C}$ are any ideals in R.

If A and B are non-empty subsets of R, then a definition of the product AB is implied by the definition of § 2. AB consists, namely, of all sums $\sum_1^n a_ib_i$, where n is an arbitrary positive integer, and $a_i \in A$, $b_i \in B$. This operation is commutative and associative. If A_1, A_2, $\cdots$, A_r are subsets of R, then $A_1A_2 \cdots A_r$ (sometimes denoted by $\prod_1^r A_i$) consists of all sums of products $a_1a_2 \cdots a_r$, where $a_i \in A_i$, $i = 1, 2, \cdots, r$. In particular A^r consists of all sums of products $\prod_1^r a_i$, where $a_i \in A$. If either A or B is an ideal, so is AB. Hence *the set of ideals is closed under the operation of multiplication.* This operation is related to sum by the easily proved distributive law:

$$\mathfrak{A}(\mathfrak{B} + \mathfrak{C}) = \mathfrak{A}\mathfrak{B} + \mathfrak{A}\mathfrak{C}, \tag{2}$$

where $\mathfrak{A}$, $\mathfrak{B}$, $\mathfrak{C}$ are any ideals of R. Moreover, for any two ideals $\mathfrak{A}$, $\mathfrak{B}$,

$$\mathfrak{A}\mathfrak{B} \subset \mathfrak{A} \cap \mathfrak{B}. \tag{3}$$

The fourth ideal-theoretic operation is the quotient.

DEFINITION 1. *If $\mathfrak{A}$ and $\mathfrak{B}$ are ideals of R, then the* QUOTIENT $\mathfrak{A} : \mathfrak{B}$ *consists of all elements c of R such that $c\mathfrak{B} \subset \mathfrak{A}$.*

It is immediate that $\mathfrak{A} : \mathfrak{B}$ is an ideal and that it contains $\mathfrak{A}$. In particular, $\mathfrak{A} : R \supset \mathfrak{A}$; but if R has an identity then $\mathfrak{A} : R = \mathfrak{A}$. If $\mathfrak{C}$ is any ideal such that $\mathfrak{C}\mathfrak{B} \subset \mathfrak{A}$, then $\mathfrak{C} \subset \mathfrak{A} : \mathfrak{B}$, and conversely. If $\mathfrak{A} \supset \mathfrak{B}$, then $\mathfrak{A} : \mathfrak{B} = R$, and conversely if R has an identity. The following relationships between quotient and the first three operations are easily checked:

$$\left(\bigcap_{i=1}^{n} \mathfrak{A}_i\right) : \mathfrak{B} = \bigcap_{i=1}^{n} (\mathfrak{A}_i : \mathfrak{B}), \tag{4}$$

$$\mathfrak{A} : \sum_{i=1}^{n} \mathfrak{B}_i = \bigcap_{i=1}^{n} (\mathfrak{A} : \mathfrak{B}_i), \tag{5}$$

$$\mathfrak{A} : \mathfrak{B}\mathfrak{C} = (\mathfrak{A} : \mathfrak{B}) : \mathfrak{C}. \tag{6}$$

For example, to prove (6), let $\mathfrak{D}$ and $\mathfrak{E}$ denote the left- and right-hand sides respectively of (6). Then by definition $\mathfrak{D}(\mathfrak{B}\mathfrak{C}) \subset \mathfrak{A}$, $(\mathfrak{D}\mathfrak{C})\mathfrak{B} \subset \mathfrak{A}$, hence $\mathfrak{D}\mathfrak{C} \subset \mathfrak{A} : \mathfrak{B}$, $\mathfrak{D} \subset (\mathfrak{A} : \mathfrak{B}) : \mathfrak{C} = \mathfrak{E}$. Likewise $\mathfrak{E}\mathfrak{C} \subset \mathfrak{A} : \mathfrak{B}$, $(\mathfrak{E}\mathfrak{C})\mathfrak{B} \subset \mathfrak{A}$, $\mathfrak{E} \subset \mathfrak{A} : \mathfrak{B}\mathfrak{C} = \mathfrak{D}$.

The four operations above were all binary; the last one, which we shall now introduce, is *unary*, that is, only one ideal is involved in this operation.

DEFINITION 2. *If $\mathfrak{A}$ is an ideal in R, the* RADICAL *of $\mathfrak{A}$, denoted by $\sqrt{\mathfrak{A}}$, consists of all elements b of R some power of which belongs to $\mathfrak{A}$.*

THEOREM 9. *The radical of $\mathfrak{A}$ is an ideal containing $\mathfrak{A}$. It satisfies the following rules:*

(7) If $\mathfrak{A}^k \subset \mathfrak{B}$, *for some positive integer k, then* $\sqrt{\mathfrak{A}} \subset \sqrt{\mathfrak{B}}$;

(8) $\sqrt{\mathfrak{A}\mathfrak{B}} = \sqrt{\mathfrak{A} \cap \mathfrak{B}} = \sqrt{\mathfrak{A}} \cap \sqrt{\mathfrak{B}}$;

(9) $\sqrt{\mathfrak{A} + \mathfrak{B}} = \sqrt{\sqrt{\mathfrak{A}} + \sqrt{\mathfrak{B}}}$;

(10) $\sqrt{\sqrt{\mathfrak{A}}} = \sqrt{\mathfrak{A}}$.

PROOF. That $\mathfrak{A} \subset \sqrt{\mathfrak{A}}$ is obvious. To show that $\sqrt{\mathfrak{A}}$ is an ideal, let $b, c \in \sqrt{\mathfrak{A}}$, so that $b^m \in \mathfrak{A}$, $c^n \in \mathfrak{A}$ for some integers m, n. In the binomial expansion of $(b - c)^{m+n-1}$ every term has a factor $b^i c^j$, with $i + j = m + n - 1$. Since either $i \geqq m$ or $j \geqq n$, either b^i or c^j is in $\mathfrak{A}$, hence $(b - c)^{m+n-1} \in \mathfrak{A}$, and $b - c \in \sqrt{\mathfrak{A}}$. If $b \in \sqrt{\mathfrak{A}}$, and $d \in R$, then $b^m \in \mathfrak{A}$ for some m; hence $(db)^m \in \mathfrak{A}$ and $db \in \sqrt{\mathfrak{A}}$. Thus $\sqrt{\mathfrak{A}}$ is indeed an ideal.

Suppose $\mathfrak{A}^k \subset \mathfrak{B}$. If $c \in \sqrt{\mathfrak{A}}$, then $c^m \in \mathfrak{A}$ for some m, $c^{mk} \in \mathfrak{A}^k \subset \mathfrak{B}$, hence $c \in \sqrt{\mathfrak{B}}$.

Since $\mathfrak{A}\mathfrak{B} \subset \mathfrak{A} \cap \mathfrak{B}$, and $\mathfrak{A} \cap \mathfrak{B}$ is contained in $\mathfrak{A}$ and in $\mathfrak{B}$, we have by (7) (with $k = 1$) that $\sqrt{\mathfrak{A}\mathfrak{B}} \subset \sqrt{\mathfrak{A} \cap \mathfrak{B}} \subset \sqrt{\mathfrak{A}} \cap \sqrt{\mathfrak{B}}$. Now if $c \in \sqrt{\mathfrak{A}} \cap \sqrt{\mathfrak{B}}$, then there exist integers m and n such that $c^m \in \mathfrak{A}$, $c^n \in \mathfrak{B}$. Then $c^m c^n \in \mathfrak{A}\mathfrak{B}$, whence $c \in \sqrt{\mathfrak{A}\mathfrak{B}}$. Thus (8) is proved.

Equation (10) is obvious. Since $\mathfrak{A} + \mathfrak{B} \subset \sqrt{\mathfrak{A}} + \sqrt{\mathfrak{B}}$,

$$\sqrt{\mathfrak{A} + \mathfrak{B}} \subset \sqrt{\sqrt{\mathfrak{A}} + \sqrt{\mathfrak{B}}};$$

since

$$\sqrt{\mathfrak{A}} + \sqrt{\mathfrak{B}} \subset \sqrt{\mathfrak{A} + \mathfrak{B}},\ \sqrt{\sqrt{\mathfrak{A}} + \sqrt{\mathfrak{B}}} \subset \sqrt{\sqrt{\mathfrak{A} + \mathfrak{B}}} = \sqrt{\mathfrak{A} + \mathfrak{B}}.$$

This proves (9).

If $\mathfrak{A} = (0)$, then $\sqrt{\mathfrak{A}}$ consists of all nilpotent elements—that is, elements some power of which is 0. This ideal is sometimes called the *radical of the ring R*.

We now consider the effect of a homomorphic mapping on the five operations defined above. Suppose, then, that R and R' are rings, T a homomorphism of R onto R' with kernel N. If $\mathfrak{A}$ and $\mathfrak{B}$ are ideals in R, then

(11) $\mathfrak{A} \subset \mathfrak{B}$ implies $\mathfrak{A}T \subset \mathfrak{B}T$;

(12) $(\mathfrak{A} + \mathfrak{B})T = \mathfrak{A}T + \mathfrak{B}T$;

(13) $(\mathfrak{A}\mathfrak{B})T = (\mathfrak{A}T)(\mathfrak{B}T)$;

(14) $(\mathfrak{A} \cap \mathfrak{B})T \subset \mathfrak{A}T \cap \mathfrak{B}T$, with equality if $\mathfrak{A} \supset N$ or if $\mathfrak{B} \supset N$;

(15) $(\mathfrak{A} : \mathfrak{B})T \subset \mathfrak{A}T : \mathfrak{B}T$, with equality if $\mathfrak{A} \supset N$;

(16) $\sqrt{\mathfrak{A}}\, T \subset \sqrt{\mathfrak{A}T}$, with equality if $\mathfrak{A} \supset N$.

If $\mathfrak{A}'$ and $\mathfrak{B}'$ are ideals in R', then:

(17) $\mathfrak{A}' \subset \mathfrak{B}'$ implies $\mathfrak{A}'T^{-1} \subset \mathfrak{B}'T^{-1}$;

(18) $(\mathfrak{A}' + \mathfrak{B}')T^{-1} = \mathfrak{A}'T^{-1} + \mathfrak{B}'T^{-1}$;

(19) $(\mathfrak{A}'\mathfrak{B}')T^{-1} \supset (\mathfrak{A}'T^{-1})(\mathfrak{B}'T^{-1})$, with equality if the right side contains N;

(20) $(\mathfrak{A}' \cap \mathfrak{B}')T^{-1} = \mathfrak{A}'T^{-1} \cap \mathfrak{B}'T^{-1}$;

(21) $(\mathfrak{A}' : \mathfrak{B}')T^{-1} = \mathfrak{A}'T^{-1} : \mathfrak{B}'T^{-1}$;

(22) $\sqrt{\mathfrak{A}'}\,T^{-1} = \sqrt{\mathfrak{A}'T^{-1}}$.

Most of the statements (11)–(16) are trivial. As an example we prove (15). If $c' \in (\mathfrak{A} : \mathfrak{B})T$, then $c' = cT$, where $c \in \mathfrak{A} : \mathfrak{B}$; then $c\mathfrak{B} \subset \mathfrak{A}$, $(cT)(\mathfrak{B}T) \subset \mathfrak{A}T$, $cT \in \mathfrak{A}T : \mathfrak{B}T$. On the other hand, suppose $c' \in \mathfrak{A}T : \mathfrak{B}T$, whence $c'(\mathfrak{B}T) \subset \mathfrak{A}T$; since T maps R *onto* R', there is an element c in R such that $c' = cT$. Then $(c\mathfrak{B})T \subset \mathfrak{A}T$, $c\mathfrak{B} \subset (\mathfrak{A}T)T^{-1}$; if $\mathfrak{A} \supset N$, then $(\mathfrak{A}T)T^{-1} = \mathfrak{A}$, hence $c\mathfrak{B} \subset \mathfrak{A}$, $c \in \mathfrak{A} : \mathfrak{B}$, $c' = cT \in (\mathfrak{A} : \mathfrak{B})T$.

To prove (17)–(22), let $\mathfrak{A} = \mathfrak{A}'T^{-1}$, $\mathfrak{B} = \mathfrak{B}'T^{-1}$, so $\mathfrak{A}' = \mathfrak{A}T$, $\mathfrak{B}' = \mathfrak{B}T$, and $\mathfrak{A}$ and $\mathfrak{B}$ contain N. To prove (19), for example, we observe that $\mathfrak{A}'\mathfrak{B}' = (\mathfrak{A}\mathfrak{B})T$, $(\mathfrak{A}'\mathfrak{B}')T^{-1} = ((\mathfrak{A}\mathfrak{B})T)T^{-1} \supset \mathfrak{A}\mathfrak{B}$; and equality holds at this last point if $\mathfrak{A}\mathfrak{B} \supset N$. Again, for (21), $\mathfrak{A}' : \mathfrak{B}' = (\mathfrak{A} : \mathfrak{B})T$ by (15), since $\mathfrak{A} \supset N$. Thus $(\mathfrak{A}' : \mathfrak{B}')T^{-1} = ((\mathfrak{A} : \mathfrak{B})T)T^{-1} = \mathfrak{A} : \mathfrak{B}$, since $\mathfrak{A} : \mathfrak{B} \supset \mathfrak{A} \supset N$. The others are similarly proved.

§ **8. Prime and maximal ideals.** So far we have considered ideals in all generality. Now we consider two important special types of ideals.

DEFINITION. *Let $\mathfrak{A}$ be an ideal in a ring R. $\mathfrak{A}$ is said to be* PRIME *if whenever a product of two elements of R is in $\mathfrak{A}$, then at least one of the factors is in $\mathfrak{A}$. An ideal $\mathfrak{A}$ is said to be* MAXIMAL *if $\mathfrak{A} \neq R$ and if there is no ideal between $\mathfrak{A}$ and R.*

Thus $\mathfrak{A}$ is prime if "$b, c \in R$, $bc \in \mathfrak{A}$" implies $b \in \mathfrak{A}$ or $c \in \mathfrak{A}$. In particular, R itself is always prime; (0) is prime if and only if R has no zero divisors.

We illustrate this definition by some examples.

1) If J is the ring of integers and $n \in J$, $n > 1$, then the principal ideal (n) is prime if and only if n is a prime number. This follows from the fact that (1) if n is a prime number then "ab divisible by n" always implies that either a or b is divisible by n and (2) if n is not a prime number then, by definition, there exist integers a, b such that $ab = n$, $0 < a < n$, $0 < b < n$.

2) The foregoing reasoning can be repeated without change for any unique factorization domain R, showing that if $w \in R$ *then the principal ideal (w) is prime if and only if w is either a unit or irreducible.* (See I, § 14, Theorem 4.)

3) Let $R = F[x_1, x_2, \cdots, x_n]$ be a polynomial ring in n variables x_i over a field F. If $a_1, a_2, \cdots, a_n$ are given elements of F, then the elements $g(x_1, x_2, \cdots, x_n)$ of R such that $g(a_1, a_2, \cdots, a_n) = 0$ form a prime ideal $\mathfrak{p}$ in R. Since R is also a polynomial ring in the elements $x_1 - a_1$, $x_2 - a_2, \cdots, x_n - a_n$, every polynomial $f(x)$ in R can be written in the form $b + \sum_i f_i(x)(x_i - a_i)$, where $b \in F$ and the $f_i(x)$ are

elements of R. Then $f(a) = 0$, that is, $f(x) \in \mathfrak{p}$, if and only if $b = 0$, showing that *the n polynomials* $x_i - a_i$ *form a basis of* $\mathfrak{p}$. If, on the other hand, $f(x)$ is not in $\mathfrak{p}$, so that $b \neq 0$, every ideal in R which contains the ideal $\mathfrak{p}$ *and* the element $f(x)$ will contain b and hence also the element 1 ($= b \cdot b^{-1}$), and the ideal is therefore the entire ring. This shows that $\mathfrak{p}$ *is a maximal ideal.* It is clear that in the canonical mapping of R onto $R/\mathfrak{p}$ the field F is mapped onto the entire ring $R/\mathfrak{p}$ (since every polynomial $f(x)$ in R is congruent, mod $\mathfrak{p}$, to an element b of F) and that this homomorphism of F onto $R/\mathfrak{p}$ is an isomorphism (since its kernel $F \cap \mathfrak{p}$ contains only the zero). Thus $R/\mathfrak{p}$ *is a field, isomorphic with the field* F.

On the other hand, if $(a_1, a_2, \cdots, a_n)$ and $(b_1, b_2, \cdots, b_n)$ are two *distinct* ordered n-tuples of elements of F, then the elements $g(x_1, x_2, \cdots, x_n)$ of R such that $g(a_1, a_2, \cdots, a_n) = 0$ *and* $g(b_1, b_2, \cdots, b_n) = 0$ do not form a prime ideal [consider, for instance, the product $(x_i - a_i)(x_i - b_i)$, assuming that $a_i \neq b_i$].

4) In the ring of integers J every prime ideal Jp (p, a prime number) is maximal, for if $m \notin Jp$, then $(m, p) = 1$ and hence $am + bp = 1$ for suitable integers a and b. This shows that every ideal which contains Jp as a proper subset also contains 1 and hence is the whole ring J. This reasoning is applicable without any change to any Euclidean domain, in particular to the ring R of example 3), *provided* $n = 1$. On the other hand if $n > 1$, then the principal ideal Rx_1 is prime (since x_1 is an irreducible element and R is a unique factorization domain), but $Rx_1 < Rx_1 + Rx_2$, and the ideal $Rx_1 + Rx_2$ contains only polynomials without constant terms and hence is not the whole ring R. Hence already in the polynomial ring $F[x_1, x_2]$ of two variables not every prime ideal is maximal.

If $\mathfrak{A}$ is prime and a product of two ideals $\mathfrak{B}$ and $\mathfrak{C}$ is in $\mathfrak{A}$, then one of the factors is in $\mathfrak{A}$. For if neither is, then $\mathfrak{B}$ and $\mathfrak{C}$ contain respectively elements b and c not in $\mathfrak{A}$. Since $\mathfrak{A}$ is prime, $bc \notin \mathfrak{A}$, hence $\mathfrak{B}\mathfrak{C} \not\subset \mathfrak{A}$, contradiction. In particular if $\mathfrak{B}^n \subset \mathfrak{A}$ for some $n > 0$, then $\mathfrak{B} \subset \mathfrak{A}$. Finally if $\mathfrak{A}$ is prime, then $\mathfrak{A}$ is equal to its radical.

On the other hand if $\mathfrak{A}$ is *not* prime, then there exist ideals $\mathfrak{B}$, $\mathfrak{C}$ such that

$$\mathfrak{A} < \mathfrak{B},\ \mathfrak{A} < \mathfrak{C},\ \mathfrak{B}\mathfrak{C} \subset \mathfrak{A}.$$

For there exist elements b, c such that $b, c \notin \mathfrak{A}$, $bc \in \mathfrak{A}$. Then take $\mathfrak{B} = (b) + \mathfrak{A}$, $\mathfrak{C} = (c) + \mathfrak{A}$.

THEOREM 10. *Let* $\mathfrak{A}$ *be an ideal different from* R. *Then* $\mathfrak{A}$ *is prime if and only if* $R/\mathfrak{A}$ *has no zero divisors. If* R *has an identity,* $\mathfrak{A}$ *is maximal* *if and only if $R/\mathfrak{A}$ is a field.* (*Hence in a ring with identity any maximal ideal is prime.*)

PROOF. Let T be the natural homomorphism of R onto $R/\mathfrak{A}$. Then $bc \in \mathfrak{A}$ if and only if $(bT)(cT) = 0$. From this the criterion for $\mathfrak{A}$ to be prime follows immediately. From Theorem 7, § 5 we see that $\mathfrak{A}$ is maximal if and only if $R/\mathfrak{A}$ has no ideals but itself and (0). Since R has an identity, so has $R/\mathfrak{A}$. It was observed in § 1 that a ring with identity is a field if and only if it has just two ideals. This proves the second statement of the conclusion. The third follows from the first two.

In rings without an identity, maximal ideals need not be prime (see example in § 1, p. 133). Prime ideals need not be maximal, as the example above shows. Later on (IV, § 2) we shall study an important class of rings in which every prime ideal different from R is maximal. In chapter V we shall study a theoretically important class of integral domains in which every *proper* prime ideal is maximal.

THEOREM 11. *Let T be a homomorphism of a ring R onto a ring R' with kernel N. If $\mathfrak{A}$ is an ideal in R containing N, then $\mathfrak{A}$ is respectively prime or maximal if and only if $\mathfrak{A}T$ is prime or maximal. If $\mathfrak{A}'$ is an ideal in R', then $\mathfrak{A}'$ is respectively prime or maximal if and only if $\mathfrak{A}'T^{-1}$ is prime or maximal.*

PROOF. We may assume $\mathfrak{A} \neq R$ since $\mathfrak{A} = R$ if and only if $\mathfrak{A}T = R'$. Then the condition for $\mathfrak{A}$ (or $\mathfrak{A}T$) to be prime is that $R/\mathfrak{A}$ (or $R'/\mathfrak{A}T$) have no zero divisors. Since, by Theorem 7, $R/\mathfrak{A} \simeq R'/\mathfrak{A}T$, $\mathfrak{A}$ is prime if and only if $\mathfrak{A}T$ is prime.

Since the correspondence between the ideals of R' and the ideals of R containing N is inclusion preserving, if there are no ideals between R and $\mathfrak{A}$ there are none between R' and $\mathfrak{A}T$; and conversely.

Since $\mathfrak{A}'T^{-1}$ contains N and $(\mathfrak{A}'T^{-1})T = \mathfrak{A}'$, the second statement follows from the first.

NOTE I. If a ring R contains an identity 1, then the set $\mathcal{I}$ of all ideals of R which contain a given ideal different from R is non-empty and is inductive if partially ordered by set-theoretic inclusion (since 1 remains outside every ideal in $\mathcal{I}$ and since it is obvious that the set-theoretic union of all the ideals which belong to a totally ordered set of ideals is itself an ideal). Hence by Zorn's lemma, $\mathcal{I}$ contains maximal elements. We have thus proved that *in a ring with identity every ideal different from R is contained in a maximal ideal.*

NOTE II. Zorn's lemma is also needed in the proof of the following general result: *in any ring R the intersection of all the prime ideals of R is the radical of the zero ideal* (*that is, the set of all nilpotent elements*). Since every ideal contains 0, it is clear that every nilpotent element is contained in every prime ideal. Hence the main point that has to be proved is that if an element u of R is not nilpotent, then there exists a prime ideal not containing u. To prove this we consider the set $\mathcal{J}$ of all ideals $\mathfrak{A}$ in R which contain no power of u. Since u is not nilpotent, the zero ideal belongs to $\mathcal{J}$, and thus $\mathcal{J}$ is non-empty. It is obvious that $\mathcal{J}$ is inductive. Let, by Zorn's lemma, $\mathfrak{p}$ be a maximal element of $\mathcal{J}$. Then $u \notin \mathfrak{p}$. We claim that $\mathfrak{p}$ is a prime ideal. For, let x and y be elements of R which do not belong to $\mathfrak{p}$. Then $\mathfrak{p} + (x) > \mathfrak{p}$ and hence some power u^m belongs to $\mathfrak{p} + (x)$. Similarly some power u^n belongs to $\mathfrak{p} + (y)$. Then u^{m+n} belongs to $\mathfrak{p} + (xy)$, and since $u^{m+n} \notin \mathfrak{p}$ it follows that $xy \notin \mathfrak{p}$, showing that $\mathfrak{p}$ is a prime ideal.

§ 9. **Primary ideals.** The general concept of a prime ideal corresponds to the concept of a prime number of ordinary arithmetic. *Primary* ideals, which we shall presently introduce, correspond in a similar fashion to *powers of prime numbers.* If p is a prime number and m is a positive integer, then $n = p^m$ has the following property: if a product ab of two integers a, b is divisible by n and if a is not divisible by n, then some power of b is divisible by n. Conversely, any integer n with this property is necessarily the power of a prime number.

DEFINITION. *Let R be an arbitrary ring and let $\mathfrak{Q}$ be an ideal in R. Then $\mathfrak{Q}$ is said to be primary if the conditions $a, b \in R$, $ab \in \mathfrak{Q}$, $a \notin \mathfrak{Q}$ imply the existence of an integer m such that $b^m \in \mathfrak{Q}$.*

In the sense of this definition, an integer n is the power of a prime number if and only if the principal ideal Jn is primary.

Clearly every prime ideal is primary, with $m = 1$. A prime ideal $\mathfrak{P}$ was characterized in the preceding section by the condition that in the residue class ring $R/\mathfrak{P}$ the only zero divisor is zero. Similarly it is easily seen that an ideal $\mathfrak{Q}$ is primary if and only if every zero divisor of $R/\mathfrak{Q}$ is nilpotent.

THEOREM 12. *Let $\mathfrak{Q}$ be a primary ideal in a ring R. If $\mathfrak{P}$ is the radical of $\mathfrak{Q}$, then $\mathfrak{P}$ is a prime ideal. Moreover, if $ab \in \mathfrak{Q}$ and $a \notin \mathfrak{Q}$, then $b \in \mathfrak{P}$. Also if $\mathfrak{A}$ and $\mathfrak{B}$ are ideals such that $\mathfrak{A}\mathfrak{B} \subset \mathfrak{Q}$, $\mathfrak{A} \not\subset \mathfrak{Q}$, then $\mathfrak{B} \subset \mathfrak{P}$.*

PROOF. The second statement is obvious, and the third follows from it. To prove $\mathfrak{P}$ is prime, suppose $ab \in \mathfrak{P}$, $a \notin \mathfrak{P}$. Since ab is in the radical of $\mathfrak{Q}$, $a^m b^m = (ab)^m \in \mathfrak{Q}$ for some m. Since $a \notin \mathfrak{P}$, $a^m \notin \mathfrak{Q}$. Since $\mathfrak{Q}$ is primary $(b^m)^n \in \mathfrak{Q}$ for some n, hence $b \in \mathfrak{P}$.

If $\mathfrak{Q}$ is a primary ideal, then its radical $\mathfrak{P}$ is called the *associated prime ideal* of $\mathfrak{Q}$, and we say that $\mathfrak{Q}$ is a primary ideal *belonging to* the prime ideal $\mathfrak{P}$, or simply that $\mathfrak{Q}$ *is primary for* $\mathfrak{P}$. It may be that there exists an integer m such that $\mathfrak{P}^m \subset \mathfrak{Q}$. (This will *always* be the case if the ring is noetherian; cf. IV, § 1, Example 2, p. 200). In this case $\mathfrak{Q}$ is said to be *strongly primary* and the least m for which $\mathfrak{P}^m \subset \mathfrak{Q}$ is called the *exponent* of $\mathfrak{Q}$. A primary ideal has exponent 1 if and only if it is prime.

The following theorem is often useful for proving that a given ideal is primary and at the same time finding its radical.

THEOREM 13. *Let $\mathfrak{Q}$ and $\mathfrak{P}$ be ideals in a ring R. Then $\mathfrak{Q}$ is primary and $\mathfrak{P}$ is its radical if and only if the following conditions are satisfied:*

(a) $\mathfrak{Q} \subset \mathfrak{P}$.

(b) *If $b \in \mathfrak{P}$, then $b^m \in \mathfrak{Q}$ for some m, (m may depend on b).*

(c) *If $ab \in \mathfrak{Q}$ and $a \notin \mathfrak{Q}$, then $b \in \mathfrak{P}$.*

PROOF. Assume (a), (b), and (c). That $\mathfrak{Q}$ is primary follows from (c) and (b). From (b) we conclude that $\mathfrak{P} \subset \sqrt{\mathfrak{Q}}$. To show $\sqrt{\mathfrak{Q}} \subset \mathfrak{P}$, suppose $b \in \sqrt{\mathfrak{Q}}$, so that $b^m \in \mathfrak{Q}$; let m be the least exponent such that $b^m \in \mathfrak{Q}$. If $m = 1$, this gives $b \in \mathfrak{Q} \subset \mathfrak{P}$; and if $m > 1$, then $b^{m-1} \cdot b \in \mathfrak{Q}$ and $b^{m-1} \notin \mathfrak{Q}$, hence $b \in \mathfrak{P}$ by (c). The proof of the converse is immediate.

An equivalent form of condition (c) is the following:

If $ab \in \mathfrak{Q}$ and $b \notin \mathfrak{P}$, then $a \in \mathfrak{Q}$.

COROLLARY 1. *Let R be a ring with identity, and let $\mathfrak{Q}$ and $\mathfrak{P}$ be ideals in R such that:*

(a′) $\mathfrak{Q} \subset \mathfrak{P}$.

(b′) *If $b \in \mathfrak{P}$, then $b^m \in \mathfrak{Q}$ for some m.*

(c′) *$\mathfrak{P}$ is a maximal ideal.*

Then $\mathfrak{Q}$ is primary and $\mathfrak{P}$ is its radical.

We need only verify (c) in the hypothesis of the above theorem. Suppose, then, that $ab \in \mathfrak{Q}$, $b \notin \mathfrak{P}$. Now $\mathfrak{P} + Rb$ contains b since R has an identity. Hence $\mathfrak{P} + Rb$ contains $\mathfrak{P}$ properly, and since $\mathfrak{P}$ is maximal, $\mathfrak{P} + Rb = R$. Hence there exist elements c, d such that

$$1 = c + db, \quad c \in \mathfrak{P}, \; d \in R. \tag{1}$$

Now by (b′), $c^m \in \mathfrak{Q}$ for some m. Raising (1) to the m-th power by the binomial theorem we obtain

$$1 = c^m + d'b, \quad \text{where} \quad d' \in R.$$

Hence

$$a = ac^m + d'(ab) \in \mathfrak{Q}. \quad \text{Q.E.D.}$$

COROLLARY 2. *In a ring with identity, a maximal ideal $\mathfrak{p}$ is prime and its powers are primary for $\mathfrak{p}$.*

We now consider further examples. Let $R = F[x, y]$ be a

polynomial ring in two indeterminates over an arbitrary field F. As was mentioned in § 8 (Example 3, p. 150) the ideal $\mathfrak{P} = Rx + Ry$ is maximal. The ideal $\mathfrak{Q} = Rx + Ry^2$ is primary by Corollary 1, since $\mathfrak{P}^2 \subset \mathfrak{Q} \subset \mathfrak{P}$. (That $\mathfrak{P}^2 \subset \mathfrak{Q}$ follows from $\mathfrak{P}^2 = Rx^2 + Rxy + Ry^2$.) But $\mathfrak{Q}$ is not a power of $\mathfrak{P}$ or of *any* prime ideal, for that matter. For suppose $\mathfrak{Q} = \mathfrak{P}_1{}^k$, where $\mathfrak{P}_1$ is prime. Since $\mathfrak{P}_1{}^k = \mathfrak{Q} \subset \mathfrak{P}$, $\mathfrak{P}^2 \subset \mathfrak{Q} \subset \mathfrak{P}_1$, and since $\mathfrak{P}$ and $\mathfrak{P}_1$ are prime we have that necessarily $\mathfrak{P} = \mathfrak{P}_1$, so that $\mathfrak{Q} = \mathfrak{P}^k$. Since $y \in \mathfrak{P}$ and $y \notin \mathfrak{Q}$, $\mathfrak{P} \neq \mathfrak{Q}$, so that $k > 1$. Thus $\mathfrak{Q} \subset \mathfrak{P}^2$; but this also cannot be since $x \in \mathfrak{Q}$, $x \notin \mathfrak{P}^2$. Thus we have proved that *a primary ideal need* NOT *be a power of a prime ideal.*

It is also true that *powers of prime ideals need not be primary.* (Thus Corollary 2 is false if "maximal" is replaced by "prime".) An example showing this is the following. Let $F[X, Y, Z]$ be a polynomial ring in three indeterminates over a field F, and let R be the residue class ring $F[X, Y, Z]/(XY - Z^2)$. We denote by x, y, and z the residue classes of X, Y, and Z respectively. The ideal generated in our polynomial ring by X and Z is prime and contains the kernel $(XY - Z^2)$ of the canonical homomorphism of $F[X, Y, Z]$ onto R. Hence the corresponding ideal $\mathfrak{P} = Rx + Rz$ in R is also prime. We have $xy = z^2 \in \mathfrak{P}^2$, the element x is not in $\mathfrak{P}^2$ (since no polynomial of the form $X + A(X, Y, Z)X^2 + B(X, Y, Z)XZ + C(X, Y, Z)Z^2$ can be divisible by $XY - Z^2$) and no power of y is in $\mathfrak{P}^2$ (no power of y is even in the prime ideal $\mathfrak{P}$ since clearly y is not in $\mathfrak{P}$). Hence $\mathfrak{P}^2$ is not a primary ideal.

The above example shows also that an ideal $\mathfrak{A}$ whose radical is prime *need not be primary.* An example of this can also be found in a polynomial ring $F[X, Y]$ of two indeterminates. Let $\mathfrak{A}$ be the ideal generated by X^2 and XY. It is immediately seen that $\sqrt{\mathfrak{A}}$ is the prime (principal) ideal (X). But $XY \in \mathfrak{A}$, $X \notin \mathfrak{A}$ and no power of Y belongs to $\mathfrak{A}$, showing that $\mathfrak{A}$ is not primary.

Various operations on primary ideals lead to primary ideals, as summarized in the following theorem:

THEOREM 14. *The intersection of a finite number of primary ideals all belonging to the same prime ideal $\mathfrak{P}$ is again primary for $\mathfrak{P}$. If $\mathfrak{P}$ is maximal, the same is true for finite sums and products. If $\mathfrak{Q}$ is primary for $\mathfrak{P}$ and $\mathfrak{A}$ is an ideal not contained in $\mathfrak{Q}$, then $\mathfrak{Q} : \mathfrak{A}$ is primary for $\mathfrak{P}$. If T is a homomorphism of a ring R onto a ring R' with kernel N, then an ideal $\mathfrak{Q}$ containing N is primary in R if and only if $\mathfrak{Q}T$ is primary in R'; and when this is so the associated prime ideal of $\mathfrak{Q}T$ is $\mathfrak{P}T$, where $\mathfrak{P}$ is the associated prime ideal of $\mathfrak{Q}$.*

These statements are easily proved using Theorems 12 and 13. For example, to show that $\mathfrak{Q}:\mathfrak{A}$ is primary for $\mathfrak{P}$, we observe that $(\mathfrak{Q}:\mathfrak{A})\,\mathfrak{A} \subset \mathfrak{Q}$ and $\mathfrak{A} \not\subset \mathfrak{Q}$, hence $\mathfrak{Q}:\mathfrak{A} \subset \mathfrak{P}$. Since also $\mathfrak{Q} \subset \mathfrak{Q}:\mathfrak{A}$, (a) and (b) of Theorem 13 are verified. To verify (c), suppose $ab \in \mathfrak{Q}:\mathfrak{A}$, $b \notin \mathfrak{P}$; then we must show that $a \in \mathfrak{Q}:\mathfrak{A}$. Now we have $b(a\mathfrak{A}) \subset \mathfrak{Q}$, and since $b \notin \mathfrak{P}$, $a\mathfrak{A} \subset \mathfrak{Q}$, so $a \in \mathfrak{Q}:\mathfrak{A}$.

NOTE. If R is a UFD and π is an irreducible element of R, then $R\pi$ is a prime ideal and $R\pi^n$ ($n \geqq 1$) is a primary ideal, with radical $R\pi$ (if π^n divides a product ab and does not divide a, then π^n divides some power of b). Conversely, every primary ideal $\mathfrak{q}$ whose radical is $R\pi$ is of the form $R\pi^n$, $n \geqq 1$. For if n is an integer such that $\mathfrak{q} \subset R\pi^n$, $\mathfrak{q} \not\subset R\pi^{n+1}$, and x is an element of $\mathfrak{q}$ of the form $y\pi^n$, $y \notin R\pi$, then necessarily $\pi^n \in \mathfrak{q}$, and hence $\mathfrak{q} = R\pi^n$.

§ **10. Finiteness conditions.** The elementary theory of vector spaces concerns itself with spaces of finite dimension; in such spaces there does not exist an infinite strictly ascending or strictly descending chain of subspaces. Similarly, the elementary theory of groups concerns itself with finite groups, or at any rate with groups which are not "too infinite" in some sense. The purpose of this section is the discussion of various finiteness conditions which can be imposed on a module.

DEFINITION. *A module M over a ring R is said to satisfy the* ASCENDING CHAIN CONDITION *if every strictly ascending chain of submodules*

(1) $$M_1 < M_2 < \cdots$$

is finite.

An obviously equivalent formulation is this: *If*

$$M_1 \subset M_2 \subset \cdots$$

is an infinite ascending sequence of submodules, then there exists an integer n such that

$$M_i = M_n \quad \text{for} \quad i = n+1, n+2 \cdots.$$

By reversing the above inclusion signs and by replacing the word "ascending" by "descending" we can similarly define the *descending chain condition.* We use the abbreviations a.c.c. and d.c.c. respectively for the ascending and descending chain conditions.

Clearly, if a group is finite, then the group (regarded as a J-module) satisfies both chain conditions. The additive group of integers is an example of a group (a J-module) satisfying the ascending but not the descending chain conditions. On the other hand, consider the field F of rational numbers and the quotient ring J_p formed by the fractions a/b, where $a, b \in J$ and b is not divisible by a given prime number p. It is easily proved (as in I, § 20) that every proper additive subgroup of F which contains J_p is of the form $p^{-n}J_p$. Thus the difference group $F - J_p$ satisfies the d.c.c. but not the a.c.c.

A module is said to satisfy the *maximum condition* if every non-empty collection C of submodules has a *maximal* element—that is, if there exists a submodule in C which is not contained in any other submodule in the collection C. The *minimum condition* is similarly defined.

THEOREM 15. *A module M satisfies the ascending (descending) chain condition if and only if it satisfies the maximum (minimum) condition.*

PROOF. If M does not satisfy the a.c.c., then there exists an *infinite* strictly ascending sequence $\{M_i\}$ of submodules, and the collection of all the M_i clearly has no maximal element. On the other hand, suppose M does satisfy the a.c.c. and let C be any non-empty collection of submodules. Since C is not empty there exists a submodule M_1 in C. If M_1 is not maximal in C, there exists an M_2 in C which contains M_1 properly. If M_2 is not maximal in C, there is an M_3 in C properly containing M_2, etc. Since M satisfies the ascending chain condition this process must stop, and thus a maximal element of C is reached.

The equivalence of the descending chain condition with the minimum condition is similarly proved.

In view of the equivalence of the chain conditions with the maximum and minimum conditions the two will be used interchangeably, depending on which is more convenient in a given context.

The following theorem is basic for determining how the chain conditions are affected by certain operations.

THEOREM 16. *Let M be a module and N a submodule. Then the ascending (descending) chain condition holds in M if and only if it holds in both N and $M - N$.*

PROOF. If the a.c.c. holds in M, then it obviously holds in N, and because of the correspondence between the submodules of $M - N$ and those of M containing N, it holds likewise in $M - N$.

Conversely, let us suppose both N and $M - N$ satisfy the a.c.c. To prove that M does also, we first note that if L and L' are two submodules of M such that

$$L \subset L', \quad L + N = L' + N, \quad L \cap N = L' \cap N,$$

then $L = L'$. Namely, $L' = L' \cap (L' + N) = L' \cap (L + N) = L + (L' \cap N)$ (by the modular law (5), § 2, p. 137) $= L + (L \cap N) = L$. Suppose now that $\{L_i\}$ is an ascending sequence of submodules of M. In order to show that this sequence remains ultimately constant, it is sufficient to show, in view of the remark just made, that each of the ascending sequences $\{L_i + N\}$, $\{L_i \cap N\}$ remains ultimately constant. For the latter sequence this follows from the a.c.c. in N. For the former, it follows from the a.c.c. in $M - N$ in view of the correspondence between submodules of $M - N$ and those of M containing N.

Thus the theorem is proved for the a.c.c. The d.c.c. is treated similarly.

COROLLARY. *Let $M_1, M_2, \cdots, M_n$ be submodules of a module M such that $M = M_1 + M_2 + \cdots + M_n$. If each M_i satisfies the ascending (descending) chain condition, then so does M.*

By induction it is sufficient to consider the case $n = 2$. In view of the theorem it is then enough to show that $M - M_1$ satisfies the chain condition in question. That it does, follows from

$$M - M_1 = (M_1 + M_2) - M_1 \cong M_2 - (M_1 \cap M_2),$$

since this last module satisfies the chain condition, by the theorem.

Let R be a ring and M an R-module. A *basis* of M is a set $\{x_\alpha\}$ of elements of M such that no proper submodule of M contains all the x_α. The module M is said to have a *finite basis*, or to be a *finite R-module*, if it has a basis consisting of a finite number of elements. It is said to be *cyclic* if it has a basis consisting of a single element.

If $\{x_\alpha\}$ is any set of elements of M, then the smallest submodule of M which contains all the x_α consists of those elements of M which can be written in the form of a finite sum

$$a_1 x_{\alpha_1} + a_2 x_{\alpha_2} + \cdots + a_n x_{\alpha_n} + m_1 x_{\alpha_1} + m_2 x_{\alpha_2} + \cdots + m_n x_{\alpha_n},$$

where the a_i are in R and the m_i are integers. This module shall be denoted by the symbol $(\{x_\alpha\})$, or by $(x_1, x_2, \cdots, x_s)$ if $\{x_\alpha\}$ is a finite set $\{x_1, x_2, \cdots, x_s\}$. If $\{x_\alpha\}$ is a basis of M then every element of M is a finite sum of the above indicated form. In particular, if M is cyclic, with basis $\{x\}$, then every element of M is of the form $ax + mx$, where $a \in R$ and m is an integer.

If R has an element 1 and M is a unitary R-module, with basis $\{x_\alpha\}$, then the integral multiples $m_\alpha x_\alpha$ can be omitted from the expression of the elements of M in terms of the x_α, since any integral multiple mx of an element x of M is itself of the form bx, $b \in R$, namely, $mx = (m \cdot 1)x$. Hence in this case, M is the sum of the modules Rx_α. In particular, if M is cyclic, with basis $\{x\}$, then we can write $M = Rx$. This shows that an abelian group as regarded a J-module in the way described in § 1 (p. 135) is a cyclic module if and only if it is a cyclic group in the usual sense. If a ring R *with identity* is regarded as a module over itself, then its cyclic submodules are its principal ideals (§ 1, p. 132).

THEOREM 17. *Let M be a module over a ring R. Then M satisfies the ascending chain condition if and only if every submodule of M has a finite basis.*

PROOF. Suppose every submodule of M has a finite basis. If $\{N_i\}$ is an ascending sequence of submodules of M, then the union N (in the set-theoretic sense) of all N_i is clearly a submodule of M. By hypothesis, N has a finite basis, say $N = (x_1, x_2, \cdots, x_h)$. Since each x_j is in N, it is in some N_i, hence there is an integer n such that $x_j \in N_n$, $j = 1, \cdots, h$. Thus $N \subset N_n$, so that $N_i = N_n$ for $i > n$. Thus the a.c.c. is proved.

Conversely, let us assume the a.c.c. If N is an arbitrary (but fixed) submodule of M, then in the collection of all submodules of N having finite basis (such exist, for example, (0)) let N' be a maximal element (Theorem 15). If x is any element of N, then $N' + (x)$ has a finite basis, since N' does. By the maximality of N', $N' + (x) = N'$, so that $x \in N'$. Thus $N = N'$, and N has a finite basis, as required.

So far the properties of the ring R have played no essential role. We now prove a theorem which relates the chain conditions in an R-module M to the chain conditions in R. Since a ring R may be regarded as an R-module, the chain conditions in R have meaning; they say simply that a strictly ascending (or descending) chain of ideals in R must be finite.

THEOREM 18. *Let R be a ring with identity, and let M be a unitary module over R having a finite basis. Then if R satisfies the ascending* (*or descending*) *chain condition, so does M.*

PROOF. If $\{x_1, \cdots, x_n\}$ is a finite basis for M, then $M = Rx_1 + \cdots + Rx_n$. To prove the theorem it is sufficient, by the corollary to Theorem 16, to consider the special case where M is a cyclic module Rx. In this case, suppose $\{N_i\}$ is an ascending chain of submodules of M. For each i, let $\mathfrak{A}_i$ be the set of all elements a of R such that $ax \in N_i$. Then $\mathfrak{A}_i$ is easily seen to be an ideal in R, and $N_i = \mathfrak{A}_i x$ since every element of M (and hence of N_i in particular) is of the form ax, $a \in R$. Moreover, it is clear the sequence of ideals $\{\mathfrak{A}_i\}$ is ascending. Since the a.c.c. is assumed in R, there is an n such that $\mathfrak{A}_i = \mathfrak{A}_n$ for $i > n$. Since $N_i = \mathfrak{A}_i x$, $N_i = N_n$ for $i > n$, and the a.c.c. is proved in M. The proof for the d.c.c. is similar.

§ **11. Composition series.** In the preceding paragraph we have considered conditions which make every increasing or decreasing sequence of modules to be finite. In the present section we consider more precisely how many modules can occur in such sequences.

Let M be an R-module. A *normal series* in M is a descending (but not necessarily strictly descending) finite chain of submodules

$$(1) \qquad M = M_0 \supset M_1 \supset M_2 \supset \cdots \supset M_r = (0),$$

beginning with M and ending with (0); the integer r is called the *length* of the normal series.

Note again that the inclusions in (1) need not be proper—that is, we may have $M_{i-1} = M_i$ for some or all i. If, however, all the inclusions are proper, that is, if we have

$$(2) \qquad M = M_0 > M_1 > M_2 > \cdots > M_r = (0),$$

then the normal series in question is said to be *without repetitions*.

A *refinement* of the normal series (1) is a normal series obtained by inserting additional terms in the series (1). In particular, if no additional terms are inserted, we speak of an *improper* refinement.

DEFINITION 1. *A* COMPOSITION SERIES *of M is a normal series without repetitions for which every proper refinement has repetitions.*

In order that a normal series (2) without repetitions be a composition series it is clearly necessary and sufficient that there exist no R-submodules between M_{i-1} and M_i, $i = 1, \cdots, r$. In other words, in view of Theorem 4, § 4, it is necessary and sufficient that each difference module $M_{i-1} - M_i$ ($i = 1, 2, \cdots, r$) be *simple*, where an R-module is said to be *simple* (or *irreducible*) if it has exactly two submodules. These must necessarily be itself and (0); the module (0) is *not* simple according to this definition. A simple module can be described as one having a composition series of length one.

Not every module has a composition series—for example, the additive group of integers.

The following theorem on composition series is fundamental:

THEOREM 19 (JORDAN). *If an R-module M has one composition series of length r, then every composition series of M has length r, and every normal series without repetitions can be refined to a composition series.*

PROOF. The theorem is trivial for $r = 1$. Hence we proceed by induction, assuming the theorem true for modules having a composition series of length less than r. For our module M we have a composition series

$$(3) \qquad M = M_0 > M_1 > M_2 > \cdots > M_r = (0).$$

By the induction hypothesis M can have no composition series of length less than r. The first statement of the theorem is proved, therefore, if we can show that *every normal series*

$$(4) \qquad M = N_0 > N_1 > N_2 > \cdots > N_s = (0)$$

without repetitions has length at most r. This will also prove the second statement, since if (4) is not already a composition series, we may insert an additional submodule without repeating any N_j; this process must lead to a composition series in exactly $r - s$ steps, if the above assertion is correct.

To prove this assertion we must show that $s \leqq r$. We note that (3) shows that M_1 has a composition series of length $r - 1$. If $N_1 = M_1$, then from (4) we get a normal series for M_1 without repetitions and of length $s - 1$; by induction hypothesis $s - 1 \leqq r - 1$, $s \leqq r$. If $N_1 < M_1$, then (4) yields a normal series for M_1 without repetitions and of length s; again by induction hypothesis we have $s \leqq r - 1$, and *a fortiori* $s \leqq r$.

We may thus confine ourselves to the case where N_1 is not contained in M_1 at all. Since there are no submodules between M_1 and M, it follows that $M_1 + N_1 = M$. Now by Theorem 5, § 4,

$$M - M_1 = (M_1 + N_1) - M_1 \cong N_1 - (M_1 \cap N_1).$$

Since $M - M_1$ is simple, so is $N_1 - (M_1 \cap N_1)$, hence there are no submodules between N_1 and $M_1 \cap N_1$. Consider the diagram

$$\begin{array}{ccccc} & & M_1 & & \\ & \nearrow & & \searrow & \\ M = M_1 + N_1 & & & & M_1 \cap N_1 > \cdots > (0). \\ & \searrow & & \nearrow & \\ & & N_1 & & \end{array}$$

Since M_1 has a composition series of length $r - 1$, and $M_1 \cap N_1 < M_1$, every normal series without repetitions of $M_1 \cap N_1$ has length at most $r - 2$, and hence $M_1 \cap N_1$ has a composition series of at most this length. Since there are no submodules between N_1 and $M_1 \cap N_1$, N_1 has a composition series of length at most $r - 1$. By induction hypothesis, $s - 1 \leqq r - 1$, $s \leqq r$. This completes the proof.

We thus see that if an R-module M has a composition series at all, then all of its composition series have the same length. This common length will be called the *length* of M and will be denoted by $l(M)$. Thus a simple module is of length one, and the module (0) is of length zero. If M has no composition series we set $l(M) = \infty$; in that case there exist normal series without repetition of arbitrarily great length. We can then state:

THEOREM 20. *If N is a submodule of the R-module M, then*

$$l(M) = l(N) + l(M - N).$$

(This is to be interpreted as meaning, in particular, that if either side is infinite, so is the other.)

PROOF. Let

$$N = N_0 > N_1 > \cdots > N_t = (0) \tag{5}$$

be any normal series of N without repetitions. Since by Theorem 4, § 4 and its corollary every submodule of $M - N$ is of the form $L - N$, where L is a submodule of M containing N, it follows that any normal series of $M - N$ without repetitions has the form

$$M - N = L_0 - N > L_1 - N > \cdots > L_s - N = (0), \qquad L_{j-1} > L_j, \quad j = 1, \cdots, s. \tag{6}$$

We thus obtain for M a normal series

$$M = L_0 > L_1 > \cdots > L_s > N_1 > \cdots > N_t = (0) \tag{7}$$

without repetitions and of length $s + t$. Hence if either $l(N)$ or $l(M - N)$ is infinite, then either t or s can be made arbitrarily large, hence $l(M) = \infty$. On the other hand if they are both finite, then we may assume (5) and (6) to be composition series. It then follows that (7) is a composition series of M, whence the theorem.

COROLLARY. *If L and N are submodules of M, then*

$$l(L) + l(N) = l(L + N) + l(L \cap N). \tag{8}$$

We make use of the relation

$$(L + N) - N \cong L - (L \cap N) \tag{9}$$

and of the evident fact that R-isomorphic modules have the same length. If either $l(L)$ or $l(N)$ is infinite, so is $l(L + N)$, and (8) is trivial. If both are finite, then the right side of (9) has finite length, hence so does the left, hence so does $L + N$, by the theorem. Equation (8) now follows from the theorem.

We have, so far, spoken of composition series and have observed that a module may not have one. Certainly any finite (commutative) group, considered as a J-module, has a composition series. More generally:

THEOREM 21. *A necessary and sufficient condition that a module M have a composition series is that it satisfy both chain conditions.*

PROOF. If M has a composition series of length r, then clearly every strictly ascending or descending chain has, at most, $r + 1$ elements. Conversely, suppose M satisfies both chain conditions. Let $M_0 = M$. If $M_0 \neq (0)$, let M_1 be maximal in the collection of submodules properly contained in M_0; if $M_1 \neq (0)$, let M_2 be similarly defined, etc. We thus get a strictly descending chain

$$M = M_0 > M_1 > M_2 > \cdots$$

such that no additional submodule can be inserted between two successive members of the chain. Since this chain cannot be infinite, we must have $M_r = (0)$ for some r. Thus we have a composition series.

Composition series will often play a role in what follows. For most of the applications the three preceding theorems are sufficient, but occasionally the stronger result contained in the *Jordan-Hölder Theorem* is needed. First we introduce the following terminology: if

$$M = M_0 \supset M_1 \supset \cdots \supset M_r = (0) \tag{10}$$

is a normal series of M, then the difference modules $M_{i-1} - M_i$ $(i = 1, 2, \cdots, r)$ are called the *normal differences* of the series. If (10) is a composition series the difference modules are called *composition differences*. Two normal series are said to be *equivalent* if the differences of one can be paired with the differences of the other so that paired differences are R-isomorphic.

Equivalent normal series have the same length, and this relation of equivalence is transitive.

THEOREM 22 (HÖLDER). *If a module has a composition series, then any two composition series are equivalent.*

PROOF. By Theorem 19 we know that any two composition series have the same length. Hence let them be

$$M = M_0 > M_1 > \cdots > M_r = (0) \tag{11}$$

$$M = N_0 > N_1 > \cdots > N_r = (0) \tag{12}$$

The proof will proceed by induction on the length of M. Since the theorem is trivial for length 1, we assume it true for all modules of length less than r. If in the above series $M_1 = N_1$, then we have two composition series for M_1, and by the induction hypothesis they are equivalent. Since $M - M_1 = M - N_1$, so are the given series for M.

Assume, then, that $M_1 \neq N_1$, so that $M = M_1 + N_1$. By taking a fixed composition series for $M_1 \cap N_1$ we obtain two composition series for M:

$$M = M_1 + N_1 > M_1 > M_1 \cap N_1 > \cdots > (0). \tag{13}$$

$$M = M_1 + N_1 > N_1 > M_1 \cap N_1 > \cdots > (0). \tag{14}$$

That these are actually composition series follows from the R-isomorphisms

$$M - M_1 \cong N_1 - (M_1 \cap N_1), \quad M - N_1 \cong M_1 - (M_1 \cap N_1),$$

and from the fact that $M - M_1$ and $M - N_1$ are simple. From these same isomorphisms it follows that the composition series (13) and (14) are equivalent. Since (11) and (13) have the member M_1 in common it follows from the preceding paragraph that they are equivalent. In like manner (12) and (14) are equivalent, whence (11) and (12) are also.

According to this theorem, then, the composition differences of an R-module are uniquely determined up to R-isomorphism.

COROLLARY. *If M has finite length and N is a submodule of M, then the composition differences of M are those of N and those of $M - N$ taken together.*

Let us assume that (5) and (6) above are composition series for N and $M - N$, whence (7) is a composition series for M. The differences for M are $N_{j-1} - N_j$ $(j = 1, \cdots, t)$ and $L_{i-1} - L_i$ $(i = 1, \cdots, s)$. Since $L_{i-1} - L_i \cong (L_{i-1} - N) - (L_i - N)$, and the latter are the differences for $M - N$, the corollary follows.

In the theory of rings the following extension of the composition series concept is often useful:

DEFINITION 2. *Let N be a submodule of an R-module M. A* NORMAL SERIES BETWEEN M AND N *is a chain*

$$M = M_0 \supset M_1 \supset \cdots \supset M_r = N.$$

It is said to be a COMPOSITION SERIES BETWEEN M AND N *if there are no repetitions and if there is no submodule between M_{i-1} and M_i, $i = 1, 2, \cdots, r$.*

Obviously a normal (or composition) series between M and N leads to a similar series of $M - N$, and conversely. Hence if there exists a composition series between M and N, all such composition series have the same length and any normal series between M and N which has no repetitions can be refined to a composition series between M and N.

§ 12. Direct sums. In this section we consider decompositions of a module into simpler components.

DEFINITION 1. *Let R be a ring, M an R-module, and $M_1, M_2, \cdots, M_r$ submodules of M. The submodules $M_1, \cdots, M_r$ are said to be* INDEPENDENT *if*

$$M_i \cap (M_1 + \cdots + M_{i-1} + M_{i+1} + \cdots + M_r) = (0), \quad i = 1, 2, \cdots, r.$$

It is immediate that this condition is equivalent to the statement that if

$$x_1 + \cdots + x_r = 0, \ x_i \in M_i,$$

then $x_i = 0$, $i = 1, 2, \cdots, r$. This criterion is often easier to apply than the definition.

DEFINITION 2. *The R-module M is said to be the* DIRECT SUM *of the* *submodules* $M_1, \cdots, M_r$ *if it is the sum of these submodules and if these submodules are independent.* *We then write*

$$M = M_1 \oplus \cdots \oplus M_r.$$

It is easily checked that M is the direct sum of $M_1, \cdots, M_r$ if and only if each x in M can be represented *uniquely* in the form

$$x = x_1 + \cdots + x_r,\ x_i \in M_i.$$

It is obvious that if $M_1, \cdots, M_r$ are any submodules of M, then $M_1 + \cdots + M_r$ is the *direct* sum of the M_i if and only if the M_i are independent. Hence we often use the expression "the sum $M_1 + \cdots + M_r$ is direct" to mean that the M_i are independent.

If $M = M_1 \oplus \cdots \oplus M_r$ and if each M_i is itself a direct sum:

$$M_i = M_{i1} \oplus M_{i2} \oplus \cdots,$$

then it is easily proved, by the criterion following the definition of independence, that M is the direct sum of all the M_{ij} taken together. Conversely, if M is the direct sum of certain submodules M_{ij} and if we define

$$M_i = M_{i1} + M_{i2} + \cdots,$$

then this sum is itself direct, and M is the direct sum of the M_i. The modular law (see (5), § 2) holds for direct sums also:

(1) If $K \supset L$, then $K \cap (L \oplus N) = L \oplus (K \cap N)$.

This is to be interpreted to mean that *if the sum on the left side is direct, so is the one on the right and vice versa.*

The statement that M is the direct sum of two submodules M_1 and M_2 is clearly equivalent to the two statements

$$M = M_1 + M_2,\quad M_1 \cap M_2 = (0).$$

From this and Theorem 5, § 4 it follows that:

If $M = M_1 \oplus M_2$, *then* M_2 *is R-isomorphic to* $M - M_1$.

From the corollary to Theorem 20, § 11, we get that $l(M) = l(M_1) + l(M_2)$, and by induction we obtain:

$$l(M_1 \oplus \cdots \oplus M_r) = l(M_1) + \cdots + l(M_r).$$

In particular if each M_i has finite length, so does their direct sum.

Direct sums are of importance, since a module is determined to within an R-isomorphism by its direct summands, as proved in the corollary to the following theorem.

THEOREM 23. *Let M and N be R-modules. For* $i = 1, \cdots, r$, *let* M_i

and N_i *be submodules of* M *and* N *respectively, and let* T_i *be an* R-*homomorphism of* M_i *into* N_i. *Finally, assume that*

$$M = M_1 \oplus \cdots \oplus M_r, \tag{2}$$

$$N = N_1 + \cdots + N_r. \tag{3}$$

Then there exists one and only one R-*homomorphism* T *of* M *into* N *which coincides with* T_i *on* M_i. *If each* T_i *is onto, so is* T. *If each* T_i *is an isomorphism and if the sum* (3) *is direct, then* T *is an isomorphism.*

PROOF. If $x \in M$, then we can write

$$x = x_1 + \cdots + x_r, \quad x_i \in M_i. \tag{4}$$

Hence if the required T exists at all we must have

$$xT = x_1T_1 + \cdots + x_rT_r, \tag{5}$$

and so T is unique. To prove T exists we define it by (5); since the representation (4) is unique, xT as defined in (5) is uniquely determined. It is easily checked that T is an R-homomorphism. If each T_i is onto, so is T, since $MT \supset M_iT = M_iT_i = N_i$ and hence $MT = N$. Now suppose each T_i is an isomorphism and (3) is direct. If, then, $xT = 0$, it follows that $\sum x_iT_i = 0$. Since (3) is direct, $x_iT_i = 0$, hence each $x_i = 0$, so $x = 0$. Thus T is an isomorphism.

COROLLARY. *If* $M = M_1 \oplus \cdots \oplus M_r$, $N = N_1 \oplus \cdots \oplus N_r$, *and if* M_i *is* R-*isomorphic to* N_i $(i = 1, \cdots, r)$, *then* M *and* N *are* R-*isomorphic.*

Despite this corollary, the structure of M cannot, in general, be concluded directly from the properties of the M_i. For example, the submodules of M cannot necessarily be determined merely because we know the submodules of the M_i. Our ignorance of the submodules is only very slightly mitigated by

THEOREM 24. *Let* $M = M_1 \oplus \cdots \oplus M_r$, *let* N_i *be a submodule of* M_i $(i = 1, \cdots, r)$, *and let*

$$N = N_1 + \cdots + N_r.$$

Then this sum is direct, and $M - N$ *is a direct sum of submodules* R-*isomorphic to the difference modules* $M_i - N_i$.

PROOF. That this sum is direct—that is, that the N_i are independent—is obvious. Let T be the natural homomorphism of M onto $M - N$. Then clearly

$$M - N = M_1T + \cdots + M_rT.$$

To show this sum is direct, suppose $0 = x_1T + \cdots + x_rT$, where $x_i \in M_i$. Since $\left(\sum_1^r x_i\right)T = 0$, $\sum_1^r x_i \in N = N_1 + \cdots + N_r$, hence

$x_i \in N_i \subset N$, so $x_i T = 0$, as was to be proved. It remains to show that $M_i T \cong M_i - N_i$. This follows from the fact that T, acting on M_i, has N_i as kernel, since $M_i \cap N = N_i$.

The following theorem, which is useful in the theory of rings, relates the direct sum concept to what we might term "direct intersection."

THEOREM 25. *Suppose that the R-module M is the direct sum of submodules* $M_1, \cdots, M_r$, *so that*

$$M = M_1 + \cdots + M_r, \tag{6}$$

$$M_i \cap (M_1 + \cdots + M_{i-1} + M_{i+1} + \cdots + M_r) = (0), \quad i = 1, \cdots, r. \tag{7}$$

If we place

$$N_i = M_1 + \cdots + M_{i-1} + M_{i+1} + \cdots + M_r, \; i = 1, \cdots, r, \tag{8}$$

then

$$(0) = N_1 \cap \cdots \cap N_r, \tag{9}$$

$$N_i + (N_1 \cap \cdots \cap N_{i-1} \cap N_{i+1} \cap \cdots \cap N_r) = M, \; i = 1, \cdots, r, \tag{10}$$

$$M_i = N_1 \cap \cdots \cap N_{i-1} \cap N_{i+1} \cap \cdots \cap N_r, \; i = 1, \cdots, r. \tag{11}$$

Conversely, if we are given submodules $N_1, \cdots, N_r$ *of M satisfying* (9) *and* (10), *and if we define* M_i *by* (11), *then* (6), (7), (8) *hold.*

(Note that (9), (10), (11) are dual to (6), (7), (8) in the sense of being obtained from them by interchanging sum and intersection, M and (0), M_i and N_i.)

PROOF. We make the preliminary observation that we have immediately

$$M_i + N_i = M, \; M_i \cap N_i = (0). \tag{12}$$

This is true whether we are given the M_i and then define the N_i in terms of them, or *vice versa*.

Suppose first that we are given $M = M_1 \oplus \cdots \oplus M_r$, and define N_i by (8). Then (11) can be proved by repeated application of the modular law, but it is easier to proceed by direct computation. Suppose, then, that $x \in \bigcap_{j \neq i} N_j$, and write $x = x_1 + \cdots + x_r$, $x_k \in M_k$. Since $x \in N_j$ $(j \neq i)$, $x_j = 0$ [by (8)]. Hence $x = x_i \in M_i$. That $M_i \subset \bigcap_{j \neq i} N_j$ is obvious. Thus (11) is proved. As for (9), we have

$$N_1 \cap \cdots \cap N_r = N_1 \cap (N_2 \cap \cdots \cap N_r) = N_1 \cap M_1 = (0), \quad \text{by (7)}.$$

Equation (10) follows from (12).

Now suppose we are given the N_i satisfying (9) and (10), and define M_i by (11). Since $M_i + N_i = M$, we may write, for any x in M:

$$x = x_i + y_i, \; x_i \in M_i, \; y_i \in N_i, \; i = 1, \cdots, r.$$

Then for any j between 1 and r,

$$x - \sum_{i=1}^{r} x_i = (x - x_j) - \sum_{i \neq j} x_i \in N_j,$$

since $x - x_j = y_j \in N_j$, and $x_i \in M_i \subset N_j$ for $i \neq j$. Thus $x - \sum x_i \in \bigcap N_j = (0)$, so $x = \sum x_i$, and $M = M_1 + \cdots + M_r$—that is, (6) holds. That (7) holds (directness of the sum) follows from

$$M_i \cap \sum_{j \neq i} M_j \subset M_i \cap N_i = (0).$$

So it remains only to prove (8). That $N_i \supset \sum_{j \neq i} M_j$ is obvious; we have indeed just used this fact. From the modular law we conclude

$$N_i = N_i \cap (\sum_{j \neq i} M_j + M_i) = \sum_{j \neq i} M_j + (N_i \cap M_i) = \sum_{j \neq i} M_j.$$

This completes the proof of the theorem.

DEFINITION 3. *If M is an R-module and N a submodule, a* COMPLEMENT OF *N is a submodule N' of M such that*

$$N \oplus N' = M.$$

If every submodule of M has a complement, M is said to be COMPLETELY REDUCIBLE.

The submodules M and (0) have (0) and M respectively as unique complements. In general, however, complements (when they exist) need not be unique. This can be seen from the situation in vector spaces (which we shall presently study in detail), where every subspace has a complement (see I, § 21) (so that they are completely reducible), and where it is well known that the complements are never unique except for M and (0). Although they are not unique, the complements of N are all R-isomorphic, since each is R-isomorphic to $M - N$. Moreover, if one complement of N contains another, they are equal. For suppose N' and N'' are complements of N, $N' \supset N''$; then

$$N' = N' \cap (N'' + N) = N'' + (N' \cap N) = N'' + (0) = N''.$$

As just observed, vector spaces are completely reducible. An example of a module which is not completely reducible is the additive group of integers. Here there exist proper subgroups and the intersection of any two is also proper, so no sum can be direct.

THEOREM 26. *If M is completely reducible, so is every submodule. If L and N are submodules such that $L \subset N$, then every complement of N is contained in a complement of L, and every complement of L contains a complement of N.*

PROOF. To show that the submodule N is completely reducible we must find a complement of L in N. Now L has a complement L' in M,

$$L \oplus L' = M.$$

Then

$$N = N \cap (L \oplus L') = L \oplus (N \cap L'),$$

so that L has $N \cap L'$ as complement in N.

Let N' be any complement of N; if L'' is a complement of L in N, then $N' + L''$ is a complement of L (in M). On the other hand, let L' be any complement of L; if N' is a complement of $N \cap L'$ in L':

$$(N \cap L') \oplus N' = L', \tag{13}$$

then N' is a complement of N (in M). For, by (13), $N + N'$ contains L'; of course it contains L, so $N + N' \supset L + L' = M$. But also

$$N \cap N' = (N \cap L') \cap N' = (0),$$

by (13). So $M = N \oplus N'$.

COROLLARY. *If a completely reducible module satisfies either chain condition, then it satisfies the other, and hence has finite length.*

For a strictly ascending chain of submodules would lead to a strictly descending chain of their complements, and *vice versa.*

THEOREM 27. *A necessary and sufficient condition that an R-module M be completely reducible and of finite length is that it be the sum of a finite number of simple submodules. When this is so, then M is, in fact, a* DIRECT *sum of simple submodules, the direct summands are uniquely determined up to R-isomorphism, and their number is $l(M)$.*

PROOF. We regard (0) as the direct sum of the empty collection of submodules.

Suppose first that M is completely reducible and of finite length, so that M satisfies both chain conditions. We say that *every* submodule of M is a direct sum of simple submodules. For if not, then in the set of those which are not, let N be one which is minimal. Now $N \neq (0)$, and also N cannot itself be simple. So N contains a submodule N' such that $(0) < N' < N$. Since M is completely reducible, so is N, hence there exists a submodule N'' such that

$$N' \oplus N'' = N. \tag{14}$$

Since $(0) < N', N'' < N$. Since N' and N'' are proper submodules of N, the minimal property of N implies that N' and N'' are both direct sums of a finite number of simple submodules. Then (14) implies that also N is such a direct sum, whence a contradiction. Hence every submodule of M is a direct sum of simple ones, as claimed.

Now suppose

(15) $$M = M_1 + \cdots + M_r,$$

where each M_i is simple. We first show that M is completely reducible. Let N be any proper submodule of M, and let M_{i_1} be the first of the modules $M_1, \cdots, M_r$ which is not contained in N. Since M_{i_1} is simple, $N \cap M_{i_1} = (0)$, so the sum $N + M_{i_1}$ is direct. If $N \oplus M_{i_1} = M$, then M_{i_1} is a complement of N; otherwise let M_{i_2} be the first M_i not contained in $N \oplus M_{i_1}$. Then, as above, the sum $(N \oplus M_{i_1}) + M_{i_2}$ is direct. Continuing in this way we obtain integers $i_1, \cdots, i_s$ such that

$$N \oplus M_{i_1} \oplus M_{i_2} \oplus \cdots \oplus M_{i_s} = M.$$

Thus N has a complement, and hence M is completely reducible. Furthermore, we have shown that N has a complement which is a direct sum of certain of the M_i involved in (15). In particular (0) has such a complement, so that

$$M = N_1 \oplus N_2 \oplus \cdots \oplus N_t,$$

where the N_j are certain of the M_i. Thus M is a *direct* sum of simple submodules, and $l(M) = t$ since each $l(N_j)$ is 1.

It remains to show that the N_j are uniquely determined up to an R-isomorphism. We assert that

$$M = N_1 \oplus N_2 \oplus \cdots \oplus N_t > N_2 \oplus \cdots \oplus N_t > \cdots > N_t > (0)$$

is a composition series. For it is certainly a normal series, and the j-th normal difference is

$$(N_j \oplus N_{j+1} \oplus \cdots \oplus N_t) - (N_{j+1} \oplus \cdots \oplus N_t),$$

which is R-isomorphic to N_j. Since each N_j is simple, the above normal series is indeed a composition series. Moreover, it has been shown that the N_j are isomorphic to the composition differences, which by the Hölder Theorem (Theorem 22, § 11) are uniquely determined up to R-isomorphisms.

We now give a decomposition theorem for modules which need not be completely reducible.

DEFINITION. *An R-module is said to be* INDECOMPOSABLE *if it is not the direct sum of two proper submodules.*

For example, the additive group of integers is indecomposable. A module $\neq (0)$ which is both indecomposable and completely reducible is clearly simple.

THEOREM 28. *An R-module M satisfying the descending chain condition is a direct sum of a finite number of indecomposable submodules.*

PROOF. We prove that every submodule is such a direct sum. For if not, let N be minimal in the set of all submodules not sums of this type. Then $N \neq (0)$, and N cannot be indecomposable, $N = N' \oplus N''$. The proof is completed as in the first half of the proof of Theorem 27.

By means of the direct sum concept we can not only decompose modules into simpler ones but also can build up big modules from little ones.

THEOREM 29. *Let $M'_1, \cdots, M'_r$ be modules over a ring R. Then there exists a module M which is the direct sum of submodules $M_1, \cdots, M_r$ such that M_i is R-isomorphic to M'_i. Moreover, M is uniquely determined up to R-isomorphism.*

PROOF. The uniqueness follows from Theorem 23, Corollary. To prove existence, define M to consist of all ordered n-tuples

$$x = (x_1, x_2, \cdots, x_r), \ x_i \in M'_i.$$

If $y = (y_1, y_2, \cdots, y_r)$ is another member of M and if $a \in R$, we define

$$x + y = (x_1 + y_1, x_2 + y_2, \cdots, x_r + y_r)$$
$$ax = (ax_1, ax_2, \cdots, ax_r).$$

Thus M clearly becomes an R-module. We define M_i to consist of all $(x_1, \cdots, x_n)$ such that $x_j = 0$ for $j \neq i$. It is obvious that

$$M = M_1 \oplus \cdots \oplus M_r$$

and that

$$x_i \rightarrow (0, \cdots, 0, x_i, 0, \cdots, 0), \ x_i \in M'_i,$$

is an R isomorphism of M'_i onto M_i.

On the basis of the results of this section we can develop very quickly the elementary properties of vector spaces. In § 1 we have observed that a vector space M over a field F is a unitary F-module. The submodules of M are then its *subspaces*. If N is subspace of M and $N \neq (0)$, then the order of N (as defined in § 6) is (0); an equivalent statement is that if $ax = 0$ (where $a \in F$ and $x \in M$), then $a = 0$ or $x = 0$. If N is a simple vector space—that is, if N has no proper subspaces—then for any $x \in N$, $x \neq 0$, it must be true that $Fx = N$, and conversely if $x \neq 0$ is in a vector space, then Fx is a simple subspace.

Let $x_1, \cdots, x_r$ be elements of M. We recall that these elements are said to be *linearly independent* over F if a relation

$$a_1x_1 + \cdots + a_rx_r = 0, \ a_i \in F,$$

implies $a_1 = \cdots = a_r = 0$; and that they are said to form a (finite) *basis* of M if they are linearly independent and if every element of M is of the form $\sum_1^r a_i x_i$, $a_i \in F$.* Equivalent formulations of these definitions in our terminology are as follows: The elements $x_1, \cdots, x_r$ are linearly independent if and only if each x_i is $\neq 0$ and the subspaces $Fx_1, \cdots, Fx_n$ are independent (in the sense of Definition 1, given in the beginning of this section); they form a finite basis for M if and only if each x_i is $\neq 0$ and

$$M = Fx_1 \oplus \cdots \oplus Fx_r.$$

As we have observed above, each Fx_i is simple, hence of length 1, so that if $x_1, \cdots, x_r$ form a basis of M, then $l(M) = r$. Thus the number of basis elements is always the same. In the usual theory of vector spaces, this number is called the *dimension* of the vector space; we have thus proved that it is the same as the length.

It follows from what we have said and from Theorem 27 that a vector space with a finite basis satisfies both chain conditions and is completely reducible. Consider now the following four properties of vector spaces:

(a) Existence of finite (vector) basis.
(b) Finite length.
(c) Ascending chain condition.
(d) Descending chain condition.

We assert that they are all equivalent. For we have just proved that (a) implies (b), and, of course, (b) implies (c) and (d). To show that (c) implies (a) we observe that (c) implies at any rate that M has a finite *module basis* over F. Since every principal module Fx is, in the present case, a simple module, we have that M is a finite sum of simple modules. Hence, by Theorem 27, M is a finite *direct* sum of simple modules, that is, (a) is satisfied. We prove now that (d) implies (c), that is, that the d.c.c. implies the a.c.c. If, namely, the a.c.c. is not satisfied, there clearly exists an infinite sequence of vectors

$$x_1, x_2, \cdots$$

in M such that every finite subset is linearly independent. If we define M_i to consist of all finite linear combinations

$$a_i x_i + a_{i+1} x_{i+1} + \cdots + a_j x_j,\ a_k \in F,\ j \geqq i,$$

j otherwise arbitrary, then clearly

$$M_1 > M_2 > M_3 > \cdots$$

* It should be carefully noted that, while any basis of a vector space M is also a module basis of M over F, the converse is not true, because of the additional condition of linear independence which we have imposed on the elements of a *vector basis*.

violates the d.c.c. Thus we have proved the equivalence of (a) to (d), so that in a vector space either chain condition implies the other and hence also finite-dimensionality and complete reducibility.

In § 3 we defined a linear transformation of one vector space into another as an F-homomorphism of the one into the other. If x and y are vectors in two vector spaces and $x \neq 0$, then the mapping

$$ax \rightarrow ay, \ (a \in F)$$

is clearly a linear transformation of Fx onto Fy. From this remark and from Theorem 23, it follows that if $x_1, \cdots, x_n$ constitute a basis for a space M and if $y_1, \cdots, y_n$ are elements of a space L, then there is one and only one linear transformation of M into L such that $x_iT = y_i$, $i = 1, \cdots, n$.

We shall see various other examples where properties of vector spaces can be deduced from theorems on modules.

§ **12**[bis]**. Infinite direct sums.** Let A be an arbitrary (finite or infinite) set of elements and let φ be a mapping of A into a set whose elements are groups. For any element a of A we shall denote by G_a the group $\varphi(a)$. We shall say then that we have a set of groups $\{G_a\}$ which *is indexed by the set A*. We do not assume that φ is univalent; it therefore may very well happen for two distinct indices a and b that $G_a = G_b$.

The *set product* of the set of groups $\{G_a\}$ indexed by A shall be by definition the set of all functions f on A such that for any element a of A the value $f(a)$ of f is an element x_a of G_a. We shall identify any such function f with the "vector" $x = \{x_a\}$, where a varies in A, and we shall call x_a the *component of* x in G_a (the term "vector" is used here in a sense which is more general than the one in which that term was used in I, § 21). The set product of the G_a is therefore to be thought of as the set of all vectors $\{x_a\}$.

The group structure of the G_a allows us to define multiplication in the set product of the G_a as follows: if $x = \{x_a\}$ and $y = \{y_a\}$, then $xy = \{x_a y_a\}$ $(x_a, y_a \in G_a)$. It is then immediately seen that the set product of the G_a becomes a group. This group will be denoted by $\tilde{\prod}_{a \in A} G_a$ and will be called the *complete direct product* of the groups G_a. The identity of the complete direct product is the vector $\{e_a\}$, where e_a is the identity of G_a, and the inverse of any element $\{x_a\}$ is the element $\{x_a^{-1}\}$.

The following assertions are straightforward and their proofs may be left to the reader:

(1) If $\{H_a\}$ is a set of groups indexed by the set A, such that for each a the group H_a is a subgroup of G_a, then the complete direct product of the H_a is a subgroup of the complete direct product of the G_a, and it is an invariant subgroup if each H_a is an invariant subgroup of G_a.

(2) If B is a subset of A and if C denotes the complement of B in A, then the complete direct product $\tilde{\prod}_{b\in B} G_b$ is isomorphic with the invariant subgroup $\tilde{\prod}_{a\in A} H_a$ of $\tilde{\prod}_{a\in A} G_a$, where we have set $H_a = G_a$ if $a \in B$ and $H_a = (e_a)$ if $a \in C$.

If each G_a is a commutative group then also the complete direct product of the G_a is commutative. In that case, if the additive notation is adopted for the group operation in each G_a, the same notation will be used for the complete direct product of the G_a and the latter will be referred to as the *complete direct sum* of the G_a and will be denoted by $\tilde{\sum}_{a\in A} G_a$.

If each G_a is a module over one and the same ring R, then the complete direct sum of the G_a can be made into an R-module by setting $\alpha\cdot\{x_a\} = \{\alpha\cdot x_a\}(\alpha \in R,\ x_a \in G_a)$. If each G_a is a module over a ring R_a which depends on a (so that the set of rings $\{R_a\}$ is itself indexed by the set A) then the complete direct sum of the G_a can be made into a module over the complete direct sum of the R_a by setting $\{u_a\}\cdot\{x_a\} = \{u_a\cdot x_a\}$ $(u_a \in R_a,\ x_a \in G_a)$. It is understood that a complete direct sum of rings R_a is viewed as a *ring* in virtue of the following definition of multiplication: $\{u_a\}\{v_a\} = \{u_a v_a\}$.

We shall seldom have occasion to use complete direct products or complete direct sums. More important for our purposes will be the concept of a *weak direct product*, or simply *direct product* (or *direct sum*, in the commutative case). We proceed to define this concept.

Let G be a group and let $\{G_a\}$ be a set of subgroups of G, indexed by a set A. We say then that G is a *weak direct sum* of the subgroups G_a if the following conditions are satisfied: (a) for $a \neq b$ each element of G_a commutes with each element of G_b; (b) for each element x of G there exists one and only one element $\{x_a\}$ of the complete direct product of the G_a such that $x_a = e_a$ for all a in A, except for a finite number of indices $a_1, a_2, \cdots, a_n$, and such that $x = x_{a_1} x_{a_2} \cdots x_{a_n}$. The element x_a is then called the *component* of x in the group G_a. We write $G = \prod_{a\in A} G_a$ to indicate that G is the weak direct product of the subgroups G_a. We shall, as a rule, omit the word "weak" and speak simply of G as being a *direct product* of the subgroups G_a. In the additive (commutative) case we shall use the term *weak direct sum* (or simply *direct sum*).

The proofs of the following assertions are straightforward and may be left to the reader:

1) If G is a direct product of subgroups G_a then G is isomorphic with a subgroup of the complete direct product of the G_a, distinct from $\tilde{\prod}_{a \in A} G_a$ if A is an infinite set. The isomorphism we allude to is the one in which to each element x of G corresponds the element $\{x_a\}$ of $\tilde{\prod}_{a \in A} G_a$, where x_a is the component of x in G_a.

2) If G is a direct product of subgroups G_a and if G'_a denotes the subgroup of G which is generated by the subgroups G_b, $b \neq a$, then

(a) each G_a is an invariant subgroup of G;
(b) G is generated by the groups G_a;
(c) $G_a \cap G'_a = (e)$, where e is the identity of G.

3) Conversely, if $\{G_a\}$ is a set of subgroups of G satisfying conditions (a), (b) and (c) then G is a direct product of the groups G_a.

4) Let $\{G_a\}$ be a set of groups indexed by a set A and let H be the subgroup of $\tilde{\prod}_{a \in A} G_a$ consisting of the elements $\{x_a\}$ such that $x_a = e_a$ for all a in A, except for a finite number of indices. Let H_a be the subgroup of H consisting of the elements $\{x_b\}$ such that $x_b = e_b$ for $b \neq a$. Then H_a and G_a are isomorphic groups, and H is a direct product of the H_a.

In the case of R-modules G_a, the group H defined in 4) is easily seen to be a submodule of the complete direct sum of the G_a.

It is immediately seen that in the case of groups (or modules) indexed by finite sets our present definitions coincide with those given in the preceding section.

§ 13. Comaximal ideals and direct sums of ideals. We now apply the results of § 12 to the theory of rings. Let R be an arbitrary ring. If we regard R as an R-module then the definitions of § 12 apply, and it is meaningful to speak of direct sums of submodules of R—that is, of ideals of R. Because of its importance we give the definition explicitly for the special case at hand:

Definition 1. *The ring R is said to be the direct sum of the ideals $R_1, R_2, \cdots, R_n$ if*

(a) $R = R_1 + R_2 + \cdots + R_n$,

(b) $R_i \cap (R_1 + \cdots + R_{i-1} + R_{i+1} + \cdots + R_n) = (0), i = 1, \cdots, n.$

We write, as before,

$$R = R_1 \oplus R_2 \oplus \cdots \oplus R_n. \tag{1}$$

We note that when (1) holds, the ideals R_i mutually annihilate one another; that is,

$$R_iR_j = (0) \quad \text{for} \quad i \neq j.$$

This follows from $R_iR_j \subset R_i \cap R_j = (0)$ if $i \neq j$. As a result, an ideal in R_i (R_i being considered as a ring) is also an ideal in R.

To make essential use of the ring structure of R we must place some restriction on it. Hence *in the remainder of this section we assume that R has an identity element*, to be denoted as usual by 1.

THEOREM 30. *Let R be a ring with identity. Let $R_1, R_2, \cdots, R_n$ be subrings of R such that*

$$R = R_1 + R_2 + \cdots + R_n, \; R_iR_j = (0) \quad \text{for } i \neq j. \tag{2}$$

Then each R_i is an ideal and the sum (2) is direct. All elements of R_i are zero divisors unless $R_j = (0)$ for all $j \neq i$. If $a_i, b_i \in R_i$ for $i = 1, 2, \cdots, n$, then

$$\begin{aligned} (a_1 + \cdots + a_n) + (b_1 + \cdots + b_n) \\ = (a_1 + b_1) + \cdots + (a_n + b_n), \\ (a_1 + \cdots + a_n)(b_1 + \cdots + b_n) = a_1b_1 + \cdots + a_nb_n. \end{aligned} \tag{3}$$

There exist uniquely determined elements e_i such that

$$1 = e_1 + \cdots + e_n, \; e_i \in R_i, \tag{4}$$

and it follows that

$$e_i^2 = e_i, \; e_i e_j = 0 \quad \text{for} \quad i \neq j, \; R_i = Re_i, \tag{5}$$

and e_i is the identity of R_i.

If $\mathfrak{A}$ *is an ideal in R, there exists a decomposition*

$$\mathfrak{A} = \mathfrak{A}_1 \oplus \cdots \oplus \mathfrak{A}_n, \; \mathfrak{A}_i \textit{ an ideal in } R_i; \tag{6}$$

this decomposition is unique, and in fact,

$$\mathfrak{A}_i = R_i\mathfrak{A} \tag{7}$$

The residue class ring $R/\mathfrak{A}$ is a direct sum of rings isomorphic (as rings) to the rings $R_i/\mathfrak{A}_i$. The ideal $\mathfrak{A}$ is a maximal or prime or primary ideal if and only if all but one of the $\mathfrak{A}_i$ coincide with the corresponding R_i and the remaining $\mathfrak{A}_i$ is respectively maximal, prime, or primary.

PROOF. From (2) and the fact that R_i is a ring it follows that $RR_i = R_i^2 \subset R_i$, so R_i is an ideal. If $c \in R_1$, then by (2) $Rc = R_1c$; and if also $c \in R_2 + \cdots + R_n$, then $Rc = R_1c = 0$, hence $c = 0$. Thus

$$R_1 \cap (R_2 + \cdots + R_n) = (0),$$

and $n - 1$ other relations of this sort together imply that the sum (2) is direct. The relation $R_iR_j = (0)$ implies that each element of R_i is a zero divisor unless $R_j = (0)$ for $j \neq i$. Relation (3) is obvious, as is the existence of unique e_i satisfying (4), and also that $e_ie_j = 0$ for $i \neq j$. Multiplying (4) by e_i we find $e_i^2 = e_i$. If $c \in R_i$, (4) implies $c = ce_i$, and hence also $R_i = Re_i$.

Suppose $\mathfrak{A}$ is an ideal in R. If a decomposition (6) exists, then $R_i\mathfrak{A} = R_i\mathfrak{A}_i = \mathfrak{A}_i$, whence (7). Existence of a decomposition follows by multiplying (2) by $\mathfrak{A}$:

$$\mathfrak{A} = R\mathfrak{A} = R_1\mathfrak{A} \oplus \cdots \oplus R_n\mathfrak{A};$$

$R_i\mathfrak{A}$ is an ideal in R_i since $R_i(R_i\mathfrak{A}) = R_i^2\mathfrak{A} = R_i\mathfrak{A}$.

If T is the natural homomorphism of R onto the ring $R/\mathfrak{A}$, then it is easily proved that $R/\mathfrak{A}$ is the direct sum of the ideals R_iT and that R_iT is isomorphic to $R_i/\mathfrak{A}_i$ (cf. proof of Theorem 24 of § 12).

Now for the last statement: Suppose $\mathfrak{A}$ is primary in R, $\mathfrak{A} \neq R$. Then not every R_iT can be (0), say, $R_1T \neq (0)$. Since no power of e_1T can be zero, it cannot be a zero-divisor in $R/\mathfrak{A}$ (since $\mathfrak{A}$ is primary), hence $R_iT = (0)$ for $i > 1$. Thus if $\mathfrak{A}$ is primary, $R_iT = 0$ for $i > 1$ and in $R_1/\mathfrak{A}_1$ ($\cong R_1T = RT$) every zero divisor is nilpotent, so that $\mathfrak{A}_i = R_i$ for $i > 1$ and $\mathfrak{A}_1$ is primary in R_1; the converse is obvious. Similarly (and even more simply) for $\mathfrak{A}$ prime or maximal.

An element e of a ring such that $e^2 = e$ is called an *idempotent*; two idempotents e and e' are said to be *orthogonal* if $ee' = 0$. Thus with a direct decomposition of the ring R we have associated a decomposition

$$1 = e_1 + \cdots + e_n$$

of the identity of R into orthogonal idempotents. Conversely if such orthogonal idempotents are given, it is easy to see that R is the direct sum of the ideals $Re_1, \cdots, Re_n$.

We point out that if S is a ring and $S = S_1 \oplus \cdots \oplus S_n$ with $R_i \cong S_i$, then $R \cong S$; cf. proof of Theorem 23 of § 12. Also, if $R'_1, \cdots, R'_n$ are arbitrary rings, there exists one and (up to isomorphism) only one ring R which is the direct sum of ideals R_i isomorphic to the R'_i. The ring R may be defined (cf. Theorem 29 of § 12) to consist of all $(a_1, \cdots, a_n)$, $a_i \in R'_i$, addition being defined in the obvious way, and multiplication by

$$(a_1, \cdots, a_n)(b_1, \cdots, b_n) = (a_1b_1, \cdots, a_nb_n)$$

DEFINITION 2. *Let R be a ring with identity. A set of ideals $\mathfrak{A}_1, \cdots, \mathfrak{A}_n$ in R is said to be* PAIRWISE COMAXIMAL *if each $\mathfrak{A}_i \neq R$ and*

$$\mathfrak{A}_i + \mathfrak{A}_j = R \quad \textit{for } i \neq j.$$

If $n = 2$, we say simply that $\mathfrak{A}_1$ and $\mathfrak{A}_2$ are comaximal.

The definition, of course, implies $\mathfrak{A}_i \neq \mathfrak{A}_j$ for $i \neq j$, also that $\mathfrak{A}_i \neq (0)$. This concept allows us to give a sort of dual decomposition to the direct sum, and in a stronger form than in Theorem 25, § 12. First:

THEOREM 31. *Let R be a ring with identity, and let $\mathfrak{A}_1, \cdots, \mathfrak{A}_n$ be ideals in R. The $\mathfrak{A}_i$ are pairwise comaximal if and only if their radicals are. If an ideal $\mathfrak{B}$ is comaximal with each $\mathfrak{A}_i$, then it is comaximal with $\mathfrak{A}_1 \cap \cdots \cap \mathfrak{A}_n$ and $\mathfrak{A}_1 \cdots \mathfrak{A}_n$. If $\mathfrak{A}_1, \cdots, \mathfrak{A}_n$ are pairwise comaximal, then*

$$\mathfrak{A}_1 \cap \cdots \cap \mathfrak{A}_n = \mathfrak{A}_1 \cdots \mathfrak{A}_n; \tag{8}$$

if, moreover, $b_1, \cdots, b_n$ are elements of R, then there exists an element b in R such that

$$b \equiv b_i(\mathfrak{A}_i), \quad i = 1, \cdots, n. \tag{9}$$

PROOF. If $\mathfrak{A}_1$ and $\mathfrak{A}_2$ are comaximal, obviously $\sqrt{\mathfrak{A}_1}$ and $\sqrt{\mathfrak{A}_2}$ are. Conversely suppose $\sqrt{\mathfrak{A}_1} + \sqrt{\mathfrak{A}_2} = R$. Then by the formulas on the radical (Theorem 9 of § 7, p. 147):

$$R = \sqrt{R} = \sqrt{\sqrt{\mathfrak{A}_1} + \sqrt{\mathfrak{A}_2}} = \sqrt{\mathfrak{A}_1 + \mathfrak{A}_2};$$

since only R has R as radical (since $1 \in R$), $\mathfrak{A}_1 + \mathfrak{A}_2 = R$. (Or directly: from $R = \sqrt{\mathfrak{A}_1} + \sqrt{\mathfrak{A}_2}$, we obtain $1 = c_1 + c_2$, $c_i \in \sqrt{\mathfrak{A}_i}$, $c_i^k \in \mathfrak{A}_i$, k an integer; since in the binomial expansion for $1 = (c_1 + c_2)^{2k-1}$ each term has a factor $c_1^i c_2^j$ with $i \geqq k$ or $j \geqq k$ and hence is in $\mathfrak{A}_1$ or $\mathfrak{A}_2$, it follows that $1 \in \mathfrak{A}_1 + \mathfrak{A}_2$.)

If $\mathfrak{B}$ is comaximal with each $\mathfrak{A}_i$, then

$$\mathfrak{B} + \mathfrak{A}_i = R, \; i = 1, \cdots, n.$$

To prove $\mathfrak{B}$ comaximal with $\prod_1^n \mathfrak{A}_i$, and hence *a fortiori* with $\bigcap_1^n \mathfrak{A}_i$, we observe that

$$R = R^n = \prod_1^n (\mathfrak{B} + \mathfrak{A}_i) \subset \mathfrak{B} + \prod_1^n \mathfrak{A}_i \subset R.$$

We have here used the fact that multiplication of ideals is distributive with respect to addition (cf. § 7, relation (2)).

Suppose the $\mathfrak{A}_i$ pairwise comaximal. If $n = 2$, then

$$\begin{aligned}(\mathfrak{A}_1 \cap \mathfrak{A}_2) = (\mathfrak{A}_1 + \mathfrak{A}_2)(\mathfrak{A}_1 \cap \mathfrak{A}_2) &= \mathfrak{A}_1(\mathfrak{A}_1 \cap \mathfrak{A}_2) + \mathfrak{A}_2(\mathfrak{A}_1 \cap \mathfrak{A}_2) \\ &\subset \mathfrak{A}_1\mathfrak{A}_2 + \mathfrak{A}_2\mathfrak{A}_1 = \mathfrak{A}_1\mathfrak{A}_2.\end{aligned}$$

Assuming (8) true for $n - 1$ factors, and observing that $\mathfrak{A}_n$ is comaximal with $\mathfrak{A}_1 \cap \cdots \cap \mathfrak{A}_{n-1}$, we have

$$(\mathfrak{A}_1 \cap \cdots \cap \mathfrak{A}_{n-1}) \cap \mathfrak{A}_n = (\mathfrak{A}_1 \cap \cdots \cap \mathfrak{A}_{n-1})\mathfrak{A}_n = (\mathfrak{A}_1 \cdots \mathfrak{A}_{n-1})\mathfrak{A}_n.$$

Suppose, in addition, the elements b_i to be given. If $n = 2$, then $1 = a_1 + a_2$ with $a_i \in \mathfrak{A}_i$, so that

$$a_1 = 0(\mathfrak{A}_1),\ a_2 \equiv 1(\mathfrak{A}_1),\ a_1 \equiv 1(\mathfrak{A}_2),\ a_2 \equiv 0(\mathfrak{A}_2).$$

Placing $b = b_1a_2 + b_2a_1$ we get

$$b \equiv b_i(\mathfrak{A}_i),\ i = 1, 2.$$

Assuming the last statement in the theorem to be true for $n - 1$, we have an element b' such that

$$b' \equiv b_i(\mathfrak{A}_i), \quad i = 1, \cdots, n - 1. \tag{10}$$

Since $\mathfrak{A}_n$ and $\mathfrak{A}_1 \cap \cdots \cap \mathfrak{A}_{n-1}$ are comaximal there is an element b such that

$$b \equiv b'(\mathfrak{A}_1 \cap \cdots \cap \mathfrak{A}_{n-1}), \quad b \equiv b_n(\mathfrak{A}_n). \tag{11}$$

From (10) and (11) $b \equiv b_i(\mathfrak{A}_i)$, $i = 1, \cdots, n - 1$.

The last statement of Theorem 31 is a generalization of the well-known fact that if $m_1, \cdots, m_n$ are integers which are relatively prime in pairs, and if $b_1, \cdots, b_n$ are arbitrary integers, then the simultaneous congruences

$$x \equiv b_i \ (\mathrm{mod}\ m_i), \quad i = 1, \cdots, n,$$

have a solution.

THEOREM 32. *Let R be a ring with identity. Let $\mathfrak{A}_1, \cdots, \mathfrak{A}_n$ be ideals such that*

$$(0) = \mathfrak{A}_1 \cap \cdots \cap \mathfrak{A}_n,\ \mathfrak{A}_i + \mathfrak{A}_j = R \quad \textit{for} \quad i \neq j. \tag{12}$$

If we place

$$R_i = \mathfrak{A}_1 \cap \cdots \cap \mathfrak{A}_{i-1} \cap \mathfrak{A}_{i+1} \cap \cdots \mathfrak{A}_n, \quad i = 1, \cdots, n, \tag{13}$$

then

$$R = R_1 \oplus \cdots \oplus R_n, \quad R_i \cong R/\mathfrak{A}_i, \tag{14}$$

$$\mathfrak{A}_i = R_1 + \cdots + R_{i-1} + R_{i+1} + \cdots + R_n. \tag{15}$$

Conversely, if R is a direct sum of ideals $R_1, \cdots, R_n$ and if we define $\mathfrak{A}_i$ by (15), *then* (12) *and* (13) *follow.*

PROOF. Suppose the $\mathfrak{A}_i$ are given, satisfying (12). From the preceding theorem it follows that $\mathfrak{A}_i + \bigcap_{j \neq i} \mathfrak{A}_j = R$, so that the second half of Theorem 25 of (§ 12) may be applied to give the first part of (14) and also (15); here the $\mathfrak{A}_i$ play the role of the N_i. In view of (14) we have

$$1 = e_1 + \cdots + e_n,\ e_i \in R_i.$$

The mapping $a \to ae_1$ is clearly a homomorphism of R onto R_1, and indeed the elements of R_1 are fixed in this mapping. The kernel consists of those a such that $ae_1 = 0$, that is, $a = \sum_{j=2}^{n} ae_j$; hence the kernel is $\mathfrak{A}_1$, and $R_1 \cong R/\mathfrak{A}_1$. Similarly, $R_i \cong R/\mathfrak{A}_i$ for $i = 2, \cdots, n$.

If R is a direct sum of the R_i and we define $\mathfrak{A}_i$ by (15), then we apply the first half of Theorem 25, with the R_i playing the role of the M_i. Then (13) and first part of (12) follow. From (15) and (14) it follows that $R = R_i + \mathfrak{A}_i$, hence *a fortiori* $R = \mathfrak{A}_j + \mathfrak{A}_i$ for $j \neq i$.

§ **14. Tensor products of rings.** All rings which are considered in this section are assumed to contain a given field k as subring. It is furthermore assumed that the element 1 of k is also an element 1 of each of the rings. We shall sometimes refer to our rings as *algebras over* k.

If A and B are subrings of a ring C, we shall denote by $[A, B]$ the smallest subring of C which contains both rings A and B.

Let A and B be two given algebras over k.

DEFINITION 1. *By a product of A and B (over k) we mean the composite concept (C, φ, ψ) consisting of an algebra C over k, a k-isomorphism φ of A into C and a k-isomorphism ψ of B into C, such that $C = [A\varphi, B\psi]$.*

DEFINITION 2. *Two products (C, φ, ψ) and (C', φ', ψ') of A and B are said to be equivalent if there exists an isomorphism f of C onto C' such that $\varphi' = \varphi f$ on A and $\psi' = \psi f$ on B.*

This relation of equivalence is clearly reflexive, symmetric, and transitive, and thus we can speak of equivalence classes of products of A and B. It is also clear that if the above isomorphism f exists at all, it is uniquely determined, for C is generated by $A\varphi$ and $B\psi$ and we must have $f = \varphi^{-1}\varphi'$ on $A\varphi$ and $f = \psi^{-1}\psi'$ on $B\psi$.

DEFINITION 3. *A product (C, φ, ψ) of A and B is called a tensor product of A and B (over k) if the rings $A\varphi$ and $B\psi$ are linearly disjoint over k* (see II, § 15).

THEOREM 33. *There exist tensor products of A and B.*

PROOF. Let $\{x_\alpha\}$ be a vector basis of A over k and let $\{y_\beta\}$ be a vector basis of B over k. We consider the set of all ordered pairs (x_α, y_β) and the set C of all formal finite sums $\sum c_{\alpha\beta}(x_\alpha, y_\beta)$, with coefficients $c_{\alpha\beta}$ in k. Then C is in a natural way a vector space over k, the set of all ordered pairs (x_α, y_β) (that is, the set-theoretic product of the two bases $\{x_\alpha\}$ and $\{y_\beta\}$ of A over k and B over k respectively) being a vector basis of C over k. We will find it convenient to use the following notation: if $x = \sum a_\alpha x_\alpha$ and $y = \sum b_\beta y_\beta$ are elements of A and B respectively ($a_\alpha, b_\beta \in k$), then $x \circ y$ shall denote the element $\sum_\alpha \sum_\beta a_\alpha b_\beta (x_\alpha, y_\beta)$ of C. In particular, we have $x_\alpha \circ y_\beta = (x_\alpha, y_\beta)$. It is clear that $x \circ y = 0$ if and only if $x = 0$ or $y = 0$. The following relations are obvious:

(1) $$(x + x') \circ y = x \circ y + x' \circ y; \quad (x, x' \in A; y \in B).$$

(2) $$x \circ (y + y') = x \circ y + x \circ y'; \quad (x \in A; y, y' \in B).$$

(3) $$c(x \circ y) = cx \circ y = x \circ cy; \quad (x \in A, y \in B; c \in k).$$

We shall now define a multiplication in C. It will be sufficient to define the products $(x_\alpha \circ y_\beta)\cdot(x_\gamma \circ y_\delta)$ of any two basis elements of C, for then the product of any two elements of C will be determined by linearity, that is, by the requirement that the multiplication be distributive and that we have $c(x_\alpha \circ y_\beta)\cdot d(x_\gamma \circ y_\delta) = cd((x_\alpha \circ y_\beta)(x_\gamma \circ y_\delta))$. We set

(4) $$(x_\alpha \circ y_\beta)\cdot(x_\gamma \circ y_\delta) = x_\alpha x_\gamma \circ y_\beta y_\delta.$$

It is obvious that the multiplication thus defined is commutative and distributive. To verify the associative law, it is only necessary to verify the validity of the following relations:

(5) $$[(x_\alpha \circ y_\beta)(x_\gamma \circ y_\delta)](x_{\alpha'} \circ y_{\beta'}) = (x_\alpha \circ y_\beta)[(x_\gamma \circ y_\delta)(x_{\alpha'} \circ y_{\beta'})].$$

These relations follow, however, from the associative laws in A and in B, for it is immediately seen that both sides of (5) are equal to $x_\alpha x_\gamma x_{\alpha'} \circ y_\beta y_\delta y_{\beta'}$. We note that the above definition implies (in view of (1) to (3)) that we have $(x \circ y)(x' \circ y') = xx' \circ yy'$, for any elements x, x' of A and any elements y, y' of B.

The mapping $\varphi : a \to a \circ 1$, $a \in A$, is obviously a homomorphism of A into C (note that we have $aa' \circ 1 = (a \circ 1)(a' \circ 1)$, for any elements a and a' in A). Since $a \circ 1 \neq 0$ if $a \neq 0$, φ is an isomorphism. Similarly the mapping $\psi : b \to 1 \circ b$, $b \in B$, is an isomorphism of B into C. The two mappings φ and ψ coincide on k, for if $c \in k$, then $c \circ 1 = c(1 \circ 1) = 1 \circ c$. We shall identify c with $c \circ 1$ for any c in k. Then (C, φ, ψ) becomes a product of A and B over k, in the sense of Definition 1, for we have $(x_\alpha \circ y_\beta) = (x_\alpha \circ 1)(1 \circ y_\beta)$ and hence $C = [A\varphi, B\psi]$. The two subrings $A' = A\varphi$ and $B' = B\psi$ of C are algebras over k and are k-isomorphic with A and B respectively.

We now prove that A' and B' are linearly disjoint over k. The elements $x_\alpha \circ 1$ form a basis of A' over k, and similarly the elements $1 \circ y_\beta$ form a basis of B' over k. We have $(x_\alpha \circ 1)(1 \circ y_\beta) = x_\alpha \circ y_\beta$, and therefore the products $(x_\alpha \circ 1)(1 \circ y_\beta)$ are linearly independent over k. The linear disjointness of A' and B' now follows from II, § 15 (Lemma 1). The proof of the theorem is now complete.

THEOREM 34. (*The universal mapping property of tensor products.*) *A necessary and sufficient condition that a product* (C, φ, ψ) *of* A *and* B

(over k) be a tensor product of A and B is that given any two k-homomorphisms g and h of A and B respectively into a ring R there should exist a homomorphism f of C into R such that $f = \varphi^{-1}g$ on $A\varphi$ and $f = \psi^{-1}h$ on $B\psi$.

PROOF. Every element ξ of C has an expression of the form $\xi = \sum_i \varphi(a_i)\psi(b_i)$, where $a_i \in A$ and $b_i \in B$. We set $f(\xi) = \sum_i g(a_i)h(b_i)$. Then f is a transformation of C into R (perhaps not single-valued) which satisfies the following two conditions: a) for any ξ in C the set $f(\xi)$ is non-empty; b) if $u \in f(\xi)$ and $v \in f(\eta)$, then $u + v \in f(\xi + \eta)$ and $uv \in f(\xi\eta)$. It follows from Lemma 2 of I, § 11, that f can be asserted to be univalent (and hence a homomorphism), provided the set $f(0)$ contains only the zero of R. We shall show now that this last condition is indeed satisfied if (C, φ, ψ) is a tensor product of A and B, and this will prove the necessity of the condition since we have $f = \varphi^{-1}g$ on $A\varphi$ and $f = \psi^{-1}h$ on $B\psi$.

Let $0 = \sum_i \varphi(a_i)\psi(b_i)$ be an expression of the zero in C. We fix a basis $\{x_\alpha\}$ of the vector space $\sum ka_i$ (over k) and a basis $\{y_\beta\}$ of the vector space $\sum kb_i$, and we express the a_i and the b_i in terms of these basis elements: $a_i = \sum_\alpha c_{i\alpha}x_\alpha$, $b_i = \sum_\beta d_{i\beta}y_\beta$ $(c_{i\alpha}, d_{i\beta} \in k)$. From the above expression of 0 and by the linear disjointness of $A\varphi$ and $B\psi$ over k it follows that

$$(6) \qquad \sum c_{i\alpha}d_{i\beta} = 0, \text{ all } \alpha \text{ and } \beta.$$

We have $g(a_i) = \sum_\alpha c_{i\alpha}g(x_\alpha)$ and $h(b_i) = \sum_\beta d_{i\beta}h(y_\beta)$. Hence $\sum_i g(a_i)h(b_i) = \sum_{\alpha,\beta}\left(\sum_i c_{i\alpha}d_{i\beta}\right)g(x_\alpha)h(y_\beta)$, and this is zero in view of (6). This completes the proof of the necessity of the condition.

Conversely, assume that the product (C, φ, ψ) satisfies the condition stated in the theorem. We fix a tensor product (C', φ', ψ') of A and B and we proceed to show that *the two products* (C, φ, ψ) *and* (C', φ', ψ') *are equivalent*. This will complete the proof of the theorem.

By assumption, there exists a homomorphism f of C into C' such that $f = \varphi^{-1}\varphi'$ on $A\varphi$ and $f = \psi^{-1}\psi'$ on $B\psi$. Since C' is a tensor product, it follows from the first part of the proof that there also exists a homomorphism f' of C' into C such that $f' = \varphi'^{-1}\varphi$ on $A\varphi'$ and $f' = \psi'^{-1}\psi$ on $B\psi'$. Then ff' is a homomorphism of C into itself which is the identity on both $A\varphi$ and $B\psi$. Since $C = [A\varphi, B\psi]$ it follows that ff' is the identity on C. Similarly, $f'f$ is the identity on C'. Consequently f is an *isomorphism* of C *onto* C', and since $f = \varphi^{-1}\varphi'$ on $A\varphi$ and $f = \psi^{-1}\psi'$

on $B\psi$, the two products (C, φ, ψ) and (C', φ', ψ') are equivalent, as was asserted.

COROLLARY 1. *Any two tensor products of A and B are equivalent.*

This has been established in the second part of the above proof.

COROLLARY 2. *If a given product (C, φ, ψ) of A and B admits a homomorphism f into a tensor product (C', φ', ψ') of A and B such that $f = \varphi^{-1}\varphi'$ on $A\varphi$ and $f = \psi^{-1}\psi'$ on $B\psi$, then (C, φ, ψ) is itself a tensor product of A and B, and f is an isomorphism of C onto C'.*

Also this has been established in the second half of the above proof.

We shall now introduce a canonical model of a tensor product of our two rings A and B whose construction is intrinsically related to these rings (the construction in the proof of Theorem 33 uses bases of A and B and is therefore not intrinsic).

We consider the set-product $A \times B$, that is, the set of all ordered pairs (a, b), $a \in A$, $b \in B$. We denote by M the set of all formal finite linear combinations $\sum_i c_i(a_i, b_i)$ of elements of $A \times B$, with coefficients c_i in k. We convert M into a vector space over k, with $A \times B$ as basis, by defining addition and scalar multiplication in an obvious way:

$$\Big(\sum_i c_i(a_i, b_i)\Big) + \Big(\sum_i d_i(a_i, b_i)\Big) = \sum_i (c_i + d_i)(a_i, b_i),$$

$$d\Big(\sum_i c_i(a_i, b_i)\Big) = \sum_i dc_i(a_i, b_i),$$

where the c_i, d_i, and d are elements of k. We now also define multiplication in M by first defining the product of any two elements (a, b) and (a', b') of $A \times B$ as follows:

$$(a, b)(a', b') = (aa', bb')$$

and then defining the product of any two elements of M by linearity. It is immediately seen that with this definition of multiplication M becomes an algebra over k (however, note that the field k is not contained in M).

Let $\mathfrak{N}$ denote the ideal generated in M by all the elements of the following form:

$$(7) \qquad \begin{cases} (a + a', b) - (a, b) - (a', b), & (ca, b) - c(a, b), \\ (a, b + b') - (a, b) - (a, b'), & (a, cb) - c(a, b), \end{cases}$$

where $a, a' \in A$; $b, b' \in B$; $c \in k$. We denote by T the residue class ring $M/\mathfrak{N}$ and by ρ the canonical homomorphism of M onto T. For any element a in A and b in B we denote by $a \otimes b$ the $\mathfrak{N}$-residue of (a, b). Finally we denote by g the mapping $a \to a \otimes 1$ of A into T, and by h the mapping $b \to 1 \otimes b$ of B into T. It is immediately seen that g and h

are ring homomorphisms. We have $\rho(c(a, b)) = \rho((ca, b)) = ca \otimes b$, and since every element of M is a sum of terms of the form $c(a, b)$, with c in k, a in A and b in B, it follows that every element of T is a sum of elements of the form $a \otimes b$. On the other hand, we have $(a, b) = (a, 1)(1, b)$, whence $a \otimes b = (a \otimes 1)(1 \otimes b)$. This shows that every element of T is a finite sum of products of elements $a \otimes 1$ of the ring Ag and elements $1 \otimes b$ of the ring Bh. In other words, we have

$$T = [Ag, Bh]. \tag{8}$$

Let us now fix a tensor product (C, φ, ψ) of A and B (over k). The mapping

$$\sigma : \sum_i c_i(a_i, b_i) \rightarrow \sum_i c_i\varphi(a_i)\psi(b_i),$$

is a ring homomorphism of M onto C. The relation $\varphi(a + a')\psi(b) = \varphi(a)\psi(b) + \varphi(a')\psi(b)$ shows that the elements of M of the form $(a + a', b) - (a, b) - (a', b)$ belong to the kernel of σ. Similarly, all the elements of the form (7) belong to the kernel of σ, that is, *the kernel of σ contains the kernel of ρ.* Hence $\sigma = \rho\tau$, where τ is a homomorphism of T onto C.

For any a in A we have $\varphi(a) = \sigma((a, 1)) = \tau(a \otimes 1) = (g\tau)(a)$. Since φ is an isomorphism, it follows that *g is an isomorphism of A onto Ag and that the restriction of τ to Ag is an isomorphism of Ag onto $A\varphi$.* Similarly we find that $\psi = h\tau$ on B, that h is an isomorphism of B onto Bh, and that the restriction of τ to Bh is an isomorphism of Bh onto $B\psi$. We note that $g = h$ on k and that consequently we can identify any element c of k with the corresponding element $g(c) = c \otimes 1 = h(c) = 1 \otimes c$. With this identification, g and h become k-isomorphisms of A and B respectively into T. *Hence, by* (8), *T is a product of A and B, over k.* In view of the existence of the homomorphism of T into C, with the properties described above, it follows, by Corollary 2 to Theorem 34, that *(T, g, h) is a tensor product of A and B.* This, canonically constructed, tensor product of A and B will be denoted by $A \underset{k}{\otimes} B$, or simply by $A \otimes B$.

The following relations are easily verified: $A \otimes B \cong B \otimes A$, $(A \otimes B) \otimes C \cong A \otimes (B \otimes C)$. The proofs may be left to the reader.

From now on we shall regard A and B as subrings of the tensor product $A \otimes B$. More precisely, we identify every element a of A with the corresponding element $g(a) = a \otimes 1$ and every element b of B with the corresponding element $1 \otimes b$. With this identification, the tensor product $A \otimes B = (T, g, h)$ is now $(T, 1, 1)$, where 1 stands both for the identity mapping of A and the identity mapping of B.

In the preceding considerations the nullring was excluded because we have always assumed that our rings contain an element 1. However, this assumption played no role whatsoever in the definition of the ring $A \otimes B$, and this ring is obviously the nullring if either A or B is the nullring.

EXAMPLES

1) If A is a field, then $A \otimes B$ is a ring containing the field A. *Any basis* $\{y_\beta\}$ *of* B *over* k *is also a basis of* $A \otimes B$ *over* A. In particular, $k \otimes B = B$.

2) If $A = k[X]$ $(= k[X_1, X_2, \cdots, X_n])$ and $B = k[Y]$ $(= k[Y_1, Y_2, \cdots, Y_m])$ are polynomial rings in n and m indeterminates respectively, then the polynomial ring $C = k[X, Y]$ in $n + m$ indeterminates is generated by A and B, and it is clear that A and B are linearly disjoint over k, in C. Hence $k[X, Y] \cong k[X] \otimes k[Y]$.

A similar result holds for polynomial rings in infinitely many variables.

Let $\mathfrak{A}$ be an ideal in A and $\mathfrak{B}$ an ideal in B. We shall denote by $(\mathfrak{A}, \mathfrak{B})$ the least ideal in $A \otimes B$ which contains $\mathfrak{A}$ and $\mathfrak{B}$. In other words, $(\mathfrak{A}, \mathfrak{B})$ is the ideal in $A \otimes B$ which is generated by the elements of $\mathfrak{A}$ and $\mathfrak{B}$. We denote by α, β, and h the canonical homomorphism of A onto $A/\mathfrak{A}$, of B onto $B/\mathfrak{B}$, and of $A \otimes B$ onto $A \otimes B/(\mathfrak{A}, \mathfrak{B})$, respectively. Since the restriction of h to A has kernel which contains $\mathfrak{A}$ it follows that $h = \alpha\varphi$ on A, where φ is a k-homomorphism of $A/\mathfrak{A}$. Similarly, $h = \beta\psi$ on B, where ψ is a k-homomorphism of $B/\mathfrak{B}$. Furthermore, since $A \otimes B$ is generated by A and B, the ring $A \otimes B/(\mathfrak{A}, \mathfrak{B})$ is generated by $(A/\mathfrak{A})\varphi$ and $(B/\mathfrak{B})\psi$.

THEOREM 35. *The rings* $A/\mathfrak{A} \otimes B/\mathfrak{B}$ *and* $(A \otimes B)/(\mathfrak{A}, \mathfrak{B})$ *are* k*-isomorphic. More precisely: the homomorphisms* φ *and* ψ *are isomorphisms, and* $(A \otimes B/(\mathfrak{A}, \mathfrak{B}), \varphi, \psi)$ *is a tensor product of* $A/\mathfrak{A}$ *and* $B/\mathfrak{B}$.

PROOF. If either $\mathfrak{A}$ or $\mathfrak{B}$ is the unit ideal, then both rings coincide with the nullring. We shall therefore assume that $\mathfrak{A} \neq A$ and $\mathfrak{B} \neq B$. Under these assumptions we first show that $(\mathfrak{A}, \mathfrak{B}) \cap A = \mathfrak{A}$ and $(\mathfrak{A}, \mathfrak{B}) \cap B = \mathfrak{B}$, and that consequently φ and ψ are isomorphisms. By Theorem 34, applied to the rings $C = A \otimes B$ and $R = A/\mathfrak{A} \otimes B/\mathfrak{B}$, there exists a homomorphism g of $A \otimes B$ into $A/\mathfrak{A} \otimes B/\mathfrak{B}$ such that $g = \alpha$ on A and $g = \beta$ on B. The kernel of g contains $\mathfrak{A}$ and $\mathfrak{B}$ and consequently also $(\mathfrak{A}, \mathfrak{B})$. Therefore $A \cap (\mathfrak{A}, \mathfrak{B}) \subset A \cap \operatorname{Ker} g = \mathfrak{A}$, showing that $(\mathfrak{A}, \mathfrak{B}) \cap A = \mathfrak{A}$. Similarly, $(\mathfrak{A}, \mathfrak{B}) \cap B = \mathfrak{B}$.

Since the kernel of g contains $(\mathfrak{A}, \mathfrak{B})$ we have $g = hf$, where f is a homomorphism of $A \otimes B/(\mathfrak{A}, \mathfrak{B})$ into the tensor product $A/\mathfrak{A} \otimes B/\mathfrak{B}$. Since $g = \alpha$ on A while $h = \alpha\varphi$ on A, it follows that φf is the identity on $A/\mathfrak{A}$, or—equivalently—that $f = \varphi^{-1}$ on $(A/\mathfrak{A})\varphi$. Similarly, $f = \psi^{-1}$ on $(B/\mathfrak{B})\psi$. By Corollary 2 of Theorem 34 this completes the proof of the theorem.

We shall now give some results concerning zero-divisors in given rings and in their tensor products. First of all, we have the following lemma:

LEMMA. *If an element a of A is not a zero-divisor in A, it is not a zero divisor in $A \otimes B$.*

The proof is immediate. For if we have $a\xi = 0$, where $\xi \in A \otimes B$ we can write ξ in the form $\xi = \sum_i a_i b_i$, where the a_i are in A, and the b_i are elements of B which are linearly independent over k (*and hence also over* A). From $\sum (aa_i)b_i = 0$ follows then $aa_i = 0$, $a_i = 0$, and hence $\xi = 0$.

COROLLARY. *The total quotient ring K of $A \otimes B$ contains the total quotient rings of A and B. More precisely: the quotient ring of A in K is a total quotient ring of A, and similarly for B. Furthermore, these total quotient rings of A and B are linearly disjoint over k (as subrings of K).*

Every regular element of A has an inverse in K (since every regular element of A is also a regular element of K, by the lemma). Hence we can speak of the quotient ring of A in K, and this quotient ring will be a total quotient ring of A (I, § 19, Corollary 3, p. 43). If $\{b_i/b\}$ is a set of elements of the quotient ring of B ($b, b_i \in B$, b regular in B) and if these elements are linearly independent over k, then also the b_i are linearly independent over k and hence also over A. From this it follows at once that also the quotients b_i/b are linearly independent over the quotient ring of A.

The following theorem includes the above lemma as a special case:

THEOREM 36. *Let A' and B' be subrings and subalgebras of A and B respectively. If no element of A' which is different from zero is a zero-divisor in A, and if, similarly, no element of B' which is different from zero is a zero-divisor in B, then every element of $A' \otimes B'$ which is a zero-divisor in $A \otimes B$ is already a zero-divisor in $A' \otimes B'$.* (Note that $A' \otimes B'$ is canonically identifiable with a subring of $A \otimes B$.)

PROOF. Let x' be an element of $A' \otimes B'$ which is not a zero-divisor in $A' \otimes B'$ and assume that we have $x'z = 0$ for some element z of $A \otimes B$. We write $z = \sum_i a_i b_i$, with $a_i \in A$ and $b_i \in B$, and we extract from the set $\{a_i\}$ a maximal subset $\{u_m\}$ of elements u_m which are linearly independent over A'. Similarly, we extract from the set $\{b_i\}$ a maximal subset $\{v_n\}$ of elements v_n which are linearly independent over B'. We note that from our assumptions it follows that both A' and B' are integral domains. Hence, there exist elements α' and β' in A' and B', both different from zero, such that $\alpha' a_i = \sum_m a'_{im} u_m$, $\beta' b_i = \sum_n b'_{in} v_n$, where the a'_{im} belong to A' and the b'_{in} belong to B'. We have therefore

$$\alpha'\beta' z = \sum_{m,n} \Big(\sum_i a'_{im} b'_{in}\Big) u_m v_n, \tag{9}$$

and

$$0 = x'\alpha'\beta' z = \sum_{m,n} \Big(x' \sum_i a'_{im} b'_{in}\Big) u_m v_n. \tag{10}$$

We set

$$y'_{mn} = x' \sum_i a'_{im} b'_{in}. \tag{11}$$

Since $y'_{mn} \in [A', B']$ we can write these elements in the form

$$y'_{mn} = \sum_{p,q} c_{mnpq} a'_p b'_q, \tag{12}$$

where the c_{mnpq} are in k, the a'_p are elements of A' which are linearly independent over k and the b'_q are elements of B' which are linearly independent over k. The linear independence of the v_n over B' and the linear independence of the b'_q over k shows easily that the products $b'_q v_n$ (elements of B) are linearly independent over k. Similarly the products $a'_p u_m$ are linearly independent over k. The linear disjointness of A and B over k implies therefore that the products $u_m v_n a'_p b'_q$ are linearly independent over k. Now relations (10), (11), and (12) yield the relations

$$\sum c_{mnpq} u_m v_n a'_p b'_q = 0.$$

Hence the elements c_{mnpq} are all zero, and therefore also the elements y_{mn} are all zero. Since x' is not a zero-divisor in $A' \otimes B'$, it follows from (11) that $\sum_i a'_{im} b'_{in} = 0$, and hence by (9), we have $\alpha'\beta' z = 0$. Since $0 \neq \alpha' \in A'$, it follows from the preceding lemma that α' is not a zero-divisor in $A \otimes B$. Hence $\beta' z = 0$. Similarly, since $0 \neq \beta' \in B'$, it follows that $z = 0$. This shows that x' is not a zero-divisor in $A \otimes B$ and completes the proof of the theorem.

COROLLARY. *Let K and K' be fields containing k and let B be a transcendence set in K (for instance, a transcendence basis). Then, in the tensor product $K \otimes_k K'$, the elements of B are also algebraically independent over K', every element of the polynomial ring $K'[B]$, different from zero, is regular in $K \otimes_k K'$, and the total quotient ring of $K \otimes_k K'$ contains the tensor product $K \otimes_{k(B)} K'(B)$.*

Any finite set of distinct monomials $x_1^{i_1}x_2^{i_2}\cdots x_n^{i_n}$, $x_i \in B$, is a set of linearly independent elements over k. Hence these monomials are also linearly independent over K', which proves the first assertion of the corollary. The rings $k[B]$ and K' are linearly disjoint over k (in $K \otimes_k K'$) and they generate the ring $K'[B]$. Hence we have $K'[B] = k[B] \otimes K'$. Since we are now dealing throughout with integral domains, the preceding theorem shows that every non-zero element of the polynomial ring $K'[B]$ is regular in $K \otimes_k K'$. Consequently the total quotient ring of $K \otimes_k K'$ contains the field $K'(B)$. *We assert that K and $K'(B)$ are linearly disjoint over $k(B)$.* Since $K'(B)$ is the quotient field of $k(B)\cdot K'$ (this latter ring being the ring of quotients of $K'[B]$ with respect to the multiplicative system of the non-zero elements of $k[B]$), it will be sufficient to show that K and $k(B)\cdot K'$ are linearly disjoint over $k(B)$. However, this follows at once from the linear disjointness of K and K' over k and from the fact that the vector space $k(B)\cdot K'$ over $k(B)$ has a basis consisting of elements of K'. The elements of such a basis are *a fortiori* linearly independent over k and hence also over K, and the linear disjointness of K and $K'(B)$ over $k(B)$ follow now from the lemma of II, § 15. The ring generated by K and $K'(B)$ in the total quotient ring of $K \otimes_k K'$ is therefore isomorphic with $K \otimes_{k(B)} K'(B)$.

§ **15. Free joins of integral domains (or of fields).** Our object in this section is to apply the concept of tensor products toward the determination of all possible ways in which two abstract *integral domains* over k (or two *fields* over k) can be *freely embedded* (in a sense that will be specified below) in a bigger field. We proceed to prepare the ground for this application.

Let R and R' be integral domains containing a given field k as subfield.

DEFINITION 1. *By a free join of two integral domains R/k and R'/k (relative to k) we mean the composite concept (Ω, τ, τ') consisting of an* INTEGRAL DOMAIN *Ω containing k, a k-isomorphism τ of R into Ω and a k-isomorphism τ' of R' into Ω, such that the following conditions are satisfied:* (1) $\Omega = [R\tau, R'\tau']$; (2) *the subrings $R\tau$ and $R'\tau'$ of Ω are free over k* (*see* II, § 16, *Definition* 2).

A similar definition can be given for the case of fields, namely, as follows:

DEFINITION 2. *By a free join of two extension fields K/k and K'/k of k we mean the composite concept (F, τ, τ') consisting of* A FIELD *F containing k, a k-isomorphism τ of K into F and a k-isomorphism τ' of K' into F such* *that the following conditions are satisfied:* (1) *F is the compositum* $(K\tau, K'\tau')$ *of the two fields $K\tau$, $K'\tau'$, that is, no proper subfield of F contains both fields $K\tau$ and $K'\tau'$*; (2) *$K\tau$ and $K'\tau'$ are free over k.*

We note, in the case of fields K and K', that if condition (1) of Definition 2 is satisfied and if we denote by S the subring $[K\tau, K'\tau']$ of F, then (F, τ, τ') is a free join of the fields K and K' if and only if $[S, \tau, \tau']$ is a free join of the integral domains K, K', in the sense of Definition 1. On the other hand, if (Ω, τ, τ') is a free join of two integral domains R/k and R'/k, and if K, K' and F denote quotient fields of R, R' and Ω respectively, then τ and τ' can be canonically extended to isomorphisms τ_1 and τ'_1 of K and K' respectively, into F, and then it is immediately seen that (F, τ_1, τ'_1) is a free (field-theoretic) join of the fields K and K', over k. Thus, Definitions 1 and 2 are essentially interchangeable. From the standpoint of tensor products it is more convenient to use Definition 1 of free joins even in the case of fields K/k and K'/k, despite the fact that the free join, in the sense of Definition 1, is itself not necessarily a field.

The *existence* of free joins of R/k and R'/k can be shown as follows: We fix a transcendence basis $B = \{x_i\}$ of R/k and a transcendence basis $B' = \{x'_j\}$ of R'/k. We then consider a pure transcendental extension $k(\{y_i\}, \{y'_j\})$ of k, where the y_i and the y'_j are "indeterminates" and where the sets $\{y_i\}$ and $\{y'_j\}$ have the same cardinal number as B and B' respectively. Let Σ denote an algebraic closure of the field $k(\{y_i\}, \{y'_j\})$. There exists a k-isomorphism τ_0 of $k(\{x_i\})$ onto $k(\{y_i\})$ such that $x_i\tau_0 = y_i$. Since the quotient field of R is an algebraic extension of $k(\{x_i\})$ and Σ is algebraically closed, τ_0 can be extended to an isomorphism τ of R onto some subring L of Σ (see II, § 14, Theorem 33). Similarly, there exists an isomorphism τ' of R' onto some subring L' of Σ such that each element of k is mapped into itself and each x'_j is mapped into y'_j. Let $\Omega = [L, L']$. Then it is immediately seen that (Ω, τ, τ') is a free join of R and R'.

Let (Ω, τ, τ') and $(\Omega^\star, \tau^\star, \tau'^\star)$ be two free joins of R/k and R'/k. Let

$$L = R\tau,\ L' = R'\tau';\ L^\star = R\tau^\star,\ L'^\star = R'\tau'^\star.$$

Then $\tau^{-1}\tau^\star$ is a k-isomorphism of L onto $L^\star$, and similarly $\tau'^{-1}\tau'^\star$ is a k-isomorphism of L' onto $L'^\star$.

DEFINITION 3. *Two free joins (Ω, τ, τ') and $(\Omega^\star, \tau^\star, \tau'^\star)$ of R/k and R'/k are said to be equivalent if there exists a k-isomorphism ψ of Ω onto $\Omega^\star$ such that ψ coincides with $\tau^{-1}\tau^\star$ on $R\tau$ and with $\tau'^{-1}\tau'^\star$ on $R'\tau'$.*

We note that if there exists an isomorphism ψ satisfying the above conditions, this isomorphism is uniquely determined, for Ω is generated by $R\tau$ and $R'\tau'$.

To find all the equivalence classes of free joins of R/k and R'/k we consider the tensor product $R \otimes R'$. Since in studying the free joins of R/k and R'/k it is permissible to replace these rings by arbitrary k-isomorphic rings, we identify R and R' with suitable subrings of $R \otimes R'$. We have now therefore: $R \otimes R' = [R, R']$, and R, R' are linearly disjoint over k.

Let (Ω, τ, τ') be a free join of R/k and R'/k. By Theorem 34 (§ 14), there exists a homomorphism f of $R \otimes R'$ into Ω such that $f = \tau$ on R and $f = \tau'$ on R'. Since $\Omega = [R\tau, R'\tau']$, f is a mapping of $R \otimes R'$ onto Ω and is uniquely determined by the above conditions. We shall call f the *canonical homomorphism* of $R \otimes R'$ onto (Ω, τ, τ').

THEOREM 37. *The kernel of the canonical homomorphism f of $R \otimes R'$ onto a free join (Ω, τ, τ') of R and R' over k is a prime ideal $\mathfrak{p}$ all elements of which are zero-divisors in $R \otimes R'$.* If f_1 *is the canonical homomorphism of $R \otimes R'$ onto another free join $(\Omega_1, \tau_1, \tau'_1)$ of R and R' over k, then (Ω, τ, τ') and $(\Omega_1, \tau_1, \tau'_1)$ are equivalent free joins of R and R' if and only if f and f_1 have the same kernel. Furthermore, if $\mathfrak{p}$ is any prime ideal in $R \otimes R'$ all elements of which are zero-divisors in $R \otimes R'$, and if φ, φ' denote the restrictions to R and R' respectively of the canonical homomorphism of $R \otimes R'$ onto $R \otimes R'/\mathfrak{p}$, then $(R \otimes R'/\mathfrak{p}, \varphi, \varphi')$ is a free join of R and R' over k.*

PROOF. Let B be a transcendence basis of R/k and let similarly B' be a transcendence basis of R'/k. The rings R and R' contain the polynomial rings $k[B]$, $k[B']$, and since R and R' are linearly disjoint over k it follows that $B \cap B'$ is the empty set and that the elements of $B \cup B'$ are algebraically independent over k. Thus $R \otimes R'$ contains the polynomial ring $k[B, B']$ ($= k[B] \otimes k[B']$; see Example 2, § 14, p. 184). Since R, R' and $k[B, B']$ are integral domains, it follows from Theorem 36, § 14, that no element of $k[B, B']$, different from zero, is a zero-divisor in $R \otimes R'$. Thus the total quotient ring of $R \otimes R'$ contains as subring the quotient field $k(B, B')$ of $k[B, B']$. Since every element of R is algebraically dependent on $k(B)$ and every element of R' is algebraically dependent on $k(B')$, it follows that every element x of $R \otimes R'$ ($= [R, R']$) is algebraic over $k(B, B')$, that is, satisfies an equation of the form

$$(1) \qquad a_0x^n + a_1x^{n-1} + \cdots + a_{n-1}x + a_n = 0, \quad a_i \in k[B, B'].$$

The proof of this assertion is the same as the proof of the similar assertion in field theory (see II, § 3, p. 60). The fact that the total quotient ring of $R \otimes R'$ may not be a field implies no changes in the proof; what matters is that we are still dealing with algebraic dependence relative to *a field*, namely, relative to the field $k(B, B')$. It all amounts to proving that if x, y are elements of $R \otimes R'$ which are algebraic over $k(B, B')$, then also every element of the ring $k[x, y]$ is algebraic over k. If x satisfies an algebraic equation of degree n, over k, and y satisfies an algebraic equation of degree m, then it is seen at once that $k[x, y]$ is a finite dimensional vector space over k, spanned by the monomials $x^i y^j$, $0 \leqq i < n$, $0 \leqq j < m$. This shows that the powers $1, z, z^2, \cdots, z^{mn}$ of any element z of $k[x, y]$ are linearly dependent over k.

Let x be an element of the kernel $\mathfrak{p}$ of the canonical homomorphism f of $R \otimes R'$ onto (Ω, τ, τ'). We consider an equation (1) of least degree n satisfied by x over $k[B, B']$. We have, then, $a_n \in \mathfrak{p} \cap k[B, B']$. Now, since (Ω, τ, τ') is a free join of R/k and R'/k, it follows that $B\tau \cap B'\tau' = \emptyset$ and that the elements of $B\tau \cup B'\tau'$ are algebraically independent over k. Hence the restriction of f to $k[B, B']$ is an isomorphism, that is, $\mathfrak{p} \cap k[B, B'] = (0)$. Thus $a_n = 0$. On the other hand, by our choice of the relation (1), we have $a_0x^{n-1} + a_1x^{n-1} + \cdots + a_{n-1} \neq 0$. Therefore x is a zero-divisor.

Assume that (Ω, τ, τ') and $(\Omega_1, \tau_1, \tau'_1)$ are two free joins of R/k and R'/k, and let $\mathfrak{p}$ and $\mathfrak{p}_1$ be the kernels of the canonical homomorphisms f and f_1 of $R \otimes R'$ onto Ω and Ω_1 respectively. If the two free joins (Ω, τ, τ'), $(\Omega_1, \tau_1, \tau'_1)$ are equivalent, let ψ be the k-isomorphism of Ω onto Ω_1 such that $\psi = \tau^{-1}\tau_1$ on $R\tau$ and $\psi = \tau'^{-1}\tau'_1$ on $R'\tau'$. Then $f\psi$ is a homomorphic mapping of $R \otimes R'$ onto $(\Omega_1, \tau_1, \tau'_1)$ such that $f\psi = \tau_1$ on R and $f\psi = \tau'_1$ on R'. Hence $f' = f\psi$, and since ψ is an isomorphism it follows that the kernels of f and f' coincide. Conversely, if $\operatorname{Ker} f = \operatorname{Ker} f'$, then it is clear that if we set $\psi = f^{-1}f'$ then ψ will be an isomorphic mapping of Ω onto Ω_1 such that $\psi = \tau^{-1}\tau_1$ on $R\tau$ and $\psi = \tau'^{-1}\tau'_1$ on $R'\tau'$, and hence (Ω, τ, τ') and $(\Omega_1, \tau_1, \tau'_1)$ are equivalent free joins of R and R' over k.

Finally, if $\mathfrak{p}$ is any prime ideal in $R \otimes R'$ consisting entirely of zero-divisors, then by the lemma proved in § 14 we know that $\mathfrak{p} \cap R = \mathfrak{p} \cap R' = (0)$, and we have pointed out above that only the zero in $k[B, B']$ is a zero-divisor in $R \otimes R'$, whence $\mathfrak{p} \cap k[B, B'] = (0)$. From this it follows that the canonical homomorphism g of $R \otimes R'$ onto $(R \otimes R')/\mathfrak{p}$ induces isomorphisms of R, R' and $k[B, B']$. From this it follows at once (after identifying kg with k) that the rings Rg and $R'g$ are free over k (in $(R \otimes R')/\mathfrak{p}$). This completes the proof of the theorem.

COROLLARY 1. *If two prime ideals* $\mathfrak{p}$ *and* $\mathfrak{p}_1$ *in* $R \otimes R'$ *consist entirely of zero-divisors and if* $\mathfrak{p} \supset \mathfrak{p}_1$, *then* $\mathfrak{p} = \mathfrak{p}_1$.

For let x be any element of $R \otimes R'$ which is not in $\mathfrak{p}_1$ and let

$$b_0x^m + b_1x^{m-1} + \cdots + b_m \equiv 0(\mathfrak{p}_1), \quad b_i \in k[B, B'],$$

be a congruence mod $\mathfrak{p}_1$, with coefficients in $k[B, B']$, of least degree m, satisfied by x (there exist such congruences since x even satisfies an exact equation of type (1)). Then $b_m \not\equiv 0$, for in the contrary case we could divide the congruence by x (since $x \notin \mathfrak{p}_1$), and therefore $b_m \notin \mathfrak{p}$ (since $\mathfrak{p} \cap k[B, B'] = (0)$). So $x \notin \mathfrak{p}$, for $b_0x^m + b_1x^{m-1} + \cdots + b_m \in \mathfrak{p}_1 \subset \mathfrak{p}$. This shows that $\mathfrak{p} = \mathfrak{p}_1$.

COROLLARY 2. *If the zero-ideal in* $R \otimes R'$ *is primary* (*or equivalently: if every zero-divisor in* $R \otimes R'$ *is nilpotent*), *then any two free joins of* R/k *and* R'/k *are equivalent.*

For in that case the radical of (0) is a prime ideal $\mathfrak{p}$ containing all the zero divisors of $R \otimes R'$, and any other prime ideal in $R \otimes R'$ must contain $\mathfrak{p}$.

Theorem 37 gives us a necessary and sufficient condition for the unicity of a free join of R/k and R'/k (up to equivalence); it is that $R \otimes R'$ contain only one prime ideal consisting entirely of zero-divisors.

We shall derive below another important necessary and sufficient condition for the unicity of a free join. We first introduce the notion of *quasi-linear disjointness*:

DEFINITION 4. *If* S *is a unitary overring of a field* k, *of characteristic* p, *two subspaces* L *and* L' *of* S *are said to be quasi-linearly disjoint over* k *if the following condition is satisfied: whenever elements* $x_1, x_2, \cdots, x_n$ *of* L *and elements* $x'_1, x'_2, \cdots, x'_m$ *of* L' *are such that for any integer* $e \geqq 0$ *the* p^e*-th powers of the* x_i *are linearly independent over* k *and the* p^e*-th powers of the* x'_j *are linearly independent over* k, *then the* mn *products* $x_ix'_j$ *are also linearly independent over* k.

Quasi-linear disjointness is clearly a symmetric relationship between subspaces L and L' and is relative to the preassigned ground field. If the characteristic is zero, then quasi-linear disjointness coincides with linear disjointness, if we set $p = 1$ in that case.

The following property is equivalent to quasi-linear disjointness and will be the one most frequently used in the sequel:

(QLD). *Whenever* $x_1, x_2, \cdots, x_n$ *are elements of* L *such that for any integer* $e \geqq 0$ *the* p^e*-th powers of the* x_i *are linearly independent over* k, *then* $x_1, x_2, \cdots, x_n$ *are linearly independent over* L'.

For assume that L and L' are quasi-linearly disjoint over k and let there be a relation of the form $\sum_i u'_ix_i = 0$, where the u'_i are in L' and the x_i satisfy the condition stated in (QLD). We denote by M_e the subspace $\sum_i k u'^{p^e}_i$ of S ($e \geqq 0$). It is clear that $\dim M_e \geqq \dim M_{e+1}$. Let s be an integer such that $\dim M_s = \dim M_e$ for all $e \geqq s$. We wish to prove that the u'_i are all zero. In the proof, we may replace the elements x_i by their p^s-th powers y_i, for also the y_i satisfy the condition stated in (QLD) and since the y_i satisfy also the relation $\sum_i u'^{p^s}_i y_i = 0$. We may therefore assume, without loss of generality, that $s = 0$, that is, that the spaces M_e all have the same dimension, say, dimension m. Let $\{x'_1, x'_2, \cdots, x'_m\}$ be a basis of M_0 over k and let $u'_i = \sum_j a_{ij} x'_j$, where the a_{ij} are in k. For any integer e the p^e-th powers of the x'_j will span the space M_e *and therefore will be independent over k*. Therefore, by the quasi-linear disjointness of L and L' over k, the products $x_i x'_j$ are also linearly independent over k. Since the relation $\sum_i u'_i x_i = 0$ yields the relation $\sum_{i,j} a_{ij} x_i x'_j = 0$, it follows that all the a_{ij} are zero and that consequently also all the u'_i are zero.

The proof of the converse, that is, that (QLD) implies quasi-linear disjointness, is straightforward (and is similar to the proof given in II, § 15, in the case of linear disjointness).

THEOREM 38. *Let (Ω, τ, τ') be a free join of two integral domains R/k and R'/k. A necessary and sufficient condition that every free join of R/k and R'/k be equivalent to (Ω, τ, τ') is that the rings $R\tau$ and $R'\tau'$ be quasi-linearly disjoint over k.*

PROOF. Assume that $R\tau$ and $R'\tau'$ are quasi-linearly disjoint over k. It will be sufficient to show that the kernel $\mathfrak{p}$ of the canonical homomorphism f of $R \otimes R'$ onto Ω consists entirely of nilpotent elements, for then it will follow that every prime ideal of $R \otimes R'$ will contain $\mathfrak{p}$ and hence that $\mathfrak{p}$ is the only prime ideal in $R \otimes R'$ which consists of zero-divisors (Corollary 1 to Theorem 37). Let z then be an element of $\mathfrak{p}$. We write z in the form $z = \sum_{i=1}^{m} x_i x'_i$, with $x_i \in R$ and $x'_i \in R'$, and we choose this expression of z in such a manner that the number m of terms $x_i x'_i$ is minimum. We denote by $\lambda(z)$ this minimum number m. In a similar way we define $\lambda(z^{p^e})$ for any integer $e \geqq 0$. We set $\lambda(z) = 0$ if $z = 0$. Since $z^{p^e} = \sum x_i^{p^e} x'^{p^e}_i$ it follows that $\lambda(z^{p^e}) \leqq \lambda(z)$, and more generally, that $\lambda(z^{p^e}) \geqq \lambda(z^{p^\rho})$ if $e \leqq \rho$. We choose one integer ρ such that $\lambda(z^{p^\rho}) \leqq \lambda(z^{p^e})$ for all e. We shall show that $z^{p^\rho} = 0$, and this will prove our assertion. Assuming the contrary, let $z^{p^\rho} = \sum_{i=1}^{h} u_i u'_i$, where

$h = \lambda(z^{p^\rho}) \neq 0$, $u_i \in R$, $u'_i \in R'$. Since $z \in \mathfrak{p}$, we have $\sum_{i=1}^{h} (u_i\tau)(u'_i\tau') = 0$. Now, since τ is an isomorphism of R and since none of the u_i is zero, the above relation implies that the $u'_i\tau'$ are linearly dependent over $R\tau$. By the quasi-linear disjointness of $R\tau$ and $R'\tau'$ over k it follows that there exists an integer e such that the elements $u'^{p^e}_i\tau'$ are linearly dependent over k. Then also the elements $u'^{p^e}_i$ are linearly dependent over k, and this obviously implies that $\lambda(z^{p^{\rho+e}}) < \lambda(z^{p^\rho})$, in contradiction with our choice of ρ.

For the proof of the necessity it will be more convenient to pass to the quotient fields K and K' of $R\tau$ and $R'\tau'$ in the quotient field F of Ω. Our assumption is now to the effect that every free join $(F^\star, \varphi, \varphi')$ of K and K' over k is equivalent to (F, τ, τ'), that is, that there exists an isomorphism ψ of F onto $F^\star$ such that $\psi = \varphi$ on K and $\psi = \varphi'$ on K'. We have to prove that K and K' are quasi-linearly disjoint over k. The proof will be divided into several parts. We shall denote by Σ an algebraic closure of the field F.

1. We shall first assume that *K is a finite separable extension of k.* Let $[K:k] = n$ and let w be a primitive element of K/k. We shall prove that in this case K and K' are not only quasi-linearly disjoint but even linearly disjoint over k. To prove this it will be sufficient to show, in view of the lemma of II, § 15, that $1, w, w^2, \cdots, w^{n-1}$ are linearly independent over K'. In other words, we have to show that the minimal polynomial $f(X)$ of w in $k[X]$ remains irreducible in $K'[X]$. Let $w_1, w_2, \cdots, w_n$ be the roots of $f(X)$ in Σ $(w_1 = w)$, and let φ_i be the k-isomorphism of $k(w_1)(= K)$ onto $k(w_i)$ such that $\varphi_i(w_1) = w_i$. If $F^\star = K'(w_i)$, then $(F^\star, \varphi_i, 1)$ (where 1 denotes the identity map of K') is a free join of K/k and K'/k (since K is an algebraic extension of k). Since, by assumption, all free joins of K/k and K'/k are equivalent, it follows that φ_i can be extended to a K'-isomorphism of $K'(w_1)$ onto $K'(w_i)$. This signifies that the n roots w_i of $f(X)$ are also conjugates over K' and that consequently $f(X)$ is irreducible over K', as was asserted.

2. The following assertion is obvious: *if every field L between k and K which is finitely generated over k has the property that L and K' are quasi-linearly disjoint (resp., linearly disjoint) over k, then also K and K' are quasi-linearly disjoint (resp., linearly disjoint) over k.* On the other hand, we assert the following: *if all the free joins of K/k and K'/k are equivalent, and if L is any field between k and K such that K is an algebraic extension of L, then also all the free joins of L/k and K'/k are equivalent.* For let $(L^\star, \varphi, \varphi')$ be a free join of L/k and K'/k, and let Δ be an algebraic

closure of $L^\star$. Since K is an algebraic extension of L, the isomorphism φ of L can be extended to an isomorphism φ_1 of K into Δ. Then if $F^\star$ denotes the compositum of $K\varphi_1$ and $K'\varphi'$ in Δ, $(F^\star, \varphi_1, \varphi')$ is a free join of K/k and K'/k. Hence there exists as isomorphism ψ of (K, K') onto $F^\star$ such that $\psi = \varphi_1$ on K and $\psi = \varphi'$ on K'. The restriction of ψ to (L, K') maps isomorphically (L, K') onto $L^\star$ and is equal to φ on L. This shows that $(L^\star, \varphi, \varphi')$ is equivalent to $((L, K'), 1, 1)$.

3. From parts 1 and 2 it follows at once that our theorem holds in the case in which K is a (finite or infinite) separable algebraic extension of k, and that in that case K and K' are linearly disjoint over k. Now, let K be an arbitrary algebraic extension of k and let L be the maximal separable extension of k in K. By part 2, all the free joins of L/k and K'/k are equivalent, and hence L and K' are linearly disjoint over k. This obviously implies that K and K' are quasi-linearly disjoint over k. (If $x'_1, x'_2, \cdots, x'_m$ are elements of K' such that, for any integer $e \geqq 0$, the p^e-th powers of these elements are linearly independent over k, then the p^e-th powers of these elements are also linearly independent over L, for all $e \geqq 0$, and therefore $x'_1, x'_2, \cdots, x'_m$ are linearly independent over K since if $c_1, c_2, \cdots, c_n$ are elements of K, then $c_1^{p^e}, c_2^{p^e}, \cdots, c_n^{p^e}$ are in L for some $e \geqq 0$.) *We have therefore completed the proof of the theorem in the case in which K is an algebraic extension of k.*

We now consider the general case. Let K_0 be a field between k and K such that K_0 is a pure transcendental extension of k and K is an algebraic extension of K_0. We denote by K'_0 the compositum (K_0, K') in F and we regard K and K'_0 as extensions of K_0. *We assert that all free joins of K/K_0 and K'_0/K_0 are equivalent.* Let $(F^\star, \sigma, \sigma')$ be a free join of K/K_0 and K'_0/K_0. We have to show that there exists an isomorphism ψ of F onto $F^\star$ such that $\psi = \sigma$ on K and $\psi = \sigma'$ on K'_0. Let σ'_1 be the restriction of σ' to K'. The field $K'_0\sigma'$ is the compositum of K_0 $(= K_0\sigma')$ and $K'\sigma'_1$ $(= K'\sigma')$. Hence $F^\star$ is the compositum of $K\sigma$ and $K'\sigma'_1$ (since $K_0 \subset K\sigma$). Since K and K' are free over k, also K_0 and K' are free over k. Therefore also K_0 and $K'\sigma'_1$ are free over k (since σ' is a k-isomorphism of (K_0, K') onto $(K_0, K'\sigma'_1)$). Since K is an algebraic extension of K_0, it follows at once that $K\sigma$ and $K'\sigma'_1$ are free over k. Thus $(F^\star, \sigma, \sigma'_1)$ is a free join of K/k and K'/k. By our assumption, there must exist an isomorphism ψ of F onto $F^\star$ such that $\psi = \sigma$ on K and $\psi = \sigma'_1$ on K'. Then we have $\psi = \sigma'$ on K'_0 (since both σ and σ' are equal to the identity on K_0 and $\sigma' = \sigma'_1$ on K'). This proves our assertion.

It follows now, by part 3 of the proof, that K and K'_0 are quasi-linearly disjoint over K_0. Now let $x_1, x_2, \cdots, x_m$ be elements of K' such that $x_1^{p^e}, x_2^{p^e}, \cdots, x_m^{p^e}$ are linearly independent over k, for all $e \geqq 0$. Then for all $e \geqq 0$ the elements $x_1^{p^e}, x_2^{p^e}, \cdots, x_m^{p^e}$ are also linearly independent over K_0, *since K_0 and K' are linearly disjoint over k* (see Lemma, Corollary 1, II, § 15). Since the x_i belong to K'_0 and since K and K'_0 are quasi-linearly disjoint over K_0, it follows that $x_1, x_2, \cdots, x_m$ are also linearly independent over K, and this establishes the fact that K and K' are quasi-linearly disjoint over k.

The proof of the theorem is now complete.

COROLLARY 1. *If K/k and K'/k are two abstract extension fields of k and if $K\tau$ and $K'\tau'$ are quasi-linearly disjoint over k for one particular free join (F, τ, τ') of K/k and K'/k, then all the free joins of K/k and K'/k are equivalent.*

In view of this corollary we can now define linear or quasi-linear disjointness of two abstract fields K/k and K'/k as follows: *we say that K and K' are linearly or quasi-linearly disjoint over k (as abstract fields) if for one particular free join (F, τ, τ') of K/k and K'/k the fields $K\tau$ and $K'\tau'$ are linearly or quasi-linearly disjoint over k (as subfields of F).* Our lemma insures that this definition is independent of the choice of the free join of K/k and K'/k (note that linear disjointness implies quasi-linear disjointness). Dealing with abstract fields K and K' which are quasi-linearly disjoint over a common subfield k, we shall frequently identify them with their isomorphic images $K\tau$ and $K'\tau'$, in a free join of K/k and $K'k$. We shall therefore often regard K and K', without further ado, as subfields of a bigger field F such that F is the compositum of K and K' and such that K and K' (as subfields of F) are free over k. Our lemma insures that this identification is not ambiguous, for in the presence of quasi-linear disjointness the free join F is uniquely determined to within equivalence.

COROLLARY 2. *A necessary and sufficient condition that all free joins of two integral domains R/k and R'/k be equivalent to each other is that the zero ideal of $R \otimes R'$ be primary.*

The sufficiency has already been proved (Corollary 2 to Theorem 37). On the other hand, if all free joins of R/k and R'/k are equivalent to each other, then, given a free join (Ω, τ, τ') of R/k and R'/k, the rings $R\tau$ and $R'\tau'$ are quasi-linearly disjoint over k. But then, by the first part of the proof of Theorem 38, every zero-divisor of $R \otimes R'$ is nilpotent.

THEOREM 39. *Given two field extensions K and K' of a field k such that one of these two fields is a separable extension of k, then the tensor product $K \otimes_k K'$ has no nilpotent elements (other than zero).*

PROOF. Let, say, K be a separable extension of k. Since every element of $K \otimes K'$ is contained in a tensor product $K_1 \otimes K'$ where K_1 is a subfield of K which is finitely generated over k, we may assume that K is finitely generated over k. Then K will be separably generated over k (II, § 15, Theorem 35). Let B be a separating transcendence basis of K/k. By the Corollary to Theorem 36 (§ 14) the elements of B are algebraically independent over K' and the total quotient ring of $K \otimes_k K'$ contains the tensor product $K \otimes_{k(B)} K'(B)$. Consequently it is sufficient to show that every zero-divisor of $K \otimes_{k(B)} K'(B)$ is nilpotent. This time the field K is a separable algebraic extension of the basic field $k(B)$. Hence we have achieved a reduction to the case in which K *is a separable algebraic extension of k.*

It is clear that any zero-divisor in $K \otimes K'$ is already a zero-divisor in some subring $K_1 \otimes K'_1$ of $K \otimes K'$, where K_1 and K'_1 are subfields of K and K' which are finitely generated over k. Hence we may now assume that K *is a finite separable extension of k.*

Let x be a primitive element of K/k and let $f(X)$ be the minimal polynomial of x over k. Since $K = k[X]/(f(X))$ and since $k[X] \otimes K' = K'[X]$ it follows from Theorem 35 that $K \otimes K' \cong K'[X]/(f(X))$. Since $f(X)$ is a separable polynomial, it is a product of *distinct* irreducible polynomials in $K'[X]$. Consequently $K \otimes K'$ is a direct sum of fields, and thus $K \otimes K'$ has no nilpotent elements.

COROLLARY. *If k is a perfect field, then $K \otimes_k K'$ has no nilpotent elements (other than zero).*

If k is a subfield of a field K, we say that k is *quasi-maximally algebraic* (q.m.a.) in K if every element of K which is separable algebraic over k is contained in k. We say that k is *maximally algebraic* (m.a.) in K if k coincides with its algebraic closure *in* K.

We shall need the following lemma in which some elementary results of V, §§ 1, 2, 3, are used.

LEMMA. *If k is m.a. (or q.m.a.) in a field K and if $K' = K(B)$ is a purely transcendental extension of K, the set B being at the same time a set of generators and a transcendence basis of K'/K, then also $k(B)$ is m.a. (or q.m.a.) in K'.*

PROOF. It is clearly sufficient to prove the lemma in the case in which B is a finite set. In that case, using an induction on the number of elements of B, we can assume that B consists of a single element, say, t.

Let α be an element of $K(t)$ which is algebraic over $k(t)$. There exists a polynomial $d(T)$ in $k[T]$ (T, an indeterminate) such that $\alpha d(t)$ is integral over $k[t]$ and therefore also over $K[t]$. Since $K[t]$ is integrally closed in $K(t)$, it follows that $\alpha d(t)$ is an element $\varphi(t)$ of $K[t]$. We have a relation of integral dependence for $\varphi(t)$ over $k[t]$:

$$(2)\qquad [\varphi(t)]^q + a_{q-1}(t)[\varphi(t)]^{q-1} + \cdots + a_0(t) = 0, \quad a_i(t) \in k[t].$$

Since t is a transcendental over K, (2) must be an "identity" in t. If we substitute for t any algebraic quantity ξ over k, (2) shows that $\varphi(\xi)$ is algebraic over k. Thus, the polynomial $\varphi(T)$ in $K[T]$ is such that for any value of T in the algebraic closure $\bar{k}$ the corresponding value of $\varphi(T)$ also belongs to $\bar{k}$. Since $\bar{k}$ contains infinitely many elements, any formula which gives the coefficients of a polynomial $\varphi(T)$ of degree q in terms of the values of that polynomial for $q + 1$ distinct values of T (for instance, the Lagrange interpolation formula) shows that the coefficients of $\varphi(T)$ are algebraic over k. Since these coefficients are in K, the assumption that k is m.a. in K implies that the coefficients of $\varphi(T)$ are in k, and that therefore $\alpha = \varphi(t)/d(t) \in k(t)$, showing that $k(t)$ is m.a. in $K(t)$. If we assume only that k is q.m.a. in K, then the coefficients of $\varphi(T)$ are purely inseparable over k. Therefore there exists a power p^s of the characteristic p such that $[\varphi(t)]^{p^s} \in k[T]$, whence $\alpha^{p^s} \in k(t)$, showing that $k(t)$ is q.m.a. in $K(t)$. This completes the proof of the lemma.

Using the above lemma we now prove the following result:

THEOREM 40. *If K and K' are two field extensions of a field k and if k is q.m.a. in one of these two fields, then K and K' are quasi-linearly disjoint over k.*

PROOF. We identify K and K' with subfields of a field F such that F is the compositum of K and K' and such that K and K' are free over k (in F). We assume that k is q.m.a. in K. We have to show that K and K' are quasi-linearly disjoint over k.

Let B' be a transcendence basis of K'/k. Since the elements of B' are algebraically independent over K, it follows easily that in order to prove that K and K' are quasi-linearly disjoint over k it is sufficient to show that $K(B')$ and K' are quasi-linearly disjoint over the field $k(B')$. Now K' is an algebraic extension of $k(B')$, and, by the above lemma, $k(B')$ is q.m.a. in $K(B')$. We have therefore achieved a reduction to the case in which k is q.m.a. in K, and K' is an algebraic extension of k. Let now $x'_1, x'_2, \cdots, x'_n$ be elements of K' such that for any integer $e \geqq 0$ the p^e-th powers of x'_i are linearly independent over k. We have to show that the x'_i are linearly independent over K. It will be sufficient to show that for some $e \geqq 0$ the p^e-th powers of the

x'_i are linearly independent over K. Since for some $e \geqq 0$ the p^e-th powers of the x'_i are separable algebraic over k, it follows that we may assume that the x'_i are separable algebraic over k. Let w be a primitive element of $k(x'_1, x'_2, \cdots, x'_n)$ over k and let $g(X)$ be the minimal polynomial of w over k. Let m be the degree of $g(X)$. To show that the x'_i are linearly independent over K, it is sufficient to show that $1, w, w^2, \cdots, w^{m-1}$ are linearly independent over K, that is, *that* $g(X)$ *remains irreducible over* K. Now, if $g_1(X)$ is an irreducible factor of $g(X)$ in $K[X]$, then the coefficients of g_1 belong to a splitting field of $g(X)$ over k, and hence are separable algebraic over k (since w is separable algebraic over k). Since k is q.m.a. in K it follows that the coefficients of $g_1(X)$ are in k, whence $g_1(X) = g(X)$. This completes the proof of the theorem.

COROLLARY 1. *If k is an algebraically closed field and K, K' are any field extensions of k, then $K \otimes K'$ is an integral domain.*

By the above theorem and by the Corollary 2 to Theorem 38, every zero-divisor in $K \otimes K'$ is nilpotent. On the other hand, by the corollary to Theorem 39, $K \otimes K'$ has no nilpotent elements other than zero.

COROLLARY 2. *If k is q.m.a. in K and if K' is a separable extension of k, then $K \underset{k}{\otimes} K'$ is an integral domain.*

Obvious.

IV. NOETHERIAN RINGS

§ 1. **Definitions. The Hilbert basis theorem.** Let R be a ring (we recall that the term "ring" always means commutative ring). When R is regarded as a module over itself, the submodules of R are identical with the ideals of R. Thus the following *finiteness conditions* are equivalent, as follows immediately from III, § 10, Theorems 15 and 17.

a) ("Ascending chain condition," or a.c.c.) *Every strictly ascending chain* $\mathfrak{A}_1 < \mathfrak{A}_2 < \mathfrak{A}_3 < \cdots$ *of ideals of* R *is finite.* Or alternatively: *Given an ascending chain* $\mathfrak{A}_1 \subset \mathfrak{A}_2 \subset \mathfrak{A}_3 \subset \cdots$ *of ideals of* R, *there exists an integer* n *such that* $\mathfrak{A}_n = \mathfrak{A}_{n+1} = \mathfrak{A}_{n+2} = \cdots$.

b) ("Maximum condition.") *In every non-empty family of ideals of* R, *there exists a maximal element*, that is, an ideal not contained in any other ideal of the family. (Of course such a maximal element need not be a maximal ideal of R.) The maximum condition implies that every ideal $\mathfrak{A} \neq R$ is contained in a maximal ideal, as is easily seen by considering the family of all ideals $\neq R$ containing $\mathfrak{A}$.*

c) ("Finite basis condition.") *Every ideal* $\mathfrak{A}$ *of* R *has a finite basis*; this means, according to III, § 10, that $\mathfrak{A}$ contains a finite set of elements $a_1, \cdots, a_n$ such that $\mathfrak{A} = Ra_1 + \cdots + Ra_n + Ja_1 + Ja_2 + \cdots + Ja_n$, where J is the set of integers (such a set is called a *basis*, or *a set of generators*, of $\mathfrak{A}$). If R has an identity then $\mathfrak{A} = Ra_1 + \cdots + Ra_n$.

The rings satisfying conditions a), b) and c) play the most important role in this book. Since these rings were first studied by Emmy Noether, we give the following definition:

DEFINITION 1. *A ring is called noetherian if it has an identity and if it satisfies the equivalent conditions* a), b), *and* c). *A noetherian domain is a noetherian ring without proper zero-divisors.*

Let us now give typical examples of reasonings about noetherian rings.

1) *Every homomorphic image* R' *of a noetherian ring* R *is noetherian.* We use here the finite basis condition: the inverse image of an ideal

* This is true in every ring with unit element (see III, § 8, Note I, p. 151).

$\mathfrak{A}'$ of R' is an ideal $\mathfrak{A}$ of R, and the image in R' of a finite set of generators of $\mathfrak{A}$ is a finite set of generators of $\mathfrak{A}'$.

2) *In a noetherian ring R, every ideal $\mathfrak{A}$ contains a power of its radical* $\sqrt{\mathfrak{A}}$. We use again the finite basis condition. If $\{c_1, \cdots, c_h\}$ is a finite basis of $\sqrt{\mathfrak{A}}$, then there exists an integer k such that $c_i^k \in \mathfrak{A}$ for $i = 1, \cdots, h$. Let $m = h(k-1)+1$. A basis for $(\sqrt{\mathfrak{A}})^m$ is then provided by the power products $\prod_i c_i^{e_i}$, where $\sum_i e_i = m$. Since $m > h(k-1)$, one at least of the exponents e_i is not less than k, and this proves that all these products are in $\mathfrak{A}$, and therefore that $\mathfrak{A}$ contains $(\sqrt{\mathfrak{A}})^m$

3) *Every non-unit a in a noetherian domain R is a product of irreducible elements* (cf. I, § 14, p. 21). We use here twice the a.c.c. We first define, by induction on n, a sequence $\{a_n\}$ of elements of R satisfying the following conditions: (1) $a_1 = a$, (2) a_n is a proper divisor of a_{n-1}. The ideals Ra_n form a strictly ascending sequence; thus the sequence $\{a_n\}$ must be finite, and its last element is irreducible. We have thus proved that every non-unit in R has an irreducible divisor. This fact provides us with a new sequence $\{b_n\}$ of elements of R defined in the following way: $b_1 = a$, $b_{n-1} = b_n p_n$ where p_n is irreducible. As above, this sequence is finite, and its last element b_m is irreducible; therefore $a = p_2 \cdots p_m b_m$ is a product of irreducible elements.*

4) *Every ideal $\mathfrak{A}$ in a noetherian ring R contains a product of prime ideals.*† We use here the maximum condition. Suppose that the family (F) of ideals in R which do not contain any product of prime ideals is non-empty. Then (F) has a maximal element $\mathfrak{A}$. The ideal $\mathfrak{A}$ cannot be prime since $\mathfrak{A} \in (F)$; hence there exist ideals $\mathfrak{B}$, $\mathfrak{C}$, properly containing $\mathfrak{A}$ and such that $\mathfrak{B}\mathfrak{C} \subset \mathfrak{A}$ (III, § 8, p. 150). Since $\mathfrak{A}$ is a maximal element of (F), $\mathfrak{B}$ and $\mathfrak{C}$ do not belong to (F) and therefore contain products of prime ideals. Therefore $\mathfrak{B}\mathfrak{C}$, and thus also $\mathfrak{A}$, contains a product of prime ideals. This contradiction shows that (F) is empty. We note that in particular the ideal (0) in a noetherian ring is a product of prime ideals.

The theorem which follows below—the celebrated *Hilbert basis theorem*—taken together with its corollaries shows that noetherian rings exist and that the class of these rings is very extensive. This theorem will not actually be used in the present chapter, except for providing examples. It is however fundamental for the theory of polynomial ideals (vol. 2, chapter VII).

* In general such a factorization is not unique.

† Note that since R itself is a prime ideal, every prime ideal $\mathfrak{p}$ is a product of prime ideals (namely $\mathfrak{p} = R\mathfrak{p}$).

THEOREM 1. *If R is a noetherian ring, then so is any polynomial ring in a finite number of indeterminates over R.*

We give two proofs of this theorem, one making use of the maximum condition and the other of the basis condition. The second proof is substantially that given by Hilbert, who stated the theorem for the case when R is a field or the ring of integers.

We note that by induction it is sufficient to consider the case of a polynomial ring S in a single indeterminate x over R. For the first proof we need the following lemma:

LEMMA. *If $\mathfrak{A}$ is an ideal in S and if i is an integer ≥ 0, let $L_i(\mathfrak{A})$ denote the set of elements of R consisting of 0 and of the coefficients of x^i of all elements of $\mathfrak{A}$ which are of degree i. Then $\{L_i(\mathfrak{A})\}$ is an increasing sequence of ideals in R. If $\mathfrak{B}$ is any other ideal in S such that $\mathfrak{A} \subset \mathfrak{B}$ and $L_i(\mathfrak{A}) = L_i(\mathfrak{B})$ for $i = 0, 1, 2, \cdots$, then $\mathfrak{A} = \mathfrak{B}$.*

PROOF OF THE LEMMA. That $L_i(\mathfrak{A})$ is an ideal and is contained in $L_{i+1}(\mathfrak{A})$ follows from the fact that if $f(x) \in \mathfrak{A}$, $g(x) \in \mathfrak{A}$ and $a \in R$, then $f(x) + g(x)$, $af(x)$ and $xf(x)$ belong to $\mathfrak{A}$. Let now $g(x)$ be an element of $\mathfrak{B}$ of degree i. Since $L_i(\mathfrak{A}) = L_i(\mathfrak{B})$, there exists an element $f_i(x)$ of $\mathfrak{A}$, of degree i, such that $g(x) - f_i(x)$ is of degree at most $i - 1$. Using the fact that $\mathfrak{A}$ is contained in $\mathfrak{B}$ we note that $g(x) - f_i(x)$ also belongs to $\mathfrak{B}$ and it follows that we can define, by induction on j, a sequence $\{f_{i+j}(x)\}$ $(j = 0, 1, 2, \cdots)$ of elements of $\mathfrak{A}$ such that $f_{i+j}(x)$ is either zero or is of degree $i-j$ and such that the polynomial $g(x) - (f_i(x) + f_{i+1}(x) + \ldots + f_{i+j}(x))$ is of degree at most $i - j - 1$. This last polynomial is necessarily 0 when $j = i$, and thus $g(x) \in \mathfrak{A}$. This completes the proof of the lemma.

FIRST PROOF OF THE THEOREM. Let $\{(\mathfrak{A}_s), s = 0,1, \cdots\}$ be an increasing sequence of ideals of S. Consider the double sequence $\{L_i(\mathfrak{A}_j)\}$ of ideals of R. When either i or j is fixed, the corresponding simple sequence $\{L_i(\mathfrak{A}_j)\}$ is increasing. Let $L_p(\mathfrak{A}_q)$ be a maximal element of the above double sequence. We have $L_p(\mathfrak{A}_q) = L_i(\mathfrak{A}_j)$ if $i \geq p$ *and* $j \geq q$, and thus $L_i(\mathfrak{A}_q) = L_i(\mathfrak{A}_j)$ for $i \geq p$ and $j \geq q$. On the other hand, if we take i fixed, the a.c.c. shows that there exists an integer $n(i)$ such that $L_i(\mathfrak{A}_j) = L_i(\mathfrak{A}_{n(i)})$ for every $j \geq n(i)$, and what we just have seen signifies that for $i \geq p$, one may take $n(i) = q$. Hence the integer $n(i)$ is bounded, and there exists an integer n_0 such that $L_i(\mathfrak{A}_j) = L_i(\mathfrak{A}_{n_0})$ for every i, and for every $j \geq n_0$. Hence, by the lemma, $\mathfrak{A}_j = \mathfrak{A}_{n_0}$ for every $j \geq n_0$, and this completes the proof.

SECOND PROOF OF THE THEOREM. Let $\mathfrak{A}$ be an ideal in S. Denote by $\mathfrak{L}$ the set of elements of R consisting of 0 and of the coefficients of the highest degree terms of all elements of $\mathfrak{A}$. The set $\mathfrak{L}$ is an ideal in R, as follows from the fact that $\mathfrak{L}$ is the union of the ideals $L_i(\mathfrak{A})$ defined in the above lemma. [This can also be proved directly: if $a, b \in \mathfrak{L}$ there exist elements $f(x)$, $g(x)$ of $\mathfrak{A}$ having respectively ax^r, bx^s as highest degree term; if, for example, $r \geq s$, then the polynomial $f(x) - x^{r-s}g(x)$ is in $\mathfrak{A}$, and (if $a \neq b$) has $(a - b)x^r$ as highest degree term; thus $a - b \in \mathfrak{L}$; if $c \in R$, $cf(x) \in \mathfrak{A}$ and $ca \in \mathfrak{L}$.] Since R is a noetherian ring, the ideal $\mathfrak{L}$ has a finite basis $\{a_1, \cdots, a_n\}$. Let $f_1(x), \cdots, f_n(x)$ be elements of $\mathfrak{A}$ having respectively $a_1, \cdots, a_n$ as highest degree term coefficients; these elements generate an ideal $\mathfrak{A}' \subset \mathfrak{A}$. Let q be the highest integer among the degrees of the $f_i(x)$, let $g(x)$ be an element of $\mathfrak{A}$ of degree $\geq q$, and let ax^s be its highest degree term. Since $a \in \mathfrak{L}$, let us write $a = \sum_i c_i a_i$ $(c_i \in R)$, and let us consider the element $g_1(x) = \sum_i c_i f_i(x) x^{s-d_i}$ of $\mathfrak{A}'$ [d_i = degree of $f_i(x)$]. The polynomial $g(x) - g_1(x)$ is an element of $\mathfrak{A}$, of degree $\leq s - 1$. By successive applications of this procedure, we get a sequence $\{g_j(x)\}(j = 1, 2, \cdots)$ of elements of $\mathfrak{A}'$, having strictly decreasing degrees, such that the polynomial $g(x) - [g_1(x) + \cdots + g_r(x)]$ is of degree $\leq s - r$. This polynomial is of degree $\leq q - 1$ as soon as $r = s - q + 1$.

Let us now take care of the elements of $\mathfrak{A}$ of degree $< q$. They form a submodule A_q of the R-module generated by $\{1, x, \cdots, x^{q-1}\}$, and since R is noetherian, A_q is finitely generated (III, § 10, Theorems 17 and 18). What we just proved shows that $\mathfrak{A} = \mathfrak{A}' + \mathfrak{A}_q$. Therefore the ideal $\mathfrak{A}$ is finitely generated.

Let us, however, complete the proof without making use of the results in chapter III. The ideal $L_i(\mathfrak{A})$ defined in the lemma has a finite basis $\{a_{ij}\}(j = 1, \cdots, n(i))$. Let $f_{ij}(x)$ be an element of $\mathfrak{A}$ having $a_{ij}x^i$ as highest degree term. We prove that the ideal $\mathfrak{A}$ is generated by the $f_k(x)$ $(k = 1, \cdots, n)$ and the $f_{ij}(x)$ [for $0 \leq i \leq q - 1$ and $1 \leq j \leq n(i)$]. In fact, given any element $g(x)$ of degree $\leq q - 1$ of $\mathfrak{A}$, we may lower its degree by adding to it a linear combination of the $f_{ij}(x)$. By at most q applications of this process we get an element of $\mathfrak{A}$ of degree 0, which is therefore a linear combination of the $f_{0j}(x)$. This completes the proof.

COROLLARY 1. *A polynomial ring in a finite number of indeterminates over a field, or over the ring of integers, is noetherian.*

COROLLARY 2. *Let R be a noetherian ring, and let S be a ring unitary over R and containing elements $y_1, \cdots, y_n$ such that $S = R[y_1, \cdots, y_n]$. Then S is also noetherian.*

This is an immediate consequence of Theorem 1 and of the fact that S is a homomorphic image of the ring of polynomials in n indeterminates over R (I, § 18, Theorem 12). We may note also that the proof of the theorem itself applies to Corollary 2 if slight changes are made.

§ 2. **Rings with descending chain condition.** A large part of ideal theory is concerned with rings satisfying the *ascending* chain condition. For many purposes the *descending* chain condition (d.c.c.) is too restrictive. For example, the only *integral domains* which satisfy the descending chain condition are the fields. For, if a is a non-zero element of such an integral domain R, then applying the d.c.c. to the descending sequence $\{Ra^n\}$ of ideals, we find that there exists an exponent n such that $Ra^n = Ra^{n+1}$. Hence there exists an element b in R such that $a^n = ba^{n+1}$, or $a^n(1 - ba) = 0$. Since $a^n \neq 0$, we deduce that $ba = 1$, so that a has an inverse and R is a field.

As a consequence it follows that *if a ring R satisfies the d.c.c. then every prime ideal $\mathfrak{P}$ of R, $\mathfrak{P} \neq R$, is maximal.* For, also the *integral domain* $R/\mathfrak{P}$ satisfies then the d.c.c. and is therefore a field, whence $\mathfrak{P}$ is maximal in R.

Any finite ring, in particular the ring of residue classes of the ring J of integers modulo an integer $\neq 0$, satisfies the d.c.c. For modules and groups we have noticed that neither chain condition implies the other. For rings, however, the d.c.c. implies the a.c.c.:

THEOREM 2. *Let R be a ring with identity. For R to satisfy the d.c.c. it is necessary and sufficient that it satisfy the a.c.c. and that every prime ideal of R different from R be maximal.*

PROOF. We first show that *if the ideal* (0) *is a product of maximal ideals,* $(0) = \mathfrak{P}_1 \cdots \mathfrak{P}_n$, *then either chain condition implies the other.* We consider the sequence $R \supset \mathfrak{P}_1 \supset \mathfrak{P}_1\mathfrak{P}_2 \supset \cdots \supset \mathfrak{P}_1\mathfrak{P}_2 \cdots \mathfrak{P}_n = (0)$ of ideals of R, and we will prove that if either chain condition is satisfied, then that sequence can be refined to a composition series and hence R satisfies both chain conditions (III, § 11, Theorem 21). The difference R-module $\mathfrak{P}_1 \cdots \mathfrak{P}_{i-1} - \mathfrak{P}_1 \cdots \mathfrak{P}_{i-1}\mathfrak{P}_i$ is clearly annihilated by $\mathfrak{P}_i$, whence it may be considered as a module over $R/\mathfrak{P}_i$ (III, § 6, p. 146), that is, as a vector space over the field $R/\mathfrak{P}_i$. Our assertion now follows from the fact that in a vector space either chain condition implies that the vector space is finite dimensional and admits a composition series (III, § 12, p. 171).

Let us now assume that R satisfies the a.c.c. and that every prime ideal of R, different from R, is maximal. We have proved in § 1 (p. 200) that every ideal of R contains a product of prime ideals. In particular, (0) is a product of prime, and therefore maximal, ideals, whence R satisfies the d.c.c.

Assume conversely that R satisfies the d.c.c. We have already seen that every prime ideal $\mathfrak{P}$ in R, different from R, is then a maximal ideal. It remains to be shown that (0) is a product of maximal (or prime) ideals. Let $\mathfrak{D}$ be a minimal element of the set of those ideals which are products of prime ideals. We suppose that $\mathfrak{D} \neq (0)$, and will derive a contradiction. We set $\mathfrak{A} = (0):\mathfrak{D}$. Since $\mathfrak{D} \neq (0)$ and R contains an identity, we have $\mathfrak{A} \neq R$, whence the family of ideals in R properly containing $\mathfrak{A}$ is non-empty. Let $\mathfrak{B}$ be a minimal element of this family. The ideal $\mathfrak{P} = \mathfrak{A}:\mathfrak{B}$ contains $\mathfrak{A}$; we claim it is *prime*. In fact, if $c \notin \mathfrak{P}$ and $d \notin \mathfrak{P}$, we have $c\mathfrak{B} + \mathfrak{A} = d\mathfrak{B} + \mathfrak{A} = \mathfrak{B}$ since both $c\mathfrak{B} + \mathfrak{A}$ and $d\mathfrak{B} + \mathfrak{A}$ contain $\mathfrak{A}$, are contained in $\mathfrak{B}$ and are distinct from $\mathfrak{A}$; thus $cd\mathfrak{B} + \mathfrak{A} = c(d\mathfrak{B} + \mathfrak{A}) + \mathfrak{A} = c\mathfrak{B} + \mathfrak{A} = \mathfrak{B}$, $cd\mathfrak{B} \not\subset \mathfrak{A}$, and $cd \notin \mathfrak{P}$, showing that $\mathfrak{P}$ is prime. Since $\mathfrak{P}\mathfrak{B} \subset \mathfrak{A}$, we have $\mathfrak{D}\mathfrak{P}\mathfrak{B} = (0)$ and $(0):\mathfrak{D}\mathfrak{P} \supset \mathfrak{B} > \mathfrak{A} = (0):\mathfrak{D}$. This proves that $\mathfrak{D}\mathfrak{P} < \mathfrak{D}$, in contradiction with the minimality property of $\mathfrak{D}$. Q.E.D.

In the last part of the proof we have actually shown that in a ring with d.c.c. (here $R/\mathfrak{A}$), the annihilator of a proper minimal ideal is prime (and thus maximal).

§ **3. Primary rings.** In this section we study a rather special class of rings, and prove that every ring with d.c.c. is a direct sum of rings of this type.

DEFINITION. *A primary ring R is a ring with identity which contains at most one prime ideal $\neq R$.*

From the existence of at least one maximal ideal in a ring with identity, we deduce that a primary ring R has *exactly one* prime ideal $\mathfrak{M} \neq R$, and that this ideal is *maximal*.

Any field is a primary ring. More generally, if $\mathfrak{Q}$ is a primary ideal of an arbitrary ring R, having a maximal ideal $\mathfrak{M}$ as associated prime ideal, then $R/\mathfrak{Q}$ is primary. For, if $\mathfrak{P}/\mathfrak{Q}$ is a prime ideal of this ring, different from $R/\mathfrak{Q}$, then $\mathfrak{P}$ is a prime ideal of R containing $\mathfrak{Q}$ and $\mathfrak{P}$ also contains the radical $\mathfrak{M}$ of $\mathfrak{Q}$. This implies $\mathfrak{M} = \mathfrak{P}$ since $\mathfrak{M}$ is maximal, showing that $\mathfrak{M}/\mathfrak{Q}$ is the only prime ideal in $R/\mathfrak{Q}$, different from $R/\mathfrak{Q}$. In particular, the ring of residue classes modulo a power of a prime integer is primary.

When we limit ourselves to noetherian rings, there is a converse to the above property: *in a noetherian primary ring R, the ideal* (0) *is primary for the unique prime ideal $\mathfrak{P}$ of R*; for, (0) is a product of prime ideals (§ 1, p. 200) and is thus a power of $\mathfrak{P}$. In this case we notice also that R satisfies the descending chain condition, since every prime ideal of R is maximal (§ 2, Theorem 2).

The following lemma will be used:

LEMMA. *In a ring R having only one maximal ideal $\mathfrak{M}$, every idempotent e is either* 0 *or* 1.

If $e \neq 0$, $e \neq 1$, then the relation $e^2 = e$ (which is equivalent to $e(1 - e) = 0$) implies that e and $1 - e$ are zero divisors and cannot therefore be units in R. Thus the ideals Re and $R(1 - e)$ are proper ideals. As they are contained in maximal ideals, they are contained in $\mathfrak{M}$. Therefore $e \in \mathfrak{M}$, $1 - e \in \mathfrak{M}$ and $1 = e + (1 - e) \in \mathfrak{M}$; a contradiction.

THEOREM 3. *A ring with identity R which satisfies the descending chain condition is the direct sum of noetherian primary rings, and this decomposition is unique.*

PROOF. As was seen in § 2, (0) is a power product $\mathfrak{P}_1^{k(1)} \cdots \mathfrak{P}_n^{k(n)}$ of maximal ideals $\mathfrak{P}_i$ which may be assumed to be distinct. For $i \neq j$ the ideals $\mathfrak{P}_i^{k(i)}$ and $\mathfrak{P}_j^{k(j)}$ are comaximal (that is, their sum is R) by Theorem 31 of III, § 13, and this theorem shows also that $(0) = \mathfrak{P}_1^{k(1)} \cap \mathfrak{P}_2^{k(2)} \cap \cdots \cap \mathfrak{P}_n^{k(n)}$. We denote by R_i the intersection $\bigcap_{j \neq i} \mathfrak{P}_j^{k(j)}$. It follows from Theorem 32, of III, § 13 that R is the direct sum of the ideals R_i $(1 \leq i \leq n)$, and that the ring R_i is isomorphic to $R/\mathfrak{P}_i^{k(i)}$. Since this last ring is primary (and noetherian) as was noticed before, the existence of the direct decomposition is proved.

As for uniqueness we observe that to a direct sum decomposition $R = R_1 \oplus R_2 \oplus \cdots \oplus R_n$ there corresponds a decomposition of the identity 1 into a sum $1 = e_1 + \cdots + e_n$ of orthogonal idempotents (that is, such that $e_ie_j = 0$ for $i \neq j$), and conversely. Now, if S is any primary ring, then no two proper ideals of S are comaximal, and hence (III, § 13, Theorem 32) S is not a direct sum of ideals in S. In other words, the element 1 of S is not the sum of two idempotents f, g distinct from 0 and 1. Applying these remarks to the rings R_i we see that if the summands R_i are primary, no e_i is the sum of two idempotents e_if, e_ig distinct from 0 and e_i. Hence, if we have two decompositions $1 = \sum_i e_i = \sum_j f_j$ of 1 into orthogonal idempotents, we deduce from the relation $e_i = \sum_j e_if_j$ that e_i is equal to one of the idempotents e_if_j, say $e_i = e_if_{j(i)}$. Similarly $f_j = f_je_{i(j)}$. Thus $e_i = e_if_{j(i)} = e_if_{j(i)}e_{i[j(i)]}$, which implies $i = i[j(i)]$ since the idempotents e_i are orthogonal and $\neq 0$. Hence the idempotents e_i and f_j are in one to one correspondence, and the relations $e_i = e_if_{j(i)}$ and $f_{j(i)} = f_{j(i)}e_i$ show that they differ only by their indexing. Q.E.D.

Note that the ideals $\mathfrak{P}_1, \cdots, \mathfrak{P}_n$ are the only prime ideals of R: for, any prime ideal $\mathfrak{P}$ in R contains $(0) = \cap \mathfrak{P}_i^{k(i)}$, thus contains some $\mathfrak{P}_i$, whence $\mathfrak{P} = \mathfrak{P}_i$. This shows that R has only a finite number of prime ideals. This last fact is also an immediate consequence of the descending chain condition. For, suppose that we have an infinite sequence $\{\mathfrak{P}_i\}$ of distinct prime ideals of R. Then the sequence of products $\{\mathfrak{A}_i = \mathfrak{P}_1 \cdots \mathfrak{P}_i\}$ is decreasing, and $\mathfrak{A}_n = \mathfrak{A}_{n+1}$ for a suitable n. From $\mathfrak{P}_1 \cdots \mathfrak{P}_n = \mathfrak{P}_1 \cdots \mathfrak{P}_n\mathfrak{P}_{n+1}$ we deduce that $\mathfrak{P}_{n+1}$ contains the product $\mathfrak{P}_1 \cdots \mathfrak{P}_n$ and hence also some $\mathfrak{P}_i$ for $i \leq n$. This is a contradiction, since all the prime ideals of R are maximal.

§ 3[bis]. Alternative method for studying the rings with d.c.c. Let R be a ring with identity satisfying the d.c.c. One shows as in § 2 (p. 203) that every prime ideal in R is maximal. Denote by $\mathfrak{r}$ the intersection of all maximal ideals of R. As R satisfies the d.c.c. $\mathfrak{r}$ is already a finite intersection of maximal ideals of R, say $\mathfrak{r} = \bigcap_{i=1}^{n} \mathfrak{m}_i$. We claim that $\mathfrak{m}_1, \mathfrak{m}_2, \cdots, \mathfrak{m}_n$ *are the only prime ideals of* R. For, assume that there is a prime ideal $\mathfrak{p}$ in R, distinct from the $\mathfrak{m}_i$. As $\mathfrak{m}_i$ is maximal, we have $\mathfrak{m}_i \not\subset \mathfrak{p}$, and there exists an element x_i in $\mathfrak{m}_i$ such that $x_i \notin \mathfrak{p}$. Since $\mathfrak{p}$ is prime, we have $y = x_1 \cdots x_n \notin \mathfrak{p}$; on the other hand, from $y \in \mathfrak{m}_i$ for $1 \leq i \leq n$, we deduce that $y \in \mathfrak{r}$, in contradiction with the fact that $\mathfrak{p}$ contains $\mathfrak{r}$. We have thus proved:

LEMMA 1. *A ring R with identity satisfying the d.c.c. has only a finite number* $\mathfrak{m}_1, \cdots, \mathfrak{m}_n$ *of prime ideals, all of them maximal.*

We note that we have also shown above that none of the ideals $\mathfrak{m}_i$ contains the intersection of others. This, however, has to do with the following general property of maximal ideals which is valid in any ring and which is precisely what we have just proved above: if $\mathfrak{m}_1, \mathfrak{m}_2, \cdots, \mathfrak{m}_n$ are maximal ideals in a ring R, then they are the only prime ideals of R which contain the intersection $\mathfrak{m}_1 \cap \mathfrak{m}_2 \cap \cdots \cap \mathfrak{m}_n$.

The following result holds in arbitrary rings with identity (and is also of importance in the non-commutative case):

LEMMA 2. *Let R be a ring with identity. The intersection $\mathfrak{r}$ of all the maximal ideals in R is the set of all elements a in R such that* $1 + xa$ *is a unit for every x in R.*

PROOF. Consider first an element a in $\mathfrak{r}$ and the principal ideal $(1 + xa)$ generated by $1 + xa$. If this ideal were contained in a maximal ideal $\mathfrak{m}$, we would have $1 + xa \in \mathfrak{m}$, $a \in \mathfrak{m}$ (since $a \in \mathfrak{r}$), and thus $1 \in \mathfrak{m}$, which is impossible, since $\mathfrak{m} \neq R$. Hence $(1 + xa) = R$* and

* We use here the fact that, in a ring R with identity, every ideal distinct from R is contained in a maximal ideal (cf. III, § 8, Note I, p. 151).

$1 + xa$ is a unit. Conversely, if $1 + xa$ is a unit for every x, let us suppose that a is not contained in some maximal ideal $\mathfrak{m}$. Then $\mathfrak{m} + (a) = R$, and there exists an element x in R such that $1 + xa \in \mathfrak{m}$—a contradiction. Therefore a belongs to all maximal ideals.

LEMMA 3. *Let R be a ring with identity satisfying the d.c.c. The intersection $\mathfrak{r}$ of all maximal ideals in R is the set of all nilpotent elements of R, that is, the radical of* (0).*

PROOF. Let first a be a nilpotent element of $R: a^h = 0$. Since a^h belongs to every ideal of R, it follows that a belongs to every prime ideal of R, hence $a \in \mathfrak{r}$. Conversely, if $a \in \mathfrak{r}$, the d.c.c. applied to the descending sequence of principal ideals (a^n) shows that there exist an integer h and an element x in R such that $a^h = xa^{h+1}$, that is, such that $(1 - xa)a^h = 0$. Since $a \in \mathfrak{r}$, $1 - xa$ is invertible by Lemma 2, and hence $a^h = 0$.

We now prove that $\mathfrak{r}$ is a *nilpotent ideal*, that is, that some power $\mathfrak{r}^s$ of $\mathfrak{r}$ is (0). More generally:

LEMMA 4. *In a ring R satisfying the d.c.c. every ideal $\mathfrak{a}$, all the elements of which are nilpotent, is nilpotent.*

PROOF. The d.c.c. applied to the descending sequence $\{\mathfrak{a}^n\}$ shows that there exists an exponent h such that $\mathfrak{b} = \mathfrak{a}^h = \mathfrak{a}^{h+1} = \cdots$. Let us suppose that $\mathfrak{b} \neq (0)$, and let us consider the family (F) of all ideals $\mathfrak{w}$ in R such that $\mathfrak{b}\mathfrak{w} \neq (0)$. Since $\mathfrak{b}^2 = \mathfrak{b} \neq (0)$, we have $\mathfrak{b} \in (F)$ and (F) is non-empty. From the d.c.c. we deduce that (F) admits at least one minimal element; let $\mathfrak{v}$ be such an element. Since $\mathfrak{b}\mathfrak{v} \neq (0)$ there exists an element c in $\mathfrak{v}$ such that $\mathfrak{b}c \neq (0)$; thus $\mathfrak{v} \supset (c) \in (F)$, where (c) denotes the ideal $Rc + Jc$ (the least ideal containing c), and $\mathfrak{v} = (c)$ since $\mathfrak{v}$ is minimal in (F). On the other hand we have $\mathfrak{b} \cdot \mathfrak{b}c = \mathfrak{b}^2 c = \mathfrak{b}c \neq (0)$, and therefore $\mathfrak{b}c \in (F)$; this implies $\mathfrak{b}c = (c)$ as (c) is minimal in (F). In particular we have $c = bc$, with $b \in \mathfrak{b}$, whence $c = bc = b^2c = \cdots = b^nc = \cdots$. Since b is a nilpotent element it follows that $c = 0$, in contradiction with $\mathfrak{b}c \neq (0)$. Therefore $\mathfrak{b} = (0)$, and $\mathfrak{a}$ is a nilpotent ideal.

We are now in position to prove the structure theorem for rings satisfying the d.c.c. (Theorem 3, § 3). In fact, since $\mathfrak{r}$ is a finite intersection $\mathfrak{m}_1 \cap \cdots \cap \mathfrak{m}_n$ of maximal ideals $\mathfrak{m}_i$, it is also the product of the $\mathfrak{m}_i$ (III, § 13, Theorem 31). From $\mathfrak{r}^s = (0)$ (Lemma 4) we deduce that $(0) = \mathfrak{m}_1{}^s \mathfrak{m}_2{}^s \cdots \mathfrak{m}_n{}^s$, and that $(0) = \mathfrak{m}_1{}^s \cap \mathfrak{m}_2{}^s \cap \cdots \cap \mathfrak{m}_n{}^s$ (again by Theorem 31 of III, § 13). The remainder of the proof is as in the proof of Theorem 3 of § 3.

* A more general version of Lemma 3 has been proved in III, § 8, Note II, pp. 151–152.

REMARK 1. For the proof of uniqueness of the decomposition $R = R_1 \oplus \cdots \oplus R_n$ of R into primary rings, one may observe that such a decomposition gives at the same time a representation of (0) as an intersection $\mathfrak{p}_1^{s(1)} \cap \cdots \cap \mathfrak{p}_n^{s(n)}$ of powers of maximal ideals (cf. III, § 13, Theorem 32) and a representation $\mathfrak{r} = \mathfrak{p}_1 \cap \cdots \cap \mathfrak{p}_n$ of the radical $\mathfrak{r}$ of (0) as an intersection of maximal ideals. By Lemma 1 this last representation is unique, the $\mathfrak{p}_i$ being all the maximal ideals in R. Concerning the first one we notice that $\mathfrak{p}_i^{s(i)} = \mathfrak{p}_i^{s(i)+1} = \cdots$; in other words, $s(i)$ is the smallest exponent for which the sequence $\{\mathfrak{p}_i^s\}$ stops decreasing, and this determines $\mathfrak{p}_i^{s(i)}$ uniquely.

REMARK 2. The burden of the second part of the proof of Theorem 2 (§ 2), namely, the proof that the d.c.c. implies the a.c.c. and the maximality of every prime ideal, rested on proving that under the assumption of the d.c.c. the zero ideal is a product of maximal ideals. In this section we have given a second proof of this assertion and hence also a new proof of Theorem 2.

§ 4. The Lasker-Noether decomposition theorem. The theorem we are going to prove in this section states that in a noetherian ring every ideal is a finite intersection of primary ideals. In many respects this theorem reduces the study of arbitrary ideals to that of primary ideals. The theorem does not extend, however, to non-noetherian rings, even if infinite intersections are allowed. The theorem was first proved, in the case of polynomial rings, by the chess master Emanuel Lasker, who introduced the notion of primary ideal; his proof was involved and computational. To Emmy Noether is due the recognition that the theorem is a consequence of the a.c.c., and the proof given here is substantially hers.

The theorem follows immediately from two lemmas. Let us call *irreducible* an ideal which is not a finite intersection of ideals strictly containing it. Observe that a prime ideal is irreducible; but a primary ideal need not be. For instance, in a polynomial ring $R = k[x, y]$ in two independent variables x and y, over a field k, the ideal $\mathfrak{m} = (x, y)$ is maximal; its square (x^2, xy, y^2) is therefore primary, but we have that $\mathfrak{m}^2$ is the intersection of the two ideals $\mathfrak{m}^2 + R\cdot x$ and $\mathfrak{m}^2 + R\cdot y$.

LEMMA 1. *In a ring R with a.c.c. every ideal is a finite intersection of irreducible ideals.*

PROOF. Suppose that there exists an ideal for which the assertion of the lemma is false. Then the family (F) of all ideals of R which are not finite intersections of irreducible ideals is non-empty, and, by the maximum condition, admits a maximal element $\mathfrak{a}$. Since $\mathfrak{a}$ cannot be irreducible, it is an intersection $\mathfrak{b} \cap \mathfrak{c}$ of two ideals strictly containing $\mathfrak{a}$. By the maximal character of $\mathfrak{a}$ in (F), $\mathfrak{b}$ and $\mathfrak{c}$ are finite intersections of irreducible ideals, and so is $\mathfrak{a}$; a contradiction.

LEMMA 2. *In a ring with a.c.c. every irreducible ideal is primary.*

PROOF. Let $\mathfrak{q}$ be an ideal of R, and suppose that it is not primary, that is, that there exist elements b, c of R, not in $\mathfrak{q}$, such that $bc \in \mathfrak{q}$ and that no power of b lies in $\mathfrak{q}$. The ideals $\{\mathfrak{q}:(b^s)\}$ form an increasing sequence, and, by the a.c.c., there exists an exponent n such that $\mathfrak{q}:(b^n) = \mathfrak{q}:(b^{n+1})$. We claim that

$$\mathfrak{q} = (\mathfrak{q} + Rb^n) \cap (\mathfrak{q} + (c)). \tag{1}$$

It is clear that the ideal on the right-hand side of (1) contains $\mathfrak{q}$. Conversely, if x is an element of that ideal, we have $x = u + yb^n = v + zc + mc (u, v \in \mathfrak{q}, y, z \in R, m$ an integer). Since $bc \in \mathfrak{q}$, we have $bx \in \mathfrak{q}$, and thus $yb^{n+1} \in \mathfrak{q}$. From $\mathfrak{q}:(b^n) = \mathfrak{q}:(b^{n+1})$ we deduce that $yb^n \in \mathfrak{q}$, $x \in \mathfrak{q}$, and this establishes (1). Together with the hypothesis on b and c, the relation (1) shows that $\mathfrak{q}$ is not irreducible: for $\mathfrak{q} + (c) > \mathfrak{q}$ since $c \notin \mathfrak{q}$, and $\mathfrak{q} + Rb^n > \mathfrak{q}$ since $b^{n+1} \in Rb^n$, $b^{n+1} \notin \mathfrak{q}$. Q.E.D.

We could now state the decomposition theorem, but we prefer to give a somewhat sharper formulation of this theorem. For this formulation, the following definition is needed: A representation $\mathfrak{a} = \bigcap_i \mathfrak{q}_i$ of an ideal $\mathfrak{a}$ as an intersection of primary ideals $\mathfrak{q}_i$ (or briefly: a *primary representation* of $\mathfrak{a}$) is said to be *irredundant* (or *reduced*) if it satisfies the following conditions:

(a) No $\mathfrak{q}_i$ contains the intersection of the other ones.

(b) The $\mathfrak{q}_i$ have distinct associated prime ideals.

Given a representation $\mathfrak{a} = \bigcap_i \mathfrak{q}_i$ of an ideal $\mathfrak{a}$ as a finite intersection of primary ideals, one can find an irredundant one as follows: First we group together all the $\mathfrak{q}_i$ which have the same associated prime ideal $\mathfrak{p}_j$, and take their intersection $\mathfrak{q}'_j$, which is primary for $\mathfrak{p}_j$ (III, § 9, Theorem 14); then $\mathfrak{a} = \bigcap_j \mathfrak{q}'_j$, and, if some $\mathfrak{q}'_j$ contains the intersection of the others, we omit it, and proceed in the same way until condition (a) is satisfied. We have therefore proved (in view of Lemmas 1 and 2) the following theorem:

THEOREM 4. *In a ring R with a.c.c. every ideal admits an irredundant representation as finite intersection of primary ideals.*

We now characterize the ideals which are their own radicals.

THEOREM 5. *Let R be a ring and let $\mathfrak{a}$ be an ideal of R admitting an irredundant primary representation $\mathfrak{a} = \bigcap_i \mathfrak{q}_i$. For $\mathfrak{a}$ to be its own radical, it is necessary and sufficient that all $\mathfrak{q}_i$ be prime ideals.*

PROOF. If the $\mathfrak{q}_i$ are prime, we deduce from $x^n \in \mathfrak{a}$, that $x^n \in \mathfrak{q}_i$, $x \in \mathfrak{q}_i$, thus $x \in \mathfrak{a}$ and $\mathfrak{a} = \sqrt{\mathfrak{a}}$. Conversely, if $\mathfrak{a} = \sqrt{\mathfrak{a}}$, let $\mathfrak{p}_i$ be the radical of $\mathfrak{q}_i$ and let x be any element of $\bigcap_i \mathfrak{p}_i$. Then a large enough power x^n of x lies in each $\mathfrak{q}_i$ and thus also in $\mathfrak{a}$, which shows that $x \in \mathfrak{a}$ and that $\mathfrak{a} = \bigcap_i \mathfrak{p}_i$. This last representation of $\mathfrak{a}$ is irredundant, since otherwise we would have for some $j: \mathfrak{a} = \bigcap_{i \neq j} \mathfrak{p}_i \supset \bigcap_{i \neq j} \mathfrak{q}_i \supset \mathfrak{a}$, whence $\mathfrak{a} = \bigcap_{i \neq j} \mathfrak{q}_i$, in contradiction with the irredundancy of the given primary representation $\bigcap_i \mathfrak{q}_i$. Now, if $y \in \mathfrak{p}_i$, there exists $z \in \bigcap_{j \neq i} \mathfrak{p}_j$ such that $z \notin \mathfrak{p}_i$; we have $yz \in \mathfrak{a} \subset \mathfrak{q}_i$, and thus $y \in \mathfrak{q}_i$ and $\mathfrak{q}_i = \mathfrak{p}_i$. This completes the proof.

The following simple properties will be useful:

A) If a prime ideal $\mathfrak{p}$ contains a finite intersection $\bigcap_i \mathfrak{q}_i$, it contains some $\mathfrak{q}_i$ (III, § 8, p. 150); if the $\mathfrak{q}_i$ are primary, then $\mathfrak{p}$ contains the associated prime ideal $\mathfrak{p}_i$ of one of them.

B) If a prime ideal $\mathfrak{p}$ is a finite intersection $\bigcap_i \mathfrak{p}_i$ of prime ideals, it contains one of them, by A), and thus is equal to it; the other $\mathfrak{p}_j$ contain then this $\mathfrak{p}_i$.

§ 5. **Uniqueness theorems.** Having proved the existence of the primary decomposition, one is naturally led to the question of the uniqueness of that decomposition. It can be shown by examples that the primary ideals $\mathfrak{q}_i$ of an irredundant representation $\mathfrak{a} = \bigcap_i \mathfrak{q}_i$ need not be uniquely determined by $\mathfrak{a}$. For instance, if $\mathfrak{a}$ is the ideal (X^2, XY) in a polynomial ring $k[X, Y]$(k, a field), then for every element c of k we have a corresponding irredundant decomposition $\mathfrak{q}_1 \cap \mathfrak{q}_{2,c}$ of $\mathfrak{a}$, where $\mathfrak{q}_1 = (X)$ and $\mathfrak{q}_{2,c} = (Y - cX, X^2)$. (See also Theorem 22 given further on, in § 11). However, we will prove that their associated prime ideals are unique (Theorem 6) and that the "most important" among the $\mathfrak{q}_i$ themselves are also uniquely determined (Theorem 8). We shall achieve this by giving intrinsic characterizations of these ideals in terms of $\mathfrak{a}$ alone.

THEOREM 6. *Let R be an arbitrary ring and $\mathfrak{a}$ an ideal of R admitting an irredundant primary representation $\bigcap_i \mathfrak{q}_i$; and let $\mathfrak{p}_i = \sqrt{\mathfrak{q}_i}$. For a prime ideal $\mathfrak{p}$ of R to be equal to some $\mathfrak{p}_i$ it is necessary and sufficient that there exist an element c of R such that the ideal $\mathfrak{a}:(c)$ is primary for $\mathfrak{p}$. The prime ideals $\mathfrak{p}_i$ are therefore uniquely determined by $\mathfrak{a}$.*

PROOF. Given an index i there exists $c \in \bigcap_{j \neq i} \mathfrak{q}_j$, $c \notin \mathfrak{q}_i$, since the representation is irredundant. For such an element c, the ideal $\mathfrak{a}:(c)$ evidently contains $\mathfrak{q}_i$ and is contained in $\mathfrak{p}_i$. On the other hand, if $xy \in \mathfrak{a}:(c)$ and $x \notin \mathfrak{p}_i$, we have $xyc \in \mathfrak{a} \subset \mathfrak{q}_i$, whence $yc \in \mathfrak{q}_i$, since $x \notin \mathfrak{p}_i$, and consequently $yc \in \mathfrak{a}$ since $yc \in (c) \subset \bigcap_{j \neq i} \mathfrak{q}_j$; this shows that $y \in \mathfrak{a}:(c)$. It follows then from III, § 9, Theorem 13, that $\mathfrak{a}:(c)$ is primary for $\mathfrak{p}_i$. Suppose conversely that for some element c, not in $\mathfrak{a}$, the ideal $\mathfrak{a}:(c)$ is primary for a given prime ideal $\mathfrak{p}$. Writing $\mathfrak{a}:(c) = \bigcap_i \{\mathfrak{q}_i:(c)\}$ and taking radicals, we get $\mathfrak{p} = \bigcap_i \sqrt{\mathfrak{q}_i:(c)}$. The first part of the proof, applied to the case $\mathfrak{a} = \mathfrak{q}_i$, shows that the radical $\sqrt{\mathfrak{q}_i:(c)}$ is $\mathfrak{p}_i$ unless $c \in \mathfrak{q}_i$, in which case that radical is R. Hence $\mathfrak{p}$ is the intersection of some of the $\mathfrak{p}_i$, and is therefore one of them [(B) of § 4, p. 210]. Q.E.D.

The (uniquely determined) associated prime ideals of the primary ideals occurring in an irredundant primary representation of an ideal $\mathfrak{a}$ are called the *associated prime ideals of* $\mathfrak{a}$, or simply the *prime ideals of* $\mathfrak{a}$. This terminology is consistent with the one used for primary ideals. A minimal element in the family of associated prime ideals of $\mathfrak{a}$ (that is, an associated prime ideal of $\mathfrak{a}$ which contains no other prime ideal of $\mathfrak{a}$) is called an *isolated prime ideal* of $\mathfrak{a}$; a prime ideal of $\mathfrak{a}$ which is not isolated is said to be *imbedded*. The isolated prime ideals of $\mathfrak{a}$ admit the following very simple characterization:

THEOREM 7. *Let R be an arbitrary ring, $\mathfrak{a}$ an ideal of R admitting a finite irredundant primary representation $\mathfrak{a} = \bigcap_i \mathfrak{q}_i$, and let $\mathfrak{p}_i = \sqrt{\mathfrak{q}_i}$. For a prime ideal $\mathfrak{p}$ of R to contain $\mathfrak{a}$ it is necessary and sufficient that $\mathfrak{p}$ contain some $\mathfrak{p}_i$. The isolated prime ideals of $\mathfrak{a}$ are the minimal elements of the family of prime ideals which contain $\mathfrak{a}$.*

PROOF. The second assertion results from the first one. It is clear that, if $\mathfrak{p}$ contains some $\mathfrak{p}_i$, it contains $\mathfrak{a}$. Conversely, if $\mathfrak{p}$ contains $\mathfrak{a} = \bigcap_i \mathfrak{q}_i$, it contains some $\mathfrak{p}_i$ by A) of § 4.

If $\mathfrak{a} = \bigcap_i \mathfrak{q}_i$ is an irredundant primary representation of $\mathfrak{a}$, the ideals $\mathfrak{q}_i$ are said to be the *primary components* of $\mathfrak{a}$ (relative to the given decomposition); and $\mathfrak{q}_i$ is called *isolated* or *imbedded* according as its associated prime ideal $\mathfrak{p}_i$ is isolated or imbedded. We now characterize the isolated primary components of $\mathfrak{a}$ in terms of $\mathfrak{a}$ alone:

THEOREM 8. *Let R be an arbitrary ring, $\mathfrak{a}$ an ideal of R admitting an irredundant finite primary representation $\mathfrak{a} = \bigcap_i \mathfrak{q}_i$, and $\mathfrak{p}_i$ the associated prime ideal of $\mathfrak{q}_i$. The set $\mathfrak{q}'_i$ of elements x of R such that $\mathfrak{a}:(x) \not\subset \mathfrak{p}_i$ is an* *ideal of R which is contained in $\mathfrak{q}_i$. If $\mathfrak{q}_i$ is an isolated primary component of $\mathfrak{a}$ then $\mathfrak{q}_i$ is equal to $\mathfrak{q}'_i$. The isolated primary components of $\mathfrak{a}$ are therefore uniquely determined by $\mathfrak{a}$.*

PROOF. The third assertion follows from the second and from the uniqueness of the isolated prime ideals of $\mathfrak{a}$ (Theorem 6 or Theorem 7). It is clear that if $x \in \mathfrak{q}'_i$ then $yx \in \mathfrak{q}'_i$ for all y in R. If x_1 and x_2 are in $\mathfrak{q}'_i$, there exist elements π_1, π_2, *not in* $\mathfrak{p}_i$, such that $\pi_1 x_1 \in \mathfrak{a}$ and $\pi_2 x_2 \in \mathfrak{a}$. We have then $\pi_1\pi_2(x_1 - x_2) \in \mathfrak{a}$ and $\pi_1\pi_2 \notin \mathfrak{p}_i$ (since $\pi_1 \notin \mathfrak{p}_i$, $\pi_2 \notin \mathfrak{p}_i$ and $\mathfrak{p}_i$ is prime), whence $x_1 - x_2 \in \mathfrak{q}'_i$. This proves that $\mathfrak{q}'_i$ is an ideal. If $x \in \mathfrak{q}'_i$, then $x\pi \in \mathfrak{a}$ for some π not in $\mathfrak{p}_i$. We have then $x\pi \in \mathfrak{q}_i$, $\pi \notin \mathfrak{p}_i$, whence $x \in \mathfrak{q}_i$, and this establishes the first assertion of the theorem. Now, suppose that $\mathfrak{q}_i$ is isolated. For $j \neq i$, we have $\mathfrak{p}_j \not\subset \mathfrak{p}_i$ and there exists $b_j \in \mathfrak{p}_j$ such that $b_j \notin \mathfrak{p}_i$; let $s(j)$ be an exponent such that $b_j^{s(j)} \in \mathfrak{q}_j$ and let $b = \prod_{j \neq i} b_j^{s(j)}$. Since $\mathfrak{p}_i$ is prime we have $b \notin \mathfrak{p}_i$; and, for any x in $\mathfrak{q}_i$, we have $bx \in \mathfrak{a}$, thus $x \in \mathfrak{q}'_i$, and $\mathfrak{q}_i \subset \mathfrak{q}'_i$. Q.E.D.

REMARK. The element b constructed above (when $\mathfrak{q}_i$ is isolated) satisfies the conditions $b \notin \mathfrak{p}_i$ and $\mathfrak{a}:(b) = \mathfrak{q}_i$. Using this element b, we see that if $\mathfrak{q}$ is an ideal which is primary for $\mathfrak{p}_i$ and contains $\mathfrak{a}$, then $\mathfrak{q}$ *must contain* $\mathfrak{q}_i$; for, we have $\mathfrak{q} \supset b\mathfrak{q}_i$ and $\mathfrak{q} \supset \mathfrak{q}_i$ since $b \notin \mathfrak{p}_i$.

The uniqueness of the isolated primary components of $\mathfrak{a}$ is a special case of a more general result. Let $\mathfrak{a} = \bigcap_i \mathfrak{q}_i$ be an irredundant primary decomposition of $\mathfrak{a}$, $\mathfrak{p}_i$ the associated prime ideal of $\mathfrak{q}_i$, and M the family of all $\mathfrak{p}_i$. A subset L of M is said to be an *isolated system of prime ideals* of $\mathfrak{a}$ if, when $\mathfrak{p}_i$ is in L, all the prime ideals of $\mathfrak{a}$ contained in $\mathfrak{p}_i$ are in L. A system L reduced to an isolated prime ideal is an isolated system. Given an isolated system L of prime ideals $(\mathfrak{p}_{i_q})$ of $\mathfrak{a}$, the intersection $\bigcap_q \mathfrak{q}_{i_q}$ of the corresponding primary components is denoted by $\mathfrak{a}_L$ and is called an *isolated ideal component* of $\mathfrak{a}$. We will prove that $\mathfrak{a}_L$ is uniquely determined by $\mathfrak{a}$ and by the isolated system L, and is independent of the given irredundant primary decomposition of $\mathfrak{a}$. Given a maximal element $\mathfrak{p}_r$ of L, the set L_r of all elements $\mathfrak{p}_j$ of L such that $\mathfrak{p}_j \subset \mathfrak{p}_r$ is obviously an isolated system, and since L is finite, it is the union of the L_r; thus $\mathfrak{a}_L = \bigcap_r \mathfrak{a}_{L_r}$, and we are reduced to proving the uniqueness of $\mathfrak{a}_{L_r}$. As in Theorem 8 one shows the existence of an element $b \notin \mathfrak{p}_r$ which lies in the intersection of all $\mathfrak{q}_j$ whose associated prime ideal $\mathfrak{p}_j$ does not lie in L_r; we then have $b\mathfrak{a}_{L_r} \subset \mathfrak{a}$. Let $\mathfrak{q}'_i$ denote (as in Theorem 8) the set of all x such that $\mathfrak{a}:(x) \not\subset \mathfrak{p}_i$. Since $b\mathfrak{a}_{L_r} \subset \mathfrak{a}$ and $b \notin \mathfrak{p}_r$, it follows that $\mathfrak{a}_{L_r} \subset \mathfrak{q}'_r$. On the other hand, if $\mathfrak{p}_j$ is any member of L_r

then $\mathfrak{q}'_r \subset \mathfrak{q}'_j$ (since $\mathfrak{p}_j \subset \mathfrak{p}_r$) and hence, by the first assertion of Theorem 8, we have $\mathfrak{q}'_r \subset \mathfrak{q}_j$. Hence $\mathfrak{q}'_r \subset \mathfrak{a}_{L_r}$, showing that $\mathfrak{q}'_r = \mathfrak{a}_{L_r}$. This equality characterizes $\mathfrak{a}_{L_r}$ in terms of $\mathfrak{a}$ and L_r alone and shows the uniqueness of $\mathfrak{a}_{L_r}$.

In a ring R in which every proper prime ideal is maximal, imbedded components do not exist, since, when the ideal (0) is prime, it can only be an associated prime ideal of itself. Thus:

THEOREM 9. *Let R be a noetherian ring in which every proper prime ideal is maximal. Then every ideal $\mathfrak{a}$ of R is, in a unique way, a finite irredundant intersection of primary ideals; $\mathfrak{a}$ is also, in a unique way, a product of primary ideals belonging to distinct prime ideals.*

PROOF. The first assertion is obvious. Now, let $\mathfrak{a} = \bigcap_i \mathfrak{q}_i$ be the irredundant primary representation of an ideal $\mathfrak{a}$ and let $\mathfrak{p}_i = \sqrt{\mathfrak{q}_i}$. If some $\mathfrak{p}_i$ is not maximal, then $\mathfrak{p}_i = \mathfrak{q}_i = (0)$, whence $\mathfrak{a} = (0)$ is prime, and our assertion is trivial. If each $\mathfrak{p}_i$ is maximal it follows that $\mathfrak{a} = \prod_i \mathfrak{q}_i$ (see III, § 13, Theorem 31, relation (8)). If $\mathfrak{a} = \prod_j \mathfrak{q}'_j$ is another representation of $\mathfrak{a}$ as a product of primary ideals $\mathfrak{q}'_j$ whose associated prime ideals are *distinct* maximal ideals, then, again by Theorem 31 of III, § 13, we have $\mathfrak{a} = \bigcap_j \mathfrak{q}'_j$, and this primary representation of $\mathfrak{a}$ is irredundant, by property A) stated at the end of § 4 (since the $\mathfrak{p}_j$ are distinct and maximal). Hence the $\mathfrak{q}'_j$ coincide with the $\mathfrak{q}_i$, except for order. Q.E.D.

REMARK CONCERNING PASSAGE TO A RESIDUE CLASS RING. Let R be a ring, $\mathfrak{a}$ and $\mathfrak{b}$ two ideals of R such that $\mathfrak{b} \subset \mathfrak{a}$. The property that $\mathfrak{a}$ be prime (or primary) is a property of the factor ring $R/\mathfrak{a}$, viz. that $R/\mathfrak{a}$ is a domain (or that every zero divisor in $R/\mathfrak{a}$ is nilpotent). Thus, if $\mathfrak{a}$ is a prime (or primary) ideal of R, the ideal $\mathfrak{a}/\mathfrak{b}$ of $R/\mathfrak{b}$ is prime (or primary). Also the radical of $\mathfrak{a}/\mathfrak{b}$ is $\sqrt{\mathfrak{a}}/\mathfrak{b}$. Consequently, if $\mathfrak{a} = \bigcap_i \mathfrak{q}_i$ is an irredundant primary representation of $\mathfrak{a}$ and if $\mathfrak{p}_i = \sqrt{\mathfrak{q}_i}$, then $\mathfrak{a}/\mathfrak{b} = \bigcap_i (\mathfrak{q}_i/\mathfrak{b})$ is an irredundant primary representation of $\mathfrak{a}/\mathfrak{b}$, and the $\mathfrak{p}_i/\mathfrak{b}$ are the associated prime ideals of $\mathfrak{a}/\mathfrak{b}$. Furthermore, to isolated (or imbedded) prime ideals and components of $\mathfrak{a}$ correspond isolated (or imbedded) prime ideals and components of $\mathfrak{a}/\mathfrak{b}$.

§ 6. Application to zero-divisors and nilpotent elements

THEOREM 10. *Let R be a ring and $\mathfrak{a}$ an ideal of R admitting a finite irredundant primary representation $\mathfrak{a} = \bigcap_i \mathfrak{q}_i$. The radical of $\mathfrak{a}$ is the intersection of the isolated prime ideals of $\mathfrak{a}$.*

PROOF. Since the radical of a finite intersection is the intersection of the radicals, the radical of $\mathfrak{a}$ is the intersection of all associated prime ideals $\mathfrak{p}_i$ of $\mathfrak{a}$. From this intersection we may delete the imbedded prime ideals.

COROLLARY. *In a noetherian ring R the set of nilpotent elements is the intersection of the isolated prime ideals of* (0) (*that is, of the minimal prime ideals of R*; cf. the last part of Theorem 7).

We observe also the following consequence of Theorem 10: $\sqrt{\mathfrak{a}}$ *is prime if and only if* $\mathfrak{a}$ *has a single isolated prime ideal.* Now, it will be proved later on (see § 11, Theorem 21) that in a noetherian domain there always exists an ideal having a preassigned (finite) set of associated prime ideals $\neq (0)$. It follows that $\sqrt{\mathfrak{a}}$ *may be prime without* $\mathfrak{a}$ *being primary.* The following example may serve as a simple illustration: in a polynomial ring $k[x, y]$ over a field k let $\mathfrak{p}_1 = (x)$, $\mathfrak{p}_2 = (x, y)$, $\mathfrak{a} = \mathfrak{p}_1 \cap \mathfrak{p}_2^2$. Then $\mathfrak{p}_1$ and $\mathfrak{p}_2$ are prime, $\mathfrak{p}_2^2$ is primary, $\mathfrak{a}$ is not primary (for, $\mathfrak{p}_1 \cap \mathfrak{p}_2^2$ is an irredundant primary representation of $\mathfrak{a}$) and $\sqrt{\mathfrak{a}} = \mathfrak{p}_1$.

THEOREM 11. *Let R be a noetherian ring,* $\mathfrak{a}$ *and* $\mathfrak{b}$ *two ideals of R such that* $\mathfrak{a} \neq R$. *Then* $\mathfrak{a} = \mathfrak{a}:\mathfrak{b}$ *if and only if* $\mathfrak{b}$ *is contained in no prime ideal of* $\mathfrak{a}$.

PROOF. We use the properties of quotient ideals given in III, § 7. Let $\mathfrak{a} = \bigcap_i \mathfrak{q}_i$ be an irredundant primary representation of $\mathfrak{a}$, and let $\mathfrak{p}_i = \sqrt{\mathfrak{q}_i}$. If $\mathfrak{b}$ is contained in no $\mathfrak{p}_i$, then, from $(\mathfrak{a}:\mathfrak{b})\mathfrak{b} \subset \mathfrak{a} \subset \mathfrak{q}_i$, we deduce $\mathfrak{a}:\mathfrak{b} \subset \mathfrak{q}_i$ and hence $\mathfrak{a}:\mathfrak{b} = \mathfrak{a}$ since the quotient $\mathfrak{a}:\mathfrak{b}$ obviously contains $\mathfrak{a}$. Conversely, if $\mathfrak{a}:\mathfrak{b} = \mathfrak{a}$, we have $\mathfrak{a}:\mathfrak{b}^s = \mathfrak{a}$ for all s. If, contrary to our assertion, $\mathfrak{b}$ is contained in some $\mathfrak{p}_i$, say $\mathfrak{b} \subset \mathfrak{p}_1$, then there exists an exponent s such that $\mathfrak{b}^s \subset \mathfrak{p}_1^s \subset \mathfrak{q}_1$ (since $\mathfrak{p}_1$ has a finite basis), and we have $\mathfrak{q}_1:\mathfrak{b}^s = R$, whence $\mathfrak{a} = \mathfrak{a}:\mathfrak{b}^s = \bigcap_i(\mathfrak{q}_i:\mathfrak{b}^s) = \bigcap_{j\neq 1}(\mathfrak{q}_j:\mathfrak{b}^s) \supset \bigcap_{j\neq 1}\mathfrak{q}_j \supset \mathfrak{a}$, whence $\mathfrak{a} = \bigcap_{j\neq 1}\mathfrak{q}_j$, contradicting irredundance.

COROLLARY 1. *For an ideal* $\mathfrak{b}$ *of a noetherian ring R to be contained in some associated prime ideal of an ideal* $\mathfrak{a}$ *of R, it is necessary and sufficient that* $\mathfrak{a}:\mathfrak{b} \neq \mathfrak{a}$.

This is a restatement of Theorem 11. Notice that this corollary gives the uniqueness of the *maximal* associated prime ideals of $\mathfrak{a}$.

COROLLARY 2. *For an element x of a noetherian ring R to belong to some associated prime ideal of an ideal* $\mathfrak{a}$ *of R, it is necessary and sufficient that there exist an element* $y \notin \mathfrak{a}$ *such that* $xy \in \mathfrak{a}$.

Apply Corollary 1 to the ideal $\mathfrak{b} = (x)$.

COROLLARY 3. *In a noetherian ring R the set of all zero-divisors is the union of all the associated prime ideals* (*isolated and imbedded*) *of* (0).

Apply Corollary 2 to the ideal $\mathfrak{a} = (0)$.

REMARK. An ideal of a noetherian ring R is entirely composed of zero-divisors if and only if it is contained in some associated prime ideal of (0). This follows from Corollary 3 and from the following fact, which is sometimes useful: *if an ideal $\mathfrak{a}$ is contained in a finite union $\bigcup_i \mathfrak{p}_i$ of prime ideals, it is contained in one of them.* In fact, we may suppose that, for $i \neq j$, we have $\mathfrak{p}_i \not\subset \mathfrak{p}_j$ for, otherwise, neither the hypothesis nor the conclusion is affected if $\mathfrak{p}_i$ is deleted. Suppose now that $\mathfrak{a}$ is contained in no $\mathfrak{p}_i$. For any i, it is then true that the ideal $\mathfrak{a} \cap \bigcap_{j \neq i} \mathfrak{p}_j$ is not contained in $\mathfrak{p}_i$ [see property A) at the end of § 4, p. 210]. If a_i is an element belonging to this ideal but not to $\mathfrak{p}_i$, then the element $\sum_i a_i$ is in $\mathfrak{a}$ without being in any $\mathfrak{p}_i$—a contradiction. This last result shows also that a finite union $\bigcup_i \mathfrak{p}_i$ of prime ideals is never an ideal, except in the trivial case where all $\mathfrak{p}_i$ are contained in one of them. However, a non-trivial finite union of ideals may be an ideal if the ideals are not prime. For example, if R is any *finite*, additive *non-cyclic* group and if we set $xy = 0$ for all x, y in R, then $(x) \neq R$ for all x in R, but $\bigcup_{x \in R} (x)$ is the unit ideal. Or also, if k is a finite field and R is the residue class ring $k[X, Y]/(X^2, XY, Y^2) = k[x, y]$ (where x and y are the residues of X and Y) then the finite union $\bigcup_{a, b \in k} (ax + by)$ is the ideal (x, y).

§ 7. Application to the intersection of the powers of an ideal.

For proving the main theorem of this section we need two lemmas:

LEMMA 1. *Let R be a noetherian ring, $\mathfrak{a}$ and $\mathfrak{m}$ two ideals of R. There exists an integer s and an ideal $\mathfrak{a}'$ of R such that $\mathfrak{m}\mathfrak{a} = \mathfrak{a} \cap \mathfrak{a}'$ and $\mathfrak{a}' \supset \mathfrak{m}^s$.*

PROOF. Let $\{\mathfrak{q}'_i\}(\{\mathfrak{q}''_j\})$ be the set of primary components of $\mathfrak{m}\mathfrak{a}$ whose associated prime ideals contain (do not contain) $\mathfrak{m}$. We take $\mathfrak{a}' = \bigcap_i \mathfrak{q}'_i$, $\mathfrak{a}'' = \bigcap_j \mathfrak{q}''_j$. Then $\mathfrak{m}\mathfrak{a} = \mathfrak{a}' \cap \mathfrak{a}''$, and there exists an integer s such that $\mathfrak{m}^s \subset \mathfrak{a}'$. On the other hand, if we fix an element y_j in $\mathfrak{m}$ such that $y_j \notin \sqrt{\mathfrak{q}''_j}$, then we have for any element x in $\mathfrak{a}$: $y_j x \in \mathfrak{m}\mathfrak{a} \subset \mathfrak{q}''_j$, which implies $x \in \mathfrak{q}''_j$; therefore $\mathfrak{a} \subset \mathfrak{a}''$. Since $\mathfrak{m}\mathfrak{a} \subset \mathfrak{a}$, we have $\mathfrak{m}\mathfrak{a} = \mathfrak{m}\mathfrak{a} \cap \mathfrak{a} = \mathfrak{a}' \cap \mathfrak{a}'' \cap \mathfrak{a} = \mathfrak{a}' \cap \mathfrak{a}$, and the lemma is proved.

LEMMA 2. *Let R be an arbitrary ring with identity, $\mathfrak{a}$ and $\mathfrak{m}$ two ideals of R such that $\mathfrak{a}$ admits a finite basis $(x_1, \cdots, x_n)$ and $\mathfrak{a} = \mathfrak{a}\mathfrak{m}$. Then there exists an element z in $\mathfrak{m}$ such that $(1 - z)\mathfrak{a} = (0)$.*

PROOF. Denote by $\mathfrak{a}_i$ the ideal $(x_i, \cdots, x_n)$ (whence $\mathfrak{a}_1 = \mathfrak{a}$) and set

$\mathfrak{a}_{n+1} = (0)$. We shall prove, by induction on i, the existence of an element z_i in $\mathfrak{m}$ such that $(1 - z_i)\mathfrak{a} \subset \mathfrak{a}_i$; then z_{n+1} will be the element z we are looking for. For $i = 1$ it suffices to take $z_1 = 0$. From $(1 - z_i)\mathfrak{a} \subset \mathfrak{a}_i$ and from $\mathfrak{a} \subset \mathfrak{m}\mathfrak{a}$, we deduce $(1 - z_i)\mathfrak{a} \subset \mathfrak{m}(1 - z_i)\mathfrak{a} \subset \mathfrak{m}\mathfrak{a}_i$; in particular we have $(1 - z_i)x_i = \sum_{j=i}^{n} z_{ij}x_j$ with $z_{ij} \in \mathfrak{m}$. Thus $(1 - z_i - z_{ii})x_i \in \mathfrak{a}_{i+1}$, and we may take $1 - z_{i+1} = (1 - z_i)(1 - z_i - z_{ii})$.

A neater proof of Lemma 2 can be obtained if we are willing to use determinants, the theory of which can be developed in any commutative ring as well as in a field. Since $\mathfrak{a} = \mathfrak{m}\mathfrak{a}$ we have relations of the form $x_i = \sum_{j=1}^{n} y_{ij}x_j$, where $y_{ij} \in \mathfrak{m}$, or $\sum_j (\delta_{ij} - y_{ij})x_j = 0$, where δ_{ij} is 0 or 1 according as i and j are distinct or equal. If d denotes the determinant $|\delta_{ij} - y_{ij}|$, the usual argument leading to Cramer's rule shows that $dx_j = 0$ for all j, that is, $d\mathfrak{a} = (0)$; and the rule for developing a determinant shows that d is of the form $1 - z$ with $z \in \mathfrak{m}$.

THEOREM 12 (KRULL). *Let R be a noetherian ring and $\mathfrak{m}$ an ideal of R. In order that $\bigcap_{n=1}^{\infty} \mathfrak{m}^n = (0)$, it is necessary and sufficient that no element of $1 - \mathfrak{m}$ be a zero-divisor in R.**

PROOF. If an element $1 - z$ of $1 - \mathfrak{m}$ is a zero-divisor, say $(1 - z)y = 0$ with $y \neq 0$, we have $y = zy = z^2y = \cdots = z^ny$, and y belongs to $\bigcap_n \mathfrak{m}^n$. Conversely, assume that no element of $1 - \mathfrak{m}$ is a zero-divisor in R, and let $\mathfrak{a} = \bigcap_n \mathfrak{m}^n$. By Lemma 1 we have $\mathfrak{m}\mathfrak{a} \supset \mathfrak{a} \cap \mathfrak{m}^s = \mathfrak{a}$ and thus $\mathfrak{m}\mathfrak{a} = \mathfrak{a}$; therefore, by Lemma 2, there exists z in $\mathfrak{m}$ such that $(1 - z)\mathfrak{a} = (0)$, and since $1 - z$ is not a zero-divisor we conclude that $\mathfrak{a} = (0)$.

We note that in the above proof we have used Lemma 1 (which will be especially useful to us in chapter VIII on local algebra) for the purpose of establishing the equality $\mathfrak{m}\mathfrak{a} = \mathfrak{a}$. This equality can also be proved in a somewhat simpler fashion, as follows:

Let $\mathfrak{a}\mathfrak{m} = \bigcap_i \mathfrak{q}_i$ be an irredundant primary decomposition of $\mathfrak{a}\mathfrak{m}$. Since $\mathfrak{a}\mathfrak{m} \subset \mathfrak{a}$, in order to prove that $\mathfrak{a}\mathfrak{m} = \mathfrak{a}$ we have only to show that $\mathfrak{a} \subset \mathfrak{q}_i$ for all i. Now, we have $\mathfrak{a}\mathfrak{m} \subset \mathfrak{q}_i$. Hence, if $\mathfrak{m} \not\subset \mathfrak{p}_i$ then certainly $\mathfrak{a} \subset \mathfrak{q}_i$. If $\mathfrak{m} \subset \mathfrak{p}_i$, then for some integer n we have $\mathfrak{m}^n \subset \mathfrak{q}_i$, and hence again $\mathfrak{a} \subset \mathfrak{m}^n \subset \mathfrak{q}_i$.

COROLLARY 1. *If R is a noetherian domain and if $\mathfrak{m}$ is an ideal of R different from R, then $\bigcap_n \mathfrak{m}^n = (0)$.*

* By $1 - \mathfrak{m}$ we mean the set of elements of the form $1 - \pi$, $\pi \in \mathfrak{m}$.

This corollary shows that if $\mathfrak{m} \neq (0)$ (and R being a noetherian domain) then $\mathfrak{m}^p \neq \mathfrak{m}^q$ for $p \neq q$; for otherwise—if, say, $p < q$—we would have $\mathfrak{m}^p = \mathfrak{m}^{p+1}$, thus, upon multiplication by $\mathfrak{m}$, $\mathfrak{m}^{p+2} = \mathfrak{m}^{p+1} = \mathfrak{m}^p$, and $\mathfrak{m}^{p+n} = \mathfrak{m}^p$ by induction on n. We would then get the contradiction $\mathfrak{m}^p = (0)$.

COROLLARY 2. *If R is a noetherian ring such that the non-units of R form an ideal $\mathfrak{m}$ (that is, if R is a "local ring"*; see § 11, p. 228) *then* $\bigcap_n \mathfrak{m}^n = (0)$.

In fact, since no element of $1 - \mathfrak{m}$ is in $\mathfrak{m}$, every element of $1 - \mathfrak{m}$ is a unit, and cannot be a zero divisor.

By Corollary 3 to Theorem 11 of § 6, the set of zero divisors in the noetherian ring R is the union $\bigcup_i \mathfrak{p}_i$ of the associated prime ideals of (0). Thus, the condition of Theorem 12 may be written "$(1 - \mathfrak{m}) \cap \mathfrak{p}_i = \emptyset$ for every i," or equivalently "$\mathfrak{m} + \mathfrak{p}_i \neq R$ for every i." Thus, by passage to a residue class ring $R/\mathfrak{a}$ (see Remark at the end of § 5) Theorem 12 yields the following result:

THEOREM 12′. *Let R be a noetherian ring and let $\mathfrak{m}$ and $\mathfrak{a}$ be two ideals of R. In order that $\bigcap_n (\mathfrak{a} + \mathfrak{m}^n) = \mathfrak{a}$, it is necessary and sufficient that $\mathfrak{m} + \mathfrak{p}_i \neq R$ for every associated prime ideal $\mathfrak{p}_i$ of $\mathfrak{a}$.*

Given an ideal $\mathfrak{m}$ of a ring R, an ideal $\mathfrak{a}$ of R is said to be *closed* (with respect to $\mathfrak{m}$) if $\bigcap_n (\mathfrak{a} + \mathfrak{m}^n) = \mathfrak{a}$.

In fact, we can define a topology on R by taking the powers $\{\mathfrak{m}^n\}$ as a neighborhood system for 0, the neighborhoods of an arbitrary x in R being the residue classes $\{x + \mathfrak{m}^n\}$. It is easy to check that R becomes in this way a topological ring, that $\bigcap_n \mathfrak{m}^n = (0)$ means that it is Hausdorff space, and that $\bigcap_n (\mathfrak{a} + \mathfrak{m}^n) = \mathfrak{a}$ means that the ideal $\mathfrak{a}$ is closed. One can then consider the question of the completeness of R with respect to this topology. These questions are of great importance in the theory of local and semi-local rings; this theory will be developed in chapter VIII. The following lemma is, actually, a well-known topological fact:

LEMMA 3. *Given a ring R, an ideal $\mathfrak{m}$ of R and a family $\{\mathfrak{a}_\lambda\}$ of ideals of R which are closed (with respect to $\mathfrak{m}$), the intersection $\bigcap_\lambda \mathfrak{a}_\lambda$ is closed.*

This follows from the obvious inclusion $(\bigcap_\lambda \mathfrak{a}_\lambda) + \mathfrak{m}^n \subset \bigcap_\lambda (\mathfrak{a}_\lambda + \mathfrak{m}^n)$ and from the associativity of intersections.

We now determine the "closure" of an ideal:

THEOREM 13. *Given a noetherian ring R and two ideals $\mathfrak{m}$ and $\mathfrak{a}$ of R,* *the intersection* $\bigcap_n (\mathfrak{a} + \mathfrak{m}^n)$ *is the intersection of those primary components* $\mathfrak{q}_i$ *of* $\mathfrak{a}$ *whose radical* $\mathfrak{p}_i$ *satisfies the relation* $\mathfrak{m} + \mathfrak{p}_i \neq R$.

PROOF. Let $\mathfrak{b} = \bigcap_i \mathfrak{q}_i$. Since each $\mathfrak{q}_i$ is closed (Theorem 12′), $\mathfrak{b}$ is closed (Lemma 3), and we have $\mathfrak{b} = \bigcap_n (\mathfrak{b} + \mathfrak{m}^n)$. Let now $\{\mathfrak{q}'_j\}$ be the other primary components of $\mathfrak{a}$ and $\{\mathfrak{p}'_j\}$ their radicals. Since $\mathfrak{m} + \mathfrak{p}'_j = R$, $\mathfrak{m}$ is comaximal with each $\mathfrak{p}'_j$, thus with each $\mathfrak{q}'_j$, thus also with the intersection $\mathfrak{b}'$ of the $\mathfrak{q}'_j$, and consequently $\mathfrak{m}^n$ is comaximal with $\mathfrak{b}'$ (III, § 13, Theorem 31); in other words, $\mathfrak{b}' + \mathfrak{m}^n = R$ for every n. We therefore have $\mathfrak{b} = (\mathfrak{b}' + \mathfrak{m}^n)\mathfrak{b} = \mathfrak{b}'\mathfrak{b} + \mathfrak{m}^n\mathfrak{b} \subset \mathfrak{a} + \mathfrak{m}^n$ for every n (in topological language this means that $\mathfrak{a}$ is dense in $\mathfrak{b}$). From $\mathfrak{a} \subset \mathfrak{b}$ and $\mathfrak{b} = \bigcap_n (\mathfrak{b} + \mathfrak{m}^n)$ we then deduce that $\mathfrak{b} = \bigcap_n (\mathfrak{a} + \mathfrak{m}^n)$. Q.E.D.

COROLLARY. *Given an ideal* $\mathfrak{m}$ *of a noetherian ring* R, $\bigcap_n \mathfrak{m}^n$ *is the intersection of those primary components* $\mathfrak{q}_i$ *of* (0) *the radical* $\mathfrak{p}_i$ *of which satisfies the relation* $\mathfrak{m} + \mathfrak{p}_i \neq R$.

We take $\mathfrak{a} = (0)$ in Theorem 13.

§ **8. Extended and contracted ideals.** We have seen that there is a quite simple relationship between the ideals of a ring and the ideals of one of its residue class rings (see III, §§ 4, 5, in particular Theorem 7 in III, § 5). The matter is much more involved if we consider a ring S and a subring R, and this problem is not essentially easier than the following more general one which we are going to study: We are given two rings R and S having identities, and a homomorphism f of R into S such that $f(1) = 1$, and we look for relations between ideals of R and ideals of S. Ideals in R will be denoted by small German letters ($\mathfrak{a}, \mathfrak{b}, \cdots$), and ideals in S by German capitals ($\mathfrak{A}, \mathfrak{B}, \cdots$). Neither S nor R need be noetherian in this discussion, since very few additional results follow from this assumption. The case where R is a *subring* of S is included in this discussion by taking for f the identity mapping of R into S.

DEFINITION. *If* $\mathfrak{A}$ *is an ideal in* S, *the ideal* $\mathfrak{A}^c = f^{-1}(\mathfrak{A})$ *is called the contracted ideal, or the contraction, of* $\mathfrak{A}$. *If* $\mathfrak{a}$ *is an ideal in* R, *the ideal* $\mathfrak{a}^e = Sf(\mathfrak{a})$ *generated by* $f(\mathfrak{a})$ *in* S *is called the extended ideal, or the extension, of* $\mathfrak{a}$.

When R is a subring of S, the ideal $\mathfrak{A}^c$ is the intersection $R \cap \mathfrak{A}$; it contains every ideal of R which is contained in $\mathfrak{A}$, and is thus the largest ideal in R contained in $\mathfrak{A}$. Similarly the ideal $\mathfrak{a}^e$ is generated by $\mathfrak{a}$ in S; it is contained in every ideal of S which contains $\mathfrak{a}$, and is thus the smallest ideal in S which contains $\mathfrak{a}$. It consists of all elements of S of the form $s_1a_1 + s_2a_2 + \cdots + s_na_n$, where n is an arbitrary positive integer, $s_i \in S$, $a_i \in \mathfrak{a}$, $i = 1, 2, \cdots, n$.

It is clear that $S^c = R$, $R^e = S$, $(R0)^e = S0$, and that $(S0)^c$ is the kernel of f. The following relations are easily proved:

(1) If $\mathfrak{A} \subset \mathfrak{B}$ then $\mathfrak{A}^c \subset \mathfrak{B}^c$; if $\mathfrak{a} \subset \mathfrak{b}$ then $\mathfrak{a}^e \subset \mathfrak{b}^e$.
(2) $\mathfrak{A}^{ce} \subset \mathfrak{A}$; $\mathfrak{a}^{ec} \supset \mathfrak{a}$.
(3) $\mathfrak{A}^{cec} = \mathfrak{A}^c$; $\mathfrak{a}^{ece} = \mathfrak{a}^e$.

(We have $\mathfrak{A}^{ce} \subset \mathfrak{A}$ by (2), thus $\mathfrak{A}^{cec} \subset \mathfrak{A}^c$ by (1); on the other hand we have $\mathfrak{A}^{cec} = (\mathfrak{A}^c)^{ec} \supset \mathfrak{A}^c$ by (2); similarly for the other formula).

(4) $(\mathfrak{A} + \mathfrak{B})^c \supset \mathfrak{A}^c + \mathfrak{B}^c$; $(\mathfrak{a} + \mathfrak{b})^e = \mathfrak{a}^e + \mathfrak{b}^e$.
(5) $(\mathfrak{A} \cap \mathfrak{B})^c = \mathfrak{A}^c \cap \mathfrak{B}^c$; $(\mathfrak{a} \cap \mathfrak{b})^e \subset \mathfrak{a}^e \cap \mathfrak{b}^e$.
(6) $(\mathfrak{A}\mathfrak{B})^c \supset \mathfrak{A}^c\mathfrak{B}^c$; $(\mathfrak{a}\mathfrak{b})^e = \mathfrak{a}^e\mathfrak{b}^e$.
(7) $(\mathfrak{A}:\mathfrak{B})^c \subset \mathfrak{A}^c:\mathfrak{B}^c$; $(\mathfrak{a}:\mathfrak{b})^e \subset \mathfrak{a}^e:\mathfrak{b}^e$.

(The second formula follows from $(\mathfrak{a}:\mathfrak{b})^e\mathfrak{b}^e = ((\mathfrak{a}:\mathfrak{b})\mathfrak{b})^e$ [by (6)] $\subset \mathfrak{a}^e$).

(8) $(\sqrt{\mathfrak{A}})^c = \sqrt{\mathfrak{A}^c}$; $(\sqrt{\mathfrak{a}})^e \subset \sqrt{\mathfrak{a}^e}$.

In (1) we cannot assert that $\mathfrak{A}^c < \mathfrak{B}^c$ if $\mathfrak{A} < \mathfrak{B}$, nor that $\mathfrak{a}^e < \mathfrak{b}^e$ if $\mathfrak{a} < \mathfrak{b}$. For instance, if R is a domain and S its quotient field, we have $\mathfrak{a}^e = R^e = S$ even if $\mathfrak{a} \neq R$, provided $\mathfrak{a} \neq (0)$. In (2), $\cdots$, (8) none of the inclusions can in general be replaced by an equality.

We notice that, in view of (3), the inclusions (2) become equalities when $\mathfrak{A}$ is an extended ideal and $\mathfrak{a}$ a contracted ideal. However, an ideal in S need not, in general, be an extended ideal, and, *a fortiori*, need not be the extension of its contraction; we may therefore have $\mathfrak{A}^{ce} < \mathfrak{A}$. Also, an ideal in R need not be a contracted ideal nor, *a fortiori*, need it be the contraction of its extension; we may therefore have $\mathfrak{a}^{ec} > \mathfrak{a}$. All that can be said, in view of (3), is that *if an ideal in S is an extended ideal, it is the extension of its contraction*, and that *if an ideal in R is a contracted ideal, it is the contraction of its extension.*

In other words, if we denote by (E) the set of all extended ideals in S and by (C) the set of all contracted ideals in R, the mappings $\mathfrak{A} \to \mathfrak{A}^c$ and $\mathfrak{a} \to \mathfrak{a}^e$ are 1-1 and are inverse mappings of (E) onto (C) and of (C) onto (E). Of course this does not preclude the possibility that members of (E) may also be extensions of ideals not in (C), and that members of (C) may also be contractions of ideals not in (E).

The 1-1 correspondence between the ideals in (C) and (E) is *an isomorphism with respect to the fundamental ideal theoretic operations* (sum, product, intersection, quotient, radical) *to the extent to which these operations do not lead to ideals outside of* (C) *or* (E). Cases where this condition is surely fulfilled are given by the equalities in formulae (4),

(5), (6), and (8). These formulae show that the set (E) is closed under addition and multiplication and that the set (C) is closed under intersection and radical formation. We now show that (C) *is also closed under quotient formation.*

PROOF. Let $\mathfrak{a}$ and $\mathfrak{b}$ be contracted ideals in R. We set $\mathfrak{A} = \mathfrak{a}^e$, $\mathfrak{B} = \mathfrak{b}^e$, whence $\mathfrak{a} = \mathfrak{A}^c$ and $\mathfrak{b} = \mathfrak{B}^c$. Our assertion that $\mathfrak{a}:\mathfrak{b} \in (C)$ will be established if we prove, more generally, the following assertion: (*) *if* $\mathfrak{A}$ *is any ideal in* S *and* $\mathfrak{B} \in (E)$ *then* $\mathfrak{A}^c:\mathfrak{B}^c = (\mathfrak{A}:\mathfrak{B})^c$. By (7), it is sufficient to prove that $\mathfrak{A}^c:\mathfrak{B}^c \subset (\mathfrak{A}:\mathfrak{B})^c$. We have: $(\mathfrak{A}^c:\mathfrak{B}^c)^e\mathfrak{B} = (\mathfrak{A}^c:\mathfrak{B}^c)^e\mathfrak{B}^{ce}$ (since $\mathfrak{B} \in (E)) = ((\mathfrak{A}^c:\mathfrak{B}^c)\mathfrak{B}^c)^e \subset \mathfrak{A}^{ce} \subset \mathfrak{A}$. Hence $(\mathfrak{A}^c:\mathfrak{B}^c)^e \subset \mathfrak{A}:\mathfrak{B}$ and $(\mathfrak{A}^c:\mathfrak{B}^c) \subset (\mathfrak{A}^c:\mathfrak{B}^c)^{ec} \subset (\mathfrak{A}:\mathfrak{B})^c$, as asserted. This established our assertion that (C) is closed under quotient formation.

The above proof shows that if $\mathfrak{a}$, $\mathfrak{b} \in (C)$ then not only does $\mathfrak{a}:\mathfrak{b}$ belong to (C) but we have also, by (*), that $\mathfrak{a}:\mathfrak{b} = (\mathfrak{a}^e:\mathfrak{b}^e)^c$.

If $\mathfrak{P}$ is a prime ideal in S and $\mathfrak{Q}$ an ideal in S which is *primary* for $\mathfrak{P}$, it is trivial to check that $\mathfrak{P}^c$ is prime and $\mathfrak{Q}^c$ primary for $\mathfrak{P}^c$. If $\mathfrak{A}$ is an ideal of S admitting a *primary representation* $\mathfrak{A} = \bigcap \mathfrak{Q}_i$, then the behavior of intersections under inverse images shows that $\mathfrak{A}^c = \bigcap \mathfrak{Q}_i{}^c$ is a primary representation of $\mathfrak{A}^c$; but this representation need not be irredundant when that of $\mathfrak{A}$ is. The behavior of prime and primary ideals of R under extension is less simple; $\mathfrak{p}$ may be prime in R without $\mathfrak{p}^e$ being prime in S. Indeed, the investigation of the character of $\mathfrak{p}^e$ is one of the central problems of ideal theory. We will study particular cases of this problem in the next section and in the next chapter.

When we are given three rings R, S, T and two homomorphisms f from R to S and g from S to T $(R \xrightarrow{f} S \xrightarrow{g} T)$ then for any ideal $\mathfrak{a}$ of R it is true that the extension (under g) of the extension (under f) of $\mathfrak{a}$ is the same ideal (in T) as the extension (under fg) of $\mathfrak{a}$; and a similar property holds for contractions. In particular, if we have a commutative diagram of rings and homomorphisms

$$\begin{array}{ccc} R & \xrightarrow{f} & S \\ {\scriptstyle g}\downarrow & & \downarrow{\scriptstyle h} \\ R' & \xrightarrow[f']{} & S' \end{array}$$

(that is, if the homomorphism fh of R into S' is the same as gf'), then, given an ideal $\mathfrak{a}$ of R, the extension under h of the extension under f of $\mathfrak{a}$ is the same ideal in S' as the extension under f' of the extension under g of $\mathfrak{a}$; and similarly for contractions. An important particular case is the one in which R and R' are subrings of S and S', f and f' are the identity mappings, and g is the restriction of h to R.

§ 9. Quotient rings. Let R be a ring with identity (not necessarily noetherian). We have seen (in I, § 19) that R admits a *total quotient ring* F, that is, a ring F which admits R as a subring and in which every regular element of R (that is, any element of R which is not a zero divisor) is a unit; furthermore every element of F may be written in the form a/b ($a, b \in R$, b regular in R). In I, § 20 we defined a *multiplicative system* in R as a non-empty subset M of R which does not contain 0 and which is closed under multiplication. When all elements of M are regular (in which case M is said to be *regular*) we have defined the *quotient ring* R_M of R with respect to M as the set of all quotients a/m where $a \in R$, $m \in M$; this is a subring of the total quotient ring F of R.

When we are given a multiplicative system M in R which contains zero divisors, a quotient ring R_M cannot be defined without further ado. In fact, the main feature of a quotient ring R_M in the regular case is that the elements of M become *units* in R_M; and a zero divisor can never be a unit. We shall now undertake a slight generalization of the concept of a quotient ring. We consider a homomorphism f of R into a ring S *such that $f(m)$ is a unit for every $m \in M$*, where M is a given multiplicative system in R. If x is an element of R such that $mx = 0$ for some m in M, we have $0 = f(xm) = f(x)f(m)$, and since $f(m)$ is a unit in S, this implies $f(x) = 0$. In other words, the kernel of f must contain the set $\mathfrak{n}$ of all elements x in R for which there exists an element m in M such that $mx = 0$. Since M is multiplicatively closed, this set $\mathfrak{n}$ is an *ideal* in R, as is readily verified; and since $0 \notin M$, we have $1 \notin \mathfrak{n}$ and $\mathfrak{n} \neq R$. Thus, the image $f(R)$ of R in S is isomorphic to a residue class ring of $R/\mathfrak{n}$, and f defines a homomorphism f' of $R/\mathfrak{n}$ into S. Now, the canonical image $\bar{M} = (M + \mathfrak{n})/\mathfrak{n}$ of M in $R/\mathfrak{n}$ is obviously closed under multiplication. Furthermore, *$\bar{M}$ does not contain any zero divisor*: for, if $\bar{x}\cdot\bar{m} = 0$ ($\bar{x} \in R/\mathfrak{n}$, $\bar{m} \in \bar{M}$) and x, m are representatives of $\bar{x}$, $\bar{m}$ in R and M, then $xm \in \mathfrak{n}$, $xmm' = 0$ for a suitable element m' in M, and since $mm' \in M$, we deduce that $x \in \mathfrak{n}$ and $\bar{x} = 0$. Thus $\bar{M}$ is a regular multiplicative system in $R/\mathfrak{n}$, and we can construct the ordinary quotient ring $(R/\mathfrak{n})_{\bar{M}}$. Since every element of $f'(\bar{M})$ $(= f(M))$ is a unit in S, the homomorphism f' may be extended to a homomorphism (still denoted by f') of $(R/\mathfrak{n})_{\bar{M}}$ into S by setting $f'(\bar{x}/\bar{m}) = f'(\bar{x})/f'(\bar{m})$ (the fact that f' is single valued and is a homomorphism is easily proved, as in I, § 19, Theorem 16). The ring $(R/\mathfrak{n})_{\bar{M}}$ is called the *quotient ring of R with respect to the multiplicative system M* and is denoted by R_M. We notice that if M is regular, we have $\mathfrak{n} = (0)$, $R/\mathfrak{n} = R$, $\bar{M} = M$, and the new terminology and notation is consistent with the old one. The quotient ring R_M has the following property:

There exists a homomorphism h of R into R_M such that:

1) *The kernel $\mathfrak{n}$ of h is the set of all elements x in R for which there exists m in M such that $xm = 0$.*

2) *The elements of $h(M)$ are units in R_M.*

3) *Every element of R_M may be written as a quotient $h(x)/h(m)$ $(x \in R, m \in M)$.*

One such homomorphism h is given by the product $\varphi\psi$, where φ is the canonical homomorphism of R onto $R/\mathfrak{n}$ and ψ is the canonical isomorphism of $R/\mathfrak{n}$ into $(R/\mathfrak{n})_{\bar{M}}$.

The preceding considerations show easily the essential uniqueness of a ring R_M and a homomorphism h satisfying conditions 1), 2), and 3). Namely, if *h is as above* (that is, if $h = \varphi\psi$), if *S is any ring and f is a homomorphism of R into S such that conditions* 1), 2), *and* 3) *are satisfied when h and R_M are replaced respectively by f and S, then there exists an isomorphism f' of R_M onto S such that $f = hf'$.* For, since the kernel of f is this time the same as the kernel of h, the homomorphism f' of $R/\mathfrak{n}$ $(= h(R))$ into S, defined by f, is an isomorphism, and hence also the extension of f' to R_M (still denoted by f') is an isomorphism into S. On the other hand, since we have also assumed that every element of S is of the form $f(x)/f(m)$ $(x \in R, m \in M)$, f' is necessarily an isomorphism onto S, and from the definition of f' it is obvious that $f = hf'$.

The particular homomorphism $h = \varphi\psi$ of R into R_M given above is called *canonical* (or *natural*).

In the course of the preceding considerations we have also proved the following "universal property" of R_M:

THEOREM 14. *Let M be a multiplicative system in a ring R with identity, and h the canonical homomorphism of R into the quotient ring R_M. For every homomorphism f of R into a ring S such that every element of $f(M)$ is a unit, there exists a homomorphism f' of R_M into S such that $f = hf'$.*

REMARK. *If M and M' are two multiplicative systems in R such that $M \subset M'$ and every element of M' is the product of an element of M and a unit in R, then $R_M = R_{M'}$.* This is obvious if M (and hence also M') is regular. In the general case we observe that our assumptions imply that the set $\mathfrak{n}$ of elements x of R for which there exists an element m in M such that $xm = 0$ coincides with the set of elements x of R for which there exists an element m' in M' such that $xm' = 0$. This shows that $R_{M'} = (R/\mathfrak{n})_{\bar{M}'}$, where $\bar{M}' = (M' + \mathfrak{n})/\mathfrak{n}$. On the other hand, we have $R_M = (R/\mathfrak{n})_{\bar{M}}$, where $\bar{M} = (M + \mathfrak{n})/\mathfrak{n}$. Now, we know that $\bar{M}$ is a regular multiplicative system, and it is clear that $\bar{M} \subset \bar{M}'$ and that every element of $\bar{M}'$ is the product of an element of $\bar{M}$ and a unit in $R/\mathfrak{n}$. Hence $(R/\mathfrak{n})_{\bar{M}'} = (R/\mathfrak{n})_{\bar{M}}$, that is, $R_{M'} = R_M$, as asserted.

§ 10. Relations between ideals in R and ideals in R_M. We now study the relations between ideals of R (denoted by $\mathfrak{a}, \mathfrak{b}, \cdots$) and ideals in R_M (denoted by $\mathfrak{a}', \mathfrak{b}', \cdots$). Extensions and contractions are with respect to the canonical homomorphism h (§ 9, p. 222).

The following terminology will be useful: an element x of a ring R is said to be *prime to an ideal* $\mathfrak{a}$ of R if $\mathfrak{a}:(x) = \mathfrak{a}$ (that is, if its residue class mod. $\mathfrak{a}$ is not a zero divisor in $R/\mathfrak{a}$). A subset E of R is said to be *prime to* $\mathfrak{a}$ if each one of its elements is prime to $\mathfrak{a}$. When $\mathfrak{a}$ is a finite intersection of primary ideals, a subset E of R is prime to $\mathfrak{a}$ if and only if it is disjoint from the union of the associated prime ideals of $\mathfrak{a}$ (cf. § 6, Theorem 11).

THEOREM 15. *Let M be a multiplicative system in a ring R with identity and let R_M be the quotient ring of R with respect to M.*

(a) *If $\mathfrak{a}$ is an ideal in R, then $\mathfrak{a}^{ec}$ consists of all elements b in R such that $bm \in \mathfrak{a}$ for some m in M.*

(b) *An ideal $\mathfrak{a}$ in R is a contracted ideal* (that is, $\mathfrak{a} = \mathfrak{a}^{ec}$) *if and only if M is prime to $\mathfrak{a}$.*

(c) *Every ideal in R_M is an extended ideal.*

(d) *The mapping $\mathfrak{a} \to \mathfrak{a}^e$ is a 1-1 mapping of the set (C) of contracted ideals in R onto the set of all ideals in R_M, and this mapping is an isomorphism with respect to the ideal theoretic operations of forming intersections, quotients and radicals.*

PROOF. (a): any element b of $\mathfrak{a}^{ec}$ is such that $h(b) \in \mathfrak{a}^e$, and by property 3) of R_M, given in § 9 (p. 222), any element of $\mathfrak{a}^e$ may be written in the form $\sum_i (h(x_i)/h(m_i))h(a_i)$ $(x_i \in R, m_i \in M, a_i \in \mathfrak{a})$. Since M is closed under multiplication, reduction to a common denominator $m = \prod_i m_i \in M$ shows that any element of $\mathfrak{a}^e$ may be written in the form $h(a)/h(m)$ $(a \in \mathfrak{a}, m \in M)$. Thus "$b \in \mathfrak{a}^{ec}$" is equivalent to "there exist a in $\mathfrak{a}$ and m in M such that $h(b) = h(a)/h(m)$," that is, to "there exist elements a and m in $\mathfrak{a}$ and M respectively such that $h(bm - a) = 0$." The characterization of the kernel $\mathfrak{n}$ of h shows that this condition is equivalent to the following one: "there exists an element a in $\mathfrak{a}$, and elements m and m' in M such that $(bm - a)m' = 0$," and this implies the existence of an element m'' $(= mm')$ in M such that $bm'' \in \mathfrak{a}$. Conversely, the existence of such an element m'' in M implies that $h(b)h(m'') \in h(\mathfrak{a})$, whence $h(b) \in \mathfrak{a}^e$ (since $h(m'')$ is a unit in R_M), that is,

$b \in \mathfrak{a}^{ec}$. This proves (a); and (b) follows from (a) and from the meaning of the expression "M is prime to $\mathfrak{a}$."

We now prove (c). If $\mathfrak{a}'$ is an ideal in R_M, any element x' of $\mathfrak{a}'$ may be written in the form $x' = h(x)/h(m)$ $(x \in R, m \in M)$. We thus have $h(x) \in \mathfrak{a}'$, $x \in \mathfrak{a}'^c$, and $x' \in \mathfrak{a}'^{ce}$; thus $\mathfrak{a}' \subset \mathfrak{a}'^{ce}$. Since the reverse inclusion is trivial (cf. (2) of § 8), (c) is proved. Then (d) follows immediately from (c) and from the discussion given in § 8, by taking into account the trivial fact that the set of all ideals in R_M is closed under all ideal theoretic operations.

COROLLARY 1. *If R is noetherian, so is R_M.*

We may use the 1-1 mapping defined in (d) and the maximum condition. Or we may use (c) and the finite basis property.

COROLLARY 2. *We have $\mathfrak{a}^e \neq R_M$ if and only if $\mathfrak{a} \cap M = \emptyset$.*

We notice that $\mathfrak{a}^e = R_M$ is equivalent to $1 \in \mathfrak{a}^{ec}$, and we use (a).

We now study the behavior of prime and primary ideals of R under extension.

THEOREM 16. *Let $\mathfrak{q}$ be a primary ideal of R disjoint from M, and let $\mathfrak{p}$ be its (prime) radical. Then:*

(a) *$\mathfrak{p}$ is disjoint from M, $\mathfrak{p}$ and $\mathfrak{q}$ are contracted ideals, and both contain the kernel $\mathfrak{n}$ of h.*

(b) *$\mathfrak{q}^e$ is primary, and $\mathfrak{p}^e$ is its associated prime.*

PROOF. If x is any element of $\mathfrak{p}$ then some power of x belongs to $\mathfrak{q}$, while if x is an element of M then any power of x belongs to M. This shows that if $\mathfrak{q}$ is disjoint from M then also $\mathfrak{p}$ must be disjoint from M. The disjointness of $\mathfrak{p}$ and M implies that M is prime to both $\mathfrak{p}$ and $\mathfrak{q}$, and thus the second assertion of (a) follows from Theorem 15, (b). The last assertion of (a) is an obvious consequence of the second assertion. As for (b) we first notice that $\mathfrak{p}^e$ is contained in the radical of $\mathfrak{q}^e$ [(8), § 8]. Let now x', y' be elements of R_M such that $x' \notin \mathfrak{p}^e$ and $x'y' \in \mathfrak{q}^e$. We may write $x' = h(x)/h(m)$ $(x \notin \mathfrak{p}, m \in M)$, $y' = h(y)/h(m')$ $(y \in R, m' \in M)$, $x'y' = h(z)/h(m'')$ $(z \in \mathfrak{q}, m'' \in M)$, and we have $h(xym'' - mm'z) = 0$. This means that there exists an element m_1 in M such that $m_1(xym'' - mm'z) = 0$; thus m_1xym'' is an element of $\mathfrak{q}$. Since M is disjoint from $\mathfrak{p}$ and $x \notin \mathfrak{p}$ we have $m_1xm'' \notin \mathfrak{p}$, whence $y \in \mathfrak{q}$ and $y' \in \mathfrak{q}^e$. In the special case $\mathfrak{q} = \mathfrak{p}$, this shows that $\mathfrak{p}^e$ is prime. In the general case, the conditions characterizing a primary ideal and its prime radical are fulfilled, and (b) is proved.

COROLLARY 1. *The mapping $\mathfrak{p} \rightarrow \mathfrak{p}^e$ is a 1-1 mapping of the set of all contracted prime ideals in R (or equivalently: the set of all prime ideals in R which are disjoint from M) onto the set of all prime ideals in R_M.*

This follows from Theorem 16 and from the following remark: Every (ideal and in particular every) prime ideal $\mathfrak{p}'$ in R_M is the extension of its contraction (Theorem 15, (c)), and the contraction of a prime ideal is a prime ideal (§ 8, p. 220).

COROLLARY 2. *Given a contracted prime ideal* $\mathfrak{p}$ *of* R, *the mapping* $\mathfrak{q} \rightarrow \mathfrak{q}^e$ *is a* 1-1 *mapping of the set of all ideals in* R *which are primary for* $\mathfrak{p}$ *onto the set of all ideals in* R_M *which are primary for* $\mathfrak{p}^e$, *and this mapping is an isomorphism for the operations* : and $\cap$.

The first assertion follows from Theorem 16 as in Corollary 1. As to the second assertion, it is sufficient to observe that if $\mathfrak{q}_1$ and $\mathfrak{q}_2$ are primary for $\mathfrak{p}$, so are $\mathfrak{q}_1 \cap \mathfrak{q}_2$ and $\mathfrak{q}_1 : \mathfrak{q}_2$ (except in the trivial case $\mathfrak{q}_2 \subset \mathfrak{q}_1$, where $\mathfrak{q}_1 : \mathfrak{q}_2 = R$; in that case we also have $\mathfrak{q}_2{}^e \subset \mathfrak{q}_1{}^e$ and $\mathfrak{q}_1{}^e : \mathfrak{q}_2{}^e = R_M = R^e$).

We now study the behavior of primary representations under extension.

THEOREM 17. *Let* $\mathfrak{a}$ *be an ideal of* R *admitting an irredundant primary representation* $\mathfrak{a} = \bigcap_{i=1}^{n} \mathfrak{q}_i$. *Suppose that, for* $1 \leq i \leq r$, *we have* $\mathfrak{q}_i \cap M = \emptyset$, *and that, for* $r + 1 \leq j \leq n$, *we have* $\mathfrak{q}_j \cap M \neq \emptyset$. *Then* $\mathfrak{a}^e = \bigcap_{i=1}^{r} \mathfrak{q}_i{}^e$ *is an irredundant primary representation of* $\mathfrak{a}^e$, *and we have* $\mathfrak{a}^{ec} = \bigcap_{i=1}^{r} \mathfrak{q}_i$, *that is,* $\mathfrak{a}^{ec}$ *is the intersection of those primary components of* $\mathfrak{a}$ *which are disjoint from* M.

PROOF. That the ideal $\mathfrak{a}^e$ is contained in $\mathfrak{a}' = \bigcap_{i=1}^{r} \mathfrak{q}_i{}^e$ follows from formula (5), § 8. Conversely, by Theorem 15, (d), any element x' of $\mathfrak{a}'$ may be written in the form $x' = h(x)/h(m)$ with $x \in \bigcap_{i=1}^{r} \mathfrak{q}_i$, since the $\mathfrak{q}_i$ are contracted ideals for $1 \leq i \leq r$. On the other hand, since M is closed under multiplication, there exists an element m' in $M \cap \left(\bigcap_{j=r+1}^{n} \mathfrak{q}_j \right)$. We then have $m'x \in \mathfrak{a}$, and $x' = h(m'x)/h(m'm) \in \mathfrak{a}^e$. This shows that $\mathfrak{a}^e = \mathfrak{a}'$ and proves the assertion about $\mathfrak{a}^e$ since the representation $\bigcap_{i=1}^{r} \mathfrak{q}_i{}^e$ is obviously irredundant, by contraction. The assertion about $\mathfrak{a}^{ec}$ also follows by contraction.

We terminate this section by giving the structure of the kernel $\mathfrak{n}$ of the canonical homomorphism h of R into R_M in the noetherian case.

THEOREM 18. *Let* R *be a noetherian ring,* M *a multiplicative system in* R. *The following ideals are equal:*

1) *The kernel* $\mathfrak{n}$ *of the canonical homomorphism of* R *into* R_M *(that is, the set of all elements* x *in* R *for which there exists an element* m *in* M *such that* $mx = 0$*).*

2) *The intersection* $\mathfrak{n}'$ *of all primary ideals in* R *which are disjoint from* M.

3) *The intersection* $\mathfrak{n}''$ *of all primary components of* (0) *in* R *which are disjoint from* M.

PROOF. The inclusion $\mathfrak{n} \subset \mathfrak{n}'$ follows from Theorem 16, (a). The inclusion $\mathfrak{n}' \subset \mathfrak{n}''$ is obvious. Now, since M is closed under multiplication, there exists an element in M which belongs to all the primary components of (0) which meet M. If m is such an element then we have, for any x in $\mathfrak{n}''$, $mx = 0$. This shows that $\mathfrak{n}'' \subset \mathfrak{n}$. Q.E.D.

We now add some properties of *transitivity* and *permutability.* Let M and M' be two multiplicative systems in a ring R such that $M \subset M'$, and let h, h' denote the canonical homomorphisms of R into R_M and $R_{M'}$ respectively. Since h' is such that all the elements of $h'(M)$ are units, there exists a homomorphism $\bar{h}$ of R_M into $R_{M'}$ such that $h' = h\bar{h}$ (§ 9, Theorem 14). We notice: (a) that $h(M')$ is a multiplicative system in R_M; (b) that the kernel of $\bar{h}$ is the set of all elements of the form $h(x)/h(m)$ $(m \in M)$ such that $m'x = 0$ for some m' in M', that is, the kernel is the set of all x' in R_M such that $y'x' = 0$ for some y' in $h(M')$; (c) that all elements of $\bar{h}(h(M'))$ $(= h'(M'))$ are units and (d) that every element of $R_{M'}$ may be written in the form $\bar{h}(x')/\bar{h}(y')$ $(x' \in R_M,$ $y' \in h(M'))$. From this we conclude (cf. characterization of quotient rings, § 9) that $R_{M'}$ *is isomorphic to the quotient ring* $(R_M)_{h(M')}$, and furthermore, that if ψ denotes the canonical homomorphism of R_M into $(R_M)_{h(M')}$ then there exists an isomorphism f of $(R_M)_{h(M')}$ onto $R_{M'}$ such that $\bar{h} = \psi f$. We have $h' = h\psi f$, and from preceding remarks concerning the transitivity of successive extensions (§ 8, p. 220) we draw at once the following consequence: if we denote by superscripts e and e' extensions of ideals in R relative to R_M and $R_{M'}$ respectively and by $\bar{e}$ extensions of ideals in R_M relative to $(R_M)_{h(M')}$, then for any ideal $\mathfrak{a}$ in R the ideal $\mathfrak{a}^{e'}$ of $R_{M'}$ corresponds to $(\mathfrak{a}^e)^{\bar{e}}$ under the isomorphism f. Note that every ideal in $R_{M'}$ is an extended ideal of an ideal $\mathfrak{a}$ of R and that consequently the above conclusion $\mathfrak{a}^{e'} = f((\mathfrak{a}^e)^{\bar{e}})$ describes fully the (1-1) correspondence which the isomorphism f induces between the ideals in $R_{M'}$ and the ideals in $(R_M)_{h(M')}$.

Let M be a multiplicative system in R and let $\mathfrak{a}$ be an ideal of R *which has no elements in common with* M. We consider now the residue class ring $R/\mathfrak{a}$ of R and we denote by f, f', and h the canonical homomorphisms of R onto $R/\mathfrak{a}$, of R_M onto $R_M/\mathfrak{a}^e$ and of R into R_M respectively. Since $\mathfrak{a} \subset \mathfrak{a}^{ec}$, h defines, by passage to the residue classes, a homomorphism $\bar{h}$ of $R/\mathfrak{a}$ into $R_M/\mathfrak{a}^e$ which satisfies the relation $hf' = f\bar{h}$. The set $f(M)$ $(=(M + \mathfrak{a})/\mathfrak{a})$ is obviously closed under multiplication, and since we have assumed that $M \cap \mathfrak{a} = \emptyset$ it is clear that the zero of $R/\mathfrak{a}$ does not belong to $f(M)$. Hence $f(M)$ is a multiplicative system in $R/\mathfrak{a}$. We shall now show that the ring $R_M/\mathfrak{a}^e$ and the homomorphism $\bar{h}$ of $R/\mathfrak{a}$ into $R_M/\mathfrak{a}^e$ satisfy the three conditions which characterize, to within "essential uniqueness," the quotient ring $(R/\mathfrak{a})_{f(M)}$ and the canonical homomorphism $\bar{h}'$ of $R/\mathfrak{a}$ into that quotient ring (these conditions were stated in § 9, p. 222). In the first place, the kernel of f' is $\mathfrak{a}^e$, whence the kernel of hf' is the inverse image of $\mathfrak{a}^e$ under h; that is, the kernel of hf' is $\mathfrak{a}^{ec}$. Since $hf' = f\bar{h}$ it follows that the kernel of $\bar{h}$ is $\mathfrak{a}^{ec}/\mathfrak{a}$. Now, let $\bar{x}$ be any element of $R/\mathfrak{a}$ and let x be a representative of $\bar{x}$ in R. Then $\bar{x}$ belongs to the kernel of the above canonical homomorphism $\bar{h}$ if and only if $\bar{x}\bar{m} = 0$ for some element $\bar{m}$ in $f(M)$—that is, if and only if $xm \in \mathfrak{a}$ for some m in M, whence—finally—if and only if $x \in \mathfrak{a}^{ec}$ (Theorem 15, (a)), that is, if and only if $\bar{x} \in \mathfrak{a}^{ec}/\mathfrak{a}$. We have thus proved that $\bar{h}$ and the canonical homomorphism $\bar{h}'$ have the same kernel. The image $\bar{h}(f(M))$ of $f(M)$ in $R_M/\mathfrak{a}^e$ consists of units, for we have $\bar{h}(f(M)) = f'(h(M))$ and $h(M)$ consists of units. Finally, it is obvious that every element of $R_M/\mathfrak{a}^e$ can be written in the form $\bar{h}(\bar{x})/\bar{h}(\bar{m})$, where $\bar{x} \in R/\mathfrak{a}$ and $\bar{m} \in f(M)$ $(= (M + \mathfrak{a})/\mathfrak{a})$. We have therefore shown *the permutability of residue class ring and quotient ring formation*:

$$R_M/\mathfrak{a}^e \cong (R/\mathfrak{a})_{(M+\mathfrak{a})/\mathfrak{a}}, \quad (M \cap \mathfrak{a} = \emptyset) \tag{1}$$

and also the existence of a particular isomorphism (1), say ψ, such that $\bar{h}\psi = \bar{h}'$, where—we repeat—$\bar{h}$ is the homomorphism of $R/\mathfrak{a}$ into $R_M/\mathfrak{a}^e$ defined by the canonical homomorphism h of R into R_M, while $\bar{h}'$ is the canonical homomorphism of $R/\mathfrak{a}$ into the quotient ring $(R/\mathfrak{a})_{(M+\mathfrak{a})/\mathfrak{a}}$.

§ 11. Examples and applications of quotient rings. The most important examples of multiplicative systems, and hence of quotient rings, are the following:

1) M is the complement, in R, of a prime ideal $\mathfrak{p}$ of R. This example will be discussed in more detail in this section.

2) M is the complement of a union $\bigcup \mathfrak{p}_i$ of prime ideals in R. Then, in R_M, the units are the elements of the complement of $\bigcup \mathfrak{p}_i^e$. When the union $\bigcup \mathfrak{p}_i$ is finite, we may suppose that $\mathfrak{p}_i \not\subset \mathfrak{p}_j$ for $i \neq j$, for we may delete p_i if $\mathfrak{p}_i \subset \mathfrak{p}_j$ without altering M. We then have $\mathfrak{p}_i^e \not\subset \mathfrak{p}_j^e$ for $i \neq j$, the ideals $\mathfrak{p}_i^e$ are maximal ideals of R_M, and they are the only maximal ideals of R_M since every element of the complement of $\bigcup \mathfrak{p}_i^e$

is a unit in R_M. A particular case is the one where R is noetherian and where the $\mathfrak{p}_i$ are the associated prime ideals of (0); in that case M is the set of regular elements of R (§ 6, Theorem 11, Corollary 3), and R_M is the total quotient ring of R.

3) M is the set of all powers of a non-nilpotent element a of R.

4) M is the set of all elements of a subring S of R which are not contained in some prime ideal $\mathfrak{p}$ of R. In particular, in a polynomial ring $F[X_1, \cdots, X_n]$ over a field, one may take for M the set of all non-zero polynomials in the first r indeterminates ($r \leqq n$).

5) M is the set of all elements x such that $x \equiv 1$ (mod. $\mathfrak{a}$), where $\mathfrak{a}$ is an ideal in R, distinct from R.

We now discuss in more detail the case where M is *the complement of a prime ideal* $\mathfrak{p}$ in R. In this case the quotient ring R_M is called the *quotient ring of* R *with respect to the prime ideal* $\mathfrak{p}$ and is denoted by $R_\mathfrak{p}$ (since $\mathfrak{p}$ is not a multiplicative system, as it contains 0, no confusion can result from the seemingly contradictory notations R_M, $R_\mathfrak{p}$ when M is the complement of $\mathfrak{p}$). Because of the importance of this case, we give a partial summary of Theorems 15, 16, 17 and 18 (and their corollaries) for this particular case.

THEOREM 19. *Let* $\mathfrak{p}$ *be a prime ideal in a ring* R. *If* $\mathfrak{a}$ *is an ideal in* R, *its extension* $\mathfrak{a}^e$ *is distinct from* $R_\mathfrak{p}$ *if and only if* $\mathfrak{a}$ *is contained in* $\mathfrak{p}$. *The mapping* $\mathfrak{a} \rightarrow \mathfrak{a}^e$ *establishes a* 1-1 *correspondence between the set of prime (primary) ideals of* R *contained in* $\mathfrak{p}$, *and the set of all prime (primary) ideals in* $R_\mathfrak{p}$. *The ideal* $\mathfrak{p}^e$ *is a maximal ideal in* $R_\mathfrak{p}$, *and contains every non-unit in* $R_\mathfrak{p}$, *as well as every proper ideal in* $R_\mathfrak{p}$. *If* $\mathfrak{a}$ *is an ideal in* R *which is a finite intersection of primary ideals, then* $\mathfrak{a}^{ec}$ *is the intersection of those primary components of* $\mathfrak{a}$ *which are contained in* $\mathfrak{p}$. *If* R *is noetherian, the kernel* $\mathfrak{n}$ *of the canonical homomorphism of* R *into* $R_\mathfrak{p}$ *is the intersection of all primary components of* (0) (*or of all primary ideals in* R) *which are contained in* $\mathfrak{p}$.

The assertion that $\mathfrak{p}^e$ is the "greatest" proper ideal of $R_\mathfrak{p}$ follows from what has been said in example 2) above, or from the fact that $\mathfrak{p}$ is the greatest proper contracted ideal (first assertion of Theorem 19).

The most important property of $R_\mathfrak{p}$ is that *its non-units form an ideal.* This property is not generally true in arbitrary rings (e.g., it is not true in the ring of integers). Rings which have the above property *and are noetherian* are called *local rings* and will be studied in chapter VIII. Local rings are of importance in the study of the geometry on an algebraic variety in the neighborhood of a point—in other words, in the study of the local properties of a variety.

If $\mathfrak{p}$ is a prime ideal in R, the passage to the quotient ring $R_\mathfrak{p}$ has the effect, we may say, of converting $\mathfrak{p}$ into a maximal ideal $\mathfrak{p}^e$. We may also say that any prime ideal, or indeed any ideal $\mathfrak{a}$ in R, which is not contained in $\mathfrak{p}$ is obliterated, or lost, in $R_\mathfrak{p}$, since then $\mathfrak{a}^e = R_\mathfrak{p}$. Since it is sometimes easier to prove a theorem for a maximal ideal than for an arbitrary prime ideal—for example, every ideal which has a maximal ideal as radical is primary (III, § 9, Theorem 13, Corollary 1), but this is not generally true for ideals with prime radical (see III, § 9, p. 154)—the technique of passage to a quotient ring may sometimes be used for getting simple proofs. Let us give an example:

THEOREM 20. *Let R be a noetherian ring and $\mathfrak{p}$ a prime ideal in R. The intersection $\mathfrak{a}$ of all ideals in R which are primary for $\mathfrak{p}$ is equal to the intersection $\mathfrak{b}$ of those primary components of* (0) *which are contained in $\mathfrak{p}$.*

PROOF. By Krull's theorem (§ 7, Theorem 12, Corollary 2) we have $\bigcap_{n=1}^{\infty} (\mathfrak{p}^e)^n = (0)$ in $R_\mathfrak{p}$. On the other hand, the ideals $(\mathfrak{p}^e)^n$ are primary for $\mathfrak{p}^e$, and every ideal in $R_\mathfrak{p}$ which is primary for $\mathfrak{p}^e$ contains some power $(\mathfrak{p}^e)^n$. Thus, since contraction maps the set of all primary ideals for $\mathfrak{p}^e$ onto the set of all primary ideals for $\mathfrak{p}$ (Theorem 19) and preserves intersections (finite and infinite), the intersection of all ideals of R which are primary for $\mathfrak{p}$ is $(0)^c$, that is, the kernel of the canonical homomorphism of R into $R_\mathfrak{p}$. By Theorem 19, this kernel is the ideal $\mathfrak{b}$. Q.E.D.

COROLLARY. *In a noetherian domain, the intersection of all primary ideals belonging to a given prime ideal $\mathfrak{p}$ is* (0).

REMARK. In the proof of the above theorem (and hence also of its corollary) we have made use of Theorem 12 (Krull's theorem). It is of interest to point out that Corollary 1 of Theorem 12 (to the effect that $\bigcap_n \mathfrak{m}^n = (0)$ if R is a noetherian domain) can be derived from the above corollary of Theorem 20, as follows: since every ideal $\mathfrak{m}$ is contained in some prime ideal, it is sufficient to prove the required relation $\bigcap_n \mathfrak{m}^n = (0)$ under the assumption that $\mathfrak{m}$ is a prime ideal, and for prime ideals $\mathfrak{m}$ this relation follows directly from the above corollary of Theorem 20 since every primary ideal belonging to $\mathfrak{m}$ contains some power of $\mathfrak{m}$. In the case of an arbitrary noetherian ring, this reasoning shows that the intersection of the powers of a prime ideal $\mathfrak{p}$ is contained in the intersection of all primary ideals belonging to $\mathfrak{p}$ (and is equal to it when $\mathfrak{p}$ is maximal). This is confirmed by a comparison between the corollary of Theorem 13 and Theorem 20: the first intersection is the intersection of those primary components $\mathfrak{q}_i$ of (0) the radical $\mathfrak{p}_i$ of which satisfies the relation $\mathfrak{p}_i + \mathfrak{p} \neq R$; the second is the intersection of those primary components $\mathfrak{q}_j$ of (0) the radical $\mathfrak{p}_j$ of which satisfies the relation $\mathfrak{p}_j \subset \mathfrak{p}$; and $\mathfrak{p}_j \subset \mathfrak{p}$ implies $\mathfrak{p}_j + \mathfrak{p} = \mathfrak{p} \neq R$, showing that the first intersection is contained in the second. In the case where $\mathfrak{p}$ is maximal, the relations $\mathfrak{p}_j \subset \mathfrak{p}$ and $\mathfrak{p}_j + \mathfrak{p} \neq R$ are equivalent, and the two intersections are equal.

We may use Theorem 20 in order to shed some light on the question of primary representation in noetherian rings. We first prove

THEOREM 21. *Let R be a noetherian ring, and let $\mathfrak{p}_1, \cdots, \mathfrak{p}_n$ be prime ideals of R none of which is an isolated prime ideal of* (0) (*in a domain they may thus be arbitrary proper prime ideals*). *There exists an ideal $\mathfrak{a}$ in R whose associated prime ideals are exactly the given ideals $\mathfrak{p}_1, \mathfrak{p}_2, \cdots, \mathfrak{p}_n$.*

PROOF. We proceed by induction on n, the case $n = 1$ being trivial (take $\mathfrak{a} = \mathfrak{p}_1$). We may suppose that $\mathfrak{p}_n$ is maximal among the given ideals. We thus suppose the existence of an ideal $\mathfrak{b}$, with irredundant primary representation $\mathfrak{b} = \bigcap_{n=1}^{n-1} \mathfrak{q}_i$, $\mathfrak{q}_i$ belonging to $\mathfrak{p}_i$. The intersection of all primary ideals in R belonging to $\mathfrak{p}_n$ cannot contain $\mathfrak{b}$, for otherwise $\mathfrak{b}$ would be contained in some isolated prime ideal $\mathfrak{p}$ of (0) (Theorem 20), and this would imply that $\mathfrak{p}$ coincides with some $\mathfrak{p}_i$ $(1 \leq i \leq n - 1)$, in contradiction with our hypothesis on the $\mathfrak{p}_i$. Thus there exists a primary ideal $\mathfrak{q}_n$ belonging to $\mathfrak{p}_n$ such that $\mathfrak{a} = \bigcap_{i=1}^{n} \mathfrak{q}_i = \mathfrak{b} \cap \mathfrak{q}_n$ is distinct from $\mathfrak{b}$. It remains to prove the irredundance of the primary representation $\mathfrak{a} = \bigcap_{i=1}^{n} \mathfrak{q}_i$. It follows from the construction that $\mathfrak{q}_n$ does not contain the intersection of the other $\mathfrak{q}_i$; and if, for example, $\mathfrak{q}_1$ contained $\bigcap_{i=2}^{n} \mathfrak{q}_i$, it would contain $\bigcap_{i=2}^{n-1} \mathfrak{q}_j$ (since $\mathfrak{q}_n$ is not contained in $\mathfrak{p}_1$ by the maximality hypothesis on $\mathfrak{p}_n$), and this contradicts the irredundance of the representation $\mathfrak{b} = \bigcap_{i=1}^{n-1} \mathfrak{q}_i$. Q.E.D.

In particular, Theorem 21 shows the existence of ideals $\mathfrak{a}$ in R admitting imbedded components, provided, of course, that R contains two distinct prime ideals, which are not isolated prime ideals of (0), and such that one of them is contained in the other. In the case of a domain R, this last proviso means that the proper prime ideals of R are not all maximal. This proviso is not fulfilled, and imbedded components are not to be expected, in the ring of integers, the rings of algebraic integers and the polynomial rings in one indeterminate over fields; on the other hand it is fulfilled, and imbedded components exist, in polynomial rings in several indeterminates.

We have proved (§ 5, Theorem 8) that the isolated components of an ideal $\mathfrak{a}$ are uniquely determined. This is not true of the imbedded components of $\mathfrak{a}$, which are *never* unique, and are even capable of infinite variation. More precisely:

THEOREM 22. *If an ideal $\mathfrak{a}$ in a noetherian ring R has an imbedded prime ideal $\mathfrak{p}$, it has infinitely many irredundant primary representations which differ only in the primary component belonging to $\mathfrak{p}$.*

PROOF. By hypothesis, there exists an associated prime ideal $\mathfrak{v}$ of $\mathfrak{a}$ strictly contained in $\mathfrak{p}$. It is then enough to prove the following statement:

LEMMA. *Given primary ideals $\mathfrak{u}$ and $\mathfrak{q}$ in a noetherian ring R, which belong to prime ideals $\mathfrak{v}$ and $\mathfrak{p}$ such that $\mathfrak{v} < \mathfrak{p} \neq R$, there exists an ideal $\mathfrak{q}'$, primary for $\mathfrak{p}$, and such that $\mathfrak{q}' < \mathfrak{q}$, $\mathfrak{q}' \cap \mathfrak{u} = \mathfrak{q} \cap \mathfrak{u}$.*

PROOF OF LEMMA. By passage to $R/\mathfrak{q} \cap \mathfrak{u}$, we may suppose that $\mathfrak{q} \cap \mathfrak{u} = (0)$ (see Remark at the end of § 5, p. 213). Then the intersection of all primary ideals belonging to $\mathfrak{p}$ is (0) (Theorem 20). Since $\mathfrak{q}$ is not contained in $\mathfrak{v}$, we have $\mathfrak{q} \neq (0)$ and there exists an ideal $\mathfrak{q}''$ primary for $\mathfrak{p}$ such that $\mathfrak{q}'' \not\supset \mathfrak{q}$. If we now set $\mathfrak{q}' = \mathfrak{q} \cap \mathfrak{q}''$, we will have $\mathfrak{q}' < \mathfrak{q}$ and since $\mathfrak{q}'$ is primary for $\mathfrak{p}$, the lemma is proved.

REMARK. In § 10 we have formulated a transitivity property of quotient ring formation (see p. 226). In the special case of quotient rings with respect to prime ideals the following slightly different formulation of the transitivity property is more useful:

Let M be a multiplicative system in R and let $\mathfrak{p}'$ be a prime ideal of R which is disjoint from M. If e denotes extension of ideals of R to R_M, then the two quotient rings $R_{\mathfrak{p}'}$ and $(R_M)_{\mathfrak{p}'^e}$ are isomorphic.

This statement is not identical with our original formulation of the transitivity property which asserts that if we set $M' = R - \mathfrak{p}'$ and if h denotes the canonical homomorphism of R into R_M then the two rings $R_{\mathfrak{p}'}$ and $(R_M)_{h(M')}$ are isomorphic. However, the two multiplicative systems $h(M')$ and $R_M - \mathfrak{p}'^e$, although not identical, are related to each other as follows: (1) $h(M') \subset R_M - \mathfrak{p}'^e$, since $\mathfrak{p}'$ and M are disjoint and since therefore $\mathfrak{p}'^{ec} = \mathfrak{p}'$ (§ 10, Theorem 16, (a)); (2) every element of $R_M - \mathfrak{p}'^e$ is the product of an element of $h(M')$ and a unit in R_M (of the form $1/h(m)$, $m \in M$). It follows, by the Remark at the end of § 9 (p. 222), that $(R_M)_{h(M')}$ and $(R_M)_{\mathfrak{p}'^e}$ are identical, whence also $R_{\mathfrak{p}'}$ and $(R_M)_{\mathfrak{p}'^e}$ are isomorphic, as asserted.

In particular, if $M = R - \mathfrak{p}$, where $\mathfrak{p}$ is a prime ideal of R, then the assumption that M and $\mathfrak{p}'$ are disjoint signifies that $\mathfrak{p}' \subset \mathfrak{p}$. In that case, then, the two quotient rings $R_{\mathfrak{p}'}$ and $(R_{\mathfrak{p}})_{\mathfrak{p}'^e}$ are isomorphic. Furthermore, if e' denotes extension of ideals of R with respect to $R_{\mathfrak{p}'}$, then for any ideal $\mathfrak{a}$ in R the ideal $\mathfrak{a}^{e'}$ in $R_{\mathfrak{p}}$ corresponds to the extension of $\mathfrak{a}^e$ in $(R_{\mathfrak{p}})_{\mathfrak{p}'^e}$.

§ 12. **Symbolic powers.** Given a ring R and a primary ideal $\mathfrak{q}$ in R, the n-th power $\mathfrak{q}^n$ of $\mathfrak{q}$ need not be primary (see III, § 9, p. 154). We can however associate with $\mathfrak{q}^n$ a certain primary ideal:

DEFINITION. *Let R be a ring with identity, $\mathfrak{p}$ a prime ideal in R ($\neq R$), $\mathfrak{q}$ a primary ideal belonging to $\mathfrak{p}$, and n a positive integer. The ideal $(\mathfrak{q}^n)^{ec}$ (extension and contraction being made with respect to the quotient ring $R_{\mathfrak{p}}$) is called the n-th symbolic power of $\mathfrak{q}$ and is denoted by $\mathfrak{q}^{(n)}$.*

The properties of symbolic powers are summarized in the following theorem:

THEOREM 23. *Let $\mathfrak{q}$ be a primary ideal belonging to $\mathfrak{p}$.*

1) *The symbolic power $\mathfrak{q}^{(n)}$ is a primary ideal belonging to $\mathfrak{p}$; it is the set of all elements x in R for which there exists $d \notin \mathfrak{p}$ such that $dx \in \mathfrak{q}^n$. If $\mathfrak{q}^n$ is primary (in particular, if $\mathfrak{p}$ is maximal) then $\mathfrak{q}^{(n)} = \mathfrak{q}^n$.*

2) *If $\mathfrak{q}^n$ is a finite intersection of primary ideals, then $\mathfrak{p}$ is its only isolated prime ideal, and $\mathfrak{q}^{(n)}$ is the corresponding primary component.*

3) *If $\mathfrak{p}$ has a finite basis, then every primary ideal belonging to $\mathfrak{p}$ contains some symbolic power of $\mathfrak{p}$. When R is noetherian, $\bigcap_{n=1}^{\infty} \mathfrak{p}^{(n)}$ is the intersection of those primary components of* (0) *which are contained in $\mathfrak{p}$. When R is a noetherian domain, we have $\bigcap_{n=1}^{\infty} \mathfrak{p}^{(n)} = (0)$, and the $\mathfrak{p}^{(n)}$ form a strictly decreasing sequence of ideals.*

4) *When R is noetherian, $\mathfrak{p}$ is the only isolated prime ideal of $\mathfrak{q}^{(n)} \cdot \mathfrak{q}^{(m)}$, and the corresponding primary component is $\mathfrak{q}^{(n+m)}$.*

PROOF OF 1). The first assertion follows from the fact that $\mathfrak{q}^e$ is primary for the *maximal* ideal $\mathfrak{p}^e$ in R and that consequently also $(\mathfrak{q}^n)^e$ ($= (\mathfrak{q}^e)^n$) is primary for $\mathfrak{p}^e$ in $R_{\mathfrak{p}}$. The second assertion is a special case of Theorem 15, (a) (§ 10), and the third follows from the second.

PROOF OF 2). This is a special case of Theorem 17 (§ 10).

PROOF OF 3). If $\mathfrak{p}$ has a finite basis, every primary ideal $\mathfrak{q}$ belonging to $\mathfrak{p}$ contains some power $\mathfrak{p}^n$. Hence, if $x \in \mathfrak{p}^{(n)}$ and if $d \notin \mathfrak{p}$ is such that $dx \in \mathfrak{p}^n$ (such an element d exists, by 1)), then $dx \in \mathfrak{q}$, whence $x \in \mathfrak{q}$ since $\mathfrak{q}$ is primary for $\mathfrak{p}$, and thus $\mathfrak{p}^{(n)} \subset \mathfrak{q}$. The second assertion follows from the first and from Theorem 20 (§ 11), and the third assertion follows from the second (cf. the Remark following Corollary 1 to Theorem 12 in § 7).

PROOF OF 4). Since $\mathfrak{p}$ is obviously the radical of $\mathfrak{q}^{(n)} \cdot \mathfrak{q}^{(m)}$, $\mathfrak{p}$ is the only isolated prime ideal of $\mathfrak{q}^{(n)} \cdot \mathfrak{q}^{(m)}$ (Theorem 10, § 6). Using the characterization of an irredundant decomposition of $\mathfrak{a}^{ec}$ in terms of that of $\mathfrak{a}$, given in Theorem 19 in § 11 we see that in order to complete the proof of 4) we have only to show that $(\mathfrak{q}^{(n)} \cdot \mathfrak{q}^{(m)})^{ec} = (\mathfrak{q}^{n+m})^{ec}$. But this is obvious, since $\mathfrak{q}^{(n)e} = (\mathfrak{q}^n)^e$ (the extended ideal $(\mathfrak{q}^n)^e$ being also the extension of its contraction $\mathfrak{q}^{(n)}$), $\mathfrak{q}^{(m)e} = (\mathfrak{q}^m)^e$, whence $(\mathfrak{q}^{(n)} \cdot \mathfrak{q}^{(m)})^e = (\mathfrak{q}^{n+m})^e$.

REMARK. We may define more generally the symbolic powers $\mathfrak{a}^{(n)}$ of an ideal $\mathfrak{a}$ which admits a primary representation without imbedded components. Instead of the quotient ring $R_\mathfrak{p}$ we consider then the quotient ring R_M, where M is the complement of the union of the associated prime ideals of $\mathfrak{a}$ and we set $\mathfrak{a}^{(n)} = (\mathfrak{a}^n)^{ec}$. The ideal $\mathfrak{a}^{(n)}$ is the set of all elements x in R for which there exists an element m in M such that $mx \in \mathfrak{a}^n$. The ideal $\mathfrak{a}^{(n)}$ has the same associated prime ideals as $\mathfrak{a}$; all these prime ideals are isolated.

§ **13. Length of an ideal.** In III, § 11 (whose content is an essential prerequisite for the reading of this section) we have defined the *length* $\mathfrak{l}$ of a module over a ring R; this length may be finite or infinite. Since an ideal $\mathfrak{a}$ of R is an R-module, its length $\mathfrak{l}(\mathfrak{a})$ is defined, but this notion has no great interest since $\mathfrak{l}(\mathfrak{a})$ is infinite in many interesting cases (for example, $\mathfrak{l}(\mathfrak{a})$ is infinite if $\mathfrak{a}$ is a proper ideal containing a regular element x, for $\mathfrak{a} > Rx > Rx^2 > Rx^3 > \cdots$ is then an infinite strictly descending chain). A more reasonable definition would be to define the length of the ideal $\mathfrak{a}$ as being the length $\mathfrak{l}(R - \mathfrak{a})$ of the difference module $R - \mathfrak{a}$; however this length would still be infinite in many important cases, for example whenever $\mathfrak{a}$ is a prime ideal which is not maximal. We therefore need a more subtle definition.

DEFINITION. *Let R be a ring with identity and let $\mathfrak{a}$ be an ideal in R having a primary representation without imbedded components. Denote by M the complement of the union of the associated prime ideals of $\mathfrak{a}$, and consider the quotient ring R_M. The length $\mathfrak{l}(R_M - \mathfrak{a}^e)$ of the difference R-module $R_M - \mathfrak{a}^e$ is called the ideal-length of the ideal $\mathfrak{a}$ and is denoted by $\lambda(\mathfrak{a})$.*

REMARK. We will in general use the simple word "length" instead of the more precise term "ideal-length." There will be no danger of confusion, since we intend always to point out the kind of length we are thinking of by using the expression "length of the ideal $\mathfrak{a}$" (or "length of $\mathfrak{a}$") and the notation $\lambda(\mathfrak{a})$ in one case, and the expression "length of $\mathfrak{a}$ considered as an R-module" (or "length of the R-module $\mathfrak{a}$") and the notation $\mathfrak{l}(\mathfrak{a})$ in the other. Notice also that the length $\lambda(\mathfrak{a})$ is not defined for ideals which do not admit primary representations, nor for ideals having a primary representation with imbedded components; however the definition of $\lambda(\mathfrak{a})$ may be extended to this latter case.

The main feature of the length $\lambda(\mathfrak{a})$ is that it is finite in an important case:

THEOREM 24. *Let R be a noetherian ring, and let $\mathfrak{a}$ be an ideal in R without imbedded components. Then the length $\lambda(\mathfrak{a})$ of the ideal $\mathfrak{a}$ is finite.*

PROOF. Let M be the complement of the union $\bigcup \mathfrak{p}_i$ of the associated prime ideals of $\mathfrak{a}$. As was seen in § 11, example 2) (p. 227), the ideals $\mathfrak{p}_i^e$ are the only maximal ideals of R_M, and, since they are the associated prime ideals of $\mathfrak{a}^e$ (Theorem 17 in § 10), they are the only prime ideals in R_M which contain $\mathfrak{a}^e$ (Theorem 7 in § 5). In other words, every prime ideal in the ring $R_M/\mathfrak{a}^e$ is maximal. On the other hand, this residue class ring is noetherian. Thus it satisfies both chain conditions (Theorem 2 in § 2), and admits a composition series. Therefore its length $\mathfrak{l}(R_M/\mathfrak{a}^e)$, that is, the length of the difference R-module $R_M - \mathfrak{a}^e$, is finite.*

The following result reduces the study of the length of an ideal to that of the length of a primary ideal:

THEOREM 25. *Let R be a ring with identity and let $\mathfrak{a}$ be an ideal in R having an irredundant primary representation $\mathfrak{a} = \bigcap \mathfrak{q}_i$ without imbedded components. Then we have $\lambda(\mathfrak{a}) = \sum_i \lambda(\mathfrak{q}_i)$.*

PROOF. We denote by $\mathfrak{p}_i$ the radical of $\mathfrak{q}_i$ and by M the complement of $\bigcup \mathfrak{p}_i$. Let the superscripts e and c denote extension and contraction of ideals with respect to the pair of rings R and R_M. The length $\lambda(\mathfrak{a})$ is equal to the length of the ring $R_M/\mathfrak{a}^e$. By III, § 13, Theorem 32, $R_M/\mathfrak{a}^e$ is isomorphic to the direct sum of the rings $R_M/\mathfrak{q}_i^e$, since $\mathfrak{a}^e = \bigcap \mathfrak{q}_i^e$ and since the associated prime ideals $\mathfrak{p}_i^e$ of the $\mathfrak{q}_i^e$ are maximal ideals in R_M (see Example 2 in § 11, p. 227). Note that since $\mathfrak{a}$ has no imbedded components, we have $\mathfrak{p}_i \not\subset \mathfrak{p}_j$ if $i \neq j$). Hence $\mathfrak{l}(R_M/\mathfrak{a}^e) = \sum_i \mathfrak{l}(R_M/\mathfrak{q}_i^e)$, and it remains to show that

$$\mathfrak{l}(R_M/\mathfrak{q}_i^e) = \lambda(\mathfrak{q}_i). \tag{1}$$

We fix an index i and denote by superscripts e', c', $(e''$, $c'')$ extension and contraction of ideals relative to the pair of rings R, $R_{\mathfrak{p}_i}$ $(R_M, (R_M)_{\mathfrak{p}_i^e})$. To prove (1) it will be sufficient to show that

$$R_M/\mathfrak{q}_i^e \cong R_{\mathfrak{p}_i}/\mathfrak{q}_i^{e'}. \tag{2}$$

The permutability of quotient ring and residue class ring formation

* Observe that the set of submodules of the *R-module* $R_M - \mathfrak{a}^e$ coincides with the set of ideals of the *ring* $R_M/\mathfrak{a}^e$, since $\mathfrak{a}^e$ annihilates $R_M - \mathfrak{a}^e$.

(§ 10, formula (1)) shows that the quotient ring $(R_M/\mathfrak{a}^e)_{\mathfrak{p}_i^e/\mathfrak{a}^e}$ is isomorphic to the ring $(R_M)_{\mathfrak{p}_i^e}/(\mathfrak{a}^e)^{e''}$. On the other hand, the transitivity of quotient ring formation (see Remark at the end of § 11, p. 231) shows that $(R_M)_{\mathfrak{p}_i^e}$ is isomorphic with $R_{\mathfrak{p}_i}$ and that in the isomorphism between these two rings the two ideals $(\mathfrak{a}^e)^{e''}$ and $\mathfrak{q}_i^{e'}$ $(= \mathfrak{a}^{e'})$ correspond to each other. Hence

$$(3) \qquad (R_M/\mathfrak{a}^e)_{\mathfrak{p}_i^e/\mathfrak{a}^e} \cong R_{\mathfrak{p}_i}/\mathfrak{q}_i^{e'}.$$

Now, the kernel of the canonical homomorphism of the ring $R_M/\mathfrak{a}^e$ into its quotient ring $(R_M/\mathfrak{a}^e)_{\mathfrak{p}_i^e/\mathfrak{a}^e}$ is $\mathfrak{q}_i^e/\mathfrak{a}^e$ (see § 10, Theorem 18; note that $\bigcap \mathfrak{q}_i^e/\mathfrak{a}^e$ is an irredundant primary representation of the zero ideal in $R_M/\mathfrak{a}^e$ and that this representation has no imbedded components), and $(R_M/\mathfrak{a}^e)/(\mathfrak{q}_i^e/\mathfrak{a}^e)$ is a ring in which the non-units form an ideal—namely, the ideal $(\mathfrak{p}_i^e/\mathfrak{a}^e)/(\mathfrak{q}_i^e/\mathfrak{a}^e)$. Hence the quotient ring $(R_M/\mathfrak{a}^e)_{\mathfrak{p}_i^e/\mathfrak{a}^e}$ coincides with the canonical map $(R_M/\mathfrak{a}^e)/(\mathfrak{q}_i^e/\mathfrak{a}^e)$ of $R_M/\mathfrak{a}^e$ into that quotient ring. Hence

$$(R_M/\mathfrak{a}^e)_{\mathfrak{p}_i^e/\mathfrak{a}^e} \cong (R_M/\mathfrak{a}^e)/(\mathfrak{q}_i^e/\mathfrak{a}^e) \cong R_M/\mathfrak{q}_i^e,$$

and this, in conjunction with (3), establishes (2).

COROLLARY. *Let $\mathfrak{a}$ be an ideal in a ring R, admitting an irredundant primary representation $\mathfrak{a} = \bigcap_i \mathfrak{q}_i$ without imbedded components. For the length $\lambda(\mathfrak{a})$ of the ideal $\mathfrak{a}$ to be finite, it is necessary and sufficient that each length $\lambda(\mathfrak{q}_i)$ be finite.*

We now characterize the length of a primary ideal $\mathfrak{q}$.

THEOREM 26. *Let $\mathfrak{q}$ be a primary ideal belonging to a prime ideal $\mathfrak{p}$ in a ring R. Consider a strictly descending chain $\mathfrak{p} = \mathfrak{q}_1 > \mathfrak{q}_2 > \cdots > \mathfrak{q}_r = \mathfrak{q}$ of primary ideals belonging to $\mathfrak{p}$ which join $\mathfrak{p}$ to $\mathfrak{q}$. The number r of terms in such a chain satisfies the inequality $r \leq \lambda(\mathfrak{q})$, where $\lambda(\mathfrak{q})$ is the length of the ideal $\mathfrak{q}$. If the length $\lambda(\mathfrak{q})$ is finite, there exists such a chain with $\lambda(\mathfrak{q})$ terms, and every other chain may be refined to a chain having exactly $\lambda(\mathfrak{q})$ terms.*

PROOF. Since $\mathfrak{p}^e$ is a maximal ideal in $R_\mathfrak{p}$, every proper ideal in $R_\mathfrak{p}$ which contains $\mathfrak{q}^e$ is a primary ideal belonging to $\mathfrak{p}^e$. On the other hand there is a 1-1 correspondence between the set of primary ideals belonging to $\mathfrak{p}$ which contain $\mathfrak{q}$, and the set of primary ideals belonging to $\mathfrak{p}^e$ which contain $\mathfrak{q}^e$ (Theorem 16, Corollary 2, § 10). The theorem follows now from Jordan's theorem (III, § 11, Theorem 19) as applied to the module $R_\mathfrak{p} - \mathfrak{q}^e$.

REMARKS. 1) Note that a composition series in $R_\mathfrak{p} - \mathfrak{q}^e$ has one more term than the corresponding chain of primary ideals, viz. $R_p - \mathfrak{q}^e$ itself. Thus r is, in this case, $\lambda(\mathfrak{q})$ and not $\lambda(\mathfrak{q}) - 1$. 2) When $\mathfrak{p}$ is not a maximal ideal, an ideal between $\mathfrak{p}$ and $\mathfrak{q}$ is not necessarily primary for $\mathfrak{p}$. Thus $\lambda(\mathfrak{q})$ is not related to the lengths $\mathfrak{l}(R - \mathfrak{q})$ and $\mathfrak{l}(\mathfrak{p} - \mathfrak{q})$ of the difference modules $R - \mathfrak{q}$ and $\mathfrak{p} - \mathfrak{q}$ (the latter ones being, in general, infinite).

COROLLARY. *If $\mathfrak{q}$ is a primary ideal in a ring R and if M is a multiplicative system disjoint from $\mathfrak{q}$ in R, then $\lambda(\mathfrak{q}) = \lambda(\mathfrak{q}^e)$ ($\mathfrak{q}^e \subset R_M$).*

We apply Theorem 16 of § 10.

THEOREM 27. *Let $\mathfrak{a}$ be an ideal in a ring R such that the difference module $R - \mathfrak{a}$ has a finite length $\mathfrak{l}(R - \mathfrak{a})$. Then $\mathfrak{a}$ admits a primary decomposition without imbedded components, and the length $\lambda(\mathfrak{a})$ of the ideal $\mathfrak{a}$ is finite and equal to $\mathfrak{l}(R - \mathfrak{a})$.*

PROOF. By Theorem 2 (§ 2) the ring $R/\mathfrak{a}$ is noetherian, and every prime ideal of this ring is maximal. This gives us a primary representation without imbedded components for (0) in $R/\mathfrak{a}$, whence for $\mathfrak{a}$ in R (see Remark at the end of § 5, p. 213). Since the associated prime ideals of $\mathfrak{a}$ are maximal, the ring $R/\mathfrak{a}$ is its own quotient ring $(R/\mathfrak{a})_{M'}$, where M' is the complement of the union of the associated prime ideals of (0) in $R/\mathfrak{a}$ (that is, M' is the set of all units in $R/\mathfrak{a}$). By the permutability of quotient ring and residue class ring formations (§ 10, formula 1), $(R/\mathfrak{a})_{M'}$ is isomorphic to $R_M/\mathfrak{a}^e$, where M denotes the complement of the union of the associated prime ideals of $\mathfrak{a}$ in R. Therefore, $R/\mathfrak{a}$ is isomorphic to $R_M/\mathfrak{a}^e$, and our theorem is proved.

COROLLARY. *If R is a ring of finite length $\mathfrak{l}(R)$ and if $\mathfrak{a}$ is an ideal in R, the length $\lambda(\mathfrak{a})$ of the ideal $\mathfrak{a}$ is defined and finite, and it satisfies the relation $\lambda(\mathfrak{a}) + \mathfrak{l}(\mathfrak{a}) = \mathfrak{l}(R)$.*

REMARKS. 1) The permutability of quotient ring and residue class ring formations shows in general (as has been seen in several particular cases) that the length $\lambda(\mathfrak{a})$ of an ideal $\mathfrak{a}$ is a property of the residue class ring $R/\mathfrak{a}$: more precisely $\lambda(\mathfrak{a})$ is equal to the length of the ideal (0) in $R/\mathfrak{a}$.

2) If two ideals $\mathfrak{a}$ and $\mathfrak{b}$ in R, admitting primary representations without imbedded components, have the same associated prime ideals and satisfy the relations $\mathfrak{a} \subset \mathfrak{b}$ and $\lambda(\mathfrak{a}) = \lambda(\mathfrak{b})$, then they are equal (notice that they are both contracted ideals of ideals in R_M, M denoting the complement of the union of the associated prime ideals of $\mathfrak{a}$ (or $\mathfrak{b}$)).

In computing the length of an ideal, one has to know when a descending chain of ideals (or of primary ideals) admits of further insertions. The following theorem and its corollaries shed some light on this question:

THEOREM 28. *Let R be a ring with identity and let N be a unitary R-module $\neq (0)$. For N to be simple it is necessary and sufficient that it be generated by one element (that is, that N be cyclic) and that there exist a* *maximal ideal* $\mathfrak{p}$ *in* R *such that* $\mathfrak{p}N = (0)$. *Then* $\mathfrak{p}$ *is the order of* N, *and* N *is* R-*isomorphic to the* R-*module* $R - \mathfrak{p}$.

PROOF. If N is simple, we have $Rx = N$ for every $x \neq 0$ in N (since Rx is a submodule of N); thus N is cyclic. Now, we notice that every cyclic module Rx is isomorphic to an R-module of the form $R - \mathfrak{a}$, where $\mathfrak{a}$ is an ideal of R: in fact the mapping $a \rightarrow ax$ $(a \in R)$ is a homomorphism of the R-module R onto N, and we may take for $\mathfrak{a}$ the kernel of this homomorphism. We also notice that, since R is commutative, $\mathfrak{a}$ is the order of N. The submodules of $R - \mathfrak{a}$ correspond to the ideals of R which contain $\mathfrak{a}$; thus $R - \mathfrak{a}$ is simple if and only if $\mathfrak{a}$ is a maximal ideal. This proves the theorem.

COROLLARY 1. *Let* $\mathfrak{a}$ *and* $\mathfrak{b}$ *be ideals in* R *such that* $\mathfrak{b} < \mathfrak{a}$. *A necessary and sufficient condition that there exist no ideal between* $\mathfrak{a}$ *and* $\mathfrak{b}$ *is that there exists a maximal ideal* $\mathfrak{p}$ *in* R *and an element* x *in* $\mathfrak{a}$ *such that* $\mathfrak{p}\mathfrak{a} \subset \mathfrak{b}$ *and* $\mathfrak{a} = \mathfrak{b} + Rx$. *When this condition is satisfied,* $\mathfrak{b}$ *is contained in* $\mathfrak{p}$.

The first assertion follows from Theorem 28 as applied to the R-module $\mathfrak{a} - \mathfrak{b}$. If $\mathfrak{b} \not\subset \mathfrak{p}$, the maximality of $\mathfrak{p}$ implies $R = \mathfrak{p} + \mathfrak{b}$, thus $\mathfrak{a} = R\mathfrak{a} = \mathfrak{b}\mathfrak{a} + \mathfrak{p}\mathfrak{a} \subset \mathfrak{b}$, a contradiction.

COROLLARY 2. *Let* $\mathfrak{q}$, $\mathfrak{q}'$ *be two primary ideals belonging to a maximal ideal* $\mathfrak{p}$ *and such that* $\mathfrak{q} < \mathfrak{q}'$. *A necessary and sufficient condition that there be no ideal between* $\mathfrak{q}$ *and* $\mathfrak{q}'$ *is that* $\mathfrak{p}\mathfrak{q}' \subset \mathfrak{q}$, *and that there exist* x *in* $\mathfrak{q}'$ *such that* $\mathfrak{q}' = \mathfrak{q} + Rx$.

§ **14. Prime ideals in noetherian rings.** Since every ideal in a noetherian ring admits a finite basis, we can roughly measure how large an ideal $\mathfrak{a}$ is by the number of elements required for constituting a basis of $\mathfrak{a}$; in this sense the principal ideals are "small." We intend to give, in this section, a more precise meaning to this vague idea, at least for prime ideals. A first step in this direction is the remark that if a principal ideal $\mathfrak{m} = Rx$ in a noetherian domain is prime and is different from R, it contains no other proper prime ideal; for if a prime ideal $\mathfrak{p} \neq (0)$ is such that $\mathfrak{p} < Rx$, then we have $\mathfrak{p} = x\mathfrak{a}$, where $\mathfrak{a}$ is a proper ideal, whence $\mathfrak{a} = \mathfrak{p}$ (since $\mathfrak{p}$ is prime and $x \notin \mathfrak{p}$), and this leads to the contradiction $\mathfrak{p} = x\mathfrak{p} = x^2\mathfrak{p} = \cdots \subset \bigcap_i \mathfrak{m}^i = (0)$ (see § 7, Theorem 12, Corollary 1).

We note that the above contradiction remains if we drop the assumption that R is noetherian but assume instead that R is a unique factorization domain. For in that case the relations $\mathfrak{p} = x^i\mathfrak{p}$, for all i, imply that any element of $\mathfrak{p}$ is divisible by any power of x, and this is impossible if $\mathfrak{p} \neq (0)$ and x is not a unit.

The next theorem is a far-reaching generalization of this fact.

DEFINITION. *In an integral domain a prime ideal is said to be minimal if it is proper and if it contains properly no prime ideal other than* (0).

As just remarked, in a noetherian domain every prime principal ideal, different from the unit ideal, is minimal. *In a unique factorization domain all minimal prime ideals are principal and are the ones generated by irreducible elements.* For, as was shown in III, § 8 (p. 149), in a unique factorization domain a proper principal ideal is prime if and only if it is generated by an irreducible element, and we have just seen that in a unique factorization domain every prime proper principal ideal is minimal; on the other hand, it is clear that in a unique factorization domain every prime ideal $\mathfrak{a}$, different from the unit ideal, contains irreducible elements. (Take a non-unit x in $\mathfrak{a}$ and factor it into irreducible factors.) It is not at all obvious that there exist minimal prime ideals in a given domain, and in fact they may fail to exist in some non-noetherian domains. However, the following theorem proves their existence and elucidates their nature in the case of a noetherian domain:

THEOREM 29 ("PRINCIPAL IDEAL THEOREM"). *In a noetherian domain R every isolated prime ideal $\mathfrak{p}$ of a proper principal ideal Ra is a minimal prime ideal. Conversely, every minimal prime ideal $\mathfrak{p}$ in R is an isolated prime ideal of some proper principal ideal Ra.*

PROOF. The second assertion follows from the characterization of the isolated prime ideals of an ideal (Theorem 7, § 5): it suffices to take for a any non-zero element of $\mathfrak{p}$. We now pass to the less trivial first assertion. By passage to the quotient ring $R_{\mathfrak{p}}$ (Theorem 19, § 11) we may suppose that $\mathfrak{p}$ is a maximal ideal, and that every element outside of $\mathfrak{p}$ is a unit in R. Suppose there exists a proper prime ideal $\mathfrak{v}$ in R such that $\mathfrak{v} < \mathfrak{p}$. We consider the infinite strictly decreasing sequence $\mathfrak{q}_1 > \mathfrak{q}_2 > \mathfrak{q}_3 > \cdots$ of primary ideals belonging to $\mathfrak{v}$, where we have set $\mathfrak{q}_n = \mathfrak{v}^{(n)}$ (cf. Theorem 23 of § 12). Then the sequence $(\mathfrak{q}_n + Ra)$ is a decreasing sequence of ideals containing Ra. But since the unique maximal ideal $\mathfrak{p}$ of R is an isolated prime ideal of Ra, Ra is a primary ideal belonging to $\mathfrak{p}$, and $\mathfrak{p}$ is the only prime ideal containing Ra; in other words, R/Ra is a primary ring (§ 3), and it therefore satisfies the descending chain condition (Theorem 2 of § 2). Therefore there exists an index n such that $\mathfrak{q}_n + Ra = \mathfrak{q}_{n+1} + Ra = \cdots$.

In particular we have $\mathfrak{q}_n \subset \mathfrak{q}_{n+1} + Ra$, and every element x of $\mathfrak{q}_n$ may be written in the form $x = y + za$, with $y \in \mathfrak{q}_{n+1}$ and $z \in R$. We have then $za = x - y \in \mathfrak{q}_n$, and on the other hand, since $\sqrt{Ra} = \mathfrak{p}$ and since $\mathfrak{v} < \mathfrak{p}$, the element a cannot belong to the radical $\mathfrak{v}$ of $\mathfrak{q}_n$. It follows that $z \in \mathfrak{q}_n$ and $\mathfrak{q}_n \subset \mathfrak{q}_{n+1} + \mathfrak{q}_n a$. Since the inverse inclusion is obvious, we have:

(1) $$\mathfrak{q}_n = \mathfrak{q}_{n+1} + \mathfrak{q}_n a.$$

At this point we pass to the residue class ring $R' = R/\mathfrak{q}_{n+1}$, we denote by a' the residue class of a, and by $\mathfrak{q}'$ the ideal $\mathfrak{q}_n/\mathfrak{q}_{n+1}$ (which is $\neq (0)$). The equality (1) gives $\mathfrak{q}' = \mathfrak{q}'a'$. Using one of the lemmas preceding Krull's theorem (Lemma 2 of § 7), we see that there exists an element x' in R' such that $(1 - x'a')\mathfrak{q}' = (0)$. But by the hypothesis on R, R' admits a unique maximal ideal, and a' belongs to that ideal; thus $1 - x'a'$ is a unit in R', and we have $\mathfrak{q}' = (0)$—a contradiction.

REMARK. The last part of the proof (after equality (1)) avoids the determinant calculation, which is customary at this point. The latter method is as follows. If $\{x_1, \cdots, x_s\}$ is a finite basis of $\mathfrak{q}_n$, equality (1) implies the existence of elements y_i in $\mathfrak{q}_{n+1}$ and z_{ij} in R such that, for every i, we have $x_i = y_i + \sum_{j=1}^{s} az_{ij}x_j$; in other words: $\sum_{j=1}^{s} (\delta_{ij} - az_{ij})x_j \in \mathfrak{q}_{n+1}$ for every i. If we denote by d the determinant $|\delta_{ij} - az_{ij}|$, the classical computation leading to Cramer's rule shows that we have $dx_j \in \mathfrak{q}_{n+1}$ for every j, that is, $d\mathfrak{q}_n \subset \mathfrak{q}_{n+1}$. But the development of d shows that d is an element of the form $1 - ba$ $(b \in R)$, that is, a unit in R. Thus $\mathfrak{q}_n \subset \mathfrak{q}_{n+1}$, a contradiction.

COROLLARY 1. *In a noetherian domain every proper prime ideal* $\mathfrak{p}$ *contains a minimal prime ideal.*

In fact, we take a non-zero element x in $\mathfrak{p}$ and we observe that some isolated prime ideal of Rx will be contained in $\mathfrak{p}$, by Theorem 7 of § 5.

COROLLARY 2. *Let R be a noetherian ring (not necessarily a domain) and let Ra be a principal ideal in R, distinct from R. If $\mathfrak{p}$ is an isolated prime ideal of Ra, there cannot exist two prime ideals $\mathfrak{p}'$ and $\mathfrak{p}''$ such that $\mathfrak{p}'' < \mathfrak{p}' < \mathfrak{p}$. Any prime ideal strictly contained in $\mathfrak{p}$ is an isolated prime ideal of* (0).

Suppose two such prime ideals $\mathfrak{p}'$ and $\mathfrak{p}''$ exist. By passage to $R/\mathfrak{p}''$, we may suppose that $\mathfrak{p}'' = (0)$ and that R is a domain. This contradicts Theorem 29. The second assertion follows from the first, if one bears in mind the fact that every prime ideal in R contains an isolated prime ideal of (0).

In connection with Theorem 29, we observe that the imbedded prime ideals of a principal ideal are certainly not minimal. That imbedded prime ideals can occur for a principal ideal (and even for *all* proper principal ideals in a suitable domain) may be shown by examples. We will prove in chapter V that in an important class of noetherian domains (the so-called "integrally closed" domains) every proper principal ideal has only isolated components.

Having proved the existence of minimal prime ideals in noetherian domains, we may inquire about the possibility of proving the stronger statement that the prime ideals satisfy the descending chain condition. We shall even prove a stronger result (see the corollary to Theorem 30 below).

DEFINITION. *Let R be an arbitrary ring with identity. A prime ideal $\mathfrak{p} \neq R$ is said to have height h (respectively, depth d) if there exists at least one chain $\mathfrak{p}_0 < \mathfrak{p}_1 < \cdots < \mathfrak{p}_{h-1} < \mathfrak{p}_h = \mathfrak{p}$ (respectively, $\mathfrak{p} = \mathfrak{p}_d < \mathfrak{p}_{d-1} < \cdots < \mathfrak{p}_1 < \mathfrak{p}_0 < R$) where the $\mathfrak{p}_i$ are prime ideals, and there exists no such chain with more than $h + 1$ (respectively, $d + 1$) ideals. We denote the height (respectively, depth) of a prime ideal $\mathfrak{p}$ by $h(\mathfrak{p})$ (respectively $d(\mathfrak{p})$).*

We point out that in the definition of the height, $\mathfrak{p}_0 = (0)$ is allowed (provided, of course, that this ideal is prime), whereas in the definition of the depth, $\mathfrak{p}_0 = R$ is *not* allowed. The prime ideals of depth 0 are the maximal ideals. The prime ideals of height 0 are the prime ideals which do not contain properly any other prime ideal: in a noetherian ring they are the isolated prime ideals of (0) (Theorem 7 of § 5), while in a domain (0) is the only prime ideal of height 0. Another way of stating the principal ideal theorem, or rather its second corollary (Corollary 2 to Theorem 29), is to say that in a noetherian ring the isolated prime ideals of principal ideals (other than R) have height 0 or 1, and that in a noetherian domain the isolated prime ideals of proper principal ideals have height 1. If $\mathfrak{p}$ and $\mathfrak{p}'$ are prime ideals such that $\mathfrak{p}' < \mathfrak{p}$, we have $h(\mathfrak{p}') < h(\mathfrak{p})$ and $d(\mathfrak{p}') > d(\mathfrak{p})$. If, for a given prime ideal $\mathfrak{p}$, there exist chains of prime ideals $\mathfrak{p}_0 < \mathfrak{p}_1 < \cdots < \mathfrak{p}_{h-1} < \mathfrak{p}$ with arbitrary large h, $\mathfrak{p}$ is said to have infinite height; and similarly for prime ideals of infinite depth.

THEOREM 30. *Let $\mathfrak{a}$ be an ideal distinct from R in a noetherian ring R. If $\mathfrak{a}$ has a basis of r elements, then every isolated prime ideal $\mathfrak{p}$ of $\mathfrak{a}$ satisfies the inequality $h(\mathfrak{p}) \leq r$ (that is, $\mathfrak{p}$ has height at most r).*

We first prove a lemma:

LEMMA. *Let $\mathfrak{p}_0 > \mathfrak{p}_1 > \cdots > \mathfrak{p}_m$ be a chain of prime ideals in a noetherian ring R, and let $(\mathfrak{v}_i)$ be a finite family of prime ideals in R, none of which contains $\mathfrak{p}_0$. Then there exists a chain $\mathfrak{p}_0 > \mathfrak{p}'_1 > \cdots > \mathfrak{p}'_{m-1} > \mathfrak{p}_m$ of prime ideals in R, with the same end terms and the same number of terms as the given one, and such that no $\mathfrak{p}'_j$ $(1 \leq j \leq m - 1)$ is contained in any $\mathfrak{v}_i$.*

PROOF. By repeated applications, it suffices to prove the lemma in the case $m = 2$; $\mathfrak{p}_0 > \mathfrak{p}_1 > \mathfrak{p}_2$. Since no $\mathfrak{v}_i$ contains $\mathfrak{p}_0$, the union

$\mathfrak{p}_2 \cup \bigcup_i \mathfrak{v}_i$ cannot contain $\mathfrak{p}_0$ (see the remark following Corollary 3 to Theorem 11 of § 6). Hence there exists an element x in $\mathfrak{p}_0$ which does not lie in any $\mathfrak{v}_i$ nor in $\mathfrak{p}_2$. We take for $\mathfrak{p}'_1$ an isolated prime ideal of $\mathfrak{p}_2 + Rx$ contained in $\mathfrak{p}_0$ (such an ideal exists by Theorem 7 of § 5). Then $\mathfrak{p}'_1$ is not contained in any $\mathfrak{v}_i$, by construction. We have $\mathfrak{p}'_1 > \mathfrak{p}_2$ since x does not lie in $\mathfrak{p}_2$, and we also have $\mathfrak{p}'_1 < \mathfrak{p}_0$ by the principal ideal theorem (Theorem 29) as applied to the principal ideal $(\mathfrak{p}_2 + Rx)/\mathfrak{p}_2$ in the ring $R/\mathfrak{p}_2$ (observe that $\mathfrak{p}_0/\mathfrak{p}_2$ is not a minimal prime ideal since $\mathfrak{p}_0/\mathfrak{p}_2 > \mathfrak{p}_1/\mathfrak{p}_2 > (0)$). Q.E.D.

We now prove Theorem 30 by induction on r. For $r = 0$, (0) is the only ideal generated by 0 elements, and its isolated prime ideals are of height 0. Now, in the general case, let $\mathfrak{a}$ be an ideal in R having a basis $\{x_1, \cdots, x_r\}$ with r elements, and let $\mathfrak{p}_0$ be an isolated prime ideal of $\mathfrak{a}$. Consider the ideal $\mathfrak{b}$ generated by $\{x_1, \cdots, x_{r-1}\}$. If $\mathfrak{p}_0$ is an isolated prime ideal of $\mathfrak{b}$, it has height at most $r - 1$ (and thus at most r) by our induction hypothesis. Assume that $\mathfrak{p}_0$ is not among the isolated prime ideals of $\mathfrak{b}$. Then $\mathfrak{p}_0$ is not contained in any isolated prime ideal $\mathfrak{v}_i$ of $\mathfrak{b}$, and hence, if $\mathfrak{p}_0$ is of height m, the lemma shows the existence of a chain $\mathfrak{p}_0 > \mathfrak{p}_1 > \mathfrak{p}_2 > \cdots > \mathfrak{p}_{m-1} > \mathfrak{p}_m$ of prime ideals in R such that $\mathfrak{p}_{m-1}$ is not contained in any $\mathfrak{v}_i$. Since $\mathfrak{p}_0$ is not an isolated prime ideal of $\mathfrak{b}$, $\mathfrak{p}_0/\mathfrak{b}$ is a prime ideal of height 1 in $R/\mathfrak{b}$, as it is an isolated prime ideal of the principal ideal generated by the $\mathfrak{b}$-residue of x_r (Corollary 2 to Theorem 29). Since $\mathfrak{p}_0/\mathfrak{b}$ contains $(\mathfrak{p}_{m-1} + \mathfrak{b})/\mathfrak{b}$ and since this last ideal is not contained in any isolated prime ideal of (0) in $R/\mathfrak{b}$, (in view of the fact that $\mathfrak{p}_{m-1}$ is not contained in any $\mathfrak{v}_i$) $\mathfrak{p}_0/\mathfrak{b}$ is an isolated prime ideal of $(\mathfrak{p}_{m-1} + \mathfrak{b})/\mathfrak{b}$ (Theorem 7 of § 5). Thus $\mathfrak{p}_0$ is an isolated prime ideal of $\mathfrak{p}_{m-1} + \mathfrak{b}$, and $\mathfrak{p}_0/\mathfrak{p}_{m-1}$ an isolated prime ideal of $(\mathfrak{b} + \mathfrak{p}_{m-1})/\mathfrak{p}_{m-1}$. Since, in $R/\mathfrak{p}_{m-1}$, this last ideal is generated by $r - 1$ elements, the induction hypothesis shows that $\mathfrak{p}_0/\mathfrak{p}_{m-1}$ is a prime ideal of height at most $r - 1$. We have therefore $m - 1 \leq r - 1$, that is, $m \leq r$, and $\mathfrak{p}_0$ is a prime ideal of height at most r. Q.E.D.

COROLLARY. *In a noetherian ring every prime ideal $\neq R$ has finite height, and the prime ideals satisfy the descending chain condition.*

REMARKS. 1) A prime ideal $\mathfrak{p}$ in a noetherian ring may very well have infinite depth. When that is the case, then the ascending chains of prime ideals starting with $\mathfrak{p}$, which, by the a.c.c., are all finite, have nevertheless lengths which are not bounded (by Theorem 30, the end-terms of these chains of unbounded lengths must include an infinite set of maximal ideals). In a noetherian ring with only a finite number of maximal ideals, the depth of any prime ideal is finite.

2) Suppose we have a prime ideal $\mathfrak{p}$ of height h, so that we have a chain $\mathfrak{p}_0 < \mathfrak{p}_1 < \cdots < \mathfrak{p}_{h-1} < \mathfrak{p}_h = \mathfrak{p}$ of prime ideals. Then no further prime ideal can be inserted in this chain, and it is easy to see that $\mathfrak{p}_i$ has height i $(0 \leq i \leq h)$. Now suppose we have a chain $\mathfrak{p}_0 < \mathfrak{p}_1 < \cdots < \mathfrak{p}_{h-1} < \mathfrak{p}_h$ of prime ideals in which no further prime ideals can be inserted. Then $\mathfrak{p}_h$ has height at least h, and one might conjecture that its height is exactly h, that is, that any ascending chain of prime ideals terminating with $\mathfrak{p}_h$ has at most h terms. An equivalent conjecture is the following: if $\mathfrak{p}$ and $\mathfrak{p}'$ are prime ideals such that $\mathfrak{p} < \mathfrak{p}'$ and that there is no prime ideal between them, then their heights differ by unity. It has been proved recently that these conjectures are false for arbitrary noetherian domains.* They can be proved to be true in an important class of rings, which includes the polynomial rings over fields.

Theorem 30 shows that a prime ideal of height h can only be an isolated prime ideal of an ideal generated by not less than h elements. That it is an isolated prime ideal of an ideal generated by exactly h elements is proved in the next theorem, which can be considered as a converse of Theorem 30. As it is much more elementary than Theorem 30, Theorem 30 will not be used in its proof.

THEOREM 31. *If $\mathfrak{p}$ is a prime ideal of height h in a noetherian ring R, then there exists an ideal $\mathfrak{a}$ in R generated by h elements and admitting $\mathfrak{p}$ as an isolated prime ideal.*

PROOF. Using induction on i we construct h elements $a_1, \cdots, a_i, \cdots, a_h$ of $\mathfrak{p}$ such that for every i, every isolated prime ideal $\mathfrak{p}_j$ of $Ra_1 + \cdots + Ra_i$ satisfies the condition $h(\mathfrak{p}_j) \geq i$ (then this height is i, by Theorem 30, but we shall not use this fact). The case $i = 0$ is trivial, and we only have to pass from i to $i + 1$ for $i < h$. Those among the $\mathfrak{p}_j$ which are of height i do not contain $\mathfrak{p}$, so that their union does not contain $\mathfrak{p}$ (see Remark following Corollary 3 to Theorem 11 of § 6). We take for a_{i+1} any element of $\mathfrak{p}$ lying outside of this union. Then every isolated prime ideal $\mathfrak{v}$ of $Ra_1 + \cdots + Ra_i + Ra_{i+1}$ contains some $\mathfrak{p}_j$ (Theorem 7 of § 5), and when this $\mathfrak{p}_j$ is of height i, we have $\mathfrak{v} > \mathfrak{p}_j$ since $a_{i+1} \notin \mathfrak{p}_j$; in any case $\mathfrak{v}$ is of height at least $i + 1$.

Now, $Ra_1 + \cdots + Ra_h$ being constructed, $\mathfrak{p}$ contains some isolated prime ideal $\mathfrak{p}'$ of this ideal. Since $h(\mathfrak{p}') \geq h$ by construction, and since $h(\mathfrak{p}) = h$ by hypothesis, we conclude that $\mathfrak{p} = \mathfrak{p}'$. Q.E.D.

§ 15. Principal ideal rings. A *principal ideal ring* (PIR) is a ring with identity in which every ideal is principal; a *principal ideal domain*

* M. Nagata, "On the Chain Problem of Prime Ideals", Nagoya Mathematical Journal, v. 10, pp. 51–64 (1956).

(PID) is a domain in which every ideal is principal, that is, a PIR without proper zero-divisors. Any PIR is obviously noetherian, and the PIR's may be considered the simplest type of noetherian rings. Very little of the general theory is needed for studying this type of ring.

Examples of PID's are the ring of integers and the polynomial rings in one variable over fields. More generally, any euclidean domain R is a PID (cf. Lemma in I, § 15, p. 24). For, $\mathfrak{a}$ being a proper ideal in R, we choose among the non-zero elements of $\mathfrak{a}$ an element x for which $\varphi(x)$ is minimum. Then, if $y \in \mathfrak{a}$, we write $y = xq + r$, with q, r in R and $\varphi(r) < \varphi(x)$, and we see that since $r(= y - xq)$ belongs to $\mathfrak{a}$ we must have $r = 0$, whence $y \in Rx$.

If R is a PIR and $\mathfrak{a}$ a proper ideal in R, then $R/\mathfrak{a}$ is obviously a PIR; this provides examples of PIR's with zero divisors.

We first study the PID's.

THEOREM 32. *Let R be a principal ideal domain. Then the proper prime ideals in R are those generated by irreducible elements, and they are maximal. The ring R is a unique factorization domain. Any two non-zero elements a and b of R have a greatest common divisor d, and we have $Rd = Ra + Rb$. If $\mathfrak{a}$ is a proper ideal in R, then $R/\mathfrak{a}$ satisfies the descending chain condition.*

PROOF. We first prove the assertion about the GCD of a and b. Since the ideal $Ra + Rb$ is principal, it is of the form Rd ($d \in R$), with $d = au + bv$ ($u, v \in R$). Since a and b are in Rd, they are multiples of d. Conversely any common divisor of a and b divides $d = au + bv$.

We now come to the assertion concerning prime ideals in R. To say that a principal ideal Rp is prime is the same as saying that if p divides a product xy it divides one of its factors. Hence, if Rp is a proper prime ideal, the element p is irreducible. Conversely, if p is an irreducible element of R, and if p divides a product xy without dividing x, the GCD of x and p is 1; we thus have $1 = ux + vp$ ($u, v \in R$), and $y = uxy + vyp$ is a multiple of p, which proves that Rp is prime. It is also maximal, since every ideal properly containing Rp is of the form Rd, where d is a divisor of p, but p is not a divisor of d. Hence d is a unit.

Let us now prove that R is a UFD (cf. I, § 14, p. 21). We have just seen that condition UF3 of I, § 14 is fulfilled. It thus remains to prove that condition UF1 of I, § 14 is also fulfilled, that is, that every element $x \neq 0$ of R is a finite product of irreducible elements. This has been proved in § 1, but, for the sake of completeness, we give here a somewhat different proof. Were UF1 false, there would exist, among the ideals Rx such that x is not a product of irreducible elements, a maximal one, say Ra. Since a cannot be irreducible, it is a product bc of elements b

and c such that $Ra < Rb$ and $Ra < Rc$. By the maximal character of Ra, b and c are both finite products of irreducible elements, whence so is a; a contradiction.

Notice that this argument about the validity of UF1 holds in every noetherian domain.

Finally we study $R/\mathfrak{a}$, where $\mathfrak{a}$ is a proper ideal of R. The ideals in $R/\mathfrak{a}$ correspond to the ideals in R which contain $\mathfrak{a}$. Writing $\mathfrak{a} = Rx$, the fact that R is a PID shows that these ideals correspond to the classes of associated divisors of x. By the unique factorization property, these classes are finite in number, and hence so are the ideals in $R/\mathfrak{a}$. Q.E.D.

COROLLARY 1. *Any UFD in which every proper prime ideal is maximal (and hence also minimal) is a PID, and conversely.*

The second ("converse") part of the corollary is contained in the first two assertions of Theorem 32. The direct part of the corollary is proved as follows:

It has already been pointed out in § 14 (p. 238) that in a UFD every minimal prime ideal is principal. Hence in the present case we are dealing with a domain R in which every prime ideal is principal. Now, let $\mathfrak{A}$ be any proper ideal in R. We consider the set of all proper principal ideals which contain $\mathfrak{A}$ (this set is not empty since $\mathfrak{A}$ is contained in at least one proper prime ideal). Since R is a UFD, R cannot contain an infinite strictly descending chain of principal ideals. Hence there exists a smallest principal ideal Ry containing $\mathfrak{A}$. From $\mathfrak{A} \subset Ry$ it follows that $\mathfrak{A} = y\mathfrak{A}_1$, where $\mathfrak{A}_1$ is an ideal $\neq (0)$. Were $\mathfrak{A}_1$ a proper ideal, we would have $\mathfrak{A}_1 \subset Rz$, where z is a non-unit $\neq 0$, and hence $\mathfrak{A} \subset Ryz < Ry$, a contradiction. Hence $\mathfrak{A}_1 = R$, $\mathfrak{A} = Ry$, and the corollary is established.

COROLLARY 2. *A necessary and sufficient condition that a domain R be a PID is that there exist a function f assigning a non-negative integer $f(x)$ to every non-zero element x of R, such that:*

(a) *If a divides b, then $f(a) \leq f(b)$, equality holding only when a and b are associates.*

(b) *If a and b are non-zero elements of R such that neither of them divides the other, then there exist elements p, q, r in R such that $r = pa + qb$ and $f(r) < \min(f(a), f(b))$.*

If R is a PID, we take for $f(x)$ the number of irreducible factors occurring in a factorization of x; then (a) is trivial, and, in (b) we may take for r the GCD of a and b. Conversely, if f is given and if $\mathfrak{a}$ is a proper ideal in R, we take, among the non-zero elements of $\mathfrak{a}$, an element x such that $f(x)$ is minimum, and we show that $\mathfrak{a} = Rx$. Let y be an element of $\mathfrak{a}$ and let us assume that y is not a multiple of x. Then, in view of (a), x cannot be a multiple of y since $f(x) \leqq f(y)$; hence, applying (b), we get the contradiction that there exists an element r of $\mathfrak{a}$ such that $f(r) < \min(f(x), f(y)) = f(x)$.

From Corollary 2 it can once more be deduced that a euclidean domain is a PID. It is natural to ask whether, conversely, every PID is euclidean; this question has been answered in the negative. Thus, the ring of integers of the quadratic number field $Q(\sqrt{-19})$ is a principal ideal ring but has no euclidean algorithm whatsoever.

We now define a type of PIR's which, together with the PID's, will enable us to construct all PIR's (see Theorem 33 below). A PIR is called *special* if it has only one prime ideal $\mathfrak{p} \neq R$ and if $\mathfrak{p}$ is nilpotent, that is, if $\mathfrak{p}^n = (0)$ for some integer $n > 0$.

EXAMPLE. If R is a PID, and if p is an irreducible element of R, then R/Rp^n is a special PIR, with Rp/Rp^n as its unique prime ideal.

When the "index of nilpotency" n is 1, the special PIR is a field; in all other cases the PIR has proper zero-divisors. At any rate, $\mathfrak{p}$ is maximal. If we place $\mathfrak{p} = Rp$, and if we denote by m the smallest integer such that $p^m = 0$, then every non-zero element x in R may obviously be written in the form $x = ep^k$, where $0 \leq k \leq m-1$, and where $e \notin Rp$. For, either x is a unit, in which case $x \notin Rp$ and so $k = 0$; or x is not a unit, in which case x must be contained in the unique maximal ideal Rp of R, and if k is the highest power of p which divides x, then $x = ep^k$, where $k \leqq m-1$ (since $x \neq 0$) and $e \notin Rp$.

We observe that the integer k in the representation $x = ep^k$ is uniquely determined by x; from $ep^k = e'p^{k'}$ and $0 \leqq k < k' < m$ we deduce $p^{k'-k} = 0$, in contradiction with the definition of m. One sees in a similar way that the unit e is uniquely determined mod. Rp^{m-k}. It follows that the only ideals in R are the Rp^k $(0 \leq k \leq m)$, and these ideals are all distinct. Conversely it is easily proved that a ring R containing a nilpotent element p such that every x in R may be written in the form $x = ep^k$ (e, a unit) is a special PIR.

We finally give a structure theorem for PIR's:

THEOREM 33. *A direct sum of PIR's is itself a PIR. Every PIR is a direct sum of PID's and of special PIR's.*

PROOF. Suppose that $R = R_1 \oplus \cdots \oplus R_n$, where each R_i is a PIR. If $\mathfrak{a}$ is an ideal in R, then $\mathfrak{a} = R\mathfrak{a} = R_1\mathfrak{a} + \cdots + R_n\mathfrak{a}$. But $R_i\mathfrak{a}$ is an ideal in R_i, and thus $R_i\mathfrak{a} = R_ix_i$ $(x_i \in R_i)$. Then clearly $\mathfrak{a} = R(x_1 + \cdots + x_n)$, and the first assertion is proved. For the proof of the second assertion, we need a lemma:

LEMMA. *Let R be a PIR. If $\mathfrak{p}$ and $\mathfrak{p}'$ are prime ideals such that $\mathfrak{p}' < \mathfrak{p} < R$, then $\mathfrak{p}$ contains no prime ideals other than $\mathfrak{p}$ and $\mathfrak{p}'$, and every* *primary ideal contained in* $\mathfrak{p}$ *contains* $\mathfrak{p}'$. *A non-maximal prime ideal of R has no primary ideal but itself. Two prime ideals in R are either comaximal, or else one of them contains the other.*

Since R is a PIR we can write $\mathfrak{p} = Rp$, $\mathfrak{p}' = Rp'$. From $\mathfrak{p}' \subset \mathfrak{p}$, we deduce $p' = rp$ $(r \in R)$. Since Rp' is prime and since $p \notin Rp'$, we have $r \in Rp'$, that is, $r = sp'$ $(s \in R)$—and thus $p' = spp'$. Let $\mathfrak{q}'$ be any primary ideal in R contained in $\mathfrak{p}$. Since $(1 - sp)p' = 0 \in \mathfrak{q}'$, and since $1 - sp$ does not belong to Rp, nor *a fortiori* to the radical of $\mathfrak{q}'$, we have $p' \in \mathfrak{q}'$. As Rp' is itself a primary ideal contained in $\mathfrak{p}$, it follows that $\mathfrak{p}'$ is the intersection of all primary ideals contained in $\mathfrak{p}$. This shows, first, that $\mathfrak{p}'$ is uniquely determined by $\mathfrak{p}$, and is contained in every primary ideal which is contained in $\mathfrak{p}$, thus proving the first assertion. This shows also that every primary ideal belonging to $\mathfrak{p}'$ contains $\mathfrak{p}'$, hence is $\mathfrak{p}'$ itself, and the second assertion is proved. Finally, if $\mathfrak{p}_1$ and $\mathfrak{p}_2$ are distinct prime ideals in R which are not comaximal, they are contained in some proper prime ideal $\mathfrak{p}$; by what has already been proved, it is impossible that both $\mathfrak{p}_1$ and $\mathfrak{p}_2$ be strictly contained in $\mathfrak{p}$. Hence one of them is $\mathfrak{p}$, and the other is strictly contained in $\mathfrak{p}$. Q.E.D.

We now complete the proof of Theorem 33. Since R is noetherian, the ideal (0) has an irredundant primary representation $(0) = \bigcap_i \mathfrak{q}_i$. Let $\mathfrak{p}_i = \sqrt{\mathfrak{q}_i}$. The ideals $\mathfrak{p}_i$ are pairwise comaximal: for if $\mathfrak{p}_i$ and $\mathfrak{p}_j$ $(i \neq j)$ were not comaximal, one would have, for example, $\mathfrak{p}_i < \mathfrak{p}_j$ (lemma); but then we would have $\mathfrak{p}_i = \mathfrak{q}_i \subset \mathfrak{q}_j$ (lemma), contradicting irredundance. It follows that the ideals $\mathfrak{q}_i$ are also pairwise comaximal; and hence, by III, § 13, Theorem 32, R is a direct sum of rings respectively isomorphic to the rings $R/\mathfrak{q}_i$. Now, each of the rings $R/\mathfrak{q}_i$ is a PIR. If $\mathfrak{p}_i$ is maximal, then $\mathfrak{q}_i$ is contained in no other prime ideal than $\mathfrak{p}_i$, so that $R/\mathfrak{q}_i$ has only one prime ideal, namely $\mathfrak{p}_i/\mathfrak{q}_i$, and is therefore a special PIR. If $\mathfrak{p}_i$ is not maximal, then $\mathfrak{p}_i = \mathfrak{q}_i$ (lemma) and $R/\mathfrak{p}_i$ is a PID. Q.E.D.

We shall conclude this section with two useful lemmas concerning finite modules over PIR's.

LEMMA 1. *If a module M over a principal ideal ring R has a basis of n elements, then every submodule N of M has a basis of n elements.*

PROOF. If $n = 1$, we have $M = Rx$ and then clearly $N = \mathfrak{A}x$, where $\mathfrak{A}$ is an ideal in R. Since R is a PIR, we have $\mathfrak{A} = Rt$, whence $N = Ry$, where $y = tx$, and this establishes the lemma in the case $n = 1$. In the general case we use induction with respect to n, assuming therefore that the lemma is true for R-modules which are generated by $n - 1$ elements. Let $M = Rx_1 + Rx_2 + \cdots + Rx_n$. We set $M_1 = Rx_2 +$

$Rx_3 + \cdots + Rx_n$, $N_1 = N \cap M_1$, and we denote by N' the difference module $N - N_1$. By the induction hypothesis, N_1 has a basis of $n - 1$ elements, say $\{y_2, y_3, \cdots, y_n\}$. By the second isomorphism theorem (III, § 4, Theorem 5) we have that N' is isomorphic to the difference module $(N + M_1) - M_1$. This last module is a submodule of the difference module $M - M_1$, the latter being a principal module (generated by the single element $x_1 + M_1$). It follows by the case $n = 1$ that N' has a basis of one element. If we denote by y_1 an element of N such that the coset $y'_1 = y_1 + N_1$ generates N' over R, then $\{y_1, y_2, \cdots, y_n\}$ is a basis of N. This completes the proof.

LEMMA 2. *If R is a PID and M is an R-module which has a basis of n elements which are linearly independent over R, then every submodule N of M has a basis of n or fewer elements which are linearly independent over R.*

PROOF. We use the notations of the proof of the preceding lemma and we first consider the case $n = 1$. In $N = (0)$ there is nothing to prove. If $N \neq (0)$, and if we have a relation $ay = 0$, $a \in R$, then $atx = 0$ and hence $at = 0$, since x is independent over R. Since we have assumed that R is an integral domain and since $t \neq 0$ (for we have $y \neq 0$) it follows that $a = 0$, and this establishes the lemma in the case $n = 1$. In the general case we again use induction with respect to n. By the induction hypothesis, the module N_1 has a basis consisting of $n - 1$ or fewer elements which are linearly independent over R. Let $\{z_2, z_3, \cdots, z_m\}$ be such a basis ($m \leqq n$). Then $\{y_1, z_2, z_3, \cdots z_m\}$ is a basis of N, and we have only to show that if $N \neq N_1$ then y_1, z_2, $z_3, \cdots z_m$ are linearly independent over R. Assume that we have a relation $a_1y_1 + a_2z_2 + a_3z_3 + \cdots + a_mz_m = 0$, $a_i \in R$. Let $y_1 = b_1x_1 + b_2x_2 + \cdots + b_nx_n$, $b_j \in R$. Since $a_1y_1 \in N_1 \subset M_1 = Rx_2 + Rx_3 + \cdots + Rx_n$, it follows from the linear independence of the x_j over R that $a_1b_1 = 0$. Since $N \neq N_1$, y_1 does not belong to N_1, and hence $b_1 \neq 0$. Consequently $a_1 = 0$, and therefore also $a_2 = a_3 = \cdots = a_m = 0$. This completes the proof of the lemma.

§ 16. Irreducible ideals. In proving the Lasker-Noether decomposition theorem (§ 4), we have seen that in a noetherian ring R, every irreducible ideal is *primary* (Lemma 2 of § 4). On the other hand, if a primary ideal $\mathfrak{q}$ belonging to a prime ideal $\mathfrak{p}$ is reducible—say $\mathfrak{q} = \mathfrak{a} \cap \mathfrak{b}$, $\mathfrak{a}$ and $\mathfrak{b}$ being distinct from $\mathfrak{q}$—then it is easily seen that we also have for $\mathfrak{q}$ a non-trivial representation of the form $\mathfrak{q} = \mathfrak{q}' \cap \mathfrak{q}''$, where $\mathfrak{q}'$ and $\mathfrak{q}''$ are suitable primary ideals belonging to $\mathfrak{p}$ and distinct from $\mathfrak{q}$. To see this, let $\mathfrak{q}'$ denote the primary component of $\mathfrak{a}$ which belongs to $\mathfrak{p}$, if such a primary component exists (in other words: if $\mathfrak{p}$ is a prime ideal of $\mathfrak{a}$); in the contrary case we set $\mathfrak{q}' = (1)$. In a similar fashion we define $\mathfrak{q}''$, using the ideal $\mathfrak{b}$ instead of $\mathfrak{a}$. From $\mathfrak{q} = \mathfrak{a} \cap \mathfrak{b}$ and from the unicity theorems concerning irredundant decompositions of an ideal into primary components, it follows at once that $\mathfrak{q} = \mathfrak{q}' \cap \mathfrak{q}''$. It now follows that neither $\mathfrak{q}'$ nor $\mathfrak{q}''$ can be the unit ideal; for if, say, $\mathfrak{q}''$ were the unit ideal, we would have $\mathfrak{q}' = \mathfrak{q}$, $\mathfrak{q} \supset \mathfrak{a}$, and hence $\mathfrak{q} = \mathfrak{a}$, in contradiction with our assumption. Hence both $\mathfrak{q}'$ and $\mathfrak{q}''$ are primary ideals belonging to $\mathfrak{p}$, both different from $\mathfrak{q}$, as asserted.

It follows that when investigating the irreducibility of a primary ideal $\mathfrak{q}$, we may restrict ourselves to representations $\mathfrak{q} = \mathfrak{a} \cap \mathfrak{b}$ in which $\mathfrak{a}$ and $\mathfrak{b}$ are primary ideals belonging to the radical of $\mathfrak{q}$. In other words, and by passage to the quotient ring $R_{\mathfrak{p}}$ (Theorem 19 of § 11), we are reduced to the following problem: given a local ring, characterize the irreducible primary ideals belonging to its maximal ideal.

THEOREM 34. *Let R be a local ring and let $\mathfrak{q}$ be a primary ideal in R belonging to the ideal $\mathfrak{m}$ of non-units of R. The following conditions are equivalent:*

1) *$\mathfrak{q}$ is irreducible.*
2) *The vector space $(\mathfrak{q}:\mathfrak{m})/\mathfrak{q}$ (over $R/\mathfrak{m}$) is one-dimensional.*
3) *The set of all ideals in R properly containing $\mathfrak{q}$ admits a smallest element (in this case that smallest ideal is $\mathfrak{q}:\mathfrak{m}$).*
4) *For every ideal $\mathfrak{a}$ containing $\mathfrak{q}$, there exists another ideal $\mathfrak{a}' \supset \mathfrak{q}$ such that $\mathfrak{a} = \mathfrak{q}:\mathfrak{a}'$.*

NOTE. Our proof that (3) $\Rightarrow$ (4) will show that (4) is satisfied with $\mathfrak{a}' = \mathfrak{q}:\mathfrak{a}$, and that we have therefore the following: if $\mathfrak{q}$ is irreducible, then $\mathfrak{q}:(\mathfrak{q}:\mathfrak{a}) = \mathfrak{a}$, for any ideal $\mathfrak{a}$ containing $\mathfrak{q}$.

PROOF. We will give a "cyclic" proof: 1) implies 2), 2) implies 3), 3) implies 4) and 4) implies 1). We first show that 1) implies 2). In fact, since $\mathfrak{q}$ is primary for $\mathfrak{m}$, we have $\mathfrak{q}:\mathfrak{m} > \mathfrak{q}$. From $\mathfrak{m}(\mathfrak{q}:\mathfrak{m}) \subset \mathfrak{q}$ we deduce that $(\mathfrak{q}:\mathfrak{m})/\mathfrak{q}$ is a vector space over $R/\mathfrak{m}$. [See III, § 6, p. 146. In the present context we have only to observe that if $(\mathfrak{q}:\mathfrak{m})/\mathfrak{q}$ is regarded as an R-module then the relation $(\mathfrak{q}:\mathfrak{m})\mathfrak{m} \subset \mathfrak{q}$ shows that the ideal $\mathfrak{m}$ is contained in the order of that R-module. Hence $(\mathfrak{q}:\mathfrak{m})/\mathfrak{q}$ can be regarded as an $R/\mathfrak{m}$-module, and since $R/\mathfrak{m}$ is a field, $(\mathfrak{q}:\mathfrak{m})/\mathfrak{q}$ is a vector space over $R/\mathfrak{m}$.] The subspaces of this vector space correspond to the ideals in R contained between $\mathfrak{q}:\mathfrak{m}$ and $\mathfrak{q}$. If this vector space were of dimension > 1, its zero element would be the intersection of two non-trivial subspaces, and $\mathfrak{q}$ would be reducible.

We now prove that 2) implies 3). We first notice that in general any ideal $\mathfrak{a}$ properly containing $\mathfrak{q}$ has with $\mathfrak{q}:\mathfrak{m}$ an intersection distinct from

$\mathfrak{q}$; for, the smallest exponent s such that $\mathfrak{a} \cdot \mathfrak{m}^s \subset \mathfrak{q}$ is ≥ 1, and we have $\mathfrak{q} \not\supset \mathfrak{a} \cdot \mathfrak{m}^{s-1} \subset (\mathfrak{q}:\mathfrak{m}) \cap \mathfrak{a}$. Now, if $(\mathfrak{q}:\mathfrak{m})/\mathfrak{q}$ is a one-dimensional vector space, there are no ideals between $\mathfrak{q}$ and $\mathfrak{q}:\mathfrak{m}$. It follows that every ideal $\mathfrak{a}$ in R, which properly contains $\mathfrak{q}$, contains $\mathfrak{q}:\mathfrak{m}$ (since the intersection of $\mathfrak{a}$ with $\mathfrak{q}:\mathfrak{m}$ is different from $\mathfrak{q}$). In other words, condition 3) is satisfied.

We observe that the implication "2) implies 3)" can also be derived from Corollary 1 of Theorem 28 (in § 13). According to that corollary we have that if $\mathfrak{a}$ is a minimal proper overideal of an ideal $\mathfrak{b}$ in a ring R, then there must exist a maximal ideal $\mathfrak{p}$ in R such that $\mathfrak{a}\mathfrak{p} \subset \mathfrak{b}$. If we apply this result to our local ring R (in which the only maximal ideal is $\mathfrak{m}$) we see that if $\mathfrak{a}$ is any minimal proper overideal of $\mathfrak{q}$, then $\mathfrak{a}\mathfrak{m} \subset \mathfrak{q}$. We have therefore $\mathfrak{q} < \mathfrak{a} \subset \mathfrak{q}:\mathfrak{m}$. If, now, 2) holds, we deduce that $\mathfrak{a} = \mathfrak{q}:\mathfrak{m}$, whence $\mathfrak{q}:\mathfrak{m}$ is the only minimal proper overideal of $\mathfrak{q}$. Since any ideal properly containing $\mathfrak{q}$ contains some minimal proper overideal of $\mathfrak{q}$ ($R/\mathfrak{q}$ being a ring satisfying the d.c.c.; see Theorem 2 of § 2), $\mathfrak{q}:\mathfrak{m}$ is the smallest ideal properly containing $\mathfrak{q}$, which proves 3).

Let us now assume that condition 3) is true. Then, since a non-trivial vector space which admits a smallest non-zero subspace must be one-dimensional, $(\mathfrak{q}:\mathfrak{m})/\mathfrak{q}$ is one-dimensional, and $\mathfrak{q}:\mathfrak{m}$ is the smallest proper overideal of $\mathfrak{q}$ as was proved above. For the proof of 4) we now need the two following lemmas:

LEMMA 1. *If* 3) *holds, and if* $\mathfrak{a}$ *is a minimal overideal of an ideal* $\mathfrak{b}$ *containing* $\mathfrak{q}$, *then* $\mathfrak{q}:\mathfrak{b}$ *is a minimal overideal of* $\mathfrak{q}:\mathfrak{a}$ *or is equal to it.*

PROOF. From our hypothesis concerning the ideal $\mathfrak{a}$ and from the above-cited Corollary 1 to Theorem 28 it follows that $\mathfrak{a}\mathfrak{m} \subset \mathfrak{b}$ and that consequently $\mathfrak{a}/\mathfrak{b}$ is a one-dimensional vector space over $R/\mathfrak{m}$. In other words, there exists an element t in $\mathfrak{a}$ such that $\mathfrak{a} = \mathfrak{b} + Rt$, $t\mathfrak{m} \subset \mathfrak{b}$. From $\mathfrak{a}\mathfrak{m} \subset \mathfrak{b}$ we deduce $(\mathfrak{q}:\mathfrak{b})\mathfrak{m}\mathfrak{a} \subset (\mathfrak{q}:\mathfrak{b})\mathfrak{b} \subset \mathfrak{q}$—that is, $(\mathfrak{q}:\mathfrak{b})\mathfrak{m} \subset \mathfrak{q}:\mathfrak{a}$. Hence also $(\mathfrak{q}:\mathfrak{b})/(\mathfrak{q}:\mathfrak{a})$ is a vector space over $R/\mathfrak{m}$. We now define the following mapping f of $\mathfrak{q}:\mathfrak{b}$ into $(\mathfrak{q}:\mathfrak{m})/\mathfrak{q}$: $f(x) =$ coset of xt mod. $\mathfrak{q}$ ($x \in \mathfrak{q}:\mathfrak{b}$; note that $t\mathfrak{m} \subset \mathfrak{a}\mathfrak{m} \subset \mathfrak{b}$, whence $xt\mathfrak{m} \subset x\mathfrak{b} \subset \mathfrak{q}$, showing that $xt \in \mathfrak{q}:\mathfrak{m}$ and that consequently f is indeed a mapping into $(\mathfrak{q}:\mathfrak{m})/\mathfrak{q}$). It is immediately seen that the kernel of f is the ideal $\mathfrak{q}:\mathfrak{a}$. Therefore f defines a mapping $\bar{f}$ of $(\mathfrak{q}:\mathfrak{b})/(\mathfrak{q}:\mathfrak{a})$ into $(\mathfrak{q}:\mathfrak{m})/\mathfrak{q}$: namely, if $x \in \mathfrak{q}:\mathfrak{b}$ and $\bar{x}$ denotes the coset of x mod. $\mathfrak{q}:\mathfrak{a}$, then we define $\bar{f}(\bar{x}) = f(x)$. Now, both $(\mathfrak{q}:\mathfrak{b})/(\mathfrak{q}:\mathfrak{a})$ and $(\mathfrak{q}:\mathfrak{m})/\mathfrak{q}$ are vector spaces over $R/\mathfrak{m}$, and it is obvious, from the definition of f, that $\bar{f}$ is a *linear* mapping, with zero kernel. Hence $(\mathfrak{q}:\mathfrak{b})/(\mathfrak{q}:\mathfrak{a})$ is isomorphic to a vector subspace of $(\mathfrak{q}:\mathfrak{m})/\mathfrak{q}$, and since $(\mathfrak{q}:\mathfrak{m})/\mathfrak{q}$ is one-dimensional, the proof of the lemma is complete.

LEMMA 2. *If* 3) *holds, and if* $\mathfrak{a}$ *is an overideal of* $\mathfrak{q}$, *the lengths of the ideals* $\mathfrak{a}$, $\mathfrak{q}:\mathfrak{a}$, *and* $\mathfrak{q}$ *are related by*

$$\lambda(\mathfrak{a}) + \lambda(\mathfrak{q}:\mathfrak{a}) = \lambda(\mathfrak{q}).$$

PROOF. Consider a composition series $\{\mathfrak{a}_i\}$ of ideals joining R to $\mathfrak{q}$ and containing $\mathfrak{a}$; let $\mathfrak{a} = \mathfrak{a}_j$, $\mathfrak{q} = \mathfrak{a}_s$. By Lemma 1, the normal series $\{\mathfrak{q}:\mathfrak{a}_{s-i}\}$ admits no proper refinement (III, § 11, p. 159). Thus, by Jordan's theorem (III, § 11, Theorem 22), its terms are all distinct, as their number is exactly s. Therefore the lengths of $\mathfrak{a}$ and $\mathfrak{q}:\mathfrak{a}$ are j and $s - j$ respectively, and the lemma is proved.

The proof that 3) implies 4) is now easy. From Lemma 2 we deduce that the two ideals $\mathfrak{a}$ and $\mathfrak{q}:(\mathfrak{q}:\mathfrak{a})$ have the same length, equal to $\lambda(\mathfrak{q}) - \lambda(\mathfrak{q}:\mathfrak{a})$. As the former is contained in the latter, the two ideals are equal, and 4) holds.

Finally we show that 4) implies 1). Suppose that we have a representation $\mathfrak{q} = \mathfrak{a} \cap \mathfrak{a}'$. We may write, by 4), $\mathfrak{a} = \mathfrak{q}:\mathfrak{b}$, $\mathfrak{a}' = \mathfrak{q}:\mathfrak{b}'$, where $\mathfrak{b}$ and $\mathfrak{b}'$ are ideals containing $\mathfrak{q}$. We then have $\mathfrak{q} = (\mathfrak{q}:\mathfrak{b}) \cap (\mathfrak{q}:\mathfrak{b}') = \mathfrak{q}:(\mathfrak{b} + \mathfrak{b}')$, and this implies $\mathfrak{b} + \mathfrak{b}' = R$, as we already have $\mathfrak{q}:\mathfrak{m} \neq \mathfrak{q}$. As $\mathfrak{m}$ is the unique maximal ideal of R, the relation $\mathfrak{b} + \mathfrak{b}' = R$ implies that either $\mathfrak{b}$ or $\mathfrak{b}'$ is equal to R, that is, that either $\mathfrak{a}$ or $\mathfrak{a}'$ is equal to $\mathfrak{q}$. In other words, the representation $\mathfrak{q} = \mathfrak{a} \cap \mathfrak{a}'$ is trivial, and $\mathfrak{q}$ is an irreducible ideal. Theorem 34 is therefore completely proved.

The notations being as in Theorem 34, suppose that the ideal $\mathfrak{q}$ is *irreducible*. Then the mapping $\mathfrak{a} \rightarrow \mathfrak{a}' = \mathfrak{q}:\mathfrak{a}$ maps the set (S) of all overideals of $\mathfrak{q}$ onto itself. In proving that 3) implies 4) we have seen that $\mathfrak{a} = \mathfrak{a}'' = \mathfrak{q}:(\mathfrak{q}:\mathfrak{a})$; in other words, the mapping $\mathfrak{a} \rightarrow \mathfrak{a}'$ is a so-called "involution," and is therefore 1-1. The general formulae

$$\mathfrak{q}:(\mathfrak{a} + \mathfrak{b}) = (\mathfrak{q}:\mathfrak{a}) \cap (\mathfrak{q}:\mathfrak{b}), \quad \mathfrak{q}:(\mathfrak{a}\mathfrak{b}) = (\mathfrak{q}:\mathfrak{a}):\mathfrak{b} = (\mathfrak{q}:\mathfrak{b}):\mathfrak{a}$$

(III, § 7, p. 147) show that under the mapping $\mathfrak{a} \rightarrow \mathfrak{a}'$ of (S) onto itself, sums of ideals are transformed into intersections, and products into quotients. More precisely we have the formulae:

$$(1) \qquad (\mathfrak{a} \cap \mathfrak{b})' = \mathfrak{a}' + \mathfrak{b}', \quad (\mathfrak{a} + \mathfrak{b})' = \mathfrak{a}' \cap \mathfrak{b}'.$$

$$(2) \qquad (\mathfrak{a}\mathfrak{b})' = (\mathfrak{a}':\mathfrak{b}) = (\mathfrak{b}':\mathfrak{a}), \quad (\mathfrak{a}:\mathfrak{b})' = \mathfrak{b} \cdot \mathfrak{a}'.$$

(The first of the two formulae (1) follows by replacing in the second of these two formulae the ideals $\mathfrak{a}$ and $\mathfrak{b}$ by the ideals $\mathfrak{a}'$ and $\mathfrak{b}'$ and by using the involutorial property of our mapping.)

THEOREM 35. *For an ideal* $\mathfrak{a}$ *in* (S) *to be irreducible, it is necessary and sufficient that* $\mathfrak{a}' = \mathfrak{q}:\mathfrak{a}$ *be principal* mod $\mathfrak{q}$.

PROOF. Relations (1) show that *an ideal* $\mathfrak{c}$ *in* (S) *is irreducible if and only if* $\mathfrak{c}'$ *is not a non-trivial sum of ideals in* (S), that is, if and only if $\mathfrak{c}'$

is not of the form $\mathfrak{a} + \mathfrak{b}$ with $\mathfrak{q} \subset \mathfrak{a} \neq \mathfrak{c}'$ and $\mathfrak{q} \subset \mathfrak{b} \neq \mathfrak{c}'$. For, if $\mathfrak{c}' = \mathfrak{a} + \mathfrak{b}$, with $\mathfrak{a}, \mathfrak{b}$ in (S), then $\mathfrak{c} = \mathfrak{c}'' = (\mathfrak{a} + \mathfrak{b})' = \mathfrak{a}' \cap \mathfrak{b}'$, and if $\mathfrak{c}$ is irreducible then either $\mathfrak{a}' = \mathfrak{c}$ or $\mathfrak{b}' = \mathfrak{c}$, whence either $\mathfrak{a} = \mathfrak{a}'' = \mathfrak{c}'$ or $\mathfrak{b} = \mathfrak{b}'' = \mathfrak{c}'$. Conversely, if $\mathfrak{c}$ is reducible—say $\mathfrak{c} = \mathfrak{a} \cap \mathfrak{b}$, $\mathfrak{a} > \mathfrak{c}$ and $\mathfrak{b} > \mathfrak{c}$—then $\mathfrak{c}' = (\mathfrak{a} \cap \mathfrak{b})' = \mathfrak{a}' + \mathfrak{b}'$, and in view of the (1, 1) character of the mapping $\mathfrak{a} \rightarrow \mathfrak{a}'$ of (S) onto itself we have that both $\mathfrak{a}'$ and $\mathfrak{b}'$ are different from $\mathfrak{c}'$, since $\mathfrak{a} \neq \mathfrak{c}$, $\mathfrak{b} \neq \mathfrak{c}$ and since $\mathfrak{a}, \mathfrak{b} \in (S)$. This proves our assertion. Now, let $\mathfrak{a}$ be any irreducible ideal in (S) and let $\{x_1, x_2, \cdots, x_n\}$ be a basis of $\mathfrak{a}'$. We have $\mathfrak{a}' = (Rx_1 + \mathfrak{q}) + (Rx_2 + \mathfrak{q}) + \cdots + (Rx_n + \mathfrak{q})$, and since each term $Rx_i + \mathfrak{q}$ belongs to (S) it follows, by what we have just proved, that at least one of the ideals $Rx_i + \mathfrak{q}$ must coincide with $\mathfrak{a}'$. In other words: *$\mathfrak{a}'$ is a principal ideal* mod $\mathfrak{q}$. Conversely, suppose that $\mathfrak{a}$ is an ideal in (S) such that $\mathfrak{a}'$ is principal mod $\mathfrak{q}$. To prove that $\mathfrak{a}$ is irreducible we have only to prove that $\mathfrak{a}'/\mathfrak{q}$ is not a non-trivial sum of ideals in $R/\mathfrak{q}$. It will be sufficient to show the following: a non-trivial sum of principal ideals in $R/\mathfrak{q}$ cannot be principal. In other words: if the ideal $R'x + R'y$ is a principal ideal $R'z$ ($R' = R/\mathfrak{q}$), then we have either $z = ux$ or $z = uy$, where u is a unit in R'. We have, by assumption, $z = ax + by$, $x = cz$, $y = dz$ (a, b, c, d in R'), thus $z(1 - ac - bd) = 0$. If both ac and bd are non-units in R', then $1 - ac - bd$ is a unit in R' (since R' is a local ring with $\mathfrak{m}/\mathfrak{q}$ as ideal of non-units), and we have $z = x = y = 0$, in which case our assertion is true. Otherwise ac (or bd) is a unit in R', and then c (or d) is a unit in R', when $z = c^{-1}x$ (or $z = d^{-1}y$). Q.E.D.

Let us now consider a primary ideal $\mathfrak{q}$ in a noetherian ring R, and let $\mathfrak{p}$ be its associated prime. The ideals $\{\mathfrak{q}:\mathfrak{p}^n\}$ (for $n = 0, \cdots, e - 1$, e being the smallest exponent n such that $\mathfrak{p}^n \subset \mathfrak{q}$—that is, e is the *exponent* of $\mathfrak{q}$; see III, § 9, p. 153) are all primary for $\mathfrak{p}$ (III, § 9, Theorem 14); and they form an ascending sequence of ideals. This sequence is called the *upper Loewy series* of $\mathfrak{q}$. Its successive factors $(\mathfrak{q}:\mathfrak{p}^n)/(\mathfrak{q}:\mathfrak{p}^{n-1})$ are modules over $R/\mathfrak{p}$. In case $\mathfrak{p}$ is a *maximal ideal*, the factor $(\mathfrak{q}:\mathfrak{p}^n)/(\mathfrak{q}:\mathfrak{p}^{n-1})$ is a vector space over the field $R/\mathfrak{p}$; its dimension (which is finite since R satisfies the a.c.c.) is called the *n-th upper Loewy invariant of* $\mathfrak{q}$. The sum of these invariants is obviously the length of $\mathfrak{q}$.

If $\mathfrak{p}$ is a maximal ideal, then it is clear that $\mathfrak{q}:\mathfrak{p}$ is the biggest overideal $\mathfrak{a}$ of $\mathfrak{q}$ such that $\mathfrak{a}/\mathfrak{q}$ is a vector space over $R/\mathfrak{p}$ in a natural way. (The condition for this to be so is that $\mathfrak{p}$ be contained in the order of the R-module $\mathfrak{a}/\mathfrak{q}$, that is, that $\mathfrak{a}\mathfrak{p} \subset \mathfrak{q}$, or $\mathfrak{a} \subset \mathfrak{q}:\mathfrak{p}$.) It is easily seen that this implies that *$(\mathfrak{q}:\mathfrak{p})/\mathfrak{q}$ is the sum of the minimal ideals of $R/\mathfrak{q}$*. We also observe that it follows from Theorem 34, 2) that a necessary and sufficient condition for $\mathfrak{q}$ to be irreducible is that its *first upper Loewy invariant be equal to* 1.

APPENDIX: PRIMARY REPRESENTATION IN NOETHERIAN MODULES

We intend to generalize to the case of *modules* some of the results which have been proved previously for ideals.

Let R be a commutative ring, M an R-module, and N a submodule of M. We define the *radical* of the submodule N as being the set $\mathfrak{r}$ of all a in R for which there exists an exponent n such that $a^n \cdot M \subset N$. The fact that this set is an ideal $\mathfrak{r}$ is easily proved, as in Chapter III, § 7 (p. 148). One can also notice that the set $\mathfrak{b}$ of all b in R such that $bM \subset N$ is an ideal in R, as it is the order (or the annihilator) of the difference module $M - N$; and $\mathfrak{r}$ is the radical of this ideal. The rules about radicals of sums and intersections (III, § 7, Theorem 9) extend to radicals of submodules (straightforward checking).

A submodule N of M is said to be *primary* if the relation $ax \in N (a \in R, x \in M)$ implies either $x \in N$ or $a \in \mathfrak{r}(N)$, ($\mathfrak{r}(N)$ denoting the radical of N). It is not difficult to see that *if N is a primary submodule of M, then the annihilator $\mathfrak{b}$ of $M - N$ is a primary ideal in R.*

PROOF. If $ab \in \mathfrak{b}$, we have $abM \subset N$; if furthermore $a \notin \mathfrak{r}(N)$ (or equivalently: if no power of a belongs to $\mathfrak{b}$), we have $bM \subset N$, that is, $b \in \mathfrak{b}$. Q.E.D. [The converse is false: in the countable direct sum $J + J + \cdots + J + \cdots = M$ the submodule $N = J + 2J + \cdots + nJ + \cdots$ is such that $\mathfrak{b} = (0)$, whence $\mathfrak{b}$ is primary; however, the element $(0, 1, 0, \cdots 0, \cdots) = x$ satisfies $2x \in N, x \notin N, 2 \notin \mathfrak{r}(N)$.] Thus, if a submodule N is primary, its radical is a *prime ideal*. We know that the converse is not true already in the case of ideals; however a submodule whose radical is a *maximal* ideal is primary.

PROOF. Let $\mathfrak{r}(N)$ be maximal. Then, from $ax \in N$ and $a \notin \mathfrak{r}(N)$, we deduce that $R = aR + \mathfrak{r}(N)$, whence $1 = ba + c$ with $c \in \mathfrak{r}(N)$. By raising to a suitably high power we get $1 = b'a + c'$, where c' belongs to the annihilator of $M - N$. Therefore $x = b'a \cdot x + c' \cdot x \in N$.

The characterization of the *pair* {primary submodule, radical} runs exactly as in the case of ideals (III, § 9, Theorem 13): let $\mathfrak{p}$ be an ideal in a ring R and let E be a submodule of an R-module M; then E is primary and $\mathfrak{p}$ is its radical if and only if the following conditions are satisfied: (a) $\mathfrak{p}$ contains the annihilator of $M - E$; (b) if $b \in \mathfrak{p}$, then $b^m \cdot M \subset E$ for some m (depending on b); (c) if $a \cdot x \in E$ and $x \notin E$, then $a \in \mathfrak{p}$.

PROOF. It is obvious that conditions (a), (b), and (c) are satisfied if E is primary and $\mathfrak{p}$ is its radical. Now suppose that (a), (b), and (c) are satisfied. Condition (b) signifies that $\mathfrak{p}$ is contained in the radical $\mathfrak{r}(E)$ of E. Therefore conditions (b) and (c) imply that E is primary. Let a be an element of $\mathfrak{r}(E)$, and let n be the least exponent such that a^n is in the annihilator of $M - E$ (that is, such that $a^nM \subset E$). If $n = 1$, we have $a \in \mathfrak{p}$ by (a); otherwise there exists x in M such that $y = a^{n-1} \cdot x \notin E$, and, since we have $a \cdot y \in E$, (c) implies $a \in \mathfrak{p}$. We have thus proved that $\mathfrak{r}(E) \subset \mathfrak{p}$, and hence $\mathfrak{p}$ is the radical of E.

It is also easily seen that a finite intersection of primary submodules E_i with the same radical $\mathfrak{p}$ is a primary submodule also having $\mathfrak{p}$ as radical.

We say that a module M over a ring R with identity is *noetherian* if it is unitary and if it satisfies the a.c.c. We have the following representation theorem: *in a noetherian module M every submodule N is a finite intersection of primary submodules.*

The proof runs as in § 4. A first lemma shows that N is a finite intersection of irreducible modules; its proof is the same as that of Lemma 1 of § 4 (that lemma actually belongs to the theory of partially ordered sets). A second lemma shows that every irreducible submodule E is primary. The proof is indirect: if E were not primary, there would exist a in R and x in M such that $ax \in E, x \notin E, a^nM \not\subset E$ for every n; one then considers the increasing sequence of submodules $E:Ra^n$ ($E:Ra^n$ denoting the set of all y in M such that $a^n \cdot y \in E$), one observes that by the a.c.c. there exists an exponent s such that $E:Ra^s = E:Ra^{s+1}$ and one then proves as in Lemma 2 of § 4 that $E = (E + R\cdot x) \cap (E + a^s\cdot M)$, in contradiction with the irreducibility of E.

Irredundant primary representations of M are defined as in § 4, and one sees immediately the existence of such a primary representation.

The radicals $\mathfrak{p}_i$ of the submodules E_i which appear in an irredundant primary representation $N = \bigcap_i E_i$ of a submodule N are characterized by the following property: they are the prime ideals $\mathfrak{p}$ in R for which there exists x in M, $x \notin N$, such that the ideal $N:Rx$ (set of all b in R such that $b\cdot x \in N$) is primary for $\mathfrak{p}$. (See Theorem 6 in § 5.) This shows the uniqueness of the prime ideals $\mathfrak{p}_i$.

PROOF. To show that each $\mathfrak{p}_i$ enjoys the above property, one takes $x \in \bigcap_{j\neq i} E_j$, $x \notin E_i$. Then $N:Rx$ contains the annihilator of $M - E_i$ (which is primary for $\mathfrak{p}_i$, as seen above), and the proof that $N:Rx$ is primary for $\mathfrak{p}_i$ as well as the proof of the converse, run exactly as in Theorem 6 of § 5.

The terminology defined in § 5 may be extended without further warning to the case of submodules. The minimal elements of the family of all prime ideals $\mathfrak{p}_i$ associated to a submodule N of M are also the minimal elements in the family of all prime ideals $\mathfrak{p}$ which contain the annihilator of $M - N$.

PROOF. We notice that, if we denote by $\mathfrak{q}_i$ the annihilator of $M - E_i$, $\mathfrak{q}_i$ is primary for $\mathfrak{p}_i$, and the annihilator $\mathfrak{a}$ of $M - N$ is the intersection of the $\mathfrak{q}_i$. Thus $\mathfrak{a} = \bigcap_i \mathfrak{q}_i$ is a (not necessarily irredundant) primary representation of the ideal $\mathfrak{a}$, and the conclusion follows by Theorem 7 of § 5.

Let $N = \bigcap_i E_i$ be an irredundant primary representation of a submodule N of M, and let $\mathfrak{p}_i$ be the associated prime ideal of E_i. The set E'_i of all elements x of M for which there exists $a \notin \mathfrak{p}_i$ such that $a\cdot x \in N$ is obviously a submodule of M. It is contained in E_i, and, when $\mathfrak{p}_i$ is an isolated prime ideal of N, we have $E'_i = E_i$. This shows the uniqueness of the isolated components of N. (Proof as in Theorem 8 of § 5. In the second part of the proof replace the relation $b_j^{s(j)} \in \mathfrak{q}_j$ by $b_j^{s(j)}\cdot M \subset E_j$.)

As an application of the preceding theory, let us give a generalization of Krull's theorem (Theorem 12 of § 7) to modules. We first state the following generalization of Lemma 1 of § 7, which is useful in local algebra:

Let E be a noetherian module over a ring R, let $\mathfrak{m}$ be an ideal in R and let F be a submodule of E. Then there exists an integer s and a submodule F' of E such that $\mathfrak{m}F = F \cap F'$ and $F' \supset \mathfrak{m}^sE$.

PROOF. As in Lemma 1 of § 7, one considers the primary components $F_i(F'_j)$ of $\mathfrak{m}F$ whose associated prime ideals contain (do not contain) $\mathfrak{m}$, and one proves that $F' = \bigcap_i F_i$ contains some $\mathfrak{m}^sE$ and that $F'' = \bigcap_j F'_j$ contains F.

Then the generalization of Krull's theorem is as follows:

Let E be a noetherian R-module and let $\mathfrak{m}$ be an ideal in R such that the relations $m \in \mathfrak{m}$, $x \in E$, $(1 + m)\cdot x = 0$ imply $x = 0$. Then we have $\bigcap_{n=1}^{\infty} \mathfrak{m}^nE = (0)$. (Notice that the condition about $\mathfrak{m}$ is automatically verified when every element of the form $1 + m$ ($m \in \mathfrak{m}$) is invertible in R).

PROOF. Let $F = \bigcap_{n=1}^{\infty} \mathfrak{m}^nE$. By the above lemma we have $F = \mathfrak{m}F$. We then take a finite basis of F, express the relation $F = mF$ in terms of this basis, and conclude from an argument about determinants (as in Lemma 2 of § 7) that $F = (0)$.

We leave to the reader the generalization of the consequences of Krull's theorem.

V. DEDEKIND DOMAINS. CLASSICAL IDEAL THEORY

All the rings in this chapter will be assumed to have an identity. Whenever two rings occur, one of which is a subring of the other, it will be tacitly assumed that the identity of the bigger ring belongs also to the subring (and therefore is the identity of the subring).

§ 1. Integral elements. Let A be a ring, let B be an overring of A, and let x be an element of B. The element x is said to be *integral over* A (or *integrally dependent on* A) if it satisfies the following condition:

(c) *There exists a finite set* $\{a_0, \cdots, a_{n-1}\}$ *of elements of* A *such that*

$$x^n + a_{n-1}x^{n-1} + \cdots + a_0 = 0. \tag{1}$$

In other words, x is integral over A if it is a root of a monic equation (1) with coefficients in A. The equation (1) is called an *equation of integral dependence* satisfied by x over A. Note that an element which is integral over A is algebraic over A. Note also that every element a of A is integral over A (a is a root of $X - a$).

We now give conditions which are equivalent to (c):

(c′) *The ring* $A[x]$ *is a finite* A*-module.*

(c″) *The ring* $A[x]$ *is contained in a subring* R *of* B *which is a finite* A*-module.*

(c‴) *There exists in* B *a finite* A*-module* M *with the following two properties:*
1) $xM \subset M$.
2) *zero is the only element* y *of* $A[x]$ *such that* $yz = 0$ *for all* z *in* M.

PROOF. We give a cyclic proof. Condition (c) implies (c′), since equation (1) signifies that $x^n \in \sum_{i=0}^{n-1} Ax^i$, whence $x^{n+q} \in \sum_{i=0}^{n-1} Ax^{i+q}$; thus $x^{n+q} \in \sum_{i=0}^{n-1} Ax^i$ by induction on q, and $\{1, x, \cdots, x^{n-1}\}$ is a finite basis of

$A[x]$ over A. It is clear that (c′) implies (c″). Also (c″) implies (c‴): we take $M = R$; then (1) is satisfied since R is a ring and (2) is satisfied since $1 \in R$. We now prove that (c‴) implies (c). This follows from the following more informative lemma:

LEMMA. *Let A be a ring, $\mathfrak{q}$ an ideal in A, M a finite A-module contained in an overring R of A, and x an element of R such that*

1′) $xM \subset M\mathfrak{q}$;

2′) *zero is the only element y of $A[x]$ such that $yz = 0$ for all z in M.*

Then x satisfies an equation of integral dependence of the form

$$x^n + q_{n-1}x^{n-1} + \cdots + q_0 = 0,$$

where all the q_i $(0 \leq i \leq n-1)$ belong to $\mathfrak{q}$.

PROOF OF THE LEMMA. Let us write $M = \sum_i Am_i$. Then $xM \subset \sum_i Am_i\mathfrak{q} = \sum_i \mathfrak{q}m_i$. In particular there exist elements q_{ji} of $\mathfrak{q}$ such that $xm_j = \sum_i q_{ji}m_i$. This is a system of n linear homogeneous equations in the m_i, and we can write it as follows:

$$\sum_i (\delta_{ji}x - q_{ji})m_i = 0,$$

where the δ_{ji} are the Kronecker symbols. Let $d = \det(\delta_{ji}x - q_{ji})$. We have then $dm_i = 0$ for every i, whence $dM = (0)$ and $d = 0$ by condition 2′). By expansion of the determinant, one sees readily that the relation $\det(\delta_{ji}x - q_{ji}) = 0$ is an equation of integral dependence of the required type. Q.E.D.

REMARK 1. Condition (2) in (c‴) is automatically satisfied if B is an integral domain and $M \neq (0)$, or if $1 \in M$. More generally, condition 2′) in Lemma 1 is automatically satisfied if $1 \in M$, in particular if M is an *overring* of A.

If A is a *noetherian* ring, then it is clear that condition (c′) is equivalent to the following condition:

(c'_n). *The ring $A[x]$ is contained in a finite A-module.*

REMARK 2. If A is not noetherian, then it may be shown by examples that condition (c'_n) is weaker than (c′). For example, we may take for A a valuation ring whose value group Δ is non-archimedean (that is, of rank > 1; see VI, § 10). Then if α and β are elements in Δ such that $\alpha > n\beta > 0$ for all positive integers n, and if x and d are elements of the quotient field of A having values $-\beta$ and α respectively, then $A[x]$ is contained in the finite A-module Ad^{-1}, without x being integral over A.

REMARK 3. If A, B and C are rings such that $A \subset B \subset C$, B is a finite A-module and C a finite B-module, then it is obvious that C is a finite A-module.

THEOREM 1. *Let A be a ring, let B be an overring of A, and let $x_1, \cdots, x_n$ be elements of B. If each of the elements x_i is integral over A, then the ring $A[x_1, \cdots, x_n]$ is a finite A-module.*

PROOF. The theorem is true for $n = 1$ by (c′). By induction on n, we may assume that the ring $B' = A[x_1, \cdots, x_{n-1}]$ is a finite A-module. On the other hand, by the case $n = 1$, the ring $A[x_1, \cdots, x_n] = B'[x_n]$ is a finite B'-module, since x_n, being integral over A, is *a fortiori* integral over B'. Hence, by Remark 3, the ring $A[x_1, \cdots, x_n]$ is a finite A-module. Q.E.D.

COROLLARY. *Let A be a ring and B an overring of A. The elements of B which are integral over A form a ring which contains A.*

In fact, if x and y are elements of B which are integral over A, then $A[x, y]$ is a finite A-module, by Theorem 1, whence $x - y$ and xy are integral over A by (c''). On the other hand, every element of A is integral over A.

The ring of all the elements of the overring B which are integral over A is called the *integral closure* of A in B. When the integral closure of A in B is A itself, then one says that A is *integrally closed in B*; this means that every element of B which is integral over A lies in A. If every element of B is integral over A, then B is said to be *integral over A* (or *integrally dependent on A*).

THEOREM 2 (TRANSITIVITY OF INTEGRAL DEPENDENCE). *Let A be a ring, B an overring of A integral over A, and C an overring of B integral over B. Then C is integral over A.*

PROOF. Let x be an element of C, and let

$$x^n + b_{n-1}x^{n-1} + \cdots + b_0 = 0 \ (b_i \in B)$$

be an equation of integral dependence for x over B. Then the ring $B' = A[b_0, \cdots, b_{n-1}]$ is a finite A-module (Theorem 1). Since x is integral over B', $B'[x]$ is a finite B'-module, and therefore a finite A-module. Therefore x is integral over A by (c′).

The above-defined notions of integral closure and of an integrally closed ring are *relative* notions and refer to a given overring B of A; the use of the words "in B" is necessary in order to avoid confusion. We, however, make the convention that the expressions "integral closure of A," "A is integrally closed," mean respectively "the integral closure of A in its total quotient ring," "A is integrally closed in its total quotient ring" (I, § 19), the role of B being played by the total quotient ring of A. The most important case is the one in which A is an integral domain, its total quotient ring being then its quotient field. When dealing with an *integrally closed integral domain A* (that is, with an integral domain which is integrally closed in its quotient field) we shall omit, as a rule, the adjective "integral" and we shall refer to A as an *integrally closed domain.*

§ 2. **Integrally dependent rings.** Let A be a ring, A' an overring of A, integrally dependent on A. We first prove that this relationship between A and A' is preserved under residue-class ring formation and under quotient ring formation. More precisely:

LEMMA 1. *Let A' be integrally dependent on A, and let $\mathfrak{J}'$ be an ideal in A'. Then $A'/\mathfrak{J}'$ is integrally dependent on $A/\mathfrak{J}' \cap A$.*

PROOF. We first recall that $A/\mathfrak{J}' \cap A$ can be canonically identified with a subring of $A'/\mathfrak{J}'$ (III, § 5, Theorem 8). Now, for any x in A', it suffices to reduce modulo $\mathfrak{J}'$ an equation of integral dependence for x over A.

LEMMA 2. *Let A' be integrally dependent on A and let S be a multiplicatively closed set of non-zero elements of A. Then A'_S is integrally dependent on A_S.*

PROOF. Let $\mathfrak{n}'$ be the ideal formed by the elements of A' which are annihilated by at least one element of S. Since A'_S and A_S are ordinary quotient rings of $A'/\mathfrak{n}'$ and of $A/\mathfrak{n}' \cap A$ (IV, § 9), we are reduced, by Lemma 1, to the case in which A and A' are subrings of A_S and A'_S respectively. Let then x/s ($x \in A'$, $s \in S$) be an element of A'_S. From an integral dependence equation

$$x^n + a_{n-1}x^{n-1} + \cdots + a_0 = 0 \quad (a_i \in A)$$

of x over A, we deduce, upon division by s^n:

$$(x/s)^n + (a_{n-1}/s)(x/s)^{n-1} + \cdots + a_0/s^n = 0,$$

and this is an integral dependence equation of x/s over A_S. Q.E.D.

THEOREM 3. *Let A be a ring, A' an overring of A, integral over A, and let $\mathfrak{p}$ be a prime ideal of A. There exists a prime ideal $\mathfrak{p}'$ in A' such that $\mathfrak{p}' \cap A = \mathfrak{p}$* (that is, $\mathfrak{p}'^c = \mathfrak{p}$).

PROOF. Let us first achieve a reduction to the case in which $\mathfrak{p}$ is the only maximal ideal of A. For this purpose, denote by S the complement of $\mathfrak{p}$ in A (which is multiplicatively closed) and consider the quotient rings A_S and A'_S. The ideal $\mathfrak{p}A_S$ is the only maximal ideal of A_S (IV, § 11, Theorem 19), and A'_S is an overring of A_S, integrally dependent on A_S (Lemma 2). Suppose that there exists a prime ideal $\mathfrak{m}'$ in A'_S such that $\mathfrak{m}' \cap A_S = \mathfrak{p}A_S$. Then the inverse image $\mathfrak{p}'$ of $\mathfrak{m}'$ in A' (i.e. the ideal $\mathfrak{m}'^c$; see Definition in IV, § 8, p. 218) is a prime ideal of A' (IV, § 8, p. 220). It is clear that $\mathfrak{p}' \cap A$ contains $\mathfrak{p}$. Conversely, if $x \in \mathfrak{p}' \cap A$, then the residue class $\bar{x}$ of x modulo the kernel $\mathfrak{n}$ of the homomorphism $A \to A_S$ lies in $\mathfrak{m}' \cap A_S$, that is in $\mathfrak{p}A_S$; thus $x \in \mathfrak{p}$ (IV, § 11, Theorem 19).

We shall now assume that $\mathfrak{p}$ is the only maximal ideal in A. If we prove that the extended ideal $A'\mathfrak{p}$ is distinct from A', then the theorem will be proved. For, if $A'\mathfrak{p} \neq A'$, then $A'\mathfrak{p}$ is contained in some maximal ideal of A'. If $\mathfrak{p}'$ is such a maximal ideal, then the ideal $\mathfrak{p}' \cap A$ is distinct from A and contains $\mathfrak{p}$, whence $\mathfrak{p}' \cap A = \mathfrak{p}$ since $\mathfrak{p}$ is maximal.

We thus have to prove that $A'\mathfrak{p}$ does not contain the identity 1. Since $1 \in A'\mathfrak{p}$ may be written as an equality $\sum_i a'_i p_i = 1$ $(a'_i \in A', p_i \in \mathfrak{p})$ which involves only a finite number of elements of A', we have a reduction to the case where A' is generated over A by a finite number of elements. By induction on the number of these elements, we have a further reduction to the case where $A' = A[x]$. Let then

(1) $$x^n = a_{n-1}x^{n-1} + \cdots + a_1 x + a_0 \quad (a_j \in A)$$

be an equation of integral dependence *of smallest possible degree* for x over A. If $1 \in A'\mathfrak{p}$ then we have a relation of the form

(2) $$1 = p_0 + p_1 x + \cdots + p_q x^q \quad (p_i \in \mathfrak{p}).$$

By using equation (1), we may suppose that $q \leq n - 1$. Furthermore, since $1 - p_0$ is not contained in the unique maximal ideal $\mathfrak{p}$ of A, it is a unit, we may divide (2) by $1 - p_0$ and we may thus assume that p_0 is zero:

(2') $$1 = p_1 x + \cdots + p_q x^q \quad (p_i \in \mathfrak{p}, q \leq n - 1).$$

This relation shows that x is a unit in A' and hence is not a zero-divisor in A'. Let us now replace a_0 in (1) by $a_0(p_1 x + \cdots + p_q x^q)$. We get

(1') $$x^n = a_{n-1}x^{n-1} + \cdots + a_1 x + a_0 x(p_1 + \cdots + p_q x^{q-1}).$$

By canceling x we get an integral dependence equation for x over A, which is at most of degree $n - 1$. This is a contradiction. Thus $\mathfrak{p}A[x]$ is a proper ideal of $A[x]$, and this proves Theorem 3.

We give a second proof of Theorem 3. Let $\mathfrak{p}$ be now an arbitrary prime ideal in R. (The reduction to the case in which $\mathfrak{p}$ is the only maximal ideal in A will not be needed in this second proof.) The set of ideals $\mathfrak{a}'$ in A' such that $\mathfrak{a}' \cap A \subset \mathfrak{p}$ is not empty (the zero ideal is in that set) and is obviously inductive. Hence Zorn's lemma provides us with a maximal element $\mathfrak{p}'$ of that set. We shall prove that $\mathfrak{p}' \cap A = \mathfrak{p}$ and that $\mathfrak{p}'$ is a prime ideal.

We shall show that the assumption $\mathfrak{p}' \cap A \neq \mathfrak{p}$ leads to a contradiction. Under this assumption we have $\mathfrak{p}' \cap A < \mathfrak{p}$. Let x be an element of $\mathfrak{p}$ which does not belong to $\mathfrak{p}'$. Then $\mathfrak{p}' + A'x > \mathfrak{p}'$, and hence, by our choice of $\mathfrak{p}'$ we will have $(\mathfrak{p}' + A'x) \cap A \not\subset \mathfrak{p}$. That means that there exists an element z' in A' and an element y in A, *not in* $\mathfrak{p}$, such that $z'x - y$ belongs to $\mathfrak{p}'$. Let $z'^n + a_1 z'^{n-1} + \cdots + a_{n-1}z' + a_n = 0$ be an equation of integral dependence for z' over A $(a_i \in A)$. Multiplying this relation by x^n and using the fact that $z'x \equiv y \pmod{\mathfrak{p}'}$ we find the following congruence: $y^n + a_1 x y^{n-1} + \cdots + a_{n-1} x^{n-1} y + a_n x^n \equiv 0 \pmod{\mathfrak{p}}$. This is in contradiction with $x \in \mathfrak{p}$, $y \notin \mathfrak{p}$.

To show that $\mathfrak{p}'$ is a prime ideal we consider any two ideals $\mathfrak{a}'$ and $\mathfrak{b}'$ in A' such that $\mathfrak{a}' > \mathfrak{p}'$ and $\mathfrak{b}' > \mathfrak{p}'$ and we prove that $\mathfrak{a}'\mathfrak{b}' > \mathfrak{p}'$. Let $\mathfrak{a} = \mathfrak{a}' \cap A$, $\mathfrak{b} = \mathfrak{b}' \cap A$. By our choice of $\mathfrak{p}'$ and from the equality $\mathfrak{p}' \cap A = \mathfrak{p}$ we deduce that $\mathfrak{a} > \mathfrak{p}$ and $\mathfrak{b} > \mathfrak{p}$. Since $\mathfrak{p}$ is a prime ideal it follows that $\mathfrak{a}\mathfrak{b} > \mathfrak{p}$. Consequently $\mathfrak{a}'\mathfrak{b}' \cap A \supset \mathfrak{a}\mathfrak{b} > \mathfrak{p}$, showing that $\mathfrak{a}'\mathfrak{b}' > \mathfrak{p}'$. This completes the proof.

An ideal $\mathfrak{p}'$ in A' such that $\mathfrak{p}' \cap A = \mathfrak{p}$ is said to "*lie over* $\mathfrak{p}$." We first give a corollary to Theorem 3:

COROLLARY. *Let A be a ring, $\mathfrak{p}$ and $\mathfrak{q}$ two prime ideals of A such that $\mathfrak{p} \subset \mathfrak{q}$, A' an overring of A integral over A, and $\mathfrak{p}'$ a prime ideal of A' lying over $\mathfrak{p}$. Then there exists a prime ideal $\mathfrak{q}'$ of A' containing $\mathfrak{p}'$ and lying over $\mathfrak{q}$.*

The residue class ring $A'/\mathfrak{p}'$ is integral over $A/\mathfrak{p}$ (Lemma 1), and the corollary follows by applying Theorem 3 to the prime ideal $\mathfrak{q}/\mathfrak{p}$ of $A/\mathfrak{p}$.

We now give two complements to Theorem 3:

1) *Two prime ideals $\mathfrak{p}'$, $\mathfrak{q}'$ of A' such that $\mathfrak{p}' < \mathfrak{q}'$ cannot lie over the same prime ideal of A.* By passage to $A'/\mathfrak{p}'$, we may suppose that $\mathfrak{p}' = (0)$ and that A' is an integral domain. We then prove that any non-zero ideal $\mathfrak{a}'$ of A' contracts to a non-zero ideal of A. We fix an element $x \neq 0$ in $\mathfrak{a}'$. There exists an equation of integral dependence $x^n + a_{n-1}x^{n-1} + \cdots + a_0 = 0$ of x over A, with $a_0 \neq 0$, since, otherwise, we could divide by x. This equation shows that $a_0 \in xA' \cap A \subset \mathfrak{a}' \cap A$. Hence $\mathfrak{a}' \cap A \neq (0)$.

2) *Let $\mathfrak{p}'$ be a prime ideal of A' lying over $\mathfrak{p}$. For $\mathfrak{p}'$ to be a maximal ideal of A', it is necessary and sufficient that $\mathfrak{p}$ be a maximal ideal of A.* For, if $\mathfrak{p}'$ is not maximal, it is contained in a maximal ideal $\mathfrak{q}'$, and $\mathfrak{q}' \cap A > \mathfrak{p} = \mathfrak{p}' \cap A$ by complement 1), showing that also $\mathfrak{p}$ is not maximal. Suppose conversely that $\mathfrak{p}$ is not maximal. Then $\mathfrak{p}$ is contained in a maximal ideal $\mathfrak{q}$; and by using the corollary, we find that also $\mathfrak{p}'$ is not maximal.

REMARK. By passage to $A'/\mathfrak{p}'$, the result we just proved is equivalent to the following special case of 2), that is, the case $\mathfrak{p}' = (0)$: let A' be an integral domain, integral over A; for A' to be a *field*, it is necessary and sufficient that A be a field. In other words: an integral domain which is integrally dependent on a *proper* integral domain (that is, on a domain which is not a field) is itself proper.

We note, however, that this result is much more elementary than Theorem 3 and can be proved very simply and directly as follows:

If A' is a field and $x \in A$, $x \neq 0$, then $1/x \in A'$, whence there is a relation of the form $(1/x)^n + a_1(1/x)^{n-1} + \cdots + a_n = 0, a_i \in A$. Therefore $1 = -x(a_1 + a_2x + \cdots + a_nx^{n-1})$, $1/x \in A$, showing that also A is a field.

Conversely, if A is a field and $x \in A'$, $x \neq 0$, then $A(x) = A[x]$, since x is algebraic over A. Hence $1/x \in A[x] \subset A'$, showing that A' is a field.

§ 3. **Integrally closed rings.** Let A be an integral domain, K its field of quotients, and L an overfield of K. If an element x of L is integral over A, it is *a fortiori algebraic* over K, by condition (c) (§ 1, p. 257). Let n be the degree of the minimal polynomial $f(X)$ of x over K, and let us denote by $\{x_1, \cdots, x_n\}$ a complete set of conjugates of x over K, that is, a set of n elements of an algebraic closure of K such that $f(X) = \prod_i (X - x_i)$ (each conjugate of x is repeated p^e times, p^e being the degree of inseparability of $f(X)$ over K; see II, § 5, Definition 2, p. 67). Since an equation of integral dependence for x over A is satisfied by all the conjugates x_i of x over K, the coefficients of the minimal polynomial $f(X) = \prod_i(X - x_i)$ are integral over A (§ 1, corollary to Theorem 1). We have thus proved the following results:

THEOREM 4. *Let A be an integral domain, K its quotient field, x an element of some extension of K. We suppose that x is integral over A. Then x is algebraic over K, and the coefficients of the minimal polynomial $f(X)$ of x over K, in particular the norm and the trace of x over K, are elements of K which are integral over A. If A is integrally closed, these coefficients are in A, and therefore already the minimal polynomial $f(X)$ yields an equation $f(x) = 0$ of integral dependence for x over A.*

A slight modification of the reasoning about conjugates leads to the following result:

THEOREM 5. *Let A be an integrally closed domain, and let K be its quotient field. If $f(X)$ and $g(X)$ are monic polynomials in $K[X]$ such that the product $h(X) = f(X)g(X)$ is in $A[X]$, then $f(X)$ and $g(X)$ are themselves in $A[X]$ (that is, have their coefficients in A).*

PROOF. Let (x_i), (y_j) be sets of elements of an algebraic closure of K such that $f(X) = \prod_i(X - x_i)$, $g(X) = \prod_j(X - y_j)$. Since $h(X) = \prod_i(X - x_i) \cdot \prod_j(X - y_j)$ is in $A[X]$, the relations $h(x_i) = 0$ and $h(y_j) = 0$ are equations of integral dependence for all the x_i and y_j over A, and these elements are therefore integral over A. Thus the coefficients of $f(X)$ and $g(X)$, which are sums of products of the x_i and the y_j respectively, are integral over A (§1, corollary to Theorem 1); they are therefore elements of A, since A is integrally closed.

REMARK. In chapter II (§ 2, p. 56) we have defined the *minimal polynomial* over a field K of an element x of some field extension L of K.

More generally, if x is an element of some overring of a field K, where the overring is not now necessarily a field, the set of all polynomials $f(X) \in K[X]$ such that $f(x) = 0$ is a principal ideal of $K[X]$. The *monic* polynomial $m(X)$ generating this ideal is called the *minimal polynomial* of x over K, or in $K[X]$. If x is not an element of some field extension of K, the minimal polynomial $m(X)$ of x need *not* be irreducible. Part of Theorem 4 may be generalized to this case: *Let A be an integrally closed domain, K its quotient field, and x an element integral over A of some overring of K. Then the minimal polynomial $m(X)$ of x over K has all its coefficients in A; in other words, the relation $m(x) = 0$ is an integral dependence equation for x over A.*

PROOF. If $f(x) = 0$ ($f(X)$, monic polynomial in $A[X]$) is an integral dependence equation for x over A, we may write $f(X) = m(X)h(X)$, where $h(X)$ is a monic polynomial in $K[X]$. Then Theorem 5 shows that $m(X) \in A[X]$. Q.E.D.

It may be noted that, if A is an integral domain and if B is an overring of A (not necessarily an overring of the quotient field K of A) such that *no non-zero element of A is a zero-divisor in B*, then the above considerations may be applied. In fact, if we denote by M the set of all non-zero elements in A, the quotient ring B_M contains both B and the quotient field K of A.

EXAMPLES OF INTEGRALLY CLOSED RINGS.

1) *Any unique factorization domain A is integrally closed.* In fact, let x/y ($x \in A, y \in A$) be integral over A; we may suppose that x and y are relatively prime. From an equation of integral dependence

$$(x/y)^n + a_{n-1}(x/y)^{n-1} + \cdots + a_0 = 0 \quad (a_i \in A)$$

we deduce

$$x^n = -y(a_{n-1}x^{n-1} + \cdots + a_0y^n).$$

Thus x^n is a multiple of y. Were y a non-unit, we would have a contradiction with the assumption that x is relatively prime to y, since any irreducible factor of y would then have to divide x. Therefore y is a unit, and x/y belongs to A. In particular, the ring J of rational integers, and the polynomial ring $k[X_1, \cdots, X_n]$ over a field k, are integrally closed.

2) If R is an integrally closed domain, and if S is a multiplicatively closed set of non-zero elements of R, then the *quotient ring* R_S is integrally closed. In fact, if an element x of the common quotient field of

R and R_S is integral over R_S, we have an integral dependence equation of the form

$$x^n + (a_{n-1}/s)x^{n-1} + \cdots + a_0/s = 0, \quad (a_i \in R)$$

since any finite number of elements of R_S have a common denominator s in S. Multiplying by s^n we see that sx is integral over R, whence $sx \in R$ since R is integrally closed. If we set $sx = z \in R$, we get $x = z/s \in R_S$, thus proving that R_S is integrally closed.

3) If R is an integrally closed domain and $\mathfrak{p}$ is a prime ideal in R, then the *residue class ring* $R/\mathfrak{p}$ is *not* integrally closed in general. In fact, any finite integral domain $k[x_1, x_2, \cdots, x_n]$ over a field k is of the form $R/\mathfrak{p}$, where R is the polynomial ring $k[X_1, X_2, \cdots, X_n]$, but finite integral domains are not in general integrally closed. In the case $n = 2$ the simplest example is the one in which $\mathfrak{p}$ is the principal ideal $(X_1^2 - X_2^3)$. In that case, x_1/x_2 does not belong to the ring $k[x_1, x_2]$, but x_1/x_2 is integral over that ring since $(x_1/x_2)^2 = x_2$.

We now prove a result which is closely related to the corollary to Theorem 3 (§ 2):

THEOREM 6. *Let A be an integrally closed domain, and A' an overring of A integral over A and such that no non-zero element of A is a zero-divisor in A'. If $\mathfrak{p}$ and $\mathfrak{q}$ are prime ideals in A such that $\mathfrak{q} \subset \mathfrak{p}$, and if $\mathfrak{p}'$ is a prime ideal of A' lying over $\mathfrak{p}$, then there exists a prime ideal $\mathfrak{q}'$ of A', contained in $\mathfrak{p}'$ and lying over $\mathfrak{q}$.*

PROOF. Let S be the multiplicatively closed set consisting of the elements of A' which may be written in the form ab', with $a \in A$, $a \notin \mathfrak{q}$, $b' \in A'$, $b' \notin \mathfrak{p}'$. The set S does not contain 0 since an element $a \notin \mathfrak{q}$ $(a \in A)$ cannot be a zero-divisor in A'. Since A and A' have an identity, S contains the complement of $\mathfrak{p}'$ in A' and the complement of $\mathfrak{q}$ in A. We are going to consider the quotient ring A'_S.

Suppose we have already proved that the ideal $\mathfrak{q}A'_S$ generated by the image of $\mathfrak{q}$ in A'_S is a *proper* ideal of A'_S. Then it is contained in a prime ideal $\mathfrak{M}$ of A'_S, for example a maximal one. The contracted ideal $\mathfrak{q}' = \mathfrak{M}^c$ of $\mathfrak{M}$ in A' is a prime ideal which does not intersect S (IV, § 10, Corollary 1 to Theorem 16), and which is therefore contained in $\mathfrak{p}'$. Now, the ideal $\mathfrak{q}' \cap A$ is obviously prime and contains $\mathfrak{q}$; but since S contains the complement of $\mathfrak{q}$ in A, and since $\mathfrak{q}'$ does not intersect S, this implies $\mathfrak{q}' \cap A = \mathfrak{q}$, and proves the theorem.

We now prove that the image of $\mathfrak{q}$ generates a proper ideal in A'_S, or, what amounts to the same thing, that the ideal $\mathfrak{q}A'$ does not intersect S. By Lemma 1 (§ 1) every element x of $\mathfrak{q}A'$ satisfies an integral dependence equation of the form

$$f(x) = x^n + q_{n-1}x^{n-1} + \cdots + q_1x + q_0 = 0, \quad \text{with } q_i \in \mathfrak{q}.$$

Let us suppose that x belongs to S, and write $x = ab'$ ($a \in A$, $a \notin \mathfrak{q}$, $b' \in A'$, $b' \notin \mathfrak{p}'$). Since $f(x) = 0$, the polynomial $f(X)$ is the product of the minimal polynomial $g(X)$ of x over the quotient field K of A and of another monic polynomial $h(X)$. Theorem 5 shows that $g(X)$ and $h(X)$ have all their coefficients in A. Denote by $\bar{f}$, $\bar{g}$, and $\bar{h}$ the polynomials obtained from f, g and h by reduction of their coefficients modulo $\mathfrak{q}$. Since $\bar{f}(X) = X^n$, $\bar{g}(X) = X^r + \cdots$, $\bar{h}(X) = X^{n-r} + \cdots$, and since $A/\mathfrak{q}$ is an integral domain, we conclude that $\bar{g}(X) = X^r$ and $\bar{h}(X) = X^{n-r}$, by inspection of the lowest degree terms. In other words, we have

$$g(x) = x^r + d_{r-1}x^{r-1} + \cdots + d_1x + d_0 = 0, \quad \text{with } d_i \in \mathfrak{q}.$$

On the other hand, the minimal polynomial of b' over K has again all its coefficients in A, by the remark following Theorem 5. Let this polynomial be

$$m(X) = X^r + e_{r-1}X^{r-1} + \cdots + e_1X + e_0, \quad (e_i \in A).$$

Since $x = ab'$, with $a \in A$, we have $d_i = e_i a^{r-i}$ for $i = 0, \cdots, r-1$. From $d_i \in \mathfrak{q}$ and $a \notin \mathfrak{q}$, we deduce $e_i \in \mathfrak{q}$ since $\mathfrak{q}$ is prime. Then the relation $m(b') = 0$ shows that $b'^r \in A'\mathfrak{q} \subset \mathfrak{p}'$, whence $b' \in \mathfrak{p}'$ since $\mathfrak{p}'$ is prime. This contradicts the hypothesis about b'. Q.E.D.

It may be shown by examples that the three conditions "A integrally closed," "A' integral over A," and "no non-zero element of A is a zero-divisor in A'" are essential for the validity of Theorem 6 (see Cohen-Seidenberg, "Prime Ideals and Integral Dependence," *Bull. Amer. Math. Soc.*, 52: 252–261, 1946).

REMARK. A simpler proof of Theorem 6 may be given in the case in which the ring A' is *noetherian*. We first prove that if A' is noetherian, then *every isolated prime ideal of* $A'\mathfrak{q}$ *contracts to* $\mathfrak{q}$ *in* A. In fact, given an isolated prime ideal $\mathfrak{q}'$ of $A'\mathfrak{q}$ and any element x of $\mathfrak{q}'$, there exists an exponent s and an element y in A', y *not* in $\mathfrak{q}'$ such that $x^sy \in A'\mathfrak{q}$: one takes y in the intersection of the primary components of $A'\mathfrak{q}$ whose radical is not $\mathfrak{q}'$, and s large enough for x^s to lie in the primary component of $A'\mathfrak{q}$ relative to $\mathfrak{q}'$ (see IV, § 5). It follows from Lemma 1 of § 1 that x^sy satisfies a relation of the form

$$f(x^sy) = (x^sy)^n + q_{n-1}(x^sy)^{n-1} + \cdots + q_0 = 0, \quad \text{with } q_i \in \mathfrak{q},$$

and, as in the last part of the proof of Theorem 6, it may be assumed that $f(X)$ is the minimal polynomial of x^sy over K. If we suppose, furthermore, that x is in A, then the comparison of $f(X)$ with the minimal polynomial $X^n + a_{n-1}X^{n-1} + \cdots + a_0$ ($a_i \in A$) of y over K shows that we have $q_i = a_i x^{s(n-i)}$ for $i = 0, \cdots, n-1$. From $x \notin \mathfrak{q}$, we would deduce that $a_i \in \mathfrak{q}$ for $i = 0, \cdots, n-1$ since $\mathfrak{q}$ is prime, whence $y^n \in A'\mathfrak{q} \subset \mathfrak{q}'$; since $\mathfrak{q}'$ is prime, this would imply $y \in \mathfrak{q}'$, in contradiction with the assumption that $y \notin \mathfrak{q}'$. Therefore $x \in \mathfrak{q}$, and since the inclusion $\mathfrak{q} \subset \mathfrak{q}' \cap A$ is evident, our assertion is proved.

For completing the proof of Theorem 6 in this case, it now suffices to observe that since $\mathfrak{p}' \supset A'\mathfrak{q}$, $\mathfrak{p}'$ contains some isolated prime ideal $\mathfrak{q}'$ of $A'\mathfrak{q}$ (IV, § 5, Theorem 7).

§ 4. Finiteness theorems

THEOREM 7. *Let A be an integrally closed domain, K its quotient field, F a finite separable algebraic extension of K, and A' the integral closure of A in F. There exists a basis $\{x_1, \cdots, x_n\}$ of F over K such that A' is contained in the A-module $\sum_i Ax_i$.*

PROOF. We first notice that if we denote by A^* the set of non-zero elements of A then we have $F = A'_{A^*}$; that is, given any element x of F there exists a non-zero element s of A such that $sx \in A'$: in fact, if $X^n + c_{n-1}X^{n-1} + \cdots + c_0$ is the minimal polynomial of x over K ($c_i \in K$), and if we take a common denominator $s \neq 0$ in A such that $sc_i = a_i \in A$, then we have $(sx)^n + a_{n-1}(sx)^{n-1} + \cdots + s^{n-1}a_0 = 0$ and sx is integral over A. It follows from this observation that there exists a basis $\{u_1, \cdots, u_n\}$ of F over K such that $u_i \in A'$ for every i.* We take any element x of A', and we write $x = \sum_i b_i u_i$ with $b_i \in K$. Since F/K is separable, there exist exactly n $(= [F:K])$ K-isomorphisms s_j $(j = 1, \cdots, n)$ of F in a least normal extension of K containing F (II, § 6, Theorem 16). The discriminant d of the basis $\{u_1, \cdots, u_n\}$ is $\neq 0$, and $d = \det(s_j(u_i))^2$ (II, § 11, p. 94). We may thus set $\sqrt{d} = \det(s_j(u_i))$. The conjugates of x over K satisfy

$$(1) \qquad s_j(x) = \sum_i b_i s_j(u_i), \quad (j = 1, \cdots, n).$$

Since x and the u_i are integral over A, $s_j(x)$ and the $s_j(u_i)$ are also integral over A. Solving the system of linear equations (1) in the b_i by Cramer's rule, we get

$$\sqrt{d}b_i = \sum_j d_{ij}s_j(x) \quad \text{and} \quad db_i = \sum_j \sqrt{d}d_{ij}s_j(x),$$

where the d_{ij} are polynomials in the $s_j(u_i)$ with ordinary integers as coefficients. Thus db_i and $\sqrt{d}b_i$ are integral over A. But since $d \in K$

* This part of the proof, and hence also the conclusion as to the existence of the above basis $\{u_1, u_2, \cdots, u_n\}$, is independent of the assumption that F/K is separable.

(II, § 11, p. 92) and since A is integrally closed, we have $db_i \in A$. Therefore, if we take $x_i = u_i/d$, then A' is contained in the A-module $\sum_i Ax_i$.

REMARK. Readers acquainted with linear algebra may prefer the following proof. Let $\{u_1, \cdots, u_n\}$ be a basis of F over K contained in A'. The bilinear function $(x, y) \rightarrow T(xy)$ is non-degenerate since F is separable over K. Thus it defines an isomorphism of F, considered as a vector space over K, onto its dual. Let $\{v_1, \cdots, v_n\}$ be the elements of F corresponding to those of the *dual basis* of $\{u_1, \cdots, u_n\}$, that is, the elements of F satisfying $T(u_i v_j) = \delta_{ij}$ for all i, j. If an element $x = \sum_j x_j v_j$ $(x_j \in K)$ is in A', we have $xu_i \in A'$ for every i, whence $T(xu_i) \in A$ (§ 3, Theorem 4). Since $T(xu_i) = \sum_j x_j T(u_i v_j) = x_i$, we have $A' \subset \sum_j Av_j$. This type of reasoning will again be used in § 11 (see the proof of Theorem 30 and Remark 2, p. 309, in § 11).

COROLLARY 1. *The assumptions being the same as in Theorem* 7, *let us furthermore assume that the ring A is noetherian. Then A' is a finite A-module and is a noetherian ring.*

In fact, A' is a submodule of the finite A-module $\sum_i Ax_i$, and is therefore a finite A-module. Thus A' satisfies the a.c.c. as an A-module (III, § 10, Theorem 18), and *a fortiori* satisfies the a.c.c. as an A'-module —that is, A' is noetherian.

COROLLARY 2. *The assumptions being the same as in Theorem* 7, *let us furthermore assume that A is a principal ideal domain. Then there exists a basis $\{y_i\}$ of F over K such that $A' = \sum_i Ay_i$.*

It was just shown that A' is contained in an A-module $\sum Ax_i$ generated by n elements x_i. Hence, by IV, § 15, Lemma 1, also A' has a basis consisting of n elements $y_1, y_2, \cdots, y_n$. Since $F = A'_{A^*}$, the set $\{y_i\}$ is necessarily also a basis of F over K.

Corollary 2 is of particular importance for the case in which A is either the ring J of rational integers, F being then an algebraic number field, or a polynomial ring $k[X]$ in one variable over a field k, F being then a field of algebraic functions of one variable. In the first case, the elements of F which are integral over J are called the *algebraic integers* of the number field F; in the second case, the elements of F which are integral over $k[X]$ are called the *integral functions* of the function field F (with respect to the element X). Corollary 2 shows that these algebraic integers (or integral functions) are the linear combinations, with ordinary integral coefficients (or with coefficients in $k[X]$), of $n(= [F:K])$ linearly independent algebraic integers y_i. Such a basis $\{y_i\}$ of F over the rational field (or over the rational function field $k(X)$) is called an *integral basis* of F.

EXAMPLE. Let x be an indeterminate over a field k of characteristic $\neq 2$, and y an algebraic element over $K = k(x)$ defined by $y^2 = P(x)$, where P is an irreducible polynomial over k. The function field $F = K(y) = k(x, y)$ admits $\{1, y\}$ as an integral basis (with respect to x). In fact, in the first place 1 and y are integral over $A = k[x]$. Furthermore, let z be an element of F which is integral over A. We write $z = a(x) + b(x)y$ $(a(x), b(x) \in k(x))$. By Theorem 4, the trace $2a(x)$ and the norm $a(x)^2 - b(x)^2P(x)$ of z over $k(x)$ belong to $k[x]$, whence both $a(x)$ and $b(x)$ are polynomials, since otherwise $P(x)$ would be divisible by the square of the denominator of $b(x)$. Consequently, the integral closure A' of $k[x]$ in $F = k(x, y)$ is the ring $k[x, y]$.

REMARK. Let A be an integrally closed domain, K its quotient field, and F a finite algebraic separable extension of K. We suppose that the integral closure A' of A in F admits an integral basis $\{y_i\}$ (that is, a basis of F over K such that $A' = \sum_i Ay_i$). Then, for any basis $\{x_i\}$ of F over K composed of elements of A', the *discriminant* $d(x_1, \cdots, x_n)$ is an element of A (by Theorem 4) and is a multiple (in A) of $d(y_1, \cdots, y_n)$ (II, § 11, formula (2)). In particular any two integral bases of A' over A have discriminants which differ only by an invertible factor in A; in other words, these discriminants generate the same principal ideal in A. This principal ideal or a canonically chosen generator of it—for example, a positive integer in the case $A = J$, or a monic polynomial in the case $A = k[X]$— is called *the discriminant* of F over K (with respect to the integral domain A). See § 11 for a generalization.

In the example given above, the discriminant of $k(x, y)$ over $k(x)$ is $P(x)$.

We now study integral domains which are generated by a finite number of elements over a field k—in other words integral domains of the form $k[x_1, \cdots, x_n]$; such domains are called *finite integral domains.*

THEOREM 8 (NORMALIZATION LEMMA). *Let $A = k[x_1, \cdots, x_n]$ be a finite integral domain over an infinite field k, and let d be the transcendence degree of $k(x_1, \cdots, x_n)$ over k. There exist d linear combinations $y_1, \cdots, y_d$ of the x_i with coefficients in k, such that A is integral over $k[y_1, \cdots, y_d]$ ($y_1, \cdots, y_d$ are then necessarily algebraically independent over k, and $k[y_1, \cdots, y_d]$ is a polynomial ring). If $k(x_1, \cdots, x_n)$ is* *separably generated over k, the y_j may be chosen in such a way that $k(x_1, \cdots, x_n)$ is a separable extension of $k(y_1, \cdots, y_d)$ ($\{y_1, \cdots, y_d\}$ thus being a separating transcendence basis of $k(x_1, \cdots, x_n)$ over k).*

PROOF. If $n = d$, we take $y_j = x_j$ and there is nothing to prove. We will proceed by induction on n, for $n > d$. Owing to the *transitivity* of integral dependence (§ 1, Theorem 2) and of separability (II, § 5, Theorem 9) we have only to prove the following result (where, for simplicity of notation, n has been changed into $n + 1$): if $k[x_1, \cdots, x_n, x_{n+1}]$ is a finite integral domain, of transcendence degree $d \leqq n$, then there exist n linear combinations $z_1, \cdots, z_n$ of the x_i such that $k[x]$ is integral over $k[z]$ (and such that $k(x)$ is separable over $k(z)$ if $k(x)$ is separably generated over k). After eventual renumbering of the x_i we may suppose that a transcendence basis of $k(x)$ over k may be found among $\{x_1, \cdots, x_n\}$ (II, § 12, Theorem 23 Corollary, 2) and that this basis is a separating transcendence basis in the separable case (II, § 13, Theorem 30). We then write $u = x_{n+1}$ and denote by $P(U, x_1, \cdots, x_n)$ the minimal polynomial of u over $k(x_1, \cdots, x_n)$. We assume that the coefficients of $P(U, x_1, \cdots, x_n)$ are in $k[x_1, \cdots, x_n]$, so that $P(U, x_1, \cdots, x_n)$ is actually the result of substituting $x_1, \cdots, x_n$ for $X_1, \cdots, X_n$ in a non-zero polynomial $P(U, X_1, \cdots, X_n)$ of $n + 1$ indeterminates $U, X_1, \cdots, X_n$, with coefficients in k.

We intend to take $z_i = x_i - a_i u (i = 1, \cdots, n)$ with suitably chosen a_i in k. Since $x_i = z_i + a_i u$, it is sufficient to prove that u is integral (and separable in the separable case) over $k[z]$. Consider the equation

$$F(u, z) = P(u, z_1 + a_1 u, \cdots, z_n + a_n u) = 0.$$

Its highest degree term in u is $u^q f(1, a_1, \cdots, a_n)$, where $f(U, X_1, \cdots, X_n)$ denotes the highest degree form of $P(U, X_1, \cdots, X_n)$ and q its degree. We will thus get an equation of integral dependence for u over $k[z]$ if $f(1, a_1, \cdots, a_n) \neq 0$.

In the separable case we have also to make sure that u is a simple root of $F(U, z)$, or in other words, that $F'_u(u, z) = P'_u(u, x) + a_1 P'_{x_1}(u, x) + \cdots + a_n P'_{x_n}(u, x)$ is not zero. But this expression is a linear function of the a_i, which is not identically zero, since it takes for $a_1 = \cdots = a_n = 0$ the value $P'_u(u, x) \neq 0$, u being separable over $k(x_1, \cdots, x_n)$. The n-tuples $\{a_1, a_2, \cdots, a_n\}$, $a_i \in k$, form an n-dimensional vector space k^n over k, and the vectors $\{a_1, a_2, \cdots, a_n\}$ such that $P'_u + a_1 P'_{x_1} + \cdots + a_n P'_{x_n} = 0$ constitute a linear variety L in k^n, distinct from k^n. Since k is infinite, we can find a vector $\{a_i\} \in k^n$ which satisfies $f(1, a_1, \cdots, a_n) \neq 0$, and which does not lie in L. Q.E.D.

THEOREM 9. *Let $A = k[x_1, \cdots, x_n]$ be a finite integral domain over* *a field k, and let F be a finite algebraic extension of the quotient field $k(x_1, \cdots, x_n)$ of A. Then the integral closure A' of A in F is a finite integral domain over k, and is a finite A-module.*

PROOF. We first achieve a reduction to the case in which F is the *quotient field of* A. For this purpose we observe that there exists a basis $\{y_1, \cdots, y_q\}$ of F over $k(x_1, \cdots, x_n)$ composed of elements which are integral over A (see footnote to proof of Theorem 7 in § 4). Then the ring $A^0 = A[y_1, \cdots, y_q]$ is a finite integral domain over k, is integral over A, and admits F as its quotient field. This achieves the desired reduction since A' is obviously the integral closure of A^0.

We now prove Theorem 9 under the additional hypotheses that k is *infinite* and that $F = k(x_1, \cdots, x_n)$ is *separably generated* over k. Under these assumptions there exist, by Theorem 8, d linear combinations $z_1, \cdots, z_d$ of the x_i such that the subring $B = k[z_1, \cdots, z_d]$ of A is a polynomial ring, over which A is integral and separable. Owing to the transitivity of integral dependence (§ 1, Theorem 2), A' is the integral closure of B in F. Since B is integrally closed and noetherian, Corollary 1 to Theorem 7 shows that A' is a finite B-module. It is, *a fortiori*, a finite A-module, and a finite integral domain over k.

In the general case, let us consider $k(x_1, \cdots, x_n)$ as a subfield of its algebraic closure, which contains the algebraic closure $\bar{k}$ of k. Since $\bar{k}$ is infinite and since $\bar{k}(x_1, \cdots, x_n)$ is separably generated over $\bar{k}$ (II, § 13, Theorem 31), we can find d linear combinations $z_i = \sum_{j=1}^{n} a_{ij}x_j$ ($a_{ij} \in \bar{k}$) such that $\bar{k}[x_1, \cdots, x_n]$ is integral and separable over $\bar{k}[z_1, \cdots, z_d]$. Let $P_j(x_j, z_1, \cdots, z_d) = 0$ be a separable and integral dependence equation for x_j over $\bar{k}[z_1, \cdots, z_d]$ (for example, the equation deduced from the minimal polynomial of x_j over $\bar{k}(z_1, \cdots, z_d)$; cf. Theorem 4 of § 3). If we denote by k' the finite algebraic extension of k generated by the coefficients a_{ij} and the coefficients of the polynomials P_j, the second part of the proof shows that the integral closure of $k'[x_1, \cdots, x_n]$ in its quotient field is a finite integral domain $k'[y_1, \cdots, y_q]$ over k'.

Now, by Theorem 1 (§ 1), $k'[y_1, \cdots, y_q]$ is a finite module over $k'[x_1, \cdots, x_n]$. On the other hand, $k'[x_1, \cdots, x_n]$ is a finite module over $A = k[x_1, \cdots, x_n]$, a finite basis of the former over the latter being given, for example, by a linear basis of k' over k. Thus $k'[y_1, \cdots, y_q]$ is a finite A-module. Since the integral closure A' of A in F is a submodule of the A-module $k'[y_1, \cdots, y_q]$ (it is $F \cap k'[y_1, \cdots, y_q]$), and since A is noetherian, A' is also a finite A-module, and *a fortiori* a finite integral domain. Q.E.D.

§ 5. The conductor of an integral closure. We have just seen an important case in which the integral closure A' of a domain A *in its quotient field* is a finite A-module $A' = \sum_i Au_i$. The elements u_i have then a common denominator $d \neq 0$ in A: $u_i = v_i/d$, with $v_i \in A$. We thus have $dA' \subset A$.

In general let A be a domain, and A' the integral closure of A in its quotient field F. The set $\mathfrak{f}$ of all elements z in A such that $zA' \subset A$ is called the *conductor* of A in A', or the conductor of the integral closure of A.

It is readily verified that $\mathfrak{f}$ is an *ideal in* A, and also an *ideal in* A'. Furthermore, if $\mathfrak{b}$ is an ideal in A which is also an ideal in A', we have $\mathfrak{b}A' \subset \mathfrak{b} \subset A$, whence $\mathfrak{b} \subset \mathfrak{f}$. Therefore $\mathfrak{f}$ is the *largest ideal* in A which is also an ideal in A'. Note that $A' = A$ if and only if the conductor $\mathfrak{f}$ is the unit ideal.

LEMMA. *Let A be an integral domain, A' its integral closure, $\mathfrak{f}$ the conductor of A in A', and S a multiplicative system in A. Then A'_S is the integral closure of A_S, and, for A_S to be integrally closed, it is sufficient that $\mathfrak{f} \cap S \neq \emptyset$. Furthermore, if A' is a finite A-module, then the conductor of A_S in A'_S is $\mathfrak{f} \cdot A_S$ and if, moreover, A_S is integrally closed, then $\mathfrak{f} \cap S$ is non-empty.*

PROOF. The ring A'_S is integrally closed (§ 3, example 2). Therefore A'_S is the integral closure of A_S (§ 2, Lemma 2). If, now, $\mathfrak{f} \cap S$ is non-empty, there exists s in $S \cap \mathfrak{f}$, and we have $A' \subset (1/s)A \subset A_S$, whence $A_S = A'_S$, and A_S is integrally closed.

From $d \in \mathfrak{f}$, we deduce $dA' \subset A$, whence $dA'_S \subset A_S$. This proves that $\mathfrak{f} \cdot A_S$ is contained in the conductor of A_S in A'_S. Conversely, if an element d/s ($d \in A$, $s \in S$) of A_S is such that $(d/s)A'_S \subset A_S$, we have $dA' \subset A_S$. Since we assume that A' is a finite A-module, and since S is a multiplicative system, there exists a common denominator s' in S such that $dA' \subset (1/s')A$. Hence $ds' \in \mathfrak{f}$, and $d/s = ds'/ss' \in \mathfrak{f} \cdot A_S$. Therefore $\mathfrak{f} \cdot A_S$ is the conductor of A_S in A'_S. The last assertion of the lemma follows since if A_S is integrally closed, its conductor $\mathfrak{f} \cdot A_S$ is the unit ideal. Q.E.D.

COROLLARY. *If A' is a finite A-module, then the prime ideals $\mathfrak{p}$ in A such that $A_\mathfrak{p}$ is not integrally closed are those which contain the conductor $\mathfrak{f}$.*

REMARK. It can be shown by examples that $A'\mathfrak{p}$ need not be a prime ideal, even if $\mathfrak{p}$ does not contain the conductor $\mathfrak{f}$. However, if $\mathfrak{p} \not\supset \mathfrak{f}$, $A'\mathfrak{p}$ is contained in the prime ideal $\mathfrak{p}' = \mathfrak{p}A_\mathfrak{p} \cap A'$ (since A' is contained in the integrally closed ring $A_\mathfrak{p}$) and $\mathfrak{p}'$ is the *only prime ideal in A' lying over $\mathfrak{p}$.* In fact, if $\mathfrak{a}'$ is an ideal in A' such that $\mathfrak{a}' \cap A = \mathfrak{p}$, we take d

in $\mathfrak{f}$ and not in $\mathfrak{p}$, and we have $d\mathfrak{a}' \subset \mathfrak{a}' \cap A = \mathfrak{p}$, whence $\mathfrak{a}' \subset \mathfrak{p}A_\mathfrak{p}$ and $\mathfrak{a}' \subset \mathfrak{p}'$. Now, if $\mathfrak{a}'$ were prime and were strictly contained in $\mathfrak{p}'$, we would have $\mathfrak{a}' \cap A < \mathfrak{p}' \cap A = \mathfrak{p}$ (§ 2, complement 1 to Theorem 3). In case A' is a noetherian ring, the ideal $\mathfrak{p}'$ is even a *primary component* of $A'\mathfrak{p}$: in fact, since $A'_{\mathfrak{p}'} = A_\mathfrak{p}$, we have $\mathfrak{p}' = \mathfrak{p}A_\mathfrak{p} \cap A' = (\mathfrak{p}A')A'_{\mathfrak{p}'} \cap A'$ (see IV, § 10, Theorem 17).

§ 6. **Characterizations of Dedekind domains.** We have seen (IV, § 1 p. 200) that, in a noetherian ring R, every ideal $\mathfrak{a}$ contains a product $\prod_i \mathfrak{p}_i^{n(i)}$ of prime ideals. It is natural to study the rings in which every ideal *is* exactly a product of prime ideals. A further reason for studying these rings, which is perhaps more important both from a historical and a conceptual point of view, is that in the first half of the nineteenth century it was noticed that the rings of algebraic integers (cf. § 4) were not in general unique factorization domains, but enjoyed the property of unique factorization of ideals into prime ideals: more precisely the notion of ideal was introduced by Kummer, Dedekind, and Kronecker in order to restore the property of unique factorization.

In this connection one may also recall Theorem 9 of IV, § 5, to the effect that if every proper prime ideal of a noetherian ring R is maximal, then every ideal of R is a unique product of primary ideals belonging to distinct prime ideals. We shall see in this section that if every ideal in a domain R is a product of prime ideals, then every proper prime ideal in R is maximal. The maximality of every proper prime ideal of a domain R does not, however, in itself ensure the possibility of factoring every ideal of R into prime ideals, for while it is true that powers of maximal ideals are primary, it is not generally true that every primary ideal belonging to a maximal ideal $\mathfrak{p}$ is a power of $\mathfrak{p}$.

DEFINITION 1. *A ring R is said to be a Dedekind domain (or also a Dedekind ring) if it is an integral domain and if every ideal in R is a product of prime ideals.*

Our first aim is to prove that in a Dedekind domain the factorization of ideals into prime ideals is *unique*. The steps taken toward the proof of this result will lead us to other important characterizations of Dedekind domains.

EXAMPLES OF DEDEKIND DOMAINS:

1) A *principal ideal domain* is a Dedekind domain (IV, § 15, Theorem 32).

2) A *quotient ring* R_M of a Dedekind domain R with respect to a

multiplicative system M is a Dedekind domain. In fact, every ideal in R_M is an extended ideal $\mathfrak{a}^e$ (IV, § 10, Theorem 15); as $\mathfrak{a} = \prod_i \mathfrak{p}_i^{n(i)}$ ($\mathfrak{p}_i$-prime ideal), we have $\mathfrak{a}^e = \prod_i (\mathfrak{p}_i^e)^{n(i)}$ (IV, § 8, p. 219), and the ideals $\mathfrak{p}_i^e$ are either prime ideals, or are equal to R_M (IV, § 10, Theorem 15, corollary 2 and Theorem 16).

3) We will prove in § 8 that if R is a Dedekind domain and if L is a finite algebraic extension of its quotient field, then the integral closure of R in L (§ 1) is also a Dedekind domain. In particular, since the ring J of rational integers and the polynomial ring $k[X]$ in one variable over a field are Dedekind domains (example 1), the ring of algebraic integers of an algebraic number field and the ring of integral functions in a field of algebraic functions of one variable are also Dedekind domains.

We introduce the useful notion of *fractionary ideal.* Given an integral domain R and its quotient field K, a sub-R-module $\mathfrak{b}$ of K is said to be a *fractionary ideal* of R if the elements of $\mathfrak{b}$ admit a common denominator $d \neq 0$ in R—more precisely, if there exists $d \neq 0$ in R such that $\mathfrak{b} \subset (1/d)R$. Then we have $\mathfrak{b} = (1/d)\mathfrak{a}$ where $\mathfrak{a}$ is an ordinary ideal in R. In contrast, the ordinary ideals in R, which are special cases of fractionary ideals ($d = 1$), are called *integral ideals.* An example of a fractionary ideal is a *principal fractionary ideal*: if $x = a/b$ ($a, b, \in R, b \neq 0$) is an element of K, the set Rx is a fractionary ideal, as it is an R-module and admits b as a common denominator; it is called the principal fractionary ideal generated by x.

The ideal theoretic *operations* $+, \cdot, \cap, :$ are defined for fractionary ideals. The operations $+, \cdot, \cap$ have already been defined for submodules or additive subgroups; and if $\mathfrak{b} \subset (1/d)R$ and $\mathfrak{b}' \subset (1/d')R$, it is clear that $\mathfrak{b} + \mathfrak{b}' \subset (1/dd')R$, that $\mathfrak{b} \cdot \mathfrak{b}' \subset (1/dd')R$, and that $\mathfrak{b} \cap \mathfrak{b}' \subset (1/d)R$. The set $(\mathfrak{b} : \mathfrak{b}')$ is defined as the set of all x in K such that $x\mathfrak{b}' \subset \mathfrak{b}$; this set is clearly an R-module, and (if $\mathfrak{b}' \neq (0)$) admits da as a common denominator, where a and d are any two non-zero elements of R such that $\mathfrak{b} \subset (1/d)R$ and $a \in \mathfrak{b}'$. These operations enjoy the same properties in the present case of fractionary ideals as those which they enjoy in the case of integral ideals.

The set $\mathscr{I}$ of all fractionary ideals of R is a partially ordered set (if ordered by inclusion); $\mathfrak{a} + \mathfrak{b}$ and $\mathfrak{a} \cap \mathfrak{b}$ are the l.u.b. and the g.l.b. of $\mathfrak{a}$ and $\mathfrak{b}$; multiplication is "compatible" with this order relation; that is, the relation $\mathfrak{a} \subset \mathfrak{a}'$ implies $\mathfrak{a} \cdot \mathfrak{b} \subset \mathfrak{a}' \cdot \mathfrak{b}$. The ring R itself is a fractionary ideal, and is the identity element of $\mathscr{I}$ for multiplication of ideals.

It is natural to inquire about the *invertible ideals* in $\mathscr{I}$, that is, about the fractionary ideals $\mathfrak{a}$ of R for which there exists an inverse—a fractionary ideal $\mathfrak{a}'$ such that $\mathfrak{a}\cdot\mathfrak{a}' = R$. A principal fractionary ideal Rx ($x \neq 0$, $x \in K$) is invertible, as it admits Rx^{-1} as an inverse.

We begin our study of Dedekind domains by proving a number of simple lemmas concerning invertible fractionary ideals. In these lemmas R denotes an integral domain, K its quotient field, $\mathscr{I}$ the set of all fractionary ideals of R, and small German letters ($\mathfrak{a}$, $\mathfrak{b}$, $\mathfrak{c}$, $\cdots$) denote elements of $\mathscr{I}$.

LEMMA 1. *If $\mathfrak{a}$ is invertible, then $\mathfrak{a}$ has a unique inverse, and this inverse is equal to $R:\mathfrak{a}$. Hence a necessary and sufficient condition for $\mathfrak{a}$ to be invertible is that $\mathfrak{a}\cdot(R:\mathfrak{a}) = R$.*

PROOF. If $\mathfrak{a}\cdot\mathfrak{a}' = R$, we have $\mathfrak{a}' \subset (R:\mathfrak{a})$. On the other hand $\mathfrak{a}\cdot(R:\mathfrak{a}) \subset R$, whence, if $\mathfrak{a}'$ is an inverse of $\mathfrak{a}$, we have $(R:\mathfrak{a}) = \mathfrak{a}'\cdot\mathfrak{a}\cdot(R:\mathfrak{a}) \subset \mathfrak{a}'\cdot R = \mathfrak{a}'$. Q.E.D.

LEMMA 2. *If every integral ideal $\neq (0)$ in R is invertible, then $\mathscr{I}$ is a group under multiplication.*

PROOF. Every fractionary ideal $\mathfrak{a}$ may be written in the form $(1/d)\mathfrak{b}$, where $\mathfrak{b}$ is an integral ideal and d is a non-zero element of R. If $\mathfrak{b}$ has an inverse $\mathfrak{b}^{-1}$, then $\mathfrak{a}$ admits $d\mathfrak{b}^{-1}$ as an inverse. Since multiplication of ideals is associative, and since every element of $\mathscr{I}$ admits an inverse, $\mathscr{I}$ is a group.

LEMMA 3. *An invertible ideal $\mathfrak{a}$, considered as an R-module, has a finite basis.*

PROOF. Since $\mathfrak{a}\cdot\mathfrak{a}^{-1} = R$, there exist two finite families $\{x_i\}$, $\{x'_i\}$ ($i = 1, \cdots, n$) of elements of $\mathfrak{a}$ and $\mathfrak{a}^{-1}$ such that $\sum_i x_i x'_i = 1$. For every x in $\mathfrak{a}$, we have $xx'_i \in R$, thus $x = \sum_i xx'_i x_i \in \sum Rx_i$, and $\{x_i\}$ is a finite basis of $\mathfrak{a}$.

LEMMA 4. *If a finite family $\{\mathfrak{a}_i\}$ of integral ideals is such that the product $\mathfrak{b} = \prod_i \mathfrak{a}_i$ is invertible, then each $\mathfrak{a}_i$ is invertible. In particular, if a product $\prod_i \mathfrak{a}_i$ of integral ideals is principal, then each $\mathfrak{a}_i$ is invertible.*

PROOF. From $\mathfrak{b}^{-1}\cdot\prod_i \mathfrak{a}_i = R$, we deduce $\mathfrak{a}_i\cdot\left(\mathfrak{b}^{-1}\cdot\prod_{j\neq i}\mathfrak{a}_j\right) = R$, and $\mathfrak{b}^{-1}\cdot\prod_{j\neq i}\mathfrak{a}_j$ is the inverse of $\mathfrak{a}_i$.

LEMMA 5. *For products of invertible prime integral ideals factorization into prime ideals is unique.*

PROOF. Let $\mathfrak{a} = \prod_i \mathfrak{p}_i$ be a product of invertible prime ideals, and suppose that we have also $\mathfrak{a} = \prod_j \mathfrak{q}_j$, where the $\mathfrak{q}_j$ are prime ideals. We take a minimal element of the set $\{\mathfrak{p}_i\}$, say $\mathfrak{p}_1$. Since $\prod_j \mathfrak{q}_j$ is contained in $\mathfrak{p}_1$, some $\mathfrak{q}_j$, say $\mathfrak{q}_1$, is contained in $\mathfrak{p}_1$. Similarly, since $\prod_i \mathfrak{p}_i$ is contained in $\mathfrak{q}_1$, some $\mathfrak{p}_i$, say $\mathfrak{p}_r$, is contained in $\mathfrak{q}_1$. Thus $\mathfrak{p}_r \subset \mathfrak{q}_1 \subset \mathfrak{p}_1$. From the minimality of $\mathfrak{p}_1$ we deduce that $\mathfrak{p}_r = \mathfrak{q}_1 = \mathfrak{p}_1$. Multiplying the relation $\prod_i \mathfrak{p}_i = \prod_j \mathfrak{q}_j$ by $\mathfrak{p}_1^{-1}$, we get $\prod_{i \neq 1} \mathfrak{p}_i = \prod_{j \neq 1} \mathfrak{q}_j$. The lemma now follows by induction on n, the case $n = 1$ being trivial.

THEOREM 10. *In a Dedekind domain R, every proper prime ideal is invertible and maximal.*

PROOF. We first show that every *invertible* proper prime ideal $\mathfrak{p}$ in R is maximal. We consider an element a of R, not in $\mathfrak{p}$, and the ideals $\mathfrak{p} + Ra$, $\mathfrak{p} + Ra^2$. As R is a Dedekind domain, we have $\mathfrak{p} + Ra = \prod_{i=1}^{n} \mathfrak{p}_i$ and $\mathfrak{p} + Ra^2 = \prod_{j=1}^{m} \mathfrak{q}_j$, where the $\mathfrak{p}_i$ and the $\mathfrak{q}_j$ are prime ideals. Let $\bar{R}$ be the residue class ring $R/\mathfrak{p}$, and $\bar{a}$ be the residue class of a modulo $\mathfrak{p}$. We have $\bar{R}\cdot\bar{a} = \prod_{i=1}^{n} (\mathfrak{p}_i/\mathfrak{p})$, $\bar{R}\cdot\bar{a}^2 = \prod_{j=1}^{m} (\mathfrak{q}_j/\mathfrak{p})$, where the ideals $\mathfrak{p}_i/\mathfrak{p}$ and $\mathfrak{q}_j/\mathfrak{p}$ are prime. By Lemma 4 these prime ideals are invertible. Thus, since $\bar{R}\cdot\bar{a}^2 = (\bar{R}\cdot\bar{a})^2 = \prod_{i=1}^{n} (\mathfrak{p}_i/\mathfrak{p})^2$, Lemma 5 shows that the ideals $\mathfrak{q}_j/\mathfrak{p}$ are the ideals $\mathfrak{p}_i/\mathfrak{p}$, each repeated twice; more precisely, we have $m = 2n$, and we can renumber the $\mathfrak{q}_j$ in such a way that $\mathfrak{q}_{2i}/\mathfrak{p} = \mathfrak{q}_{2i-1}/\mathfrak{p} = \mathfrak{p}_i/\mathfrak{p}$. Thus $\mathfrak{q}_{2i} = \mathfrak{q}_{2i-1} = \mathfrak{p}_i$, and we have $\mathfrak{p} + Ra^2 = (\mathfrak{p} + Ra)^2$. This implies $\mathfrak{p} \subset (\mathfrak{p} + Ra)^2 \subset \mathfrak{p}^2 + Ra$. Thus any element x of $\mathfrak{p}$ may be written in the form $x = y + za$ with $y \in \mathfrak{p}^2$ and $z \in R$. We then have $za \in \mathfrak{p}$, whence $z \in \mathfrak{p}$, since $a \notin \mathfrak{p}$; in other words, $\mathfrak{p}$ is contained in $\mathfrak{p}^2 + \mathfrak{p}a$. As the inclusion $\mathfrak{p} \supset \mathfrak{p}^2 + \mathfrak{p}a$ is obvious, we conclude that $\mathfrak{p} = \mathfrak{p}^2 + \mathfrak{p}a = \mathfrak{p}(\mathfrak{p} + Ra)$. Since $\mathfrak{p}$ is invertible by hypothesis, we can multiply this equality by $\mathfrak{p}^{-1}$, and we get $R = \mathfrak{p} + Ra$. Since a is an arbitrary element of the complement of $\mathfrak{p}$ in R, this proves that $\mathfrak{p}$ is maximal.

This being so, to prove the theorem we need only prove that every proper prime ideal $\mathfrak{p}$ in R is invertible. We take a non-zero element b in $\mathfrak{p}$, and write $Rb = \prod_i \mathfrak{p}_i$, where the $\mathfrak{p}_i$ are prime ideals. Since $\mathfrak{p}$ contains $\prod_i \mathfrak{p}_i$, it contains some $\mathfrak{p}_i$. But, by Lemma 4, every $\mathfrak{p}_i$ is invertible. Thus every $\mathfrak{p}_i$ is maximal, by the first part of the proof. Since $\mathfrak{p}$ contains one of them, say $\mathfrak{p}_1$, we have $\mathfrak{p} = \mathfrak{p}_1$, and $\mathfrak{p}$ is invertible. Q.E.D.

COROLLARY. *In a Dedekind domain the factorization of any ideal into prime ideals is unique.*

This follows immediately from Theorem 10 and from Lemma 5.

We can state a result which is both more general and more precise than Theorem 10.

THEOREM 11. *Let R be a Dedekind domain. Every fractionary ideal* $\mathfrak{a} \neq (0)$ *of R is invertible and may be written, in a unique way, in the form*

$$\mathfrak{a} = \prod_{\mathfrak{p},\ \text{prime}} \mathfrak{p}^{n_{\mathfrak{p}}(\mathfrak{a})}, \tag{1}$$

where the $n_{\mathfrak{p}}(\mathfrak{a})$ *are integers (positive, negative, or zero) such that, for given* $\mathfrak{a}$, *the integers* $n_{\mathfrak{p}}(\mathfrak{a})$ *which are* $\neq 0$ *are finite in number. In order that* $\mathfrak{a} \subset \mathfrak{b}$, *it is necessary and sufficient that* $n_{\mathfrak{p}}(\mathfrak{a}) \geq n_{\mathfrak{p}}(\mathfrak{b})$ *for every* $\mathfrak{p}$. *We have the relations:*

$$n_{\mathfrak{p}}(\mathfrak{a} + \mathfrak{b}) = \min(n_{\mathfrak{p}}(\mathfrak{a}),\ n_{\mathfrak{p}}(\mathfrak{b})), \tag{2}$$

$$n_{\mathfrak{p}}(\mathfrak{a} \cap \mathfrak{b}) = \max(n_{\mathfrak{p}}(\mathfrak{a}),\ n_{\mathfrak{p}}(\mathfrak{b})), \tag{3}$$

$$n_{\mathfrak{p}}(\mathfrak{a} \cdot \mathfrak{b}) = n_{\mathfrak{p}}(\mathfrak{a}) + n_{\mathfrak{p}}(\mathfrak{b}). \tag{4}$$

The ideals $\mathfrak{a}:\mathfrak{b}$ *and* $\mathfrak{a} \cdot \mathfrak{b}^{-1}$ *are equal, and we have*

$$n_{\mathfrak{p}}(\mathfrak{a}:\mathfrak{b}) = n_{\mathfrak{p}}(\mathfrak{a} \cdot \mathfrak{b}^{-1}) = n_{\mathfrak{p}}(\mathfrak{a}) - n_{\mathfrak{p}}(\mathfrak{b}). \tag{5}$$

PROOF. Since a fractionary ideal $\mathfrak{a}$ may be written in the form $\mathfrak{b} \cdot \mathfrak{c}^{-1}$ where $\mathfrak{b}$ and $\mathfrak{c}$ are integral ideals (for example, if $\mathfrak{a} = (1/d)\mathfrak{b}$ with $d \in R$ and $\mathfrak{b} \subset R$, we take $\mathfrak{c} = Rd$) and since by definition we can express $\mathfrak{b}$ and $\mathfrak{c}$ as products of prime ideals, Theorem 10 shows that we can write $\mathfrak{a} = \prod_i \mathfrak{p}_i \cdot \prod_j \mathfrak{q}_j^{-1}$, where the $\mathfrak{p}_i$ and the $\mathfrak{q}_j$ are prime ideals in R. Thus $\mathfrak{a}$ is invertible, by Theorem 10. We may evidently assume that $\mathfrak{p}_i \neq \mathfrak{q}_j$ for all i and j. If we have another factorization of $\mathfrak{a}$, say $\mathfrak{a} = \prod_s \mathfrak{p}'_s \cdot \prod_t \mathfrak{q}'^{-1}_t$, with $\mathfrak{p}'_s \neq \mathfrak{q}'_t$ for every s and every t, then the relation $\prod_i \mathfrak{p}_i \cdot \prod_t \mathfrak{q}'_t = \prod_s \mathfrak{p}'_s \cdot \prod_j \mathfrak{q}_j$ holds true, and the uniqueness of factorization for integral ideals (corollary to Theorem 10) shows that we have $\prod_i \mathfrak{p}_i = \prod_s \mathfrak{p}'_s$, and $\prod_t \mathfrak{q}'_t = \prod_j \mathfrak{q}_j$. This proves the uniqueness of factorization for fractionary ideals, and also formula (1).

Since $\mathfrak{b}$ is invertible, the relation $\mathfrak{a} \subset \mathfrak{b}$ is equivalent to $\mathfrak{a} \cdot \mathfrak{b}^{-1} \subset \mathfrak{b} \cdot \mathfrak{b}^{-1}$—that is, to $\mathfrak{a} \cdot \mathfrak{b}^{-1} \subset R$. This is equivalent to $n_{\mathfrak{p}}(\mathfrak{a} \cdot \mathfrak{b}^{-1}) \geq 0$ for all prime ideals $\mathfrak{p}$ of R, since the integral ideals $\mathfrak{c}$ are those characterized by $n_{\mathfrak{p}}(\mathfrak{c}) \geq 0$ for all $\mathfrak{p}$. In other words, the relation $\mathfrak{a} \subset \mathfrak{b}$ is equivalent to $n_{\mathfrak{p}}(\mathfrak{a}) - n_{\mathfrak{p}}(\mathfrak{b}) \geq 0$, that is, to $n_{\mathfrak{p}}(\mathfrak{a}) \geqq n_{\mathfrak{p}}(\mathfrak{b})$, for all $\mathfrak{p}$. This characterization of inclusion shows immediately that $\prod_{\mathfrak{p}} \mathfrak{p}^{\nu(\mathfrak{p})}$, with $\nu(\mathfrak{p}) = \min(n_{\mathfrak{p}}(\mathfrak{a}), n_{\mathfrak{p}}(\mathfrak{b}))$ is the smallest ideal containing $\mathfrak{a}$ and $\mathfrak{b}$, and that

$\prod_{\mathfrak{p}} \mathfrak{p}^{\mu(\mathfrak{p})}$, with $\mu(\mathfrak{p}) = \max(n_{\mathfrak{p}}(\mathfrak{a}), n_{\mathfrak{p}}(\mathfrak{b}))$ is the greatest ideal contained in $\mathfrak{a}$ and $\mathfrak{b}$. This proves formulae (2) and (3). Formula (4) is trivial.

Finally, since $\mathfrak{b}^{-1} = (R:\mathfrak{b})$ (Lemma 1), we have $\mathfrak{a}\cdot\mathfrak{b}^{-1} = \mathfrak{a}\cdot(R:\mathfrak{b}) \subset \mathfrak{a}:\mathfrak{b}$. On the other hand, we have $R = (R:\mathfrak{b})\,\mathfrak{b}$, hence $\mathfrak{a}:\mathfrak{b} = (\mathfrak{a}:\mathfrak{b})\mathfrak{b}(R:\mathfrak{b}) \subset \mathfrak{a}\cdot(R:\mathfrak{b})$. Therefore $\mathfrak{a}\cdot(R:\mathfrak{b}) = \mathfrak{a}:\mathfrak{b}$. This proves the assertion about $\mathfrak{a}:\mathfrak{b}$, and formula (5) follows immediately. Q.E.D.

If, for any non-zero element x of K, we denote by $v_{\mathfrak{p}}(x)$ the integer $n_{\mathfrak{p}}(Rx)$, we have $v_{\mathfrak{p}}(xy) = v_{\mathfrak{p}}(x) + v_{\mathfrak{p}}(y)$ by formula (4), and $v_{\mathfrak{p}}(x + y) \geq \min(v_{\mathfrak{p}}(x), v_{\mathfrak{p}}(y))$ by formula (2), since the ideal $R(x + y)$ is contained in $Rx + Ry$. We will see in the next chapter that this means that $v_{\mathfrak{p}}$ is a *valuation* of the field K.

The following theorem gives a characterization of Dedekind domains:

THEOREM 12. *Let R be an integral domain. In order that R be a Dedekind domain, it is necessary and sufficient that the set $\mathscr{I}$ of fractionary ideals of R be a group under multiplication (that is, that every ideal in R be invertible).*

PROOF. The necessity is clear, since every fractionary ideal of a Dedekind domain is invertible by Theorem 11. Conversely, if $\mathscr{I}$ is a group, every ideal in R has a finite basis (Lemma 3), and R is noetherian. Using the fact that R is noetherian, we can now prove, by an indirect argument, that every proper integral ideal in R is a product of maximal ideals, and this will complete the proof of the theorem. Assuming the contrary, there exists, among the ideals (different from zero) which are not products of maximal ideals, a maximal one, say $\mathfrak{a}$ (since R is noetherian). The ideal $\mathfrak{a}$ is not a maximal ideal of R, by hypothesis. Thus it is strictly contained in some maximal ideal $\mathfrak{m}$. The ideal $\mathfrak{m}^{-1}\mathfrak{a}$, which exists since $\mathscr{I}$ is a group, is an integral ideal which strictly contains $\mathfrak{a}$: in fact, from $\mathfrak{a} = \mathfrak{m}^{-1}\mathfrak{a}$, we would deduce $\mathfrak{m}\mathfrak{a} = \mathfrak{a}$, in contradiction with Lemma 2 in IV, § 7. Therefore $\mathfrak{m}^{-1}\mathfrak{a}$ is a product of maximal ideals, in virtue of the maximality of $\mathfrak{a}$, and $\mathfrak{a} = \mathfrak{m}\cdot\mathfrak{m}^{-1}\mathfrak{a}$ is also a product of maximal ideals. This contradicts our assumption, and proves Theorem 12.

The following characterization of Dedekind domains often yields the simplest method of checking whether a given integral domain is or is not a Dedekind domain:

THEOREM 13. *Let R be an integral domain. In order that R be a Dedekind domain, it is necessary and sufficient that it satisfy the following conditions:*

1) *R is noetherian.*
2) *Every proper prime ideal of R is maximal.*
3) *R is integrally closed.*

PROOF. The necessity of 1) follows from Theorem 11 and Lemma 3, and that of 2) from Theorem 10. As for 3), consider an element x of the quotient field K of R which is integral over R. There exists a common denominator $d \neq 0$ in R such that $dx^n \in R$ for every $n \geq 0$. Then, for every prime ideal $\mathfrak{p}$ in R, we have $v_\mathfrak{p}(dx^n) = v_\mathfrak{p}(d) + nv_\mathfrak{p}(x) \geq 0$ for every n. As $v_\mathfrak{p}(d)$ and $v_\mathfrak{p}(x)$ are ordinary integers, this implies $v_\mathfrak{p}(x) \geq 0$, that is, $v_\mathfrak{p}(Rx) \geq 0$ for every prime ideal $\mathfrak{p}$; that is, $x \in R$. Thus R is integrally closed.

For proving the converse, we first observe that, in the proof of Theorem 12, the assumption that *every* ideal in R is invertible has been used only for the purpose of establishing that R is noetherian, while the rest of the proof was based exclusively on the established fact that R is noetherian and on the assumption that every proper *prime* ideal is invertible; in fact, in that proof we only needed the fact that the maximal ideal $\mathfrak{m}$ is invertible. Since, in the present case, we are *given* that R is noetherian, it follows that in order to prove that R is a Dedekind domain, we have only to show that every proper *prime* ideal $\mathfrak{p}$ of R is invertible. We observe that if y is some non-zero element of $\mathfrak{p}$, then $\mathfrak{p}$ must contain some prime ideal of the principal ideal Ry, and hence $\mathfrak{p}$ itself must be a prime ideal of Ry since all proper prime ideals in R are maximal. The proof of Theorem 13 will therefore be complete if we prove the following lemma:

LEMMA 6. *Let R be a noetherian integrally closed domain, and let $\mathfrak{p} \neq (0)$ be a maximal ideal in R. If $\mathfrak{p}$ is a prime ideal of a principal ideal Ry, then $\mathfrak{p}$ is invertible.*

PROOF. By assumption we have $Ry:\mathfrak{p} \neq Ry$ (IV, § 6, Theorem 11). If, then, x is some element of $Ry:\mathfrak{p}$ not in Ry, then $(x/y)\mathfrak{p} \subset R$ and $x/y \notin R$. We have therefore shown that $R:\mathfrak{p} \neq R$. Now let us assume that $\mathfrak{p}$ is not invertible. Then we have $\mathfrak{p} \subset \mathfrak{p}(R:\mathfrak{p}) < R$, and since $\mathfrak{p}$ is maximal, it follows that $\mathfrak{p} = \mathfrak{p}(R:\mathfrak{p})$. Now, $\mathfrak{p} \neq (0)$, and $\mathfrak{p}$ is a finite R-module since R is noetherian; furthermore R is an integral domain. It follows therefore from $\mathfrak{p}(R:\mathfrak{p}) \subset \mathfrak{p}$, in view of condition ($c'''$) of § 1, p. 254 (see also Remark after the proof of Lemma 1 in § 1) that every element of $R:\mathfrak{p}$ is integral over R, and hence belongs to R, since R is integrally closed. In other words we have $R:\mathfrak{p} \subset R$, in contradiction with the inequality $R:\mathfrak{p} \neq R$ proved above. Q.E.D.

REMARK. It follows from the proof of Lemma 6 that the assumption that $\mathfrak{p}$ is a prime ideal of a principal ideal could be replaced by the assumption that $R:\mathfrak{p} \neq R$.

Lemma 6 can be used to prove a result on integrally closed noetherian domains, which is of importance in itself:

THEOREM 14. *In an integrally closed noetherian domain R, the prime ideals of any proper principal ideal Ra ($a \neq 0$, a not a unit) are minimal prime ideals of R (and consequently a principal ideal has no imbedded components).*

PROOF. Let $\mathfrak{p}$ be a prime ideal of Ra. If R' denotes the quotient ring $R_{\mathfrak{p}}$, then it is clear that the maximal ideal $\mathfrak{p}' = R'\mathfrak{p}$ of R' is a prime ideal of $R'a$ (IV, § 10, Theorem 17), and that $\mathfrak{p}$ will be minimal in R if and only if $\mathfrak{p}'$ is minimal in R' (IV, § 11, Theorem 19). Since also R' is noetherian and integrally closed, it follows that we may assume in the proof that $\mathfrak{p}$ is a maximal ideal. Under this assumption, Lemma 6 shows that $\mathfrak{p}$ is invertible. We assert that every ideal $\mathfrak{q} \neq 0$ contained in $\mathfrak{p}$ admits $\mathfrak{p}$ as an associated prime ideal. To see this, we observe that we have $\mathfrak{q}(R:\mathfrak{p})\mathfrak{p} \subset \mathfrak{q}$, whence $\mathfrak{q}(R:\mathfrak{p}) \subset \mathfrak{q}:\mathfrak{p}$. On the other hand, we have $(\mathfrak{q}:\mathfrak{p})\mathfrak{p} \subset \mathfrak{q}$; and since $\mathfrak{p}$ is invertible it follows that $\mathfrak{q}:\mathfrak{p} = (\mathfrak{q}:\mathfrak{p})\mathfrak{p}\mathfrak{p}^{-1} \subset \mathfrak{q}\mathfrak{p}^{-1} = \mathfrak{q}(R:\mathfrak{p})$. Hence $\mathfrak{q}:\mathfrak{p} = \mathfrak{q}(R:\mathfrak{p}) = \mathfrak{q}\cdot\mathfrak{p}^{-1}$. The ideal $\mathfrak{q}\cdot\mathfrak{p}^{-1}$ is distinct from $\mathfrak{q}$, since, otherwise, we would have $\mathfrak{q} = \mathfrak{q}\cdot\mathfrak{p}$, and such an equality is impossible in our noetherian domain since $\mathfrak{q} \neq (0)$ and $\mathfrak{p} \neq R$ (IV, § 7, Lemma 2). The inequality $\mathfrak{q}:\mathfrak{p} \neq \mathfrak{q}$ implies that $\mathfrak{q}$ admits $\mathfrak{p}$ as an associated prime ideal (IV, § 6, Theorem 11), as asserted. In particular, $\mathfrak{p}$ cannot contain any proper prime ideal distinct from $\mathfrak{p}$, and is therefore minimal. Q.E.D.

COROLLARY. *Let R be a noetherian integrally closed domain. If $\mathfrak{p}$ is a minimal prime ideal in R, then the only primary ideals belonging to $\mathfrak{p}$ are its symbolic powers $\mathfrak{p}^{(n)}$.*

For, the quotient ring $R' = R_{\mathfrak{p}}$ is noetherian, is integrally closed, and contains only one proper prime ideal, namely, the maximal ideal $\mathfrak{p}' = R'\mathfrak{p}$. Hence, by Theorem 13, R' is a Dedekind domain, and every proper ideal in R' is therefore a power of $\mathfrak{p}'$ and is a primary ideal belonging to $\mathfrak{p}'$, since $\mathfrak{p}'$ is maximal. Since the primary ideals of R' which belong to $\mathfrak{p}'$ are in 1-1 correspondence with the primary ideals of R which belong to $\mathfrak{p}$, the corollary follows from the definition of symbolic powers (IV, § 12, p. 232).

THEOREM 15. *Let R be a noetherian integrally closed domain having only one maximal ideal $\mathfrak{m}$ (whence R is a local ring). If the fractionary ideal $(R:\mathfrak{m})$ is distinct from R, then $\mathfrak{m}$ is a principal ideal Rm; every non-zero element x of R may be written, and in a unique way, as $x = em^k$ where e is a unit; and the only proper ideals of R are the ideals Rm^k.*

PROOF. From the remark following Lemma 6 we deduce that $\mathfrak{m}$ is invertible. Then the proof of Theorem 14 shows that $\mathfrak{m}$ is a minimal prime ideal of R, whence $\mathfrak{m}$ is the only proper prime ideal in R, since it is the only maximal ideal in R. Thus R is a Dedekind domain (Theorem

13), and every proper ideal of R is a power of $\mathfrak{m}$. It remains to be shown that $\mathfrak{m}$ is a principal ideal. To show this, we observe that since $\mathfrak{m}\cdot(R:\mathfrak{m}) = R$ there exist finite families $\{m_i\}$, $\{m'_i\}$ of elements of $\mathfrak{m}$ and $(R:\mathfrak{m})$ such that $1 = \sum_i m_i m'_i$; since all products $m_i m'_i$ are in R; and since they cannot all lie in $\mathfrak{m}$, one of them, say $m_1 m'_1$, is outside of $\mathfrak{m}$ and is therefore a unit; in other words, there exist m in $\mathfrak{m}$ and m' in $(R:\mathfrak{m})$ such that $mm' = 1$. Hence, for every x in $\mathfrak{m}$, we have $x = (xm')m \in Rm$, and this proves that $\mathfrak{m} = Rm$. Q.E.D.

In the next chapter we shall see that R is a *discrete valuation ring*. As discrete valuation rings will often be encountered in the present chapter, we temporarily define them as being Dedekind domains with *only one* proper prime ideal (a definition equivalent to the one that will be given in chapter VI). Theorem 15 is thus a theorem concerning discrete valuation rings. Notice also that if A is a Dedekind domain and $\mathfrak{p}$ a proper prime ideal of A, then the quotient ring $A_\mathfrak{p}$ is a discrete valuation ring.

§ 7. Further properties of Dedekind domains

THEOREM 16. *A Dedekind domain R with only a finite number of proper prime ideals $(\mathfrak{p}_i)$ $(i = 1, \cdots, n)$ is a principal ideal domain.*

PROOF. It is sufficient to show that every $\mathfrak{p}_i$ is principal, and for this we have only to show that there exists an element p_i in $\mathfrak{p}_i$ such that $p_i \notin \mathfrak{p}_i^2$ and $p_i \notin \mathfrak{p}_j$ for $j \neq i$, since in that case the factorization of Rp_i into prime ideals can only be $Rp_i = \mathfrak{p}_i$. Since R is a noetherian domain, we have $\mathfrak{p}_i^2 < \mathfrak{p}_i$ (IV, § 7, p. 217), and there exists therefore an element a_i of $\mathfrak{p}_i$ which does not lie in $\mathfrak{p}_i^2$. As element p_i we may then take a solution of the system of n congruences $x \equiv a_i(\mathfrak{p}_i^2)$, $x \equiv 1\ (\mathfrak{p}_j)$ $(j \neq i)$. Since the ideals $\mathfrak{p}_i^2$, $\mathfrak{p}_j$ are pairwise comaximal, this system has a solution by Theorem 31, III, § 13. Q.E.D.

COROLLARY 1. *A residue class ring $R/\mathfrak{a}$ of a Dedekind domain R by a proper ideal $\mathfrak{a}$ is a principal ideal ring.*

Let $\mathfrak{a} = \prod_i \mathfrak{p}_i^{n(i)}$ be the factorization of $\mathfrak{a}$ into prime ideals, and let M be the complement of $\bigcup_i \mathfrak{p}_i$ in R. The ring $R/\mathfrak{a}$ is the direct sum of rings isomorphic to the rings $R/\mathfrak{p}_i^{n(i)}$. The set M is a multiplicative system in R, and the quotient ring R_M is a Dedekind domain (Example 2, § 6). The only prime ideals of R_M are the ideals $\mathfrak{p}_i^e$ (IV, § 11), whence R_M is a principal ideal domain (Theorem 16). Hence also $R_M/\mathfrak{a}^e$ is a principal ideal ring. By the permutability of residue class ring and quotient ring formation (IV, § 10, formula 1), $R_M/\mathfrak{a}^e$ is isomorphic to

$(R/\mathfrak{a})_{(M+\mathfrak{a})/\mathfrak{a}}$, and this last ring is $R/\mathfrak{a}$ itself since every element of M is invertible modulo $\mathfrak{a}$. (If $x \in M$ then $Rx + \mathfrak{a}$ is not contained in any proper prime ideal of R; hence it is the unit ideal.) This proves the corollary.

The following alternate proof of Corollary 1 is more direct and is independent of Theorem 16.

Since $\mathfrak{a}$ is an intersection (product) of pairwise comaximal powers of prime ideals, the ring $R/\mathfrak{a}$ is a direct sum of rings of the type $R/\mathfrak{p}^n$, where $\mathfrak{p}$ is a prime ideal in R. It is therefore sufficient to consider the case in which $\mathfrak{a}$ is a power $\mathfrak{p}^n$. In this case we fix an element t in $\mathfrak{p}$, not in $\mathfrak{p}^2$. Then for $s = 1, 2, \cdots, n$ we have $\mathfrak{p}^s = Rt^s + \mathfrak{p}^n$. This implies that the ideals $\mathfrak{p}/\mathfrak{p}^n, \mathfrak{p}^2/\mathfrak{p}^n, \cdots, \mathfrak{p}^{n-1}/\mathfrak{p}^n$ are principal ideals in $R/\mathfrak{p}^n$, and since these are the only proper ideals in $R/\mathfrak{p}^n$, the corollary is proved.

COROLLARY 2. *In a Dedekind domain R, every proper ideal $\mathfrak{a}$ has a basis consisting of two elements.*

We take a non-zero element a in $\mathfrak{a}$. As R/Ra is a principal ideal ring (Corollary 1), the ideal $\mathfrak{a}/Ra$ is principal. Let b be an element of $\mathfrak{a}$ whose residue class modulo Ra generates $\mathfrak{a}/Ra$. Then it is clear that $\{a, b\}$ is a basis of $\mathfrak{a}$.

In the proof of Theorem 16, we encountered a simple case of the problem of solving a finite system of simultaneous congruences. The next theorem treats the general case of this problem.

THEOREM 17 (CHINESE REMAINDER THEOREM).* *A Dedekind domain R possesses the following property:*

(CRT) *Given a finite number of ideals $\mathfrak{a}_i$ and of elements x_i of R $(i = 1, \cdots, n)$, the system of congruences $x \equiv x_i \pmod{\mathfrak{a}_i}$ admits a solution x in R if and only if these congruences are pairwise compatible, that is, if and only if we have $x_i \equiv x_j \pmod{\mathfrak{a}_i + \mathfrak{a}_j}$ for $i \neq j$.*

PROOF. The property (CRT) is related to the fact that in the set of ideals of a Dedekind domain R, each of the operations $\cap$ and $+$ is *distributive* with respect to the other; that is, that given three ideals $\mathfrak{a}, \mathfrak{b}, \mathfrak{b}'$ in R, we have:

$$\mathfrak{a} \cap (\mathfrak{b} + \mathfrak{b}') = (\mathfrak{a} \cap \mathfrak{b}) + (\mathfrak{a} \cap \mathfrak{b}')$$
$$\mathfrak{a} + (\mathfrak{b} \cap \mathfrak{b}') = (\mathfrak{a} + \mathfrak{b}) \cap (\mathfrak{a} + \mathfrak{b}').$$

* A rule for the solution of simultaneous linear congruences, essentially equivalent with Theorem 17 in the case of the ring J of integers, was found by Chinese calendar makers between the fourth and the seventh centuries A.D. It was used for finding the common periods to several cycles of astronomical phenomena.

In the case of a Dedekind domain R, these distributivity relations are easily verified by using Theorem 11: they are equivalent to:

$$\max \{n_\mathfrak{p}(\mathfrak{a}), \min (n_\mathfrak{p}(\mathfrak{b}), n_\mathfrak{p}(\mathfrak{b}'))\} = \min \{\max (n_\mathfrak{p}(\mathfrak{a}), n_\mathfrak{p}(\mathfrak{b})), \max (n_\mathfrak{p}(\mathfrak{a}), n_\mathfrak{p}(\mathfrak{b}'))\},$$

$$\min \{n_\mathfrak{p}(\mathfrak{a}), \max (n_\mathfrak{p}(\mathfrak{b}), n_\mathfrak{p}(\mathfrak{b}'))\} = \max \{\min (n_\mathfrak{p}(\mathfrak{a}), n_\mathfrak{p}(\mathfrak{b})), \min (n_\mathfrak{p}(\mathfrak{a}), n_\mathfrak{p}(\mathfrak{b}'))\}$$

(for every prime ideal $\mathfrak{p}$ in R); and these relations in their turn follow immediately from the fact that in the set of ordinary integers, each of the operations *min* and *max* is distributive with respect to the other, a fact whose verification is straightforward.

This being so, Theorem 17 follows immediately from:

THEOREM 18. *Given an arbitrary ring R, the property (CRT) is equivalent to the distributivity of each of the operations $+$ and $\cap$ with respect to the other in the set of all ideals of R.*

PROOF. We notice that in (CRT) the part "only if" is trivially true in any ring; we shall therefore disregard this part of (CRT) as being irrelevant to the proof.

We first consider *the case $n = 2$*. If $x_1 \equiv x_2 \pmod{(\mathfrak{a}_1 + \mathfrak{a}_2)}$, we have $x_1 - x_2 = a_1 - a_2$ with a_i in $\mathfrak{a}_i$. We may then take $x = x_1 - a_1 = x_2 - a_2$ as a solution of the congruences $x \equiv x_1 \pmod{\mathfrak{a}_1}$, $x \equiv x_2 \pmod{\mathfrak{a}_2}$. Thus (CRT) holds unconditionally for $n = 2$.

Now we prove that the distributivity conditions imply (CRT) for any number n of congruences. Using induction with respect to n, we need only examine *the step from $n - 1$ to n*. We have to solve n congruences $x \equiv x_i \pmod{\mathfrak{a}_i}$ such that $x_i \equiv x_j \pmod{(\mathfrak{a}_i + \mathfrak{a}_j)}$, and we know that any system of $n - 1$ such pairwise compatible congruences is solvable. We then know a solution x' of the system of the first $n - 1$ congruences: $x' \equiv x_i \pmod{\mathfrak{a}_i}$ $(i = 1, \cdots, n - 1)$. Then the given system of n congruences is equivalent to $x \equiv x' \pmod{\mathfrak{a}_i}$ $(i = 1, \cdots, n - 1)$, $x \equiv x_n \pmod{\mathfrak{a}_n}$; in other words, it is equivalent to the system of two congruences $x \equiv x' \pmod{\bigcap_{i=1}^{n-1} \mathfrak{a}_i}$, $x \equiv x_n \pmod{\mathfrak{a}_n}$. As was proved before, this system is solvable if we have $x' \equiv x_n \pmod{(\mathfrak{a}_n + \bigcap_i \mathfrak{a}_i)}$. Suppose that the distributive law

$$(D_1) \qquad \mathfrak{a} + (\mathfrak{b} \cap \mathfrak{b}') = (\mathfrak{a} + \mathfrak{b}) \cap (\mathfrak{a} + \mathfrak{b}')$$

holds for ideals in R. Then our condition of solvability may be written as follows:

$$x' \equiv x_n \left(\operatorname{mod} \bigcap_{i=1}^{n-1} (\mathfrak{a}_n + \mathfrak{a}_i)\right)$$

and this condition is indeed fulfilled, for we have $x' \equiv x_i \pmod{\mathfrak{a}_i}$ and, by hypothesis, $x_i \equiv x_n \pmod{\mathfrak{a}_i + \mathfrak{a}_n}$ for $i = 1, 2, \cdots, n-1$. The solvability condition is thus fulfilled, and the given system of n congruences admits a solution. Therefore the distributive law (D_1) implies the validity of (CRT).

Conversely we prove that both distributive laws follow from the validity of (CRT) *for* $n = 3$. As to (D_1), the left-hand side is obviously contained in the right-hand side, and hence it is sufficient to prove that any element d of $(\mathfrak{a} + \mathfrak{b}) \cap (\mathfrak{a} + \mathfrak{b}')$ belongs to $\mathfrak{a} + (\mathfrak{b} \cap \mathfrak{b}')$. By hypothesis we have $d = a + b = a' + b'$ with a and a' in $\mathfrak{a}$, b in $\mathfrak{b}$, b' in $\mathfrak{b}'$. We now try to write d in the form $x + y$ with x in $\mathfrak{a}$ and y in $\mathfrak{b} \cap \mathfrak{b}'$. This is equivalent to looking for an element x in $\mathfrak{a}$ such that $x - d \in \mathfrak{b} \cap \mathfrak{b}'$—that is, for a solution of the three congruences $x \equiv 0 \pmod{\mathfrak{a}}$, $x \equiv d \pmod{\mathfrak{b}}$, $x \equiv d \pmod{\mathfrak{b}'}$. As $d - 0 = d \in \mathfrak{a} + \mathfrak{b}$, $d - 0 = d \in \mathfrak{a} + \mathfrak{b}'$, and $d - d = 0 \in \mathfrak{b} + \mathfrak{b}'$, these congruences are pairwise compatible, and the solution x exists by (CRT). Therefore (D_1) is proved.

From what we have proved above, we immediately deduce that (CRT) holds for every n.

For the proof of the other distributive law

$$(D_2) \qquad \mathfrak{a} \cap (\mathfrak{b} + \mathfrak{b}') = (\mathfrak{a} \cap \mathfrak{b}) + (\mathfrak{a} \cap \mathfrak{b}')$$

we notice again that the right-hand side is contained in the left-hand side, and so it suffices to prove that any element d of $\mathfrak{a} \cap (\mathfrak{b} + \mathfrak{b}')$ is an element of $(\mathfrak{a} \cap \mathfrak{b}) + (\mathfrak{a} \cap \mathfrak{b}')$. Trying to write d in the form $x + y$ with x in $\mathfrak{a} \cap \mathfrak{b}$ and y in $\mathfrak{a} \cap \mathfrak{b}'$, is equivalent to looking for an element x of $\mathfrak{a} \cap \mathfrak{b}$ such that $x \equiv d \pmod{\mathfrak{a} \cap \mathfrak{b}'}$, that is, to solving the system of four congruences $x \equiv 0 \pmod{\mathfrak{a}}$, $x \equiv 0 \pmod{\mathfrak{b}}$, $x \equiv d \pmod{\mathfrak{a}}$, $x \equiv d \pmod{\mathfrak{b}'}$. As the six compatibility conditions $0 \in \mathfrak{a} + \mathfrak{b}$, $d \in \mathfrak{a} + \mathfrak{a}$, $d \in \mathfrak{a} + \mathfrak{b}'$, $d \in \mathfrak{a} + \mathfrak{b}$, $d \in \mathfrak{b} + \mathfrak{b}'$, and $0 \in \mathfrak{a} + \mathfrak{b}'$ are fulfilled, the system has a solution x, by (CRT).

REMARKS.

1) We actually proved that the distributive law (D_1) implies the distributive law (D_2), and that (CRT) for $n = 3$ implies (CRT) for every n.

2) Examples in which (D_1) or (D_2) does not hold may already be constructed in the polynomial ring $k[X, Y]$ in two variables over a field.

§ 8. Extensions of Dedekind domains

THEOREM 19. *Let R be a Dedekind domain, and L a finite algebraic extension of the quotient field K of R. Then the integral closure R' of R in L (that is, the set of elements of L which are integral over R) is a Dedekind domain.*

PROOF. We first study the case in which L is *separable* over K, and in this case we use the characterization of Dedekind domains given in Theorem 13 of § 6. In this case, R' is noetherian by Corollary 1 to Theorem 7 of § 4, every proper prime ideal of R' is maximal by complement 2) to Theorem 3 of § 2, and R' is integrally closed by construction. Therefore R' is a Dedekind domain.

In the general case, L is a purely inseparable extension of a separable extension L_s of K (II, § 5, p. 71). If we denote by R_s the integral closure of R in L_s, then R_s is a Dedekind domain, and R' is obviously the integral closure of R_s in L. In other words, it is now permissible to assume that L is a *purely inseparable* extension of K. Then the minimal equation of any element x of R' over K is of the form $x^{p^e} - a = 0$, where p is the characteristic of L, and where a is in K. But since R is integrally closed, this is an integral dependence equation (Theorem 4, § 3) and we have $a \in R$. Since L is a finite extension of K, the exponents e are bounded, and there exists a power q of p such that R' is the set of all elements x in L such that $x^q \in R$.

We introduce the field M consisting of all the q-th roots of all the elements of K—that is, $M = K^{q^{-1}}$—and the subring S of M consisting of all the elements x in M such that $x^q \in R$. As the mapping $x \to x^q$ is an isomorphism of M onto K (II, § 4, p. 64), it is an isomorphism of S onto R. Thus S is a Dedekind domain.

In order to show that R' is a Dedekind ring it will be sufficient to show that every proper ideal $\mathfrak{A}$ in R' is invertible (Theorem 12 of § 6). Since S is a Dedekind ring, the ideal $S\mathfrak{A}$ is invertible, and hence there exist elements s_i in $S:S\mathfrak{A}$ and elements a_i in $\mathfrak{A}$ such that $\sum a_i s_i = 1$. Since q is a power of the characteristic, we have $\sum a_i^q s_i^q = 1$, with $s_i^q \in K \subset L$. Let us write this relation in the form $\sum a_i \cdot a_i^{q-1} s_i^q = 1$. The elements $b_i = a_i^{q-1} s_i^q$ are in L. On the other hand, we have $b_i \mathfrak{A} \subset \mathfrak{A}^q s_i^q \subset S$, since $\mathfrak{A} s_i \subset S$. Therefore $b_i \mathfrak{A} \in S \cap L = R'$, that is, $b_i \in R':\mathfrak{A}$; and since $\sum b_i a_i = 1$, it follows that $\mathfrak{A}(R':\mathfrak{A}) = R'$, showing that $\mathfrak{A}$ is invertible. Q.E.D.

In the purely inseparable case which was treated in the last part of the above proof, we have established the following result: *If M is a purely inseparable extension of a field L and if there exists a power q of the characteristic of L such that $M^q \subset L$, then every Dedekind ring S contained in M contracts in L to a Dedekind ring.* This result can be generalized as follows:

LEMMA. *Let L be a field and M a finite normal and separable extension of a purely inseparable extension M' of L. Suppose that there exists a power q of the characteristic of L such that $M'^q \subset L$. If S is a Dedekind* *domain contained in M and integral over $R' = S \cap L$, then R' is also a Dedekind domain.*

PROOF. The lemma has already been proved in the special case $M = M'$. It is therefore sufficient to prove that $S \cap M'$ is a Dedekind domain. In other words, *we may assume that* $M' = L$. Under this assumption we take a proper ideal $\mathfrak{a}$ in R', and prove that it is invertible. Since the ideal $S\mathfrak{a}$ is invertible, there exist elements a_i in $\mathfrak{a}$ and x_i in M such that $\sum_i a_i x_i = 1$ and $x_i \mathfrak{a} \subset S$. We denote by $x_i^{(j)}$ the conjugates of x_i over L, and we consider the relation $\prod_j \left(\sum_i a_i x_i^{(j)}\right) = 1$. Denoting by n the degree $[M:L]$, we may write this relation as follows:

$$\sum_m m(a) P_m(x_i^{(j)}) = 1, \tag{1}$$

where the $m(a)$ are the monomials of degree n in the a_i, and where the $P_m(x_i^{(j)})$ are symmetric functions and homogeneous polynomials of degree n in the $x_i^{(j)}$, with ordinary integers as coefficients. By Galois theory (II, § 7), $P_m(x_i^{(j)})$ is an element of L. In each monomial $m(a)$ let us factor out some a_i; relation (1) may then be written in the form $\sum_i a_i b_i = 1$, where b_i is a sum of products of monomials of degree $n - 1$ in the a_j by symmetric functions $P_m(x_i^{(j)})$, and is therefore an element of L. On the other hand, from the relation $x_i \mathfrak{a} \subset S$, we deduce $x_i^{(j)} \mathfrak{a} \subset S$ since $\mathfrak{a}$ is contained in L and since S, which is the integral closure of R' in M, is invariant under any L-automorphism of M. Hence $b_i \mathfrak{a} \subset \sum_m P_m(x_i^{(j)}) \mathfrak{a}^{n-1} \cdot \mathfrak{a} \subset S$, and thus $b_i \mathfrak{a} \subset S \cap L = R'$. This proves that $\mathfrak{a}$ is invertible, and the lemma is proved.

Theorem 19 admits a converse:

THEOREM 20. *Let R be a Dedekind domain, and L a subfield of the quotient field K of R, over which K is finite algebraic. Then if R is integral over $T = L \cap R$, T is also a Dedekind domain.*

PROOF. Theorem 20 follows immediately from the lemma if K is a normal extension of L, since K is then a normal and separable extension of a purely inseparable extension of finite degree of L (that is, of the fixed field of all L-automorphisms of K). The general case may be reduced to this one by replacing K by the least normal extension K' of L containing K, and R by its integral closure R' in K': we have evidently $T = L \cap R'$, and R' is a Dedekind domain by Theorem 19.

In the next chapter, by using valuation theory, we shall be able to prove Theorem 19 without any reference to separability or inseparability, and without having to give a proof in three steps (up, up, down) in the inseparable case [see VI, § 13, Theorem 30, c)].

The most important cases of application of Theorem 19 are the following ones: R is the ring of integers, or R is a polynomial ring in one variable over a field. In the first case Theorem 19 shows that the algebraic integers of an algebraic number field form a Dedekind ring (but not always a principal ideal domain); this answers the classical problem of unique factorization for algebraic integers. In the second case Theorem 19 shows that the integral functions of a function field F of one variable x form a Dedekind ring. This is of importance in the study of affine normal curves.

All Dedekind domains which occur in number theory or in algebraic geometry are obtainable from PID's by application of the process described in Theorem 19, that is, as the integral closure of a suitable PID R in a finite algebraic extension of the quotient field of R. It is not known whether all Dedekind domains are obtainable in this way. It is *not* true that, given a Dedekind domain R which is not a field, the polynomial ring $R[X]$ is a Dedekind domain; in fact it admits proper prime ideals which are not maximal, e.g., $R[X]\mathfrak{p}$ where $\mathfrak{p}$ is a proper prime ideal in R.

§ 9. Decomposition of prime ideals in extensions of Dedekind domains. Let R be a Dedekind domain, L a finite algebraic extension of degree n of the quotient field K of R, and R' the integral closure of R in L; by Theorem 19, R' is a Dedekind domain. We denote ideals of R by small German letters ($\mathfrak{a}, \mathfrak{b}, \mathfrak{p}, \cdots$), and ideals of R' by capital German letters ($\mathfrak{A}, \mathfrak{B}, \mathfrak{P}, \cdots$). Let $\mathfrak{p}$ be a proper prime ideal in R. Since R' is a Dedekind domain, the ideal $\mathfrak{p}^e = R'\mathfrak{p}$ is a product of prime ideals; let us write

$$\mathfrak{p}^e = \prod_i \mathfrak{P}_i^{e_i},$$

where the prime ideals $\mathfrak{P}_i$ are all distinct. We have $\mathfrak{P}_i \cap R = \mathfrak{P}_i^c = \mathfrak{p}$ since $\mathfrak{p}$ is a maximal ideal. The integer e_i is called the *reduced ramification index** of $\mathfrak{P}_i$ over $\mathfrak{p}$.

Since $\mathfrak{P}_i \cap R = \mathfrak{p}$, the residue field $R/\mathfrak{p}$ may be identified with a subfield of the residue field $R'/\mathfrak{P}_i$. The field $R'/\mathfrak{P}_i$ is a *finite algebraic* extension of $R/\mathfrak{p}$, as follows from the following more general result.

LEMMA 1. *Let R be a Dedekind domain, L a finite algebraic extension of the quotient field K of R, $\mathfrak{p}$ a proper prime ideal in R, R' an overring of R contained in L, and $\mathfrak{A}$ an ideal of R' such that $\mathfrak{A} \cap R = \mathfrak{p}$. Then the dimension of $R'/\mathfrak{A}$, considered as a vector space over $R/\mathfrak{p}$, is $\leq [L:K]$.*

PROOF. We denote by M the complement of $\mathfrak{p}$ in R. Since the elements of M are invertible modulo $\mathfrak{p}$, and consequently modulo $\mathfrak{A}$, the permutability of residue class ring and quotient ring formation (IV,

* The ramification index of $\mathfrak{P}_i$ over $\mathfrak{p}$ will be defined later as being e_i times the inseparable factor of the degree $[R'/\mathfrak{P}_i:R/\mathfrak{p}]$.

§ 10, formula (1)) shows that $R/\mathfrak{p}$ is isomorphic to $R_M/\mathfrak{p}R_M$ and $R'/\mathfrak{A}$ to $R'_M/\mathfrak{A}R'_M$. On the other hand, $\mathfrak{A}R'_M \cap R_M = \mathfrak{p}R_M$: in fact, if z is any element of $\mathfrak{A}R'_M \cap R_M$, then we have $z = a/m$ $(a \in \mathfrak{A}, m \in M)$ and also $z = r/m'$ $(r \in R, m' \in M)$; then $am' = rm$ is an element of $\mathfrak{A} \cap R = \mathfrak{p}$, whence $a/m = am'/m'm \in \mathfrak{p}R_M$. Since R_M is a Dedekind domain (§ 6, Example 2, p. 271), we may therefore replace R, R', $\mathfrak{p}$ and $\mathfrak{A}$ by R_M, R'_M, $\mathfrak{p}R_M$ and $\mathfrak{A}R'_M$. By Theorem 16, R_M is a PID. Thus we may prove Lemma 1 under the additional assumption that the ideal $\mathfrak{p}$ is *principal*, say $\mathfrak{p} = Rp$. This being so, consider a finite family $\{\bar{x}_i\}$ of elements of $R'/\mathfrak{A}$ which are linearly independent over $R/\mathfrak{p}$, and denote by x_i a representative in R' of the residue class $\bar{x}_i$. If we had a non-trivial linear relation $\sum_i a_i x_i = 0$ with a_i in K, we could suppose that all the a_i are in R (since K is the quotient field of R) and furthermore, by dividing them by a suitable power of p, we could also suppose that they are not all in $\mathfrak{p}$; thus, by reducing the relation $\sum_i a_i x_i = 0$ modulo $\mathfrak{A}$, we would get a non-trivial linear relation $\sum_i \bar{a}_i \bar{x}_i = 0$, with $\bar{a}_i$ in $R/\mathfrak{p}$; and this is a contradiction. Therefore the elements x_i are linearly independent over K, and their number, which is also the number of the $\bar{x}_i$, cannot exceed $[L:K]$. This proves Lemma 1.

Coming back to the situation described before Lemma 1 (that is, R' is now the integral closure of R in L), the degree $[R'/\mathfrak{P}_i:R/\mathfrak{p}]$ is called the *relative degree of* $\mathfrak{P}_i$ over $\mathfrak{p}$ and is denoted by f_i. It will be tacitly understood from now on that $\mathfrak{P}_i$ ranges over the set of all prime ideals in R' which are factors of the extended ideal $\mathfrak{p}^e$. The notations e_i and f_i are classical and will be used without further warning in this section.

THEOREM 21. *The integer* $\sum_i e_i f_i$ *is equal to the dimension of the ring* $R'/\mathfrak{p}^e$ *considered as a vector space over* $R/\mathfrak{p}$. *We have the inequality* $\sum_i e_i f_i \leq [L:K]$, *with equality if and only if* R'_M *is a finite* R_M*-module,* M *denoting the complement of* $\mathfrak{p}$ *in* R.

PROOF. Theorem 21 is an easy consequence of the following more general lemma:

LEMMA 2. *Let* R *be a Dedekind domain,* K *its quotient field,* $\mathfrak{p}$ *a proper prime ideal in* R, L *a finite algebraic extension of* K, *and* R' *a noetherian overring of* R *contained in* L. *Let* $\mathfrak{p}^e = \mathfrak{Q}_1 \cap \cdots \cap \mathfrak{Q}_g$ *be an irredundant decomposition of* $\mathfrak{p}^e = R'\mathfrak{p}$ *into primary components; let* $\mathfrak{P}_i$ *be the radical of* $\mathfrak{Q}_i$, e_i *the length of* $\mathfrak{Q}_i$ (IV, § 13), *and* f_i *the degree* $[R'/\mathfrak{P}_i:R/\mathfrak{p}]$. *Then the integer* $\sum_i e_i f_i$ *is equal to the dimension of the ring* $R'/\mathfrak{p}^e$ *considered as a* *vector space over $R/\mathfrak{p}$. If L is the quotient field of R' we have the inequality $\sum_i e_i f_i \leq [L:K]$, with equality if and only if R'_M is a finite R_M-module, M denoting the complement of $\mathfrak{p}$ in R.*

We first show how Theorem 21 may be deduced from Lemma 2. Under the assumptions of Theorem 21, the primary components $\mathfrak{Q}_i$ of $\mathfrak{p}^e$ are the prime ideal powers $\mathfrak{P}_i^{e_i}$, and we need only to prove that e_i is the length of $\mathfrak{P}_i^{e_i}$; but this follows from the fact that the only primary ideals belonging to $\mathfrak{P}_i$ are its powers since R' is a Dedekind ring.

PROOF OF THE LEMMA. We first notice that since $\mathfrak{p}$ is maximal, we have $\mathfrak{P}_i \cap R = \mathfrak{p}^e \cap R = \mathfrak{p}$; thus $R/\mathfrak{p}$ may be identified with subfields of $R'/\mathfrak{P}_i$ and of $R'/\mathfrak{p}^e$. By Lemma 1, the dimensions of the vector spaces $R'/\mathfrak{P}_i$ and $R'/\mathfrak{p}^e$ over $R/\mathfrak{p}$ are finite. In particular, by the remark made at the end of § 2 (p. 259), $R'/\mathfrak{P}_i$ is a field, since it is an integral domain which is a finite module over $R/\mathfrak{p}$ and therefore integrally dependent on the field $R/\mathfrak{p}$. Hence $\mathfrak{P}_i$ is a maximal ideal in R'. Thus the statement of Lemma 2 is meaningful.

Let d be the dimension of the ring $R'/\mathfrak{p}^e$ considered as a vector space over $R/\mathfrak{p}$. The inequality $d \leq [L:K]$ follows immediately from Lemma 1. We now prove that we have $d = \sum_i e_i f_i$. By III, § 13, Theorem 32, the vector space $R'/\mathfrak{p}^e$ over $R/\mathfrak{p}$ is isomorphic to the direct sum of the spaces $R'/\mathfrak{Q}_i$, the latter being themselves regarded as vector spaces over $R/\mathfrak{p}$. But $R'/\mathfrak{Q}_i$, considered as an R'-module, admits a composition series of length e_i whose successive difference modules are one-dimensional vector spaces over $R'/\mathfrak{P}_i$ (IV, § 13, Theorem 28). Since $[R'/\mathfrak{P}_i : R/\mathfrak{p}] = f_i$, it follows easily that the dimension of $R'/\mathfrak{Q}_i$ over $R/\mathfrak{p}$ is $e_i f_i$ Therefore $R'/\mathfrak{p}^e$ has dimension $\sum_i e_i f_i$ over $R/\mathfrak{p}$.

We suppose now that the equality $\sum_i e_i f_i = [L:K]$ holds, and we wish to prove that R'_M is then a finite R_M-module. By IV, § 10 (p. 225), neither the hypotheses in Lemma 2 nor the integers e_i, f_i d and $[L:K]$ are changed if we replace R by R_M, $\mathfrak{p}$ by $\mathfrak{p}R_M$ and R' by R'_M. In other words, we may suppose, as in Lemma 1, that R is a PID with $\mathfrak{p}$ as unique proper prime ideal; we write $\mathfrak{p} = Rp$. It follows from the proof of Lemma 1 that if $\{x_j\}$ is a set of elements of R' whose residue classes modulo $\mathfrak{p}$ constitute a basis of the vector space $R'/\mathfrak{p}^e$ over $R/\mathfrak{p}$, then the elements x_j are linearly independent over K. Since by hypothesis L has the same dimension over K as $R'/\mathfrak{p}^e$ over $R/\mathfrak{p}$, these elements x_j constitute a basis of L over K. We shall now prove that we have $R' = \sum_j Rx_j$. In fact, take an element x of R'. As $\{x_j\}$ is a basis of L

over K, we can write $x = \sum_j a_j x_j$ with $a_j \in K$. If the elements a_j were not all in R, there would exist an integer $n > 0$ such that all the elements $p^n a_j$ belong to R, and not all of them to $Rp = \mathfrak{p}$; then, reducing modulo $\mathfrak{p}^e = R'p$ the equality $p^n x = \sum_j (p^n a_j) x_j$, we would get a non-trivial linear relation $\sum_j \bar{b}_j \bar{x}_j = 0$ with coefficients $\bar{b}_j$ in $R/\mathfrak{p}$; and this contradicts the fact that the $\bar{x}_j$ are linearly independent over $R/\mathfrak{p}$. Thus our assertion is proved.

Conversely, suppose that R'_M is a finite R_M-module. As R_M is a PID, and as L is the quotient field of R'_M, R'_M can be generated by exactly n $(= [L:K])$ elements x_j (see proof of Corollary 2 to Theorem 7 in § 4, p. 265), and these elements are linearly independent over K. Then (denoting again by p a generator of the principal ideal $\mathfrak{p}R_M$) we deduce from the relation $R'_M = \sum_j R_M x_j$ that we have $pR'_M = \sum_j (pR_M) x_j$. As the elements x_j are linearly independent over K, this implies that R'_M/pR'_M is isomorphic to the product of R_M/pR_M n times with itself. Therefore the dimension of R'_M/pR'_M considered as a vector space over R_M/pR_M is equal to n. Since (see proof of Lemma 1) R'_M/pR'_M is isomorphic to $R'/\mathfrak{p}^e$ and R_M/pR_M to $R/\mathfrak{p}$, the dimension of $R'/\mathfrak{p}^e$ over $R/\mathfrak{p}$ is also equal to n, and this latter dimension is $\sum_i e_i f_i$, as has been seen in the beginning of the proof. This completes the proof of Lemma 2, and consequently of Theorem 21.

Note that the hypothesis that R' is noetherian is automatically verified if L is separable over K, and R' integral over R (Corollary 1 to Theorem 7 of § 4) or if R is a finite integral domain, and R' is integral over R (Theorem 9 of § 4).

COROLLARY. *The hypotheses and notations being as in Theorem* 21, *we suppose that L is a separable extension of K or that R is a finite integral domain. Then* $\sum_i e_i f_i = [L:K]$.

In fact, R'_M is then a finite R_M-module by Corollary 1 to Theorem 7 in § 4, or by Theorem 9 in § 4.

EXAMPLE: *Gaussian integers.* We take for R the ring J of rational integers, and for L the quadratic field obtained by adjunction of $i = \sqrt{-1}$ to the rational number field K. Any element z of L may be written, and in a unique way, as $z = x + iy$, with x and y in K. For z to be integral over $R = J$, it is necessary and sufficient that its trace $2x$ and its norm $x^2 + y^2$ be rational integers (§ 3); we then have $x = a/2$ and $y^2 = (4b - a^2)/4$ with a and b in J, whence $y = c/2$ with c in J;

this means that $4b = a^2 + c^2$, and this is possible only if a and c are even, since the sum of the squares of two odd integers is congruent to 2 modulo 4. Thus x and y are rational integers, and the integral closure R' of J in L is the ring

$$R' = J + i \cdot J,$$

which is called the ring of *gaussian integers*. It is a Dedekind domain (Theorem 19).

Given a prime number p, the relation $\sum_i e_i f_i = 2$ (Corollary to Theorem 21) shows that the factorization of $R'p$ into prime ideals is either $R'p = \mathfrak{P}_1 \cdot \mathfrak{P}_2$ (with $R'/\mathfrak{P}_1 = R'/\mathfrak{P}_2 = J/(p)$), or $R'p = \mathfrak{P}$ (with $[R'/\mathfrak{P} : J/(p)] = 2$), or $R'p = \mathfrak{P}^2$ (with $R'/\mathfrak{P} = J/(p)$). The prime number p is then said to be *decomposed* in the first case, *inertial* in the second, and *ramified* in the third (with respect to the quadratic field L). Notice that this classification into three cases holds for any quadratic field.

From $R' = J + i \cdot J$, and from $\mathfrak{P}_1 \cap J = (p)$ (or $\mathfrak{P}_2 \cap J = (p)$, or $\mathfrak{P} \cap J = (p)$), it follows that $R'/\mathfrak{P}_1$ is generated over $J/(p)$ by the $\mathfrak{P}_1$-residue of i, that is, by a root of $X^2 + 1$ in some extension field of $J/(p)$. We must then study whether $X^2 + 1$ does or does not have a root in $J/(p)$ or, equivalently, whether -1 is or is not a square modulo p. If $p = 2$, -1 is a square modulo 2. If p is an odd prime, the multiplicative group of $J/(p)$ is a cyclic group of order $p - 1$ (II, § 8, Theorem 18); if we denote by x a generator of this group, we have $-1 = x^{(p-1)/2}$ since $(-1)^2 = 1$ and since $-1 \neq 1$ in $J/(p)$. Thus -1 is a square modulo p if $(p-1)/2$ is even (that is, if $p = 4n + 1$), and is not a square modulo p when $(p-1)/2$ is odd (that is, if $p = 4n - 1$). Therefore *the only odd primes which are inertial are the primes of the form* $4n - 1$. Any such prime is an irreducible element of R'.

We now use the well-known fact that R' is a *euclidean domain* (I, § 15), hence a PID. In fact, with the notation of I, § 15, we take for the function φ the function defined by $\varphi(z) = \varphi(x + iy) = x^2 + y^2$. As quotient q of the division of z by z' we may then take any one of the gaussian integers $a + bi$ whose distance to z/z' in the complex plane is < 1 (such a gaussian integer exists, since, in the complex plane, the gaussian integers are the vertices of a lattice of squares of side 1). From the multiplicativity of the norm, and from the formula $(a + bi)^{-1} = (a - bi)/N(a + bi)$, it follows that the only units in R' are the gaussian integers $a + bi$ whose norm $a^2 + b^2$ is 1 or -1; in other words, these units are $1, -1, i, -i$. For a prime $p = 4n + 1$, we consider the decomposition $R'p = \mathfrak{P}_1 \cdot \mathfrak{P}_2$, and we denote by $a + bi$ a generator of

$\mathfrak{P}_1$. Since $x + yi \to x - yi$ is an automorphism of L, the ideal $R'(a - bi)$ is also a prime ideal lying over (p). This latter ideal is different from $\mathfrak{P}_1$, for otherwise $(a + bi)/(a - bi)$ would be a unit $1, -1, i$, or $-i$. It cannot be 1 or -1, since in the contrary case either a or b would be zero, whence the ideal $(a + bi)$ would be generated by a rational integer and it would then follow that $\mathfrak{P}_1 = R'p$, a contradiction. It cannot be i or $-i$, since this would imply that $a + bi$ is an integral multiple of $1 + i$ or $1 - i$, whence that it is $\pm(1 + i)$ or $\pm(1 - i)$ since $a + bi$ is irreducible; but then, since $-i(1 + i)^2 = i(1 - i)^2 = 2$, we would have $2 \in \mathfrak{P}_1{}^c = (p)$, and this contradicts the fact that p is odd. We therefore have $\mathfrak{P}_1 = (a + bi)$, $\mathfrak{P}_2 = (a - bi)$ with $\mathfrak{P}_1 \neq \mathfrak{P}_2$. It follows that all the primes $p = 4n + 1$ are decomposed. Furthermore, we have $R'p = R'(a^2 + b^2)$, and since $1, -1, i$, and $-i$ are the only units in R', it follows that $p = a^2 + b^2$. In other words, *a prime* $4n + 1$ *is a sum of two squares.*

We find directly that $R' \cdot 2 = \mathfrak{P}^2$ where $\mathfrak{P} = R' \cdot (1 + i)$. Hence 2 *is the only ramified prime.*

REMARKS.

1) In the general quadratic field, generated by $\sqrt{d}$, where d is a rational integer without square factors, the question whether a prime number p is decomposed, inertial, or ramified is, by an analogous reasoning, closely related to the question whether d is or is not a square modulo p, or, in the standard terminology of number theory, "whether d is or is not a quadratic residue." (See § 12.) This question is a cornerstone of number theory. We will prove that the ramified primes are finite in number (§ 11). On the contrary, the sets of decomposed and of inertial primes are both infinite; and one can prove that they have the same "asymptotic density"; more precisely, that the number of decomposed or inertial primes which are $\leq n$ is asymptotic to $n/2 \log n$. Generalizations to other algebraic number fields are also important in number theory. For these questions, see H. Weyl, "Algebraic Theory of Numbers," *Ann. Math. Studies*, I (Princeton, 1940).

2) We have proved the "two squares theorem" ("every prime of the form $4n + 1$ is a sum of two squares") by investigating the divisibility properties of the ring $R' = J + i \cdot J$ of gaussian integers. An analogous method holds for the "four squares theorem" ("every prime is a sum of four squares"): one studies then the divisibility properties of the (non-commutative) ring of integral quaternions $a + bi + cj + dk$. See Hardy-Wright, *Theory of Numbers* (Oxford, 1938) Chap. 20.

When L is a normal extension of K, Theorem 21 admits a useful complement:

THEOREM 22. *Let R be an integrally closed domain, and R' the integral closure of R in a finite normal extension L of the quotient field K of R. If $\mathfrak{p}$ is a prime ideal in R, then the prime ideals $\mathfrak{P}_i$ of R' which lie over $\mathfrak{p}$ are all conjugates of any one of them. If, furthermore, R is a Dedekind domain, then the $\mathfrak{P}_i$ are the prime factors of $R'\mathfrak{p}$, the integers e_i (or f_i) are all equal* *to the same integer e (or f); and denoting by g the number of prime ideals* $\mathfrak{P}_i$, *we have* $efg \leq n = [L:K]$. *If L is a separable extension of K, we have* $efg = n$.

PROOF. By Theorem 3 of § 2, there exists a prime ideal $\mathfrak{P}$ of R' which lies over $\mathfrak{p}$. We denote by $\mathfrak{P}^{(j)}$ ($1 \leq j \leq q$) the conjugates of $\mathfrak{P}$, that is, the ideals of the form $s(\mathfrak{P})$, where s is a K-automorphism of L. Since R' is the integral closure of a subring of K, we have $s(R') = R'$ for any K-automorphism s of L. Hence the set $s(\mathfrak{P})$ is also a prime ideal in R' which lies over $\mathfrak{p}$. Suppose we have a prime ideal $\mathfrak{Q}$ of R' lying over $\mathfrak{p}$ and distinct from any of the ideals $\mathfrak{P}^{(j)}$. Then $\mathfrak{Q}$ cannot be contained in any $\mathfrak{P}^{(j)}$ by complement 1) to Theorem 3 of § 2, and there exists an element x of $\mathfrak{Q}$ which is not contained in any $\mathfrak{P}^{(j)}$ (see IV, § 6, Remark at the end of section, p. 215). But then none of the conjugates of x is in $\mathfrak{P}$; hence neither is any power of their product. Some such power, however, is in K, hence also in R (since R is integrally closed), and hence, finally, also in $\mathfrak{p} = \mathfrak{Q} \cap R$. Since $\mathfrak{p}$ is contained in $\mathfrak{P}$, this is a contradiction, and our first statement is proved.

In the case of a Dedekind domain R it is clear that the $\mathfrak{P}_i$ are the prime factors of $\mathfrak{p}R'$. The equality of the ramification indices e_i on the one hand, and of the relative degrees f_i on the other, is evident by automorphism. Then the inequality $efg \leq n$ follows from Theorem 21, and the equality $efg = n$ in the separable case follows from the corollary of Theorem 21. Q.E.D.

§ 10. Decomposition group, inertia group, and ramification groups. In this section R denotes a Dedekind domain, R' the integral closure of R in a finite, *normal and separable* extension L of the quotient field K of R, and G the Galois group of L over K (II, § 7). The notations are as in Theorem 22. Given a proper prime ideal $\mathfrak{p}$ of R and a prime ideal $\mathfrak{P}$ of R' lying over $\mathfrak{p}$, the automorphisms $s \in G$ such that $s(\mathfrak{P}) = \mathfrak{P}$ form obviously a subgroup G_Z of G; this subgroup is called the *decomposition group* of $\mathfrak{P}$. By Theorem 22 the order of G_Z is equal to (order of G)/g—that is, to ef. Given another prime ideal $\mathfrak{P}'$ of R' lying over $\mathfrak{p}$, we have, by Theorem 22, $\mathfrak{P}' = t(\mathfrak{P})$ with t in G, and the decomposition group of $\mathfrak{P}'$ is obviously $t^{-1} \cdot G_Z \cdot t$, therefore a *conjugate subgroup* of G_Z.

If G is *abelian* (in which case one says that L is an *abelian extension* of K), then the decomposition groups of the prime ideals of R' lying over $\mathfrak{p}$ are all equal. One then says that G_Z is the decomposition group *of* $\mathfrak{p}$.

The fixed field K_Z of G_Z is called the *decomposition field* of $\mathfrak{P}$. The field K_Z is an extension of K contained in L, and by Galois theory (II, § 7) L is a normal and separable extension of K_Z, with G_Z as Galois group. We thus have

(1) $$[K_Z:K] = g, \quad [L:K_Z] = ef.$$

If G_Z is an invariant subgroup of G (in particular, if L is an abelian extension of K), then K_Z is a normal and separable extension of K, admitting G/G_Z as Galois group.

THEOREM 23. *Let K_Z be the decomposition field of the prime ideal $\mathfrak{P}$ in R', R_Z the integral closure of R in K_Z, and $\mathfrak{P}_Z$ the contracted ideal $\mathfrak{P} \cap R_Z$. Then $\mathfrak{P}$ is the only prime ideal of R' lying over $\mathfrak{P}_Z$; its relative degree is f, and its reduced ramification index is e: $R'\mathfrak{P}_Z = \mathfrak{P}^e$. If the decomposition group G_Z of $\mathfrak{P}$ is an invariant subgroup of G, then K_Z is a normal and separable extension of K, and the factorization of the ideal $R_Z\mathfrak{p}$ in R_Z consists of g distinct and conjugate prime factors, all of them with relative degree* 1.

PROOF. By definition of G_Z, the conjugate prime ideals of $\mathfrak{P}$ over K_Z consist of $\mathfrak{P}$ only. Thus, by Theorem 22, $\mathfrak{P}$ is the only prime ideal of R' lying over $\mathfrak{P}_Z$. (Note that R_Z is integrally closed and that consequently Theorem 22 is applicable to the pair of rings R_Z, R'.) Therefore $R'\mathfrak{P}_Z$ is a power $\mathfrak{P}^{e(Z)}$ of $\mathfrak{P}$. Since $R/\mathfrak{p} \subset R_Z/\mathfrak{P}_Z \subset R'/\mathfrak{P}$, the relative degree $f(Z)$ of $\mathfrak{P}$ over $\mathfrak{P}_Z$ is a divisor of f. On the other hand, consider the factorization $R_Z\mathfrak{p} = \mathfrak{P}_Z{}^q \cdot \prod_j \mathfrak{Q}_j{}^{r(j)}$ of $R_Z\mathfrak{p}$ in R_Z. Since extension of ideals preserves products (IV, § 8), the factorizations $R'\mathfrak{P}_Z = \mathfrak{P}^{e(Z)}$, $R'\mathfrak{Q}_j = \prod_{i=1}^{h_j} \mathfrak{P}_{ji}{}^{s_j(i)}$ give the factorization

$$R'\mathfrak{p} = \mathfrak{P}^{qe(Z)} \prod_j \prod_{i=1}^{h_j} \mathfrak{P}_{ij}{}^{s_j(i)r(j)}.$$

Since $\mathfrak{P}$ has exponent e in the factorization of $R'\mathfrak{p}$, this implies $qe(Z) \leq e$. By Theorem 22 applied to $\mathfrak{P}_Z$, we have $e(Z)f(Z) = [L:K_Z] = ef$, and this together with the inequalities $qe(Z) \leq e$ and $f(Z) \leq f$, implies $e(Z) = e$, $f(Z) = f$, $q = 1$; thus our first assertion is proved. From this we deduce that 1 is the exponent of $\mathfrak{P}_Z$ in the factorization of $R_Z\mathfrak{p}$, and that $R/\mathfrak{p} = R_Z/\mathfrak{P}_Z$, that is, that the relative degree of $\mathfrak{P}_Z$ over $\mathfrak{p}$ is 1. The assertion relative to the case where G_Z is an invariant subgroup of G follows at once from this and from Theorem 22, if one takes into account the relation $[K_Z:K] = g$.

In other words, *if K_Z is normal over K* then the passage of K to K_Z involves only a decomposition of $\mathfrak{p}$ into distinct prime factors, without ramification and without increase of residue field; this is the reason for the names "decomposition group" and "decomposition field". The index Z is customary and is the initial of *Zerlegung*, the German word for "decomposition."

We now show that the phenomenon "extension of the residue field" may also be isolated in a further step of the field extension. Given a prime ideal $\mathfrak{P}$ of R' lying over $\mathfrak{p}$, the automorphisms $s \in G$ such that $s(x) \equiv x \pmod{\mathfrak{P}}$ for every x in R', form obviously a subgroup of G; this subgroup is called the *inertia group* of $\mathfrak{P}$, and is denoted by G_T. For x in $\mathfrak{P}$ and s in G_T, we have $s(x) \in x + \mathfrak{P}$, that is, $s(x) \in \mathfrak{P}$; this shows that $s(\mathfrak{P}) \subset \mathfrak{P}$, whence $s(\mathfrak{P}) = \mathfrak{P}$ [since we have, by a similar argument: $s^{-1}(\mathfrak{P}) \subset \mathfrak{P}$, that is, $\mathfrak{P} \subset s(\mathfrak{P})$]. Therefore *the inertia group G_T of $\mathfrak{P}$ is a subgroup of the decomposition group G_Z of $\mathfrak{P}$.* Furthermore, for s in G_T and t in G_Z, and for any x in R', we have $t^{-1}(x) \in R'$, whence $st^{-1}(x) - t^{-1}(x) \in \mathfrak{P}$, and $tst^{-1}(x) - x = t(st^{-1}(x) - t^{-1}(x)) \in t(\mathfrak{P}) = \mathfrak{P}$; therefore $tst^{-1} \in G_T$, and the inertia group G_T is an *invariant subgroup* of the decomposition group G_Z. The fixed field of G_T is called the *inertia field* of $\mathfrak{P}$, and is denoted by K_T. We have $K \subset K_Z \subset K_T \subset L$. By Galois theory (II, § 7), L is a normal and separable extension of K_T admitting G_T as Galois group, and K_T is a normal and separable extension of K_Z admitting G_Z/G_T as Galois group. (The index T is customary and is the initial of *Trägheit* (the German word for "inertia")).

THEOREM 24. *Let K_Z and K_T be the decomposition and the inertia fields of the prime ideal $\mathfrak{P}$, R_Z and R_T the integral closures of R in K_Z and K_T, $\mathfrak{P}_Z$ and $\mathfrak{P}_T$ the contracted ideals $\mathfrak{P} \cap R_Z$ and $\mathfrak{P} \cap R_T$. Then $R'/\mathfrak{P}$ is a normal extension of $R/\mathfrak{p}$, and its Galois group is isomorphic to G_Z/G_T. If $f = f_0 p^s$, where f_0 is the degree over $R/\mathfrak{p}$ of the maximal separable extension of $R/\mathfrak{p}$ in $R'/\mathfrak{P}$* (II, § 5, p. 71) *and where p is the characteristic of $R/\mathfrak{p}$, then K_T is a normal and separable extension, of degree f_0, of K_Z, and $\mathfrak{P}_T$ is the only prime ideal of R_T lying over $\mathfrak{P}_Z$; its relative degree is f_0 and its reduced ramification index is 1: $\mathfrak{P}_Z R_T = \mathfrak{P}_T$. We have $R/\mathfrak{p} = R_Z/\mathfrak{P}_Z$, and $R_T/\mathfrak{P}_T$ is the maximal separable extension of $R/\mathfrak{p}$ in $R'/\mathfrak{P}$. The field L is a normal and separable extension of K_T, of degree ep^s, and $\mathfrak{P}$ is the only prime ideal of R' lying over $\mathfrak{P}_T$; its relative degree (over $\mathfrak{P}_T$) is p^s, and its reduced ramification index is e: $\mathfrak{P}_T R' = \mathfrak{P}^e$.*

PROOF. Let s be an element of G_Z. Since $s(R') = R'$ and $s(\mathfrak{P}) = \mathfrak{P}$, the element s defines an automorphism $\bar{s}$ of $R'/\mathfrak{P}$ over $R/\mathfrak{p}$. By definition of G_T, $\bar{s}$ is the identity if and only if s belongs to G_T. Thus G_Z/G_T can be identified with a group of automorphisms of $R'/\mathfrak{P}$ over $R/\mathfrak{p}$. We will now investigate whether the extension $R'/\mathfrak{P}$ of $R/\mathfrak{p}$ is normal, and whether G_Z/G_T is its full Galois group.

Consider any element $\bar{x}$ of $R'/\mathfrak{P}$, and a representative $x \in R'$ of the residue class $\bar{x}$. The minimal polynomial of x over K, say $X^q + a_{q-1}X^{q-1} + \cdots + a_0$, has its coefficients in R (Theorem 4, § 3).

Since L is normal over K, this polynomial factors into linear factors in R':

$$X^q + a_{q-1}X^{q-1} + \cdots + a_0 = \prod_i (X - x_i)$$

with $x = x_1$ (II, § 6). If we denote by bars residue classes modulo $\mathfrak{P}$, the polynomial $X^q + \bar{a}_{q-1}X^{q-1} + \cdots + \bar{a}_0$ has its coefficients in $R/\mathfrak{p}$ and factors into linear factors $\prod_i (X - \bar{x}_i)$ in $R'/\mathfrak{P}$. Since this polynomial admits $\bar{x}$ as a root, it is a multiple of the minimal polynomial of $\bar{x}$ over $R/\mathfrak{p}$; thus this minimal polynomial has all its roots in $R'/\mathfrak{P}$. In other words, all the conjugates of $\bar{x}$ over $R/\mathfrak{p}$ are in $R'/\mathfrak{P}$. Therefore $R'/\mathfrak{P}$ is a *normal* extension of $R/\mathfrak{p}$.

In order to prove that G_Z/G_T is the Galois group of $R'/\mathfrak{P}$ (over $R/\mathfrak{p}$) it is sufficient to prove that it is the Galois group of the maximal separable extension S of $R/\mathfrak{p}$ in $R'/\mathfrak{P}$, that is, that every automorphism s' of S over $R/\mathfrak{p}$ comes from some element s of G_Z. If we take a primitive element $\bar{x}$ of S over $R/\mathfrak{p}$ (II, § 9, Theorem 19), the automorphism s' is completely determined if one knows which one of the conjugates of $\bar{x}$ is $s'(\bar{x})$. But the preceding reasoning (applied to K_Z, R_Z, $\mathfrak{P}_Z$ instead of to K, R, $\mathfrak{p}$), together with the equality $R_Z/\mathfrak{P}_Z = R/\mathfrak{p}$ established in the proof of the preceding theorem, shows that if we denote by x an element of R' whose $\mathfrak{P}$-residue is $\bar{x}$, and by x_i its conjugates over K_Z, then the conjugates of $\bar{x}$ over $R/\mathfrak{p}$ are among the $\mathfrak{P}$-residues $\bar{x}_i$ of the x_i. Thus, there exists an index j such that $s'(\bar{x}) = \bar{x}_j$. Since x_j is a conjugate of x over K_Z, there exists s in G_Z such that $x_j = s(x)$. Since the automorphism $\bar{s}$ of $R'/\mathfrak{P}$ determined by s is such that $\bar{s}(\bar{x}) = s'(\bar{x})$, s' and $\bar{s}$ coincide on S, whence also $s' = \bar{s}$ on $R'/\mathfrak{P}$, since $R'/\mathfrak{P}$ is a purely inseparable extension of S. Therefore G_Z/G_T is the Galois group of $R'/\mathfrak{P}$ over $R/\mathfrak{p}$.

From this we first deduce that the order of G_Z/G_T is equal to the degree f_0 of S over $R/\mathfrak{p}$ (II, § 7), and hence $[K_T:K_Z] = f_0$. We deduce also, by applying this result to K_T (instead of K) as ground field (in this case $G_Z = G_T = G$), that $R'/\mathfrak{P}$ is a purely inseparable extension of $R_T/\mathfrak{P}_T$ and that $S = R_T/\mathfrak{P}_T$. Thus the relative degree of $\mathfrak{P}_T$ over $\mathfrak{P}_Z$ is $f_0 = [S:R_Z/P_Z]$, and Theorem 21 shows that its reduced ramification index is 1.

On the other hand, from $[L:K_Z] = ef = ef_0p^s$ and from $[K_T:K_Z] = f_0$, we deduce that $[L:K_T] = ep^s$. Since $R_T/\mathfrak{P}_T = S$, the relative degree of $\mathfrak{P}$ over $\mathfrak{P}_T$ is p^s. Hence its reduced ramification index is e, by Theorem 21. This completes the proof of Theorem 24.

Corollary. *The assumptions and notations being as in Theorem* 24, *we suppose furthermore that* $R'/\mathfrak{P}$ *is a separable extension of* $R/\mathfrak{p}$. *Then* $R'/\mathfrak{P} = R_T/\mathfrak{P}_T$, $[L:K_T] = e$, *and the relative degree of* $\mathfrak{P}$ *over* $\mathfrak{P}_T$ is 1.

The proof of the corollary is immediate.

The conclusions of Theorem 24 may be summarized in Table I.

TABLE I

Fields	K		K_Z		K_T		L
Degrees		g		f_0		ep^s	
Prime ideals	$\mathfrak{p}$		$\mathfrak{P}_Z$		$\mathfrak{P}_T$		$\mathfrak{P}$
Relative degrees		1		f_0		p^s	
Reduced ramification indices		1		1		e	

If $R'/\mathfrak{P}$ is separable over $R/\mathfrak{p}$, we take, in this table, $f_0 = f$ and $p^s = 1$.

It may be shown by examples that $R'/\mathfrak{P}$ is not necessarily separable over $R/\mathfrak{p}$. But this separability condition is fulfilled in the important cases of the rings of algebraic integers, and of the rings of integral functions of one variable over a finite ground field: in fact, in these cases the residue fields are finite, and hence perfect (II, § 4, Theorem 5).

The integers f_0 and ep^s are respectively called the *reduced relative degree* and the *ramification index* of $\mathfrak{P}$ over $\mathfrak{p}$. A prime ideal $\mathfrak{P}$ of R' whose ramification index is > 1 is said to be *ramified*.

We can go farther than the inertia group G_T. For every integer $n \geq 1$, the set of all automorphisms s in G such that $s(x) \equiv x \pmod{\mathfrak{P}^n}$ for every x in R' is obviously a subgroup of G_T; this subgroup is called the n-th *ramification group* of $\mathfrak{P}$ over $\mathfrak{p}$, and is denoted by G_{V_n}. We have $G_{V_1} = G_T$. (The index V is customary. It stands for the initial of *Verzweigung*, the German word for "ramification.") It is clear that the subgroups G_{V_n} form a decreasing sequence of subgroups of G. Since $\bigcap_{n=0}^{\infty} \mathfrak{P}^n = (0)$, their intersection is reduced to the identity. Thus, since G is a finite group, there are only a finite number of distinct G_{V_n}, and G_{V_n} is reduced to the identity for n large enough. The indices n (finite in number) for which $G_{V_{n+1}} < G_{V_n}$ are called the *ramification numbers* of $\mathfrak{P}$ over $\mathfrak{p}$.

For any s in G_{V_n}, any t in G_Z and any x in R', we have $t^{-1}(x) \in R'$, whence $st^{-1}(x) - t^{-1}(x) \in \mathfrak{P}^n$. Since $t(\mathfrak{P}) = \mathfrak{P}$ and $t(\mathfrak{P}^n) = \mathfrak{P}^n$, this implies that $tst^{-1}(x) - x = t(st^{-1}(x) - t^{-1}(x)) \in \mathfrak{P}^n$. Therefore we have $tst^{-1} \in G_{V_n}$, and hence the ramification group G_{V_n} is an *invariant subgroup* of G_Z.

We now take s in G_{V_q} and t in G_{V_r}, and study the *commutator* $sts^{-1}t^{-1}$. We first consider $s(y) - y$ for y in $\mathfrak{P}^r$ and prove that $s(y) - y \in \mathfrak{P}^{q+r-1}$. It is sufficient to consider the case in which $y = x_1 \cdot x_2 \cdots x_r$ with x_j in

$\mathfrak{P}$, since every element in $\mathfrak{P}^r$ is a finite sum of such products. Then the element $s(y) - y = s(x_1 \cdots x_r) - x_1 \cdots x_r = \sum_{j=1}^{r} s(x_1) \cdots s(x_{j-1}) \cdot [s(x_j) - x_j] \cdot x_{j+1} \cdots x_r$ is an element of $\mathfrak{P}^{q+r-1}$ since $s(x_i) \in \mathfrak{P}$, $x_k \in \mathfrak{P}$, and $s(x_j) - x_j \in \mathfrak{P}^q$. Similarly, we have $t(z) - z \in \mathfrak{P}^{q+r-1}$ for z in $\mathfrak{P}^q$. This being so, we take any x in R' and set $y = t(x) - x$ and $z = s(x) - x$. Since $y \in \mathfrak{P}^r$ and $z \in \mathfrak{P}^q$, we have

$$s(y) - y = st(x) - s(x) - t(x) + x \in \mathfrak{P}^{q+r-1},$$
$$t(z) - z = ts(x) - t(x) - s(x) + x \in \mathfrak{P}^{q+r-1};$$

whence, by subtraction, $st(x) - ts(x) \in \mathfrak{P}^{q+r-1}$. Replacing x by $s^{-1}t^{-1}(x)$, we get $sts^{-1}t^{-1}(x) - x \in \mathfrak{P}^{q+r-1}$. We have thus proved:

LEMMA 1. *The commutator $sts^{-1}t^{-1}$ of elements s of G_{V_q} and t of G_{V_r} belongs to $G_{V_{q+r-1}}$. In particular it follows (in the case $q = r = n$) that the factor groups $G_{V_n}/G_{V_{2n-1}}$ are abelian, and that consequently also the groups $G_{V_n}/G_{V_{n+1}}$, for $n \geq 2$, are abelian.*

More precise results about the structure of these last factor groups can be given:

THEOREM 25. *The groups $G_1 = G_T/G_{V_2}$ and $G_n = G_{V_n}/G_{V_{n+1}}$ ($n \geq 2$) contain invariant subgroups G'_1 and G'_n whose orders are powers of the characteristic p of $R/\mathfrak{p}$. The factor group G_1/G'_1 is isomorphic with a multiplicative subgroup of $R'/\mathfrak{P}$, and is therefore cyclic. The factor groups G_n/G'_n ($n \geqq 2$) are isomorphic with additive subgroups of $R'/\mathfrak{P}$. The subgroups G'_1, G'_n are reduced to the identity if $R'/\mathfrak{P}$ is separable over $R/\mathfrak{p}$.*

PROOF. We can replace R and R' by the quotient rings R_M and R'_M (M = complement of $\mathfrak{p}$ in R) without changing anything. In other words, we may suppose that $\mathfrak{P}$ is a principal ideal (Theorem 16 of § 7). Let u be a generator of this ideal.

For s in G_T, we have $s(u) \in \mathfrak{P}$; furthermore $s(u) \notin \mathfrak{P}^2$, for in the contrary case $u = s^{-1}(s(u))$ would be in $\mathfrak{P}^2$. We may thus write $s(u) = x_s u$, with x_s in R', x_s not in $\mathfrak{P}$. For t in G_T, we have $st(u) = s(x_t u) = s(x_t)x_s u$, whence $x_{st} = s(x_t)x_s$. Since s is in G_T, we have $s(x_t) \equiv x_t \pmod{\mathfrak{P}}$, whence $x_{st} \equiv x_t x_s \pmod{\mathfrak{P}}$, and by passage to the residue classes mod $\mathfrak{P}$, we find $\bar{x}_{st} = \bar{x}_s \bar{x}_t$. Therefore the mapping $s \to \bar{x}_s$ is a homomorphism of G_T into the multiplicative subgroup of $R'/\mathfrak{P}$. Its kernel is the group H_1 of all automorphisms s in G_T such that $x_s \equiv 1 \pmod{\mathfrak{P}}$, that is, such that $s(u) - u \in \mathfrak{P}^2$.

Similarly, for s in G_{V_n} ($n \geq 2$), we have $s(u) - u \in \mathfrak{P}^n$, and we can write $s(u) - u = y_s u^n$ with y_s in R'. For t in G_{V_n}, we have $y_{st}u^n =$

$st(u) - u = s(y_t u^n + u) - u = s(y_t)s(u^n) + s(u) - u = s(y_t)(u + y_s u^n)^n + y_s u^n$. Dividing by u^n we get $y_{st} = s(y_t)(1 + y_s u^{n-1})^n + y_s$, since $n \geq 2$. Since $s(y_t) \equiv y_t \pmod{\mathfrak{P}^n}$, and since all the terms of the expansion of $(1 + y_s u^{n-1})^n$ are in $\mathfrak{P}$ except the first one, it follows that $y_{st} \equiv y_t + y_s \pmod{\mathfrak{P}}$. By passage to the residue classes mod $\mathfrak{P}$, we thus have $\bar{y}_{st} = \bar{y}_s + \bar{y}_t$, and the mapping $s \to \bar{y}_s$ is a homomorphism of G_{V_n} into the additive group of $R'/\mathfrak{P}$. Its kernel H_n is the set of all s in G_{V_n} such that $y_s \equiv 0 \pmod{\mathfrak{P}}$, that is, such that $s(u) - u \in \mathfrak{P}^{n+1}$.

The kernels H_1, H_n contain G_{V_2} and $G_{V_{n+1}}$ respectively, and we intend to compare these groups. Suppose that s is an element of G_{V_n} $(n \geq 1)$ such that $s(u) - u \in \mathfrak{P}^{n+1}$. Then we have $s(z) - z \in \mathfrak{P}^{n+1}$ for all z *in* $\mathfrak{P}$: in fact, we have $z = au$ $(a \in R')$, whence $s(z) - z = s(au) - au = (s(a) - a)u + s(a)(s(u) - u)$; here we have $s(a) - a \in \mathfrak{P}^n$ (since $s \in G_{V_n}$), $u \in \mathfrak{P}$, and $s(u) - u \in \mathfrak{P}^{n+1}$, and our assertion is proved. We now take any x in R' (not necessarily in $\mathfrak{P}$), and write $s^p(x) - x = s^{p-1}(s(x) - x) + s^{p-2}(s(x) - x) + \cdots + s(s(x) - x) + (s(x) - x)$. Here $s(x) - x$ is an element z of $\mathfrak{P}^n$ and, *a fortiori*, of $\mathfrak{P}$. From what has been proved above, we know that z is congruent modulo $\mathfrak{P}^{n+1}$ to $s(z)$, whence z is also congruent modulo $\mathfrak{P}^{n+1}$ to each of the terms $s^2(z), \cdots, s^{p-1}(z)$ of the above sum $s^p(x) - x$. Hence $s^p(x) - x \equiv pz \pmod{\mathfrak{P}^{n+1}}$. We have $p \cdot 1 \in \mathfrak{P}$ in R', since p is the characteristic of $R'/\mathfrak{P}$, and we also have $z \in \mathfrak{P}^n$. Therefore $s^p(x) - x \in \mathfrak{P}^{n+1}$, whence $s^p \in G_{V_{n+1}}$. In other words, in the factor group $G_{V_n}/G_{V_{n+1}}$ $(n \geq 1)$, all the elements of the subgroup $G'_n = H_n/G_{V_{n+1}}$ are *of order* p. Thus the order of G'_n is a power of p. From what has been seen above, the factor group G_T/H'_1 $(= G_{V_1}/H'_1)$ is isomorphic with a multiplicative subgroup of $R'/\mathfrak{P}$, and the factor group G_{V_n}/H_n $(n \geq 2)$ is isomorphic with an additive subgroup of $R'/\mathfrak{P}$. This proves our assertion in the general case.

In the case where $R'/\mathfrak{P}$ is separable over $R/\mathfrak{p}$, it remains to be proved that we have $H_n = G_{V_{n+1}}$ for every $n \geq 1$, that is, that the relation $s \in H_n$ implies $s(x) - x \in \mathfrak{P}^{n+1}$ for every x in R'. We already know that this is true if x is in $\mathfrak{P}$. But, in the separable case, the fields $R'/\mathfrak{P}$ and $R_T/\mathfrak{P}_T$ are equal (Corollary to Theorem 24). Hence any element x of R' may be written in the form $x = y + z$, with $y \in R_T$ and $z \in \mathfrak{P}$. Then $s(x) - x = s(y) - y + s(z) - z$ is in $\mathfrak{P}^{n+1}$, since $s(y) = y$ (s being in G_T) and since $s(z) - z \in \mathfrak{P}^{n+1}$. This completes the proof.

COROLLARY. *If $R'/\mathfrak{P}$ is a field of characteristic* 0, *then G_{V_n} is reduced to the identity for $n \geq 2$.*

In fact we are in the separable case, whence G'_n is reduced to the identity. Since (0) is the only finite additive subgroup of $R'/\mathfrak{P}$, we have $G_{V_n} = G_{V_{n+1}}$, and from this the corollary follows, since the intersection of the groups $G_{V_{n+q}}$ is reduced to the identity.

REMARKS.

1) In the inseparable case, the p-groups G'_n are *abelian* for $n \geq 2$, since they are subgroups of the *abelian* group $G_{V_n}/G_{V_{n+1}}$. We now prove that G'_1 is also *abelian*. Any element of G'_1 is the (G_{V_2})-residue of an element s of G_T such that $s(u) - u \in \mathfrak{P}^2$. We have seen in the proof of Theorem 25 that we then have $s(z) - z \in \mathfrak{P}^2$ for every z in $\mathfrak{P}$. Furthermore, since any element y of $\mathfrak{P}^2$ may be written in the form $y = zz'$ with z and z' in $\mathfrak{P}$, and since $s(y) - y = s(z)(s(z') - z') + z'(s(z) - z)$, we have $s(y) - y \in \mathfrak{P}^3$ for any y in $\mathfrak{P}^2$. Consider now two elements s and t of G_T such that $s(u) - u$ and $t(u) - u$ lie in $\mathfrak{P}^2$. Then, for any z in $\mathfrak{P}$, the difference $st(z) - ts(z)$ is the difference of the two elements $s(t(z) - z) - (t(z) - z)$ and $t(s(z) - z) - (s(z) - z)$; since $y = t(z) - z$ is in $\mathfrak{P}^2$, the first element $s(y) - y$ is in $\mathfrak{P}^3$, and similarly the second also. Thus $st(z) - ts(z) \in \mathfrak{P}^3$ for all z in $\mathfrak{P}$, whence $sts^{-1}t^{-1}(z) - z \in \mathfrak{P}^3$ for all z in $\mathfrak{P}$. Let us denote by c the commutator $sts^{-1}t^{-1}$. We need only to prove that c is in G_{V_2}, that is, that $c(x) - x \in \mathfrak{P}^2$ for every x in R'. This is already true for x in $\mathfrak{P}$. If x is not in $\mathfrak{P}$, it is a unit in R'. (We recall that we have replaced R' by a quotient ring having only one prime ideal.) We may write $x = z'/z$, with z and z' in $\mathfrak{P}$ but not in $\mathfrak{P}^2$. Then $c(x) - x$ is equal to $(z(c(z') - z') - z'(c(z) - z))/zc(z)$. The numerator is in $\mathfrak{P}^4$, and the denominator is in $\mathfrak{P}^2$ but not in $\mathfrak{P}^3$. Therefore $c(x) - x$ is in $\mathfrak{P}^2$, and our assertion is proved.*

2) The homomorphism $s \to \bar{x}_s$ of G_T into the multiplicative group of $R'/\mathfrak{P}$ defined in the proof of Theorem 25 is *independent* of the choice of the generator u of $\mathfrak{P}$. In fact, any other generator u' of $\mathfrak{P}$ may be written in the form $u' = au$, where a is a unit in R'. If we set $s(u) = x_s u$ and $s(u') = x'_s u'$, an easy computation shows that $x'_s = x_s \cdot s(a) \cdot a^{-1}$. Since $s(a) - a \in \mathfrak{P}$, we have $s(a)a^{-1} \equiv 1 \pmod{\mathfrak{P}}$, whence $x'_s \equiv x_s \pmod{\mathfrak{P}}$, and $\bar{x} = \overline{x'}_s$.

3) On the contrary, the homomorphism $s \to \bar{y}_s$ of G_{V_n} $(n \geq 2)$ into the additive group of $R'/\mathfrak{P}$, is only determined modulo a multiplication in $R'/\mathfrak{P}$ if one changes the generator u of $\mathfrak{P}$. In fact, taking another generator $u' = au$ of $\mathfrak{P}$ (a, a unit), and setting $s(u) - u = y_s u^n$ and $s(u') - u' = y'_s u'^n$, an easy computation shows that

$$y'_s = (s(a)-a)\cdot u^{-(n-1)} + y_s \cdot s(a) \cdot a^{-n}.$$

The first term is in $\mathfrak{P}$, as $s(a) - a \in \mathfrak{P}^n$. Thus $\bar{y}'_s = \bar{b} \cdot \bar{y}_s$, where

* It can be shown by examples that $G_T/G_{V_2} = G_1$ need not be abelian.

b is the $\mathfrak{P}$-residue of $s(a)a^{-n}$, that is, of $a^{-(n-1)}$, since $s(a) \equiv a \pmod{\mathfrak{P}}$.

§ **11. Different and discriminant.** Let R be an integrally closed ring, K its quotient field, K' a finite algebraic *separable* extension of K, and R' an integral extension of R admitting K' as quotient field. We denote by T, or $T_{K'/K}$, the mapping of K' into K defined by the *trace* (II, § 10). The set $\mathscr{C}$ of all z in K' such that $T(zR') \subset R$ is obviously an R'*-module*; it is called the *complementary module* of R' with respect to R. Since R is integrally closed, the trace of any element of R' lies in R (§ 3, Theorem 4), and the complementary module $\mathscr{C}$ contains R'.

THEOREM 26. *The complementary module $\mathscr{C}$ of R' with respect to R is a fractionary ideal of R'.*

PROOF. By definition of a fractionary ideal (§ 6, p. 271), we need only show that $\mathscr{C}$ is contained in a finite R'-module, and for this it will be sufficient to see that it is contained in a finite R-module. Since $K' = R'_R$ (see proof of Theorem 7, § 4), there exists a K-basis $\{e_1, \cdots, e_n\}$ of K' all the elements of which are in R'. We take an element z of $\mathscr{C}$, and write $z = \sum_i a_i e_i$ $(a_i \in K)$. We have $T(ze_i) = \sum_j a_j T(e_j e_i) \in R$ for $i = 1, \cdots, n$. As K' is separable over K, the determinant $d = \det(T(e_j e_i))$ is different from 0 (II, § 11). By the usual computation leading to Cramer's rule, and since $T(e_j e_i) \in R$, we get $da_i \in R$ for $i = 1, \cdots, n$, whence $\mathscr{C} \subset \sum_i (e_i/d)R$. Q.E.D.

The *different* of R' over R is the set of all elements x in K' such that $zx \in R'$ whenever $T(zR') \subset R$. In other words, the different is the ideal $(R' : \mathscr{C})$. Since $\mathscr{C}$ is a fractionary ideal containing R', we have:

COROLLARY. *The different of R' over R is an ideal $\neq (0)$ contained in R'.*

The different of R' over R is denoted by $\mathfrak{D}$, or by $\mathfrak{D}_{R'/R}$ (or by $\mathfrak{D}_{K'/K}$, whenever it is clear from the context which rings play the role of R and R').

In the case where R is a *Dedekind domain* and if one takes for R' its integral closure in K', R' is also a Dedekind domain, and the different $\mathfrak{D}_{R'/R}$ may be factored into prime ideals:

$$\mathfrak{D}_{R'/R} = \prod_{\mathfrak{P}} \mathfrak{P}^{m(\mathfrak{P})}. \tag{1}$$

The exponent $m(\mathfrak{P})$ is called the *differential exponent* of $\mathfrak{P}$ over R. It is positive or zero. The prime ideals $\mathfrak{P}$ for which $m(\mathfrak{P}) \neq 0$ are finite in number. We show that the different $\mathfrak{D}_{R'/R}$ is determined by "local data"; more precisely:

THEOREM 27. *Let R be a Dedekind domain, and R' its integral closure in a finite algebraic separable extension K' of the quotient field K of R. Let $\mathfrak{p}$ be a proper prime ideal in R, M the complement of $\mathfrak{p}$ in R, $\mathfrak{P}$ a prime ideal in R' lying over $\mathfrak{p}$, and $m(\mathfrak{P})$ its differential exponent. Then the differential exponent of $\mathfrak{P}R'_M$ over R_M is $m(\mathfrak{P})$.*

PROOF. Under an equivalent form, our assertion is to the effect that the different of R'_M over R_M is the ideal $\prod_i \mathfrak{P}_i^{m(\mathfrak{P}_i)}R'_M$, (where the $\mathfrak{P}_i$ denote the prime ideals of R' lying over $\mathfrak{p}$), that is, the ideal $\mathfrak{D}_{R'/R}R'_M$. We take any element x of $\mathfrak{D}_{R'/R}R'_M$: $x = x'/m$ with x' in $\mathfrak{D}$ and m in M. If z is an element of the complementary module of R'_M (with respect to R_M), we have $T(zR'_M) \subset R_M$; in particular, for any r' in R', the element $T(zr')$ may be written in the form r/m', with r in R and m' in M, whence $T(m'zr') = r$ since $T(m'zr') = m'T(zr')$ as $m' \in K$. Since R' is a finite R-module (Corollary 1 to Theorem 7 of § 4), a common denominator m' in M may be found for all elements $T(zr')$ $(r' \in R')$, and if m'_0 is such a common denominator then we have $T(m'_0zR') \subset R$. Therefore m'_0z is an element of the complementary module of R', whence $x'm'_0z \in R'$ since $x' \in \mathfrak{D}$. We thus have $xz = x'm'_0z/mm'_0 \in R'_M$, and this proves that x belongs to the different of R'_M over R_M. In other words, $\mathfrak{D}_{R'/R}R'_M \subset \mathfrak{D}_{R'_M/R_M}$.

Conversely, we show that every element x of $\mathfrak{D}_{R'_M/R_M}$ is in $\mathfrak{D}_{R'/R}R'_M$. We take an element z such that $T(zR') \subset R$, and we study zx. Since M is contained in K, we have $T(zR'_M) \subset R_M$, and z is in the complementary module of R'_M. Thus, by definition of the different $\mathfrak{D}_{R'_M/R_M}$, we have $zx \in R'_M$, and there exists an element $m(z)$ of M such that $zm(z)x \in R$ for every z in the complementary module $\mathscr{C}$ of R'. Since $\mathscr{C}$ is a finite R-module, we may suppose that $m(z)$ is an element m of M independent of z. Then, from $z(mx) \in R'$ for every z in $\mathscr{C}$, we deduce that $mx \in \mathfrak{D}_{R'/R}$, whence $x \in \mathfrak{D}_{R'/R}R'_M$. Q.E.D.

Before proving an important relation between reduced ramification indices and differential exponents, we need a useful formula about traces:

LEMMA 1. *Let R be a Dedekind domain, K its quotient field, K' a finite algebraic separable extension of K, and R' the integral closure of R in K'. Let $\mathfrak{p}$ be a proper prime ideal in R, and $\{\mathfrak{P}_i\}$ the finite family of prime ideals of R' lying over $\mathfrak{p}$. Denote by e_i the reduced ramification index of $\mathfrak{P}_i$ over $\mathfrak{p}$, by k the residue field $R/\mathfrak{p}$, by h the canonical homomorphism of R onto k, by k_i the residue field $R'/\mathfrak{P}_i$, and by h_i the canonical homomorphism of R' onto k_i. Then we have, for x in R',*

$$h(T_{K'/K}(x)) = \sum_i e_i \cdot T_{k_i/k}(h_i(x)),$$

$$h(N_{K'/K}(x)) = \prod_i (N_{k_i/k}(h_i(x)))^{e_i}.$$

PROOF. Let $f(X)$ be the field polynomial of x, relative to K (see II, § 10, p. 87):

$$f(X) = X^n + a_1X^{n-1} + \cdots + a_n, \quad n = [K':K].$$

Since x is integral over R and R is integrally closed in K, the coefficients of the minimal polynomial of x over K belong to R (§ 3), and as $f(X)$ is a power of this minimal polynomial (see II, § 10, relation (10)) it follows that also the a_i belong to R. Let $\bar{a}_i = h(a_i), \bar{f}(X) = X^n + \bar{a}_1X^{n-1} + \cdots + \bar{a}_n$. We have $T_{K'/K}(x) = -a_1$, $N_{K'/K}(x) = (-1)^n a_n$, and hence

$$h(T_{K'/K}(x)) = -\bar{a}_1,$$
$$h(N_{K'/K}(x)) = (-1)^n \bar{a}_n.$$

Let $f_i(X)$ be the field polynomial of $h_i(x)$, relative to k, $h_i(x)$ being regarded as an element of k_i. To prove the lemma we shall prove the following stronger result:

$$\bar{f}(X) = \prod_{i=1}^{g} [f_i(X)]^{e_i}.$$

We recall the definition of the field polynomial $f(X)$. The mapping M: $z \to zx$, $z \in K'$, of K' into itself is a K'-linear additive transformation. If K' is regarded as a vector space over K, then M is also a linear transformation of K'/K into itself, *and the field polynomial $f(X)$ is the characteristic polynomial of M.*

The ring $R'/R'\mathfrak{p}$ is a vector space over the field $k = R/\mathfrak{p}$ (*of dimension $n = \sum e_i f_i$*; see Theorem 21, § 9). The transition from R' to $R'/R'\mathfrak{p}$ leads from M to a k-linear transformation $\bar{M}$ of the vector space $R'/R'\mathfrak{p}$, defined as follows: if $\bar{z}$ is the $R'\mathfrak{p}$ residue of an element z of R' then $\bar{M}(\bar{z}) = R'\mathfrak{p}$-residue of $M(z)$ (note that since $x \in R'$, the element $M(z) = zx$ belongs to R' if z belongs to R'). If $\bar{x}$ is the $R'\mathfrak{p}$-residue of x then we have for any element $\bar{z}$ of $R'/R'\mathfrak{p}$: $\bar{M}(\bar{z}) = \overline{zx}$. In the proof one may replace R by R_p; when that is done, then R' will have an R-basis consisting precisely on n elements (Corollary 2, p. 265). If $\{z_1, z_2, \cdots, z_n\}$ is an R-basis of R', then $\{z_1, z_2, \cdots, z_n\}$ is a vector basis of K'/K, and if $\bar{z}_\nu$ denotes the $R'\mathfrak{p}$-residue of z_ν, then $\{\bar{z}_1, \bar{z}_2, \cdots, \bar{z}_n\}$ is a vector basis of $R'/R'\mathfrak{p}$. If $M(z_\nu) = \sum c_{\nu\mu} z_\mu$, $c_{\nu\mu} \in K$, then the $c_{\nu\mu}$ belong to R (since $M(R') \subset R' = \sum R z_\nu$) and we have $f(X) = |\delta_{\nu\mu} X - c_{\nu\mu}|$. On the other hand, we aiso have $\bar{M}(\bar{z}_\nu) = \sum \bar{c}_{\nu\mu} \bar{z}_\mu$, where $\bar{c}_{\nu\mu} = \mathfrak{p}$-residue of $c_{\nu\mu}$ ($= R'\mathfrak{p}$-residue of $c_{\nu\mu}$), and $\bar{f}(X) = |\delta_{\nu\mu}X - \bar{c}_{\nu\mu}|$. It follows that $\bar{f}(X)$ *is the characteristic polynomial of the linear transformation* $\bar{M}$, and the lemma will be proved if we show that *the characteristic polynomial of* $\bar{M}$ *is equal to the power product*

$$\prod_{i=1}^{g} [\bar{f}_i(X)]^{e_i}$$

of the field polynomials of the elements $h_i(x)$.

The vector space $R'/R'\mathfrak{p}$ is the direct sum $S_1 + S_2 + \cdots + S_g$ of the subspaces $S_i = (\bigcap_{j \neq i} \mathfrak{P}_j^{e_j})/R'\mathfrak{p}$, and each of these subspaces S_i is an invariant space of $\bar{M}$ (since each prime ideal $\mathfrak{P}_i$ is invariant under M). Hence if $\bar{M}_i$ denotes the restriction of $\bar{M}$ to S_i, then $\bar{M}$ is the direct sum $\bar{M}_1 + \bar{M}_2 + \cdots + \bar{M}_g$ of the linear transformations $\bar{M}_i$, and the characteristic polynomial of $\bar{M}$ is the product of the characteristic polynomials of the $\bar{M}_i$. Hence, in order to prove our lemma, it will be sufficient to show that *the characteristic polynomial of* $\bar{M}_i$ *is equal to the* e_i*-th power* $(\bar{f}_i(X))^{e_i}$ *of the field polynomial* $\bar{f}_i(X)$ *of* $h_i(x)$.

If $\bar{x}_1 + \bar{x}_2 + \cdots + \bar{x}_g$ is the direct decomposition of $\bar{x}$ ($\bar{x} = R'\mathfrak{p}$-residue of x; $\bar{x}_i \in S_i$), then it is clear that for any $\bar{z}_i$ in S_i we have $\bar{M}_i(\bar{z}_i) = \bar{z}_i\bar{x}_i$ (since $\bar{M}_i(\bar{z}_i) = \bar{M}(\bar{z}_i) = \bar{z}_i\bar{x} = \bar{z}_i\bar{x}_i$). We now replace the ring S_i by the canonically isomorphic replica $L_i = R'/\mathfrak{P}_i^{e_i}$. The canonical isomorphism φ_i of S_i onto L_i is as follows: if $\bar{z} = \bar{z}_1 + \bar{z}_2 + \cdots + \bar{z}_g$ ($\bar{z}_i \in S_i$) is the $R'\mathfrak{p}$-residue of an element z of R', then $\varphi_i(\bar{z}_i) = \mathfrak{P}_i^{e_i}$-residue of z (see II, § 13, Theorem 32, relation (14)). Also L_i is a vector space over $k = R/\mathfrak{p}$, and the isomorphism φ_i is k-linear. We may now replace $\bar{M}_i$ by the similar linear transformation M'_i, of L_i into itself, where $M'_i = \varphi_i^{-1}M_i\varphi_i$, since $\bar{M}_i$ and M'_i have the same characteristic polynomial. It is clear that M'_i is the transformation which carries every element ζ of $R'/\mathfrak{P}_i^{e_i}$ into the element $\zeta\tilde{x}$, where $\tilde{x}$ denotes the $\mathfrak{P}_i^{e_i}$-residue of x.

We denote by L_{ij} the subspace $\mathfrak{P}_i^j/\mathfrak{P}_i^{e_i}$, $0 \leq j \leq e_i$ of L_i ($L_{i0} = L_i$). The L_{ij} are invariant spaces of M'_i and they form a descending chain: $L_i = L_{i0} > L_{i1} > \cdots > L_{i,e_i-1} > L_{i,e_i} = 0$. For each j, the linear transformation M'_i determines in a natural fashion a linear transformation M'_{ij} in the factor space $L_{ij}/L_{i,j+1}$ ($0 \leq j \leq e_{i-1}$); M'_{ij} sends each coset $u + L_{i,j+1}$ ($u \in L_{ij}$) into the coset $M'_i(u) + L_{i,j+1}$. If we choose as basis of L_i a set of elements u_{j,q_j} ($j = 0, 1, \cdots, e_{i-1}$; $q_j = 1, 2, \cdots$, dim $L_{ij}/L_{i,j+1}$) such that, for each j, the cosets $u_{j,q_j} + L_{i,j+1}$ form a basis of $L_{ij}/L_{i,j+1}$, and use this basis for the purpose of finding the characteristic polynomial of M'_i, we see at once that this polynomial is equal to the product of the characteristic polynomials of the e_i linear transformations $M'_{i0}, M'_{i1}, \cdots, M'_{i,e_i-1}$. *We shall now complete the proof of the lemma by showing that the characteristic polynomial of each M'_{ij} is equal to the field polynomial of $h_i(x)$.*

The factor space $L_{ij}/L_{i,j+1}$ is canonically isomorphic to $\mathfrak{P}_i^j/\mathfrak{P}_i^{j+1}$, and we may identify $M'_{i,j}$ with the linear transformation in $\mathfrak{P}_i^j/\mathfrak{P}_i^{j+1}$ which carries every element u of this ring into the element $u\tilde{x}_j$, where $\tilde{x}_j$ denotes the $\mathfrak{P}_i^{j+1}$-residue of x. On the other hand, the vector space $\mathfrak{P}_i^j/\mathfrak{P}_i^{j+1}$ (over k) is isomorphic with $R'/\mathfrak{P}_i$, and a particular isomorphism ψ between these spaces is obtained as follows:

We fix an element u_0 in $\mathfrak{P}_i^j$, not in $\mathfrak{P}_i^{j+1}$. Then $\mathfrak{P}_i^{j+1} < \mathfrak{P}_i^{j+1} + R'u_0 \subset \mathfrak{P}_i^j$, and hence $\mathfrak{P}_i^{j+1} + R'u_0 = \mathfrak{P}_i^j$ (since there are no ideals between $\mathfrak{P}_i^j$ and $\mathfrak{P}_i^{j+1}$). Consequently, if $u \in \mathfrak{P}_i^j$, we can write $u \equiv u'u_0 \pmod{\mathfrak{P}_i^{j+1}}$, $u' \in R'$, and it is clear that the element u' is uniquely determined mod $\mathfrak{P}_i$ by the element u. The mapping ψ: $\mathfrak{P}_i^{j+1}$-residue of $u \to \mathfrak{P}_i$-residue of u', is a k-linear isomorphism of $\mathfrak{P}_i^j/\mathfrak{P}_i^{j+1}$ onto $R'/\mathfrak{P}_i$. The ψ-transform $\psi^{-1}M'_{ij}\psi$ of M'_{ij} is the linear transformation in $R'/\mathfrak{P}_i$ given by the multiplication $z' \to z'h_i(x)$. Consequently, the characteristic polynomial of $\psi^{-1}M'_{ij}\psi$, and hence also of M'_{ij}, is precisely the field polynomial of the element $h_i(x)$ of the field $k_i = R'/\mathfrak{P}_i$, relative to the ground field $k = R/\mathfrak{p}$. This completes the proof of the lemma.

THEOREM 28. *The notations and hypotheses being as in Theorem 27, and $e(\mathfrak{P})$ denoting the reduced ramification index of $\mathfrak{P}$ over $\mathfrak{p}$, we have the inequality $m(\mathfrak{P}) \geq e(\mathfrak{P}) - 1$. In this formula the equality holds if and only if*

a) *$e(\mathfrak{P})$ is not a multiple of the characteristic of $R/\mathfrak{p}$, and*
b) *$R'/\mathfrak{P}$ is separable over $R/\mathfrak{p}$.*

PROOF. By Theorem 27 we may assume that $\mathfrak{p}$ is the only proper prime ideal in R. Then $\mathfrak{p}$ is a principal ideal $\mathfrak{p} = Ru$, and R' has only a finite number of proper prime ideals $\mathfrak{P}_i$, all of them lying over $\mathfrak{p}$; we set $\mathfrak{P} = \mathfrak{P}_1$, $m_i = m(\mathfrak{P}_i)$. As the complementary module $\mathscr{C}$ of R' is $\prod_i \mathfrak{P}_i^{-m_i}$, the inequality $m_i \geq e_i - 1$ will be proved if we show that $\prod_i \mathfrak{P}_i^{1-e_i} \subset \mathscr{C}$. Let then z be any element in $\prod_i \mathfrak{P}_i^{1-e_i}$. As $R'u = \prod_i \mathfrak{P}_i^{e_i}$, we have $zu \in \prod_i \mathfrak{P}_i$, whence $zu \in \mathfrak{P}_i$ for every i. From this conclusion that zu belongs to all the prime ideals lying over $\mathfrak{p}$ it follows at once that the same conclusion remains valid if K' is replaced by the least normal extension of K containing K' and zu is replaced by any of its conjugates over K. Therefore $T_{K'/K}(zu)$, which is a sum of such conjugates, is an element of $\mathfrak{p}$, and we have $uT(z) = T(zu) \in Ru$, whence $T(z) \in R$. As this is true for every element z of $\prod_i \mathfrak{P}_i^{1-e_i}$, we also have $T(zR') \subset R$ for every element z of $\prod_i \mathfrak{P}_i^{1-e_i}$ (since $\prod_i \mathfrak{P}_i^{1-e_i}$ is an R-module). Hence the inclusion $\prod_i \mathfrak{P}_i^{1-e_i} \subset \mathscr{C}$ is proved.

If a) and b) are fulfilled, there exists (by b)) an element $\bar{y}$ of $R'/\mathfrak{P}$ whose trace in $R/\mathfrak{p}$ is not zero. Since the ideals $\mathfrak{P}_i^{e_i}(i \geqq 2)$, $\mathfrak{P}_1$ are pairwise comaximal, there exists a representative y of $\bar{y}$ in R' such that $y \in \mathfrak{P}_i^{e_i}$ for $i \geq 2$. Then, by Lemma 1, the $\mathfrak{p}$-residue of $T_{K'/K}(y)$ is $e(\mathfrak{P}) \cdot T_{(R'/\mathfrak{P})/(R/\mathfrak{p})}(\bar{y})$; it is $\neq 0$ by a) and by the choice of $\bar{y}$. Therefore the element y/u, which belongs to $\mathfrak{P}^{-e(\mathfrak{P})}$, admits a trace $T(y)/u$ which is not in R. Thus $y/u \notin \mathscr{C}$, and $m(\mathfrak{P}) = e(\mathfrak{P}) - 1$.

Conversely, suppose that either a) or b) is not true. Take then any element z of the fractional ideal $\mathfrak{B}' = \mathfrak{P}^{-e(\mathfrak{P})} \cdot \prod_{i \neq 1} \mathfrak{P}_i^{1-e_i}$. The element zu is in $\mathfrak{P}_i$ for $i \neq 1$. Then, by Lemma 1, the $\mathfrak{p}$-residue of the trace of zu is $e(\mathfrak{P}) \cdot T_{(R'/\mathfrak{P})/(R/\mathfrak{p})}(\bar{z}\bar{u})$. Under either hypothesis this $\mathfrak{p}$-residue is zero, trivially if a) is false, by the Corollary to Theorem 20 of II, § 10, if b) is false. Hence $uT(z)(= T(uz))$ is an element of the ideal Ru $(= \mathfrak{p})$, and consequently $T(z)$ is in R. Since this holds for every element z of the R'-module $\mathfrak{B}'$ it follows that $\mathfrak{B}' \subset \mathscr{C}$, and consequently $m(\mathfrak{P}) \geq e(\mathfrak{P})$. Q.E.D.

COROLLARY. *The ramified prime ideals in R' are those which divide the different $\mathfrak{D}_{R'/R}$.*

In fact, if $\mathfrak{P}$ divides the different, we have $m(\mathfrak{P}) \geqq 1$. This implies $e(\mathfrak{P}) > 1$, and thus that $\mathfrak{P}$ is ramified, unless either a) or b) is false. In the first case $e(\mathfrak{P})$ is a multiple of the characteristic of $R/\mathfrak{p}$, and is therefore > 1. In the second case $R'/\mathfrak{P}$ is inseparable over $R/\mathfrak{p}$, and we made the convention to call $\mathfrak{P}$ ramified in that case. Conversely, if $\mathfrak{P}$ is ramified, we have either $e(\mathfrak{P}) > 1$, or $R'/\mathfrak{P}$ is inseparable over $R/\mathfrak{p}$; in either case this implies $m(\mathfrak{P}) \geqq 1$.

The ramified primes in R' are therefore finite in number.

THEOREM 29. *Let R be a noetherian integrally closed integral domain, K its quotient field, K' a finite algebraic and separable extension of K, and R' the integral closure of R in K'. Let y be an element of R' such that $K' = K(y)$, and let $F(X)$ be the minimal polynomial of y over K. Let $R'_0 = R[y]$ and let $\mathfrak{C}$ be the conductor of R' in R'_0. Then we have*

$$\mathscr{C}_{R'_0/R} = R'_0 \cdot 1/F'(y), \qquad (3)$$

and, if R is a Dedekind domain, then we have also

(4) $$R'F'(y) = \mathfrak{C}\mathfrak{D}_{R'/R}.$$

[*It follows in particular that the derivative $F'(y)$ belongs to the different $\mathfrak{D}_{R'/R}$ and that we have $\mathfrak{D}_{R'/R} = R'F'(y)$ if and only if $R' = R[y]$, that is, if and only if the set $\{1, y, y^2, \cdots, y^{n-1}\}$ (where $n = [K':K]$) is a basis of R' over R.*]

PROOF. Let z be any element of K'. As $\{1, y, \cdots, y^{n-1}\}$ is a basis of K' over K, we have $z = g(y)$, where $g(X)$ is a polynomial of degree $\leq n - 1$ over K, uniquely determined by z. We denote by y_i $(i = 1, \cdots, n)$ the conjugates of y over K. We have the interpolation formula of Lagrange:

$$g(X) = \sum_i g(y_i)F(X)/F'(y_i)(X - y_i).$$

[The right-hand side is a polynomial of degree $\leq n - 1$ since $F(X)$ is a multiple of $X - y_i$. Its coefficients are in K by Galois theory, and for $X = y_i$, its value reduces to one term—namely, to the result of the substitution $X = y_i$ in $g(y_i)F(X)/F'(y_i)(X - y_i)$; but an easy computation shows that $F'(y_i) = \prod_{y_j \neq y_i} (y_i - y_j)$, that is, that $F'(y_i)$ is the value of the polynomial $F(X)/(X - y_i)$ for $X = y_i$. Hence the value of the right-hand side for $X = y_i$ is $g(y_i)$. This establishes the above expression of $g(X)$.] If we define the trace of a polynomial coefficient-wise, the above expression of $g(X)$ may be written as follows:

$$g(X) = T_{K'/K}(zF(X)/F'(y)(X - y)).$$

Take any element z' in the complementary module $\mathscr{C}_{R'_0/R}$ and take for z the element $z'F'(y)$. By the division algorithm, all the coefficients of $F(X)/(X - y)$ are in R'_0, and hence, by the definition of $\mathscr{C}_{R'_0/R}$ and by our choice of z, all the coefficients of $g(X)$ are in R. Thus $z'F'(y) = z = g(y) \in R'_0$. As this is true for every z' in $\mathscr{C}_{R_0'/R}$, we have

$$\mathscr{C}_{R'_0/R} \subset R'_0 \cdot 1/F'(y).$$

To prove (3), we shall now establish the opposite inclusion, i.e., $R'_0 \cdot 1/F'(y) \subset \mathscr{C}_{R'_0/R}$. Since for any element ξ of $R'_0 \cdot 1/F'(y)$ the set $\xi R'_0$ consists of elements of $R'_0 \cdot 1/F'(y)$, it is sufficient to prove that $T(y^k/F'(y)) \in R$, for $k = 0, 1, \cdots, n - 1$. We shall prove that

(a) $$T(y^k/F'(y)) = 0, \qquad k = 0, 1, \cdots, n - 2$$

and that

(b) $$T(y^{n-1}/F'(y)) = 1.$$

If we apply the above expressions of $g(X)$ to the case in which $z = y^{k+1}$ $(k = 0, 1, \cdots, n-2)$, we find the identity

$$X^{k+1} = \sum_{i=1}^{n} (y_i^{k+1}F(X)/F'(y_i)(X - y_i), \qquad k = 0, 1, \cdots, n-2,$$

and upon setting $X = 0$ in this identity (and observing that $F(0) \neq 0$, since we naturally exclude the trivial case $n = 1$), we find (a). If, on the other hand, we apply the above expressions of $g(X)$ to the case in which $z = y^n$ and if we observe that in this case the polynomial $g(X)$ coincides with the polynomial $X^n - F(X)$, we find the identity

$$X^n - F(X) = \sum_{i=1}^{n} (y_i^n F(X)/F'(y_i)(X - y_i).$$

Upon setting $X = 0$ in this last identity we find that

$$-a_n = -a_n \sum \frac{y_i^{n-1}}{f'(y_i)},$$

where $a_n = F(0)$. Since $a_n \neq 0$, this establishes (b) and formula (3) of the theorem.

To prove formula (4) of the theorem we first observe that since $\mathscr{C}_{R'/R} \subset \mathscr{C}_{R'_0/R}$, it follows from (3) that $\mathscr{C}_{R'/R} \subset R'_0 \cdot 1/F'(y)$. It follows (in view of the definition $\mathfrak{D}_{R'/R} = \mathscr{C}^{-1}{}_{R'/R}$ of $\mathfrak{D}_{R'/R}$) that $F'(y)\mathfrak{D}^{-1}{}_{R'/R} \subset R'_0$. Thus, the R'-module $F'(y)\mathfrak{D}^{-1}{}_{R'/R}$ is an ideal in R' (since $R'_0 \subset R'$) which is contained in R'_0. Therefore we have $F'(y)\mathfrak{D}^{-1}{}_{R'/R} \subset \mathfrak{C}$, i.e.,

$$R'F'(y) \subset \mathfrak{C}\mathfrak{D}_{R'/R}.$$

We shall now prove the opposite inclusion: $\mathfrak{C}\mathfrak{D}_{R'/R} \subset R'F'(y)$, and this will complete the proof of the theorem. If ξ is any element of $\mathfrak{C}$ and η is any element of R', then $\xi\eta \in \subset R'_0$, and therefore, by (3), we have $T(\xi\eta \cdot 1/F'(y)) \in R$. This implies that $\xi/F'(y) \in \mathscr{C}_{R'/R} = \mathfrak{D}^{-1}{}_{R'/R}$ (since η is *any* element of R'), i.e., $\xi\mathfrak{D}_{R'/R} \subset R'F'(y)$. This being true for any element ξ of $\mathfrak{C}$, the inclusion $\mathfrak{C}\mathfrak{D}_{R'/R} \subset R'F'(y)$ is established, and the proof of the theorem is now complete.

[NOTE. In the proof of (4) we have noted the validity of the inclusion $F'(y)\mathscr{C}_{R'/R} \subset R'_0$ (without using the assumption that R is a Dedekind domain). Since $F'(y)\mathscr{C}_{R'/R}$ is an R'-module, it follows that $F'(y)\mathscr{C}_{R'/R}$ is an ideal in R', contained in R'_0, and hence contained in the conductor $\mathfrak{C}$. We have therefore $F'(y)\mathscr{C}_{R'/R} \subset \mathfrak{C}$. In particular, it follows that $F'(y)$ *always belongs to the conductor of* R' in R'_0. We also observe that $\mathscr{C}_{R'/R} \subset \mathscr{C}_{R'_0/R}$ and that consequently $\mathfrak{D}_{R'/R} \supset \mathfrak{D}_{R'_0/R}$.

Now, every element z of R' is contained in the complementary module $\mathscr{C}_{R'_0/R}$, since for any element x of R'_0 we have that zx is integral over R and hence $T(zx) \in R$. We have therefore $\mathfrak{D}_{R'_0/R} \cdot R' \subset R'_0$; in other words, *the different of R'_0 over R is contained in the conductor $\mathfrak{C}$, in R'_0, of the integral closure R' of R'_0.* It should be pointed out, however, that in spite of the fact that $\mathfrak{D}_{R'_0/R}$ is contained in $\mathfrak{C}$, it need not be an ideal *in* R'.]

Again let R be a Dedekind domain, K its quotient field, K' a finite algebraic and separable extension of K, and R' the integral closure of R in K'. For every proper prime ideal $\mathfrak{P}$ in R' we denote by $f(\mathfrak{P})$, $e(\mathfrak{P})$, and $m(\mathfrak{P})$ its relative degree, its reduced ramification index, and its differential exponent. Given an ideal $\mathfrak{A}$ of R', we factor it:

$$\mathfrak{A} = \prod_{\mathfrak{P}} \mathfrak{P}^{n(\mathfrak{P})}.$$

The ideal $\mathfrak{a} = \prod_{\mathfrak{P}} (\mathfrak{P} \cap R)^{n(\mathfrak{P})f(\mathfrak{P})}$, which is defined since there is only a finite number of exponents $n(\mathfrak{P})$ which are different from zero, is called the *norm* of $\mathfrak{A}$ (with respect to R) and is denoted by $N_{K'/K}(\mathfrak{A})$, or simply by $N(\mathfrak{A})$.

We have the following formulae:

$$(5) \qquad N(\mathfrak{A} \cdot \mathfrak{B}) = N(\mathfrak{A}) \cdot N(\mathfrak{B}); \quad \text{if } \mathfrak{A} \subset \mathfrak{B} \text{ then } N(\mathfrak{A}) \subset N(\mathfrak{B}).$$

$$(6) \qquad N_{K'/K}(N_{K''/K'}(\mathfrak{A})) = N_{K''/K}(\mathfrak{A}) \quad \text{if } K \subset K' \subset K''.$$

$$(7) \qquad N_{K'/K}(R'\mathfrak{a}) = \mathfrak{a}^n \ (\mathfrak{a}\text{-ideal of } R,\ n = [K':K]).$$

In fact, (5) is evident. Formula (6) follows from the multiplicativity of the relative degrees. Finally, it suffices to prove (7) for a prime ideal $\mathfrak{p}$ of R; in this case we have $R'\mathfrak{p} = \prod_{\mathfrak{P}} \mathfrak{P}^{e(\mathfrak{P})}$, and (7) follows from the formula $\sum e(\mathfrak{P})f(\mathfrak{P}) = n$ (Corollary to Theorem 21 of § 9).

LEMMA 2. *For x in R', we have $N(R'x) = R \cdot N(x)$.*

PROOF. We first suppose that K' is normal over K. We write $R'x = \prod_{\mathfrak{P}} \mathfrak{P}^{v(\mathfrak{P}; x)}$. Let G denote the Galois group of K' over K. We have $v(\mathfrak{P};\ N(x)) = \sum_{s \in G} v(\mathfrak{P};\ s(x)) = \sum_{s \in G} v(s^{-1}(\mathfrak{P});\ x)$. Denoting by $\mathfrak{P}_1, \cdots, \mathfrak{P}_g$ the distinct conjugates of $\mathfrak{P}$ (that is, the prime ideals of R' lying over $\mathfrak{P} \cap R$), each $\mathfrak{P}_i$ occurs $e(\mathfrak{P})f(\mathfrak{P})$ times among the $s^{-1}(\mathfrak{P})$ (Theorem 22, § 9). Thus $v(\mathfrak{P}; N(x)) = e(\mathfrak{P})f(\mathfrak{P})\left(\sum_i v(\mathfrak{P}_i; x)\right)$. If we denote by $\mathfrak{p}$ the ideal $\mathfrak{P} \cap R$, the exponent $v(\mathfrak{p}; N(x))$ in the factorization of $R \cdot N(x)$ is therefore $\sum_{\mathfrak{P} \cap R = \mathfrak{p}} v(\mathfrak{P}; x)f(\mathfrak{P})$; and the exponent of $\mathfrak{p}$ in $N(R'x)$ is the same integer, by definition. This proves our assertion in the normal case.

In the general case, we introduce the least normal extension K'' of K containing K', and the integral closure R'' of R (or R') in K''. Since K'' is normal over both K and K', we have, for any x in R'', $N_{K''/K}(R''x) = R \cdot N_{K''/K}(x)$ and $N_{K''/K'}(R''x) = R'N_{K''/K'}(x)$. If we take x in R', we have $(R'x)^q = N_{K''/K'}(R''x)$, where $q = [K'':K']$, by (7). Then $(N_{K'/K}(R'x))^q = N_{K'/K}(R'x^q)$ (by (5)) $= N_{K'/K}(N_{K''/K'}(R''x)) = N_{K''/K}(R''x)$ (by (6)) $= R \cdot N_{K''/K}(x) = R \cdot N_{K'/K}(N_{K''/K'}(x)) = R \cdot N_{K'/K}(x^q) = (R \cdot N_{K'/K}(x))^q$ (II, § 10). Comparing the extreme terms of these equalities, we conclude that $N_{K'/K}(R'x) = R \cdot N_{K'/K}(x)$ by the unique factorization of ideals in R.

LEMMA 3. *The ideal $N(\mathfrak{A})$ is generated by the norms of the elements of $\mathfrak{A}$.*

PROOF. By Lemma 2, and formula (5), we have, for any a in $\mathfrak{A}$, $R \cdot N(a) = N(R'a) \subset N(\mathfrak{A})$. It remains to show, for every prime ideal $\mathfrak{p}$ in R, the existence of an element a of $\mathfrak{A}$ such that the exponent of $\mathfrak{p}$ in the factorization of $R \cdot N(a)$ is equal to the exponent of $\mathfrak{p}$ in $N(\mathfrak{A})$. If $\mathfrak{A} = \prod_{\mathfrak{P}} \mathfrak{P}^{n(\mathfrak{P})}$, the exponent of $\mathfrak{p}$ in $N(\mathfrak{A})$ is $\sum_{\mathfrak{P} \cap R = \mathfrak{p}} f(\mathfrak{P})n(\mathfrak{P})$. By the Chinese remainder theorem (Theorem 17, § 7), there exists an element a of $\mathfrak{A}$ which does not lie in any $\mathfrak{P}^{n(\mathfrak{P})+1}$ for $\mathfrak{P}$ lying over $\mathfrak{p}$: in fact, for every $\mathfrak{P}$ which lies over $\mathfrak{p}$ there exists an element $x_{\mathfrak{P}}$ in $\mathfrak{A}$ which does not lie in $\mathfrak{P}^{n(\mathfrak{P})+1}$, and the congruences $x \equiv x_{\mathfrak{P}} \pmod{\mathfrak{P}^{n(\mathfrak{P})+1}}$ (for $\mathfrak{P} \cap R = \mathfrak{p}$) and $x \equiv 0 \pmod{\mathfrak{A}}$ are obviously pairwise compatible. Thus the exponent of $\mathfrak{P}$ in $R'a$ is $n(\mathfrak{P})$ for $\mathfrak{P}$ lying over $\mathfrak{p}$, whence the exponent of $\mathfrak{p}$ in $R \cdot N(a)$ is $\sum_{\mathfrak{P} \cap R = \mathfrak{p}} f(\mathfrak{P})n(\mathfrak{P})$ by Lemma 2. Q.E.D.

With the same hypotheses and notations, the *norm of the different* $\mathfrak{D}_{R'/R}$ is an ideal in R. It is called *the discriminant of R' over R*, and is denoted by $\mathfrak{d}_{R'/R}$, or simply by $\mathfrak{d}$. The exponent of $\mathfrak{p}$ in $\mathfrak{d}$ is $\sum_{\mathfrak{P} \cap R = \mathfrak{p}} f(\mathfrak{P})m(\mathfrak{P})$. By the Corollary to Theorem 28 the prime ideals $\mathfrak{p}$ of R which "*ramify in R'*" (that is, those for which there exists a ramified prime ideal $\mathfrak{P}$ in R' such that $\mathfrak{P} \cap R = \mathfrak{p}$) are those which contain the discriminant $\mathfrak{d}_{R'/R}$; their number is therefore *finite*. The discriminant $\mathfrak{d}_{R'/R}$ is closely related to the discriminants of the bases of K' over K (II, § 11); more precisely:

THEOREM 30. *For any basis $\{u_1, \cdots, u_n\}$ of K' over K which is contained in R', the element $d(u)$, discriminant of the basis $\{u_1, \cdots, u_n\}$, is contained in the ideal $\mathfrak{d}_{R'/R}$, discriminant of R' over R. The ideal $\mathfrak{d}_{R'/R}$ is generated by the elements $d(u)$. In order that $\mathfrak{d}_{R'/R} = R \cdot d(u)$, it is necessary and sufficient that $\{u_1, \cdots, u_n\}$ be a basis of R' over R.*

PROOF. By definition of the discriminant, and by the analogous property of the different, the discriminant is determined by local data.

More precisely, if we denote by M the complement of a prime ideal $\mathfrak{p}$ in R, then the discriminant $\mathfrak{d}_M$ of R'_M over R_M is $\mathfrak{d}R_M$. In particular, the exponents of $\mathfrak{p}$ in $\mathfrak{d}$ and of $\mathfrak{p}R_M$ in $\mathfrak{d}_M$ are both equal to $\sum_{\mathfrak{P}\cap R=\mathfrak{p}} f(\mathfrak{P})m(\mathfrak{P})$.

Since R_M is a PID, R'_M admits a basis $\{u_1, \cdots, u_n\}$ over R_M (Corollary 2 to Theorem 7, § 4). We consider n elements $v_1, \cdots, v_n$ of K', and the equations $T(u_iv_j) = \delta_{ij}$ (δ_{ij} being the Kronecker symbols). If we set $v_j = \sum_s a_{js}u_s$ $(a_{js} \in K)$, this system of equations becomes $\sum_s a_{js}T(u_iu_s) = \delta_{ij}$, and for fixed j we get n linear equations between the n elements a_{js}. Since K' is separable, the determinant $\det (T(u_iu_s))$ of this system is $\neq 0$, (II, § 11) and the above system admits one and only one solution $\{a_{js}\}$ in K. In other words, there exists one and only one system of elements $v_1, \cdots, v_n$ of K' which satisfies the relations $T(u_iv_j) = \delta_{ij}$. We show that if $\{u_1, \cdots, u_n\}$ is a basis of R'_M over R_M, then $\{v_1, \cdots, v_n\}$ is a *basis of the complementary module* $\mathscr{C}_M$ of R'_M. In fact, we have $\sum_j T(u_iu_j)v_j = \sum_{j,s} T(u_iu_j)a_{js}u_s = \sum_s \delta_{is}u_s = u_i$. In particular,. $\{v_1, \cdots, v_n\}$ is a basis of K' over K. For an element $z = \sum_j a_jv_j$ $(a_j \in K)$ to belong to $\mathscr{C}_M$ it is necessary and sufficient that $T(zx) \in R_M$ for every x in R'_M, that is, that $T\left(\left(\sum_j a_jv_j\right)\cdot\left(\sum_i b_iu_i\right)\right) \in R_M$ for all systems $\{b_1, \cdots, b_n\}$ of n elements of R_M. But since $T(u_iv_j) = \delta_{ij}$, this condition may be written $\sum_j a_jb_j \in R_M$ for every system $\{b_1, \cdots, b_n\}$ of n elements of R_M. If we take $b_j = 1$ and $b_{j'} = 0$ for $j' \neq j$, this condition implies that $a_j \in R_M$; and conversely, if $a_j \in R_M$ for every j, then our condition is obviously verified. This proves that $\{v_1, \cdots, v_n\}$ is a basis of $\mathscr{C}_M$.

We now notice two simple facts, which will be useful later in the proof:

(a) If $\{u'_1, \cdots, u'_n\}$ and $\{u''_1, \cdots, u''_n\}$ are two bases of K' over K such that every u''_i is in the R-module (or R_M-module) generated by the u'_j then the relation $d(u'') \in R\cdot d(u')$ [or $d(u'') \in R_M d(u')$] holds between their discriminants. This is an immediate consequence of formula (2) in II, § 11, relating the discriminants of two bases of K' over K.

(b) In particular if the above two bases generate the same R-module (or R_M-module), the quotient $d(u')/d(u'')$ is a unit in R (or in R_M)—that is, $d(u')$ and $d(u'')$ generate the same principal fractionary ideal.

This being so, the complementary module $\mathscr{C}_M$ is a principal fractionary ideal, since R'_M is a PID. If we set $\mathscr{C}_M = R'_My$, then $\{yu_1, \cdots, yu_n\}$ is a basis of $\mathscr{C}_M$ over R_M. Using the expression of the discriminant of a basis as the square of a determinant, given in II, § 11, formula (5), the discriminant of this basis is $(N(y))^2 d(u)$. Thus, by (b), $(N(y))^2 d(u)/d(v)$ is a unit in R_M. But the above proved formula $u_i = \sum_j T(u_i u_j) v_j$ shows (formula (2), II, § 11) that $d(u) = (\det (T(u_i u_j)))^2 d(v)$—that is, that $d(u)d(v) = 1$, since $\det (T(u_i u_j)) = d(u)$. Hence $(N(y)d(u))^2$ is a unit in R_M, whence also $N(y)d(u)$ is a unit in R_M. By Lemma 3 and the definition of the discriminant, $N(y)^{-1}$ generates $\mathfrak{d}_M$. Thus $\mathfrak{d}_M = R_M d(u)$.

Now let $\{u'_1, \cdots, u'_n\}$ be a basis of K' over K composed of elements of R'. By (a), its discriminant $d(u')$ is an R_M-multiple of $d(u)$, whence $d(u') \in \mathfrak{d}_M$ for every prime ideal $\mathfrak{p}$. Thus the exponent of $\mathfrak{p}$ in $Rd(u')$ is at least equal to the exponent of $\mathfrak{p}$ in $\mathfrak{d}$. This proves that $d(u') \in \mathfrak{d}$.

The basis $\{u_1, \cdots, u_n\}$ of R'_M over R_M may be chosen, after multiplication by a unit in R_M, in such a way that $u_i \in R'$ for every i. If we take one such basis for every $\mathfrak{p}$, the discriminants of these bases generate an ideal $\mathfrak{b}$ in R. It is contained in $\mathfrak{d}$ as we just saw. On the other hand, the exponent of $\mathfrak{p}$ in $\mathfrak{b}$ is at most equal to the exponent of $\mathfrak{p}$ in $Rd(u)$, that is, to the exponent of $\mathfrak{p}$ in $\mathfrak{d}$. Therefore $\mathfrak{b} = \mathfrak{d}$, and our second assertion is proved.

Suppose that we have a basis $\{u_1, \cdots, u_n\}$ of R' over R. Then it is a basis of R'_M over R_M, and we have $R_M d(u) = \mathfrak{d}_M$ as has been proved above. Since this holds for every prime ideal $\mathfrak{p}$, we conclude that $R \cdot d(u) = \mathfrak{d}$.

Suppose conversely that we have a basis $\{u'_1, \cdots, u'_n\}$ of K' over K, which is contained in R', and such that $R \cdot d(u') = \mathfrak{d}$. Then, if $\{u_1, \cdots u_n\}$ is a basis of R'_M over R_M, the elements $d(u)$ and $d(u')$ generate the same ideal $\mathfrak{d}_M$ in R_M. If we set $u'_i = \sum_j a_{ij} u_i$ $(a_{ij} \in R_M)$, the formula $d(u') = (\det (a_{ij}))^2 d(u)$ (II, § 11 (2)) shows that $(\det (a_{ij}))^2$ is a unit in R_M. Thus $\det (a_{ij})$ is also a unit in R_M, and Cramer's formulae show that every u_i belongs to the R_M-module generated by the u'_j. In other words, $\{u'_1, \cdots, u'_n\}$ is also a basis of R'_M over R_M. Since this holds for every prime ideal $\mathfrak{p}$, $\{u'_1, \cdots, u'_n\}$ is a basis of R' over R: in fact, if, for x in R', we write $x = \sum_i a_i u'_i$, we have $a_i \in R_M$ for every $\mathfrak{p}$ since $x \in R'_M$; thus the exponent of $\mathfrak{p}$ in $a_i R$ is ≥ 0 for every $\mathfrak{p}$, and this proves that $a_i \in R$. The proof of Theorem 30 is now complete.

REMARKS.

1) Suppose that $\{u_1, \cdots, u_n\}$ is a basis of R' over R. Then it is a basis of R'_M over R_M for every $\mathfrak{p}$; and hence the basis $\{v_1, \cdots, v_n\}$ of K' over K constructed in the proof of Theorem 30 is a basis of the

complementary module $\mathscr{C}_M$ of R'_M for every $\mathfrak{p}$. As the complementary module $\mathscr{C}$ is determined by local data, we see, as at the end of the proof of Theorem 30, that $\{v_1, \cdots, v_n\}$ is a basis of $\mathscr{C}$ over R.

2) The assumption of separability in Theorem 30—that is, the assumption that det $(T(u_iu_j)) \neq 0$ for every basis of K' over K—means that the *bilinear form* $T(xy)$ on K' considered as a vector space over K, is *non-degenerate*. It establishes therefore a duality between the vector space K' and itself, that is, an isomorphism between K' and its dual vector space. The basis $\{v_1, \cdots, v_n\}$ constructed in the proof of Theorem 30 is the *dual basis* of the basis $\{u_1, \cdots, u_n\}$.

3) The discriminant d of a basis $\{1, y, \cdots, y^{n-1}\}$ (y: primitive element of K' over K) is $N(F'(y))$, where F is the minimal polynomial of y over K: this follows from formula (6) of II, § 11, which gives, by expansion of the Vandermonde determinant,

$$d = \prod_{i \neq j} (y_i - y_j) = N((y_1 - y_2) \cdots (y_1 - y_n)) = N(F'(y_1)) = N(F'(y)).$$

This establishes a link between Theorems 29 and 30.

THEOREM 31 (TRANSITIVITY FORMULAE). *Let R be a Dedekind domain, K' and K'' two finite algebraic and separable extensions of the quotient field K of R, such that $K' \subset K''$, and R' and R'' the integral closures of R in K' and K''. Then we have:*

$$\mathfrak{D}_{R''/R} = \mathfrak{D}_{R''/R'}(R''\mathfrak{D}_{R'/R}), \quad \mathfrak{d}_{R''/R} = N_{K'/K}(\mathfrak{d}_{R''/R'})\cdot(\mathfrak{d}_{R'/R})^{[K'':K']}.$$

PROOF. We first prove the formula for the differents, or rather, the formula relating their inverses—that is, the complementary modules: $\mathscr{C}_{R''/R} = \mathscr{C}_{R''/R'}\cdot(R''\mathscr{C}_{R'/R})$. For z in R'' the following relations are equivalent: $z \in \mathscr{C}_{R''/R}$, $T_{K''/K}(zR'') \subset R$, $T_{K'/K}(T_{K''/K'}(zR'')) \subset R$, $T_{K'/K}(T_{K''/K'}(zR'')\cdot R') \subset R$ (if one multiplies an element of K'' by $x \in R'$, its trace $T_{K''/K'}$ is multiplied by x), $T_{K''/K'}(zR'') \subset \mathscr{C}_{R'/R}$, $(\mathscr{C}_{R'/R})^{-1}\cdot T_{K''/K'}(zR'') = \mathfrak{D}_{R'/R}\cdot T_{K''/K'}(zR'') = T_{K''/K'}(z\cdot\mathfrak{D}_{R'/R}R'') \subset R'$, $z\mathfrak{D}_{R'/R} \subset \mathscr{C}_{R''/R'}$, $z \in \mathscr{C}_{R''/R'}\cdot\mathscr{C}_{R'/R}$. Thus the formula for the different is proved. For getting the formula for discriminants, we take the norm $N_{K''/K}$ of both sides in the formula for the differents. We get $\mathfrak{d}_{R''/R} = N_{K''/K}(\mathfrak{D}_{R''/R'}(R''\mathfrak{D}_{R'/R})) = N_{K''/K}(\mathfrak{D}_{R''/R'}) \cdot N_{K''/K}(R''\mathfrak{D}_{R'/R})$ [formula (5)] $= N_{K'/K}(\mathfrak{d}_{R''/R'})\cdot N_{K'/K}(N_{K''/K'}(R''\mathfrak{D}_{R'/R}))$ [formula (6)] $= N_{K'/K}(\mathfrak{d}_{R''/R'})\cdot N_{K'/K}((\mathfrak{D}_{R'/R})^n)$ [by formula (7)] $= N_{K'/K}(\mathfrak{d}_{R''/R'})\cdot(\mathfrak{d}_{R'/R})^n$, where $n = [K'':K']$, by formula (5). Q.E.D.

In the case of a *normal and separable extension* K' of K, the differential exponent $m(\mathfrak{P})$ of a prime ideal $\mathfrak{P}$ of R' may be computed if one knows the *orders of the ramification groups* of $\mathfrak{P}$. As in § 10 we denote by

G_Z, G_T, G_{V_i} the decomposition, inertia, and i-th ramification groups ($G_{V_1} = G_T$) and by K_Z, K_T, K_i the corresponding subfields of K'; let n_i be the degree $[K':K_i]$, that is, the order of G_i. We suppose that $R'/\mathfrak{P}$ is *separable* over $R/\mathfrak{p}$ ($\mathfrak{p} = R \cap \mathfrak{P}$). Then $[K_Z:K] = g$, $[K_T:K_Z] = f$, $[K':K_T] = e = n_1$. For any two extensions L, L' of K such that $K \subset L \subset L' \subset K'$, we denote by $m(L', L)$ the differential exponent over L of the prime ideal $\mathfrak{P}_{L'}$, which is the contraction of $\mathfrak{P}$ in the integral closure $R_{L'}$ of R in L'.

For any $i \geq 1$, the residue field $R_{K_i}/\mathfrak{P}_{K_i}$ is equal to $R'/\mathfrak{P}$, and $\mathfrak{P}$ is the only prime ideal in R' lying over $\mathfrak{P}_{K_i}$ (Corollary to Theorem 24, § 10). By "localization" we may suppose that R_{K_i} has only one prime ideal $\mathfrak{P}_{K_i}$, which is then principal; then $\mathfrak{P}$ is the only prime ideal of R'; it is a principal ideal, say $\mathfrak{P} = R'u$; we have $R'\mathfrak{P}_{K_i} = \mathfrak{P}^{n_i}$. For any x in K', we denote by $v(x)$ the exponent of $\mathfrak{P} = R'u$ in the factorization of $R'x$. We assert that $\{1, u, \cdots, u^{n_i-1}\}$ constitutes *a basis of R' over R_{K_i}*. This is a consequence of the following lemma:

LEMMA 4. *Let R be a discrete valuation ring, and R' the integral closure of R in a finite algebraic and separable extension K' of the quotient field K of R. We suppose that the prime ideal $\mathfrak{p}$ of R is completely ramified in R', that is, that there is only one prime ideal $\mathfrak{P}$ of R' lying over $\mathfrak{p}$, and that $R'/\mathfrak{P} = R/\mathfrak{p}$. Then, if u denotes a generator of $\mathfrak{P}$ and n the degree of K' over K, $\{1, u, \cdots, u^{n-1}\}$ is a basis of R' over R.*

PROOF. By Corollary to Theorem 21 of § 9, we have $R'\mathfrak{p} = \mathfrak{P}^n$. Since $\mathfrak{P}$ is the only prime ideal of R', $\mathfrak{P}$ is actually a principal ideal $R'u$ (Theorem 15 of § 6). For any x in K', we denote by $v(x)$ the exponent of $\mathfrak{P}$ in $R'x$. Then if a is in K, $v(a)$ is a multiple of n since, if we denote by s the exponent of $\mathfrak{p}$ in Ra, we have $R'a = R'\mathfrak{p}^s = \mathfrak{P}^{ns}$. In order to prove that the elements $\{1, u, \cdots, u^{n-1}\}$ form a basis of R' over R, we first prove that they are *linearly independent*. In fact, if we have a non-trivial linear relation $0 = a_0 + a_1u + \cdots + a_{n-1}u^{n-1} (a_j \in K)$ the integers $v(a_ju^j) = j + v(a_j)$ are all distinct, since the integers $v(a_j)$ are multiples of n. Thus, if $v(a_qu^q) = r$ is the smallest one, the sum of the other terms is in $\mathfrak{P}^{r+1}$, in contradiction with the fact that it is equal to $-a_qu^q$. Therefore $\{1, u, \cdots, u^{n-1}\}$ is a basis of K' over K. For any x in R' we write $x = \sum_{j=1}^{n-1} b_ju^j$ with b_j in K. As above the integers $v(b_ju^j)$ are all distinct. Thus, if r denotes the smallest one, the sum $\sum_j b_ju^j$ is in $\mathfrak{P}^r$ but not in $\mathfrak{P}^{r+1}$. Since x is in R', we have $r \geq 0$, whence $j + v(b_j) = v(b_ju^j) \geq 0$ for $j = 1, \cdots, n-1$. As the integers $v(b_j)$

are multiples of n, this implies $v(b_j) \geq 0$, whence $b_j \in R$. This proves the lemma.

We may thus use Theorem 29 for computing the different of R' over R_{K_i}: it is the ideal $R'F'(u)$. But $F'(u) = \prod(u - u_j)$ where the u_j are the conjugates of u which are distinct from u. As K' is a normal and separable extension of K_i with G_{V_i} as Galois group we have $F'(u) = \prod_{s \in G_{V_i}, s \neq 1} (u - s(u))$. Thus the differential exponent $m(K', K_i)$ is equal to $\sum_{s \in G_{V_i}, s \neq 1} v(u - s(u))$. Therefore we have

$$m(K', K_{i-1}) - m(K', K_i) = \sum_{s \in G_{V_{i-1}}, s \notin G_{V_i}} v(u - s(u)).$$

But, since $R'/\mathfrak{P}$ is separable over $R/\mathfrak{p}$, we have seen in the proof of Theorem 25, § 10, that, for an element s of $G_{V_{i-1}}$ which is not in G_{V_i}, we have $v(u - s(u)) = i - 1$. Therefore $m(K', K_{i-1}) - m(K', K_i) = (i - 1)(n_{i-1} - n_i)$. But the transitivity formula for the differents (Theorem 31) shows, by repeated applications, that we have

$$m(K', K_T) = (n_1 - n_2) + 2(n_2 - n_3) + \cdots + j(n_j - n_{j+1}) + \cdots,$$

the sum having only a finite number of non-zero terms, since $n_j = 1$—that is, $G_{V_j} = (1)$—for j large enough. Furthermore the prime ideal $\mathfrak{p}$ of R does not ramify in the inertia field K_T (Corollary to Theorem 24). Hence the differential exponent $m(K_T, K)$ is equal to 0 by Theorem 28. Therefore the transitivity formula for the differents gives the following relation, called *Hilbert's formula*:

$$(8) \quad m(K', K) = m(\mathfrak{P}) = n_1 - n_2 + 2(n_2 - n_3) + \cdots + j(n_j - n_{j+1}) + \cdots.$$

We can now compute the exponent of $\mathfrak{p}$ in the *discriminant* $\mathfrak{d}_{R'/R}$. As seen before, this exponent is $\sum_{\mathfrak{P} \cap R = \mathfrak{p}} f(\mathfrak{P})m(\mathfrak{P})$. But the prime ideals $\mathfrak{P}$ of R' lying over $\mathfrak{p}$ are the conjugates of one of them; therefore all their differential exponents $m(\mathfrak{P})$ are equal, and so are their relative degrees $f(\mathfrak{P})$. Since there are $g = n/ef$ $(e = n_1)$ of them, the exponent of $\mathfrak{p}$ in $\mathfrak{d}_{R'/R}$ is $gf \cdot m(\mathfrak{P}) = ne^{-1}m(\mathfrak{P})$, and therefore:

$$(9) \quad n \cdot n_1^{-1}((n_1 - n_2) + 2(n_2 - n_3) + \cdots + j(n_j - n_{j+1}) + \cdots).$$

§ 12. Application to quadratic fields and cyclotomic fields. A quadratic field K is an extension of degree 2 of the rational number field Q. The classroom solution of the quadratic equation shows that it is generated over Q by the square root of a rational number. Multiplying (or dividing) this rational number by the square of a suitable integer, we may assume that it is a "square-free" integer m, that is, an integer m without square factors. Let $K = Q(e)$ where $e^2 = m$. Any element x of K is of the form $x = a + be$ with a and b in Q. The mapping $a + be \to a - be$ is an automorphism of K over Q. Thus K is a *normal* extension of Q (with a cyclic group of order 2 as Galois group). For $x = a + be$ to be an algebraic integer, it is necessary and sufficient that its trace $2a$ and its norm $a^2 - mb^2$ be ordinary integers. This implies that $a = \frac{1}{2}a'$ (a' an integer) and that $(2b)^2m$ is an integer. As m is square-free, $2b$ must be an integer, whence $b = \frac{1}{2}b'$ where b' is an integer. Our condition now reduces to $a'^2 - mb'^2 \equiv 0 \pmod 4$. If m is congruent to 2 or to 3 modulo 4, a simple examination of cases shows that our condition is satisfied if and only if a' and b' are both even. If m is congruent to 1 modulo 4, then it is easily verified that our condition is satisfied if and only if a' and b' are either both even or both odd. Note that m cannot be $\equiv 0 \pmod 4$ as it is square-free. To summarize: the ring R of algebraic integers has the following bases over the ring J of rational integers:

$m \equiv 2$ or $3 \pmod 4$: $\{1, e\}$ *is a basis*;
$m \equiv 1 \pmod 4$: $\{1, \frac{1}{2}(1 + e)\}$ *is a basis.*

By Theorems 29 and 30 the *different* and the *discriminant* of R over J are the ideals:

$m \equiv 2$ or $3 \pmod 4$: *different* $= 2eR$, *discriminant* $= 4mJ$
$m \equiv 1 \pmod 4$: *different* $= eR$, *discriminant* $= mJ$.

Extending the terminology introduced for the gaussian integers (§ 9, p. 288), we first see that the *ramified* prime numbers p are those which divide the discriminant. In particular, the prime 2 is always ramified if $m \equiv 2$ or $3 \pmod 4$, and in no other case.

Among the unramified odd primes, some are decomposed and some are inertial. As in the case of gaussian integers, we see that $\mathfrak{p}$ is decomposed if m is a square modulo p, and inertial if m is not a square modulo p. We introduce the *Legendre symbol* $\left(\frac{s}{p}\right)$ for every integer s which is not a multiple of p: by definition it is $+1$ if the p-residue of s is a square in $J/(p)$, and -1 if the p-residue of s is not a square in $J/(p)$. As the multiplicative group of $J/(p)$ is a cyclic group of order $p - 1$ (II, § 9), the relation $\left(\frac{s}{p}\right) = 1$ is equivalent to $s^{\frac{1}{2}(p-1)} \equiv 1 \pmod p$, and the relation $\left(\frac{s}{p}\right) = -1$ to $s^{\frac{1}{2}(p-1)} \equiv -1 \pmod p$. It follows that $\left(\frac{s}{p}\right)\left(\frac{t}{p}\right) = \left(\frac{st}{p}\right)$.

It remains to investigate, in the case where 2 is unramified—that is, in the case $m \equiv 1 \pmod 4$—whether the prime 2 is inertial or decomposed. Since $\{1, \frac{1}{2}(1 + e)\}$ is a basis of R over J, the residues modulo $2R$ of these elements form a basis of $R/2R$ over $J/(2)$. As the minimal polynomial of $\frac{1}{2}(1 + e)$ over Q is $X^2 - X - (m - 1)/4$, we have to investigate whether this polynomial is irreducible or not over $J/(2)$. It is clear that it is reducible over $J/(2)$ if and only if $(m - 1)/4$ is even.

We can now state our results:

THEOREM 32. *In the quadratic field* $K = Q(\sqrt{m})$ *(m, a square-free integer),*

a) *the ramified primes are the odd prime divisors of m, and also 2 if* $m \equiv 2$ *or* $3 \pmod 4$;

b) *the inertial primes are the odd primes p which do not divide m and which are such that* $\left(\frac{m}{p}\right) = -1$, *and also 2 if* $m \equiv 5 \pmod 8$;

c) *the decomposed primes are the odd primes p which do not divide m and which are such that* $\left(\frac{m}{p}\right) = 1$, *and also 2 if* $m \equiv 1 \pmod 8$.

One usually sets $\left(\frac{m}{2}\right) = 1$ if $m \equiv 1, 7 \pmod 8$, and $\left(\frac{m}{2}\right) = -1$ if $m \equiv 3, 5 \pmod 8$. In other words, for an odd m, we have $\left(\frac{m}{2}\right) = (-1)^{(m^2-1)/8}$.

We now study, for an *odd prime p*, the *cyclotomic field* of the p-th roots of unity, that is, the splitting field of the polynomial $Z^p - 1$ over the field Q of rational numbers. The roots of this polynomial form a group under multiplication, of prime order p. Hence if z is any root of $Z^p - 1$, other than 1, then the other roots of Z^{p-1} will be then $z^2, z^3, \cdots, z^{p-1}$ and $z^p = 1$. Therefore the cyclotomic field of p-th roots of unity *is the simple extension* $Q(z)$ of Q. For computing the *degree* $[Q(z):Q]$, we first notice that $z, z^2, \cdots, z^{p-1}$ are the roots of the polynomial $F(Z) = Z^{p-1} + \cdots + Z + 1$, whence we have $F(Z) = (Z - z)(Z - z^2) \cdots (Z - z^{p-1})$. Let R' be the ring of algebraic integers in $Q(z)$; we have $z \in R'$. For $r = 2, 3, \cdots, p - 1$, the element $(1 - z^r)/(1 - z)$ is in R', since it is equal to $1 + z + \cdots + z^{r-1}$. Its inverse $(1 - z)/(1 - z^r)$ is also in R', for if we denote by r' an integer such that $rr' \equiv 1 \pmod p$, we have $z = (z^r)^{r'}$, whence $(1 - z)/(1 - z^r)$

$= 1 + z^r + z^{2r} + \cdots + z^{(r'-1)r}$. Therefore $(1 - z^r)/(1 - z)$ is a *unit* in R'. This being so, the formula $p = F(1) = (1 - z)(1 - z^2) \cdots (1 - z^{p-1})$ shows that the ideal $R'p$ is equal to $R'(1 - z)^{p-1}$. The formula $\sum e_i f_i = [Q(z):Q]$ (Corollary to Theorem 21 of § 9) about the decomposition of $R'p$ in R' thus shows that $p - 1 \leq [Q(z):Q]$. As z is a root of the polynomial $F(Z)$ of degree $p - 1$, *the degree* $[Q(z):Q]$ *is exactly* $p - 1$. Hence the polynomial $F(Z) = Z^{p-1} + \cdots + Z + 1$ is *irreducible*. Also the formulae $e(p)f(p)g(p) = p - 1$ (§ 9, Theorem 22) and $R'p = R'(1 - z)^{p-1}$ show that $e(p) = p - 1$, $f(p) = g(p) = 1$ (one says then that p is "*completely ramified*" in R') and that $R'(1 - z)$ is a *prime ideal*.

Since $Q(z)$ is a normal extension of degree $p - 1$ of Q, there are $p - 1$ conjugates of z over Q, and these conjugates are obviously $z, z^2, \cdots, z^{p-1}$. The Galois group of $Q(z)$ over Q, whose elements s_j are defined by $s_j(z) = z^j$ for $j = 1, \cdots, p - 1$ and obviously satisfy the relations $s_i s_j = s_k$ if $k \equiv ij \pmod{p}$, is therefore isomorphic to the multiplicative group of $J/(p)$; hence it is a *cyclic group of order* $p - 1$.

Let us now compute the *discriminant* of R' over J. We have $(Z - 1)F(Z) = Z^p - 1$ and hence $F'(z) = pz^{p-1}/(z - 1)$. As z is in R', this number $pz^{p-1}/(z - 1)$ is contained in the different of R' over J (Theorem 29). As the minimal polynomial of $z - 1$ over Q is $(X + 1)^{p-1} + \cdots + (X + 1) + 1$, the norm $N(z - 1)$ is p; since $N(z) = 1$, we have $N(F'(z)) = p^{p-1}/p = p^{p-2}$. Thus the discriminant of R' over J divides p^{p-2}, and p is the only prime number which may ramify in R'. On the other hand we have seen that p is "completely ramified" in R', and that $R'(1 - z)$ is the unique prime divisor of $R'p$ in R'. Then Lemma 4 (§ 11) shows that the powers $(1 - z)^j$ $(j = 0, 1, \cdots, p - 2)$, whence also the powers z^j, form a basis of $R'_{R'(1-z)}$ over $J_{(p)}$. Thus, by Theorem 30, the discriminant of $R'_{R'(1-z)}$ over $J_{(p)}$ is the discriminant of this basis, that is, p^{p-2}, as has already been computed. Therefore, since p is the only prime number which ramifies in R', p^{p-2} *is also the discriminant of* R' *over* J; and $\{1, z, \cdots, z^{p-2}\}$ *is an integral basis of* R' (Theorem 30).

Since p is odd, the Galois group of $Q(z)$, which is a cyclic group of order $p - 1$, contains one and only one subgroup of index 2. To this subgroup corresponds a *quadratic subfield* K of $Q(z)$, uniquely determined by p. By the transistivity formula for discriminants (Theorem 31), the discriminant of the ring R of algebraic integers of K over J divides p^{p-2}. Then the formulae for the discriminant of a quadratic field imply that $K = Q(\sqrt{p})$ if $p \equiv 1 \pmod{4}$ and $K = Q(\sqrt{-p})$ if $p \equiv 3 \pmod{4}$. At any rate the discriminant of R over J is p.

We now study the *decomposition of a prime* $q \neq p$ in $Q(z)$. Let g be the number of its prime factors, and f their common relative degree (Theorem 22 of § 9). We have $fg = p - 1$, as q is unramified. If $\mathfrak{Q}$ is any prime ideal of R' lying over (q), then $R'/\mathfrak{Q}$ is obtained by adjoining to $J/(q)$ the p-th roots of unity. This implies that f is the smallest positive integer for which $q^f \equiv 1 \pmod{p}$: in fact if a p-th root $\bar{z}$ of unity belongs to the field with q^f elements, we have $\bar{z}^{q^f-1} = 1$, whence p divides $q^f - 1$; conversely, if $q^{f'} \equiv 1 \pmod{p}$ for $f' < f$, any p-th root $\bar{z}$ of unity (over $J/(q)$) satisfies $\bar{z}^{q^{f'}-1} = 1$, hence belongs to the field with $q^{f'}$ elements. The decomposition field L of q is of degree $g = (p-1)/f$ over Q. Since $Q(z)$ is an abelian extension of Q, all prime ideals of R' lying over (q) have the same decomposition group and the same decomposition field; we can thus talk of the decomposition group and the decomposition field of (q), or of q, by a remark made in the beginning of § 10. As the quadratic field K has only Q as proper subfield, we have either $K \cap L = Q$, or $K \cap L = K$, that is, $K \subset L$. If q is inertial in K, then K cannot be contained in L, as is easily seen by examining the residue fields; thus $K \cap L = Q$ in this case. Conversely if $K \cap L = Q$, then the compositum $L(K)$ is a quadratic extension of L; if $\mathfrak{Q}$ is a prime ideal of L lying over (q), then $\mathfrak{Q}$ is inertial in $L(K)$ since it can be neither decomposed nor ramified (Theorem 24, § 10); if we denote by $\mathfrak{Q}'$ the only prime ideal of $L(K)$ lying over $\mathfrak{Q}$, by $\mathfrak{Q}''$ the prime ideal $\mathfrak{Q}' \cap K$, and by k_0, k, k', k'' the residue fields corresponding to the prime ideals (q), $\mathfrak{Q}$, $\mathfrak{Q}'$, $\mathfrak{Q}''$, then $k = k_0$, $[k':k] = 2$, and k' is the compositum of k and k''; therefore $k'' = k'$, and q is inertial in K. If the prime q is decomposed into two prime ideals $\mathfrak{q}'$, $\mathfrak{q}''$ in the quadratic field K, $\mathfrak{q}'$ and $\mathfrak{q}''$ are conjugate over Q; thus they decompose into the same number of prime ideals in $Q(z)$, and the number g must be *even*. Conversely, if g is even, the decomposition group of q is a subgroup of even index in the cyclic Galois group G of $Q(z)$ over Q; therefore it must be contained in the unique subgroup of index 2 of G; in other words the decomposition field L contains the quadratic field K, and the prime q is decomposed in K.

We now compare these results with those given in Theorem 32. In the notation of this theorem, we have $m = (-1)^{\frac{1}{2}(p-1)}p$. If q is decomposed in K, we have $\left(\frac{(-1)^{\frac{1}{2}(p-1)}p}{q}\right) = 1$; on the other hand g is even, whence $q^{\frac{1}{2}(p-1)} = (q^f)^{\frac{1}{2}g} \equiv 1 \pmod{p}$, and therefore $\left(\frac{q}{p}\right) = 1$ as has been seen just after the definition of Legendre's symbol. If q is inertial in K, then $\left(\frac{(-1)^{\frac{1}{2}(p-1)}p}{q}\right) = -1$; on the other hand, since g is odd and $p-1$ even, f must be even, whence $q^{\frac{1}{2}(p-1)} = (q^{\frac{1}{2}f})^g$, but by the above given characterization of f, we have $q^{\frac{1}{2}f} \equiv -1 \pmod p$, whence $q^{\frac{1}{2}(p-1)} \equiv -1 \pmod p$ since g is odd, and $\left(\frac{q}{p}\right) = -1$. We have now almost proved the following theorem:

THEOREM 33 ("QUADRATIC RECIPROCITY LAW"). *If p and q are distinct odd primes, we have*

$$\left(\frac{p}{q}\right)\cdot\left(\frac{q}{p}\right) = (-1)^{\frac{1}{2}(p-1)\frac{1}{2}(q-1)}.$$

If p is an odd prime, we have $\left(\frac{2}{p}\right) = (-1)^{(p^2-1)/8}$.

PROOF. Just before stating Theorem 33, we saw that

$$\left(\frac{q}{p}\right) = \left(\frac{(-1)^{\frac{1}{2}(p-1)}p}{q}\right)$$

for any prime q and any odd prime p. Hence

$$\left(\frac{p}{q}\right)\left(\frac{q}{p}\right) = \left(\frac{p}{q}\right)\left(\frac{(-1)^{\frac{1}{2}(p-1)}p}{q}\right) = \left(\frac{(-1)^{\frac{1}{2}(p-1)}p^2}{q}\right) = \left(\frac{(-1)^{\frac{1}{2}(p-1)}}{q}\right).$$

If q is also odd, we get by permutation of p and q,

$$\left(\frac{p}{q}\right)\left(\frac{q}{p}\right) = \left(\frac{(-1)^{\frac{1}{2}(q-1)}}{p}\right).$$

Taking $p = 3$, and comparing these equalities, we get

$$\left(\frac{-1}{q}\right) = \left(\frac{(-1)^{\frac{1}{2}(q-1)}}{3}\right) = \left(\frac{-1}{3}\right)^{\frac{1}{2}(q-1)} = (-1)^{\frac{1}{2}(q-1)}.$$

Therefore

$$\left(\frac{p}{q}\right)\left(\frac{q}{p}\right) = \left(\frac{(-1)^{\frac{1}{2}(p-1)}}{q}\right) = \left(\frac{-1}{q}\right)^{\frac{1}{2}(p-1)} = (-1)^{\frac{1}{2}(p-1)\frac{1}{2}(q-1)},$$

and our first formula is proved.

On the other hand, for $q = 2$, we have

$$\left(\frac{2}{p}\right) = \left(\frac{(-1)^{\frac{1}{2}(p-1)}p}{2}\right).$$

Since, by definition, $\left(\frac{m}{2}\right) = (-1)^{(m^2-1)/8}$ for any odd number m, and since $m = (-1)^{\frac{1}{2}(p-1)}\,p$ in the case which now interests us, we have

$m^2 - 1 = p^2 - 1$, whence $\left(\frac{2}{p}\right) = (-1)^{(p^2-1)/8}$. Q.E.D.

§ 13. A theorem of Kummer. In the last section, we have seen that, both in the quadratic and in some cyclotomic fields, the rings of algebraic integers admit integral bases of the form $\{1, y, \cdots, y^{n-1}\}$ over the ring J of rational integers. We have also been able to get useful information about the decomposition of prime numbers in these fields. We shall now prove a theorem which shows how some information about the decomposition of prime ideals may be derived from the existence of an integral basis of the above-mentioned type:

THEOREM 34 (KUMMER). *Let R be an integrally closed domain, K its quotient field, K' a finite algebraic extension of K, R' the integral closure of R in K'. We suppose that there exists an element y of R' such that $R' = R + Ry + \cdots + Ry^{n-1}$ $(n = [K':K])$ (y is then a primitive element of K' over K). Let $F(Y)$ be the minimal polynomial of y over K. ($F(Y)$ has its coefficients in R, by Theorem 4 of § 3.) Let $\mathfrak{p}$ be a maximal ideal in R; for every polynomial $G(X)$ over R, we denote by $\bar{G}(X)$ the polynomial over $R/\mathfrak{p}$ whose coefficients are the $\mathfrak{p}$-residues of the corresponding coefficients of G. Let $\bar{F}(X) = \prod_{i=1}^{g} (f_i(X))^{e(i)}$ be the factorization of $\bar{F}(X)$ into distinct irreducible factors $f_i(X)$ over $R/\mathfrak{p}$; for $i = 1, \cdots, g$ we denote by $F_i(X)$ a polynomial over R such that $\bar{F}_i(X) = f_i(X)$. Then the ring R' has exactly g maximal ideals $\mathfrak{P}_i$ which lie over $\mathfrak{p}$, and we have*

$$\mathfrak{P}_i = R'\mathfrak{p} + R'F_i(y).$$

Furthermore we have

$$R'\mathfrak{p} = \mathfrak{Q}_1 \cap \mathfrak{Q}_2 \cdots \cap \mathfrak{Q}_g = \mathfrak{Q}_1 \cdot \mathfrak{Q}_2 \cdot \cdots \mathfrak{Q}_g,$$

where $\mathfrak{Q}_i = R'\mathfrak{p} + R' \cdot (F_i(y))^{e(i)}$.

PROOF. We denote by k the field $R/\mathfrak{p}$. We consider the homomorphisms

$$R[X] \to k[X] \to k[X]/(f_i(X))$$

(the first homomorphism being defined by $G(X) \to \bar{G}(X)$). The kernel of the composite homomorphism is $(\mathfrak{p}R[X], F_i(X))$ and obviously contains $F(X)$. By passage to residue class rings, we have thus defined a homomorphism h_i of $R' = R[y] = R[X]/F(X)$ onto the ring $k[X]/(f_i(X))$, this ring being a field, since $f_i(X)$ is irreducible over k. The kernel $\mathfrak{P}_i$ of h_i is therefore a maximal ideal of R'. As $\mathfrak{P}_i$ contains $\mathfrak{p}$, and as $\mathfrak{p}$ is maximal, we have $\mathfrak{P}_i \cap R = \mathfrak{p}$. Since the g irreducible polynomials $f_i(X)$ are distinct, the fields $k[X]/(f_i(X))$ are also distinct, and so are the g maximal ideals $\mathfrak{P}_i$. The kernel $\mathfrak{P}_i$ is clearly equal to $R'\mathfrak{p} + F_i(y)R'$.

When the coefficients of the product of the g polynomials $(F_i(X))^{e(i)}$ are reduced modulo $\mathfrak{p}$, the resulting polynomial is equal to $\bar{F}(X)$. Since $F(y) = 0$, it follows that the product of the h elements $(F_i(y))^{e(i)}$ belongs to $R'\mathfrak{p}$. Hence the product $\mathfrak{Q}_1 \cdot \mathfrak{Q}_2 \cdot \cdots \cdot \mathfrak{Q}_g$ is contained in $R'\mathfrak{p}$, if we set $\mathfrak{Q}_i = R'\mathfrak{p} + R' \cdot (F_i(y))^{e(i)}$. On the other hand the intersection $\mathfrak{Q}_1 \cap \cdots \cap \mathfrak{Q}_g$ contains $R'\mathfrak{p}$. In order to prove our second assertion, it will be sufficient to show that the intersection and the product of the $\mathfrak{Q}_i$ coincide, and for this it suffices to prove that $\mathfrak{Q}_i$ and $\mathfrak{Q}_j$ are comaximal for $i \neq j$. But in this case we have an identity of the form $a_i(X)(f_i(X))^{e(i)} + a_j(X)(f_j(X))^{e(j)} = 1$, where $a_i(X)$ and $a_j(X)$ are polynomials over k. Hence if $A_i(X)$ and $A_j(X)$ are polynomials over R such that $\bar{A}_i(X) = a_i(X)$ and $\bar{A}_j(X) = a_j(X)$, then

$$A_i(y)(F_i(y))^{e(i)} + A_j(y)(F_j(y))^{e(j)}$$

is congruent to 1 modulo $R'\mathfrak{p}$. But this is an element of $\mathfrak{Q}_i + \mathfrak{Q}_j$. As $R'\mathfrak{p}$ is contained in $\mathfrak{Q}_i + \mathfrak{Q}_j$, this proves that $\mathfrak{Q}_i + \mathfrak{Q}_j = R'$, and consequently our second assertion is proved.

Every maximal ideal $\mathfrak{P}$ of R' which contracts to $\mathfrak{p}$ contains $R'\mathfrak{p}$. Thus it must contain one of the ideals $\mathfrak{Q}_i$. But, since $\mathfrak{P}_i^{e(i)} \subset \mathfrak{Q}_i$, $\mathfrak{P}$ must also contain $\mathfrak{P}_i$. As the $\mathfrak{P}_i$ are maximal ideals, $\mathfrak{P}$ must be one of them. This completes the proof of Theorem 34.

EXAMPLES.

1) *Case of a Dedekind ring.* The degree $f(i)$ of $f_i(X)$ is obviously the relative degree of $\mathfrak{P}_i$. Since $\mathfrak{P}_i^{e(i)} \subset \mathfrak{Q}_i$, we have $e(i) \geq e'(i)$, where $e'(i)$ is the reduced ramification index of $\mathfrak{P}_i$. But as R' is a finite R-module, we have $\sum_i e'(i)f(i) = n$ (Theorem 21 of § 9); since we obviously have $\sum_i e(i)f(i) = n$, we conclude that $e(i)$ is the reduced ramification index of $\mathfrak{P}_i$.

2) *Case of a quadratic field* $Q(\sqrt{m})$ ($m \equiv 2$ or 3 (mod 4)). We know that $\{1, e\}$ ($e^2 = m$) is an integral basis of this field. Hence $F(X) = X^2 - m$, and we have to study its decomposition in $J/(p)$ (p:prime). The only cases in which it is a square are $p = 2$ (since $J/(2)$ is perfect), and $p|m$. In the other cases (p an odd prime, $p \nmid m$), the polynomial $X^2 - m$ factors into two distinct factors over $J/(p)$ or is irreducible over $J/(p)$, according as m is or is not a square modulo p.

3) *Case of a quadratic field* $Q(\sqrt{m})$ ($m \equiv 1$ (mod 4)). We know that $\{1, \frac{1}{2}(1 + e)\}$ ($e^2 = m$) is an integral basis for this field. Hence

$F(X) = X^2 - X - (m-1)/4$. For $F(X)$ to be a square in $(J/(p))[X]$ it is necessary and sufficient that the only root s of its derivative $F'(X) = 2X - 1$ be a root of $f(X)$; this implies $p \neq 2$, and, in this case, we must have $4F(s) = 2(2s) - (m-1) = -m = 0$, whence $p \mid m$; in other words the ramified primes are those which divide m. Given any other prime p, for deciding whether it is inertial or decomposed, we have to decide whether the polynomial $X^2 - X - (m-1)/4$ is or is not irreducible over $J/(p)$. In case $p = 2$ it is irreducible if and only if $(m-1)/4$ is even. If p is odd, the classroom method for solving quadratic equations reduces the question about the irreducibility of $X^2 - X - (m-1)/4$ to the same question about $Z^2 - m$. Thus p is inertial if and only if m is a square modulo p.

4) *Case of a cyclotomic field* $Q(z)$ ($z^p = 1$, $z \neq 1$). To make the method more elementary, we shall prove that $\{1, z, \cdots, z^{p-2}\}$ is an integral basis without using the theory of differents and discriminants. We set $F(Z) = Z^{p-1} + \cdots + Z + 1 = (Z - z) \cdots (Z - z^{p-1})$. We first see that $p = F(1)$ is a multiple of $(1 - z)$ in the ring R' of integers of $Q(z)$, and is equal to $N(1 - z)$. Thus $(1 - z)$ cannot be a unit in R', and $R'(1 - z)$ contracts to (p) in J. Now, if $x \in R'$, $T((z-1)x) = (z-1)x + (z^2-1)x_2 + \cdots + (z^{p-1}-1)x_{p-1}$ (where x_i is the conjugate of x defined by $s_i(x) = x_i$, s_i being the automorphism of $Q(z)$ defined by $s_i(z) = z^i$) belongs to $R'(z-1) \cap J$, and is therefore a multiple of p. Finally, if we write any x in R' in the form $x = a_0 + a_1 z + \cdots + a_{p-2} z^{p-2}$ a simple computation, using $T(z) = T(z^2) = \cdots = T(z^{p-1}) = -1$ and $T(1) = p - 1$, shows that $T((z-1)x) = -pa_0$, whence a_0 is a rational integer. Replacing x by $xz^2, \cdots, xz^{p-1}$, which are also algebraic integers, we see that $a_{p-2}, a_{p-3}, \cdots, a_1$ are also rational integers. Thus $\{1, z, \cdots, z^{p-2}\}$ is a basis of R' over J.

Now, for studying the decomposition of a prime q in R', we have to study the factorization of $X^{p-1} + \cdots + X + 1$ over the field $J/(q)$, or what amounts trivially to the same thing, the factorization of $X^p - 1$. The only case in which this polynomial has a multiple factor (that is, in which it is not relatively prime to its derivative pX^{p-1}) is the case $q = p$. For $q \neq p$ the p-th roots of unity over $J/(q)$ lie in the field with q^f elements, where f is the smallest positive integer such that $q^f \equiv 1 \pmod{p}$ (proof as in §12, p. 315). Then the polynomical $X^{p-1} + \cdots + X + 1$ factors over $J/(q)$, into $(p-1)/f$ distinct irreducible factors of degree f. The remaining part of the study is as in § 12.

INDEX OF NOTATIONS

INDEX OF DEFINITIONS

The numbers opposite each entry refer to chapter, section, and page respectively. Thus the entry "Perfect field, II, 4, 64" means that a definition of perfect fields may be found in Chapter II, § 4, page 64. In the text, all newly defined terms are usually either introduced in a formal DEFINITION or italicized.

Graduate Texts in Mathematics

continued from page ii

119 ROTMAN. An Introduction to Algebraic Topology.
120 ZIEMER. Weakly Differentiable Functions: Sobolev Spaces and Functions of Bounded Variation.
121 LANG. Cyclotomic Fields I and II. Combined 2nd ed.
122 REMMERT. Theory of Complex Functions. *Readings in Mathematics*
123 EBBINGHAUS/HERMES et al. Numbers. *Readings in Mathematics*
124 DUBROVIN/FOMENKO/NOVIKOV. Modern Geometry—Methods and Applications. Part III.
125 BERENSTEIN/GAY. Complex Variables: An Introduction.
126 BOREL. Linear Algebraic Groups.
127 MASSEY. A Basic Course in Algebraic Topology.
128 RAUCH. Partial Differential Equations.
129 FULTON/HARRIS. Representation Theory: A First Course. *Readings in Mathematics*
130 DODSON/POSTON. Tensor Geometry.
131 LAM. A First Course in Noncommutative Rings.
132 BEARDON. Iteration of Rational Functions.
133 HARRIS. Algebraic Geometry: A First Course.
134 ROMAN. Coding and Information Theory.
135 ROMAN. Advanced Linear Algebra.
136 ADKINS/WEINTRAUB. Algebra: An Approach via Module Theory.
137 AXLER/BOURDON/RAMEY. Harmonic Function Theory.
138 COHEN. A Course in Computational Algebraic Number Theory.
139 BREDON. Topology and Geometry.
140 AUBIN. Optima and Equilibria. An Introduction to Nonlinear Analysis.
141 BECKER/WEISPFENNING/KREDEL. Gröbner Bases. A Computational Approach to Commutative Algebra.
142 LANG. Real and Functional Analysis. 3rd ed.
143 DOOB. Measure Theory.
144 DENNIS/FARB. Noncommutative Algebra.
145 VICK. Homology Theory. An Introduction to Algebraic Topology. 2nd ed.
146 BRIDGES. Computability: A Mathematical Sketchbook.
147 ROSENBERG. Algebraic *K*-Theory and Its Applications.
148 ROTMAN. An Introduction to the Theory of Groups. 4th ed.
149 RATCLIFFE. Foundations of Hyperbolic Manifolds.
150 EISENBUD. Commutative Algebra with a View Toward Algebraic Geometry.
151 SILVERMAN. Advanced Topics in the Arithmetic of Elliptic Curves.
152 ZIEGLER. Lectures on Polytopes.
153 FULTON. Algebraic Topology: A First Course.
154 BROWN/PEARCY. An Introduction to Analysis.
155 KASSEL. Quantum Groups.
156 KECHRIS. Classical Descriptive Set Theory.
157 MALLIAVIN. Integration and Probability.
158 ROMAN. Field Theory.
159 CONWAY. Functions of One Complex Variable II.
160 LANG. Differential and Riemannian Manifolds.
161 BORWEIN/ERDÉLYI. Polynomials and Polynomial Inequalities.
162 ALPERIN/BELL. Groups and Representations.
163 DIXON/MORTIMER. Permutation Groups.

9 780387 900896